Chemistry and the Environment

SVEN E. HARNUNG

University of Copenhagen

MATTHEW S. JOHNSON

University of Copenhagen

CAMBRIDGE
UNIVERSITY PRESS

CAMBRIDGE UNIVERSITY PRESS
Cambridge, New York, Melbourne, Madrid, Cape Town,
Singapore, São Paulo, Delhi, Mexico City

Cambridge University Press
32 Avenue of the Americas, New York, NY 10013-2473, USA

www.cambridge.org
Information on this title: www.cambridge.org/9781107682573

First published 2012

Printed in the United States of America

A catalog record for this publication is available from the British Library.

Library of Congress Cataloging in Publication Data
Harnung, Sven E.
Chemistry and the environment / Sven E. Harnung, Matthew S. Johnson.
p. cm.
Includes bibliographical references and index.
ISBN 978-1-107-02155-6 (hardback)
1. Environmental chemistry. I. Johnson, Matthew S. (Matthew Stanley), 1966– II. Title.
TD193.H366 2012
551.9–dc23 2011053164

ISBN 978-1-107-02155-6 Hardback
ISBN 978-1-107-68257-3 Paperback

Additional resources for this publication at www.cambridge.org/harnung.

Chemistry and the Environment

From the origin of the Earth to climate change, this textbook presents the chemistry of the environment using the full strength of physical, inorganic, and organic chemistry, in addition to the necessary mathematics and physics, using modern notation and terminology. It provides a broad yet thorough description of the environment and the environmental impact of human activity using scientific principles.

Chemistry and the Environment describes the chemistry of Earth's atmosphere, hydrosphere, and lithosphere (including soils) and the biogeochemical cycles. The book presents a variety of industrial processes, from paper and steel to energy and pesticide production, focusing discussion on the environmental impact of these processes and showing how increasing environmental awareness has led to improved methods. The text provides an accessible account of environmental chemistry while paying attention to the fundamental basis of the science, showing derivations of formulas and giving primary references and historical insight. The authors make consistent use of professionally accepted nomenclature (IUPAC and SI), allowing transparent access to the material by students and scientists from other fields.

The authors created this textbook primarily for their own courses, and it has been developed through many years of feedback from students and colleagues. The book will be invaluable for advanced undergraduate and graduate students in environmental chemistry courses, and for professionals in chemistry and allied fields.

Sven E. Harnung is a Senior Lecturer in the Department of Chemistry at the University of Copenhagen and was Head of Department for 12 years. He teaches courses on environmental, inorganic, physical, and analytical chemistry, including pharmaceutical applications. His current research concentrates on magnetic studies of single-molecule magnets. He is the author of three chemistry textbooks in Danish. Dr. Harnung has organized several congresses, including an International Union of Pure and Applied Chemistry (IUPAC) General Assembly. He has been a member of the Danish National Committee for Chemistry for more than 30 years, and he is a Fellow of IUPAC. He has served as a board member of the journals *Acta Chemica Scandinavica* and *Physical Chemistry Chemical Physics*.

Matthew S. Johnson is a Senior Lecturer at the Department of Chemistry at the University of Copenhagen. He teaches courses on environmental chemistry, physical and quantum chemistry, and scientific writing. His main research interest is atmospheric chemistry, including kinetics, spectroscopy, and stable isotopes in atmospheric trace gases. He is a coauthor of more than 70 articles in peer-reviewed journals. He has invented and patented a method for efficient emissions control and improving building energy efficiency. He was awarded a Fulbright Fellowship to study stratospheric chemistry at the Max-lab electron storage ring in Lund, Sweden. He has worked as a researcher for Honeywell and Medtronic and has research collaborations with many groups around the world, including Ford Motor Company and the Tokyo Institute of Technology.

Advance praise for *Chemistry and the Environment*

"This outstanding text brings together fundamental information about the natural chemistry of the Earth and its atmosphere and the environmental impacts of anthropogenic chemicals. It is well suited for upper-level undergraduate and graduate students and researchers in chemistry, Earth sciences, and atmospheric science."

> – Mark Jacobson, Department of Civil and Environmental Engineering, Stanford University, author of *Air Pollution and Global Warming*

"Both authors have excellent scientific standing and complementary backgrounds. They have combined well on this excellent textbook, based on their long experience of teaching environmental chemistry to undergraduate students at the University of Copenhagen. There are many textbooks on environmental chemistry aimed at undergraduate and graduate courses, but this is one of the best that I have come across. It will be adopted for courses in every university for the next decade and beyond due to its logical and comprehensive content. I strongly recommend this excellent textbook for environmental chemistry and related courses at the graduate and undergraduate levels."

> – Naohiro Yoshida, Department of Environmental Chemistry & Engineering, Tokyo Institute of Technology

"Harnung and Johnson have produced a textbook on environmental chemistry that is firmly rooted in physical and chemical principles and follows a strict quantitative and analytical approach. Nevertheless, the accessible style and informative footnotes make it a joy to read and explore for graduate students and professionals alike. It perfectly fills the gap left by more phenomenological introductions to the field."

> – Jan Kaiser, Department of Environmental Sciences, University of East Anglia

Contents

Preface

Chemical species made by humans affect many naturally occurring processes and organisms. The observation of an anthropogenic substance in Nature raises a series of questions: Where did it come from? How and why was it produced and released? How does it move around within the environment? What is its chemistry, including the reaction rate, mechanism, and products, and how does it influence living organisms and the Earth system? More generally, and perhaps not within the focus of scientific chemistry, there are questions such as: Who is entitled to make use of Nature and to what extent? Are there limits to growth? A rational discussion of these questions involves the scientific method and results from the physical sciences, as well as law, economy, and the humanities.

Turning to chemistry: there is no doubt that success in the field of environmental chemistry requires mastering fundamental disciplines such as analytical chemistry, thermodynamics, and modern experimental and theoretical chemistry. The important role of environmental chemistry as a field in its own right is recognized internationally: the International Union of Pure and Applied Chemistry, IUPAC,[a,233] has organized scientific investigations of the environmental impact of chemistry for many years; examples include its series of reports on pesticides,[179] starting long before environmental issues were of political and public interest, and the White Book on chlorine.[191] Recognition of the importance of the subject has driven IUPAC to rename its Applied Chemistry Division the Chemistry and the Environment Division. The significance of this change is underlined by the fact that the word *Applied* is part of the very name of the Union.

Despite the central role of environmental chemistry in sustainable development, we have often wished that there was a textbook that would address the subject using the full strength of physical, inorganic, and organic chemistry, in addition to the necessary mathematics and physics. The target audience for this book is interested professionals and advanced undergraduate and graduate students in chemistry and allied fields.

Scientists with very different backgrounds have contributed to environmental sciences, and various traditions regarding nomenclature are found in the literature. Accordingly, much time and effort are sometimes required in order to interpret scientific papers. For this reason we have emphasized the use of standard ISO-IUPAC nomenclature throughout. The overall objective of a nomenclature is the safe exchange of scientific and technical information among people in different disciplines

---

[a] The international organizations are discussed in the Introduction.

and between nations.[a] For example, public safety demands that chemists and nonexperts (e.g., customs authorities, emergency and health services) be able to communicate clearly concerning the identities of chemical species involved.

Environmental chemistry is driven by specific examples such as detection of pesticide residues and characterization of the ozone hole. The background knowledge needed to understand these subjects in depth has been put into a separate chapter on environmental dynamics, which includes derivations and formulas related to fluid dynamics, thermodynamics, and reaction kinetics. The intention is that this material not be taught from start to finish, but rather taken up when it is relevant. Similarly, teachers are encouraged to choose the specific sections of the book that are most relevant to their educational programs. The purpose of the forward references in the text is to help the reader during the final reading of the book; they may not be helpful in the first reading. We have included dates of significant events in the history of chemistry and the environment. Dates prior to 1950, mainly of chemical history, may be found in the treatise *A History of Chemistry*,[73a,b] while more recent events are referenced directly. Dates of historical interest for other disciplines are provided without explicit references.

Together we have taught environmental chemistry at the University of Copenhagen for more than 30 years, and this book has grown out of our classes. We are indebted to the many gifted students it has been an honor to teach and who have helped us refine our methods and the material.

---

[a] The loss of the Mars Climate Orbiter on September 23, 1999, because of confusion of the nonstandard pound force (lbf) with the SI newton (N), illustrates the point.

Acknowledgments

We thank the following:

- Professional colleagues for help and advice, in particular Thomas Blunier, Carl Meusinger, Ole Mønsted, Ole John Nielsen, Yuichiro Ueno, and Högni Weihe.
- Naohiro Yoshida and the Tokyo Institute of Technology for hosting a sabbatical for MSJ.
- The environmental chemistry students at the University of Copenhagen.
- The Department of Chemistry, University of Copenhagen.

Chemistry and the Environment

Introduction

The environment is a fascinating subject for a chemist, with a seemingly endless variety of conditions combining dramatic changes in temperature, pressure, phase, and composition. It is also an extremely useful object of study from the practical viewpoints of economy and health. Environmental chemistry is the science in which the methods and results of chemistry are applied to processes involving chemical species in the environment.[a]

Overview

More than 2,000 years ago in the Middle East, it was established that the Earth was a sphere, and the physical and chemical laws discovered since then allow the present description of the Earth as a physicochemical system. In a strictly chemical sense, the "elements" of the environment are the elements of the periodic table. However, it is convenient to add other kinds of structural units, for example, the spheres, which together with special branches of physics, for example, the laws of fluid dynamics, are needed to understand environmental systems.

Human activity

The agricultural and industrial revolutions (beginning around 1750) have allowed the human population to increase exponentially (Figure 1). In the recent past, the human population has been doubling about every 40 years, and 20 % of all humans born in the past 6,000 years are alive today. Population growth and advances in quality of life have been made possible by knowledge concerning how to use Earth's resources. Specific examples include the production of fertilizer and cement, the invention of refrigeration, medical advances, and the invention and mass production of consumer goods ranging from clothes to personal electronic devices. A key challenge facing human society today is dealing with the consequences of our success.

There are many examples of negative effects that industrial activity and population growth have had on the environment. One is an air pollution episode in London in 1952 that resulted in more than 4,000 deaths. Another is the contamination of the Arctic environment with mercury and persistent organic pollutants. Phenomena such as ozone depletion and anthropogenic climate change take place on a global scale. In order to mitigate the negative effects of humanity on the environment, it is necessary to understand how the environment works; in the end, knowledge is the key to sustainable development.

---

[a] This is not a circular definition, because the term *chemical species* is defined using more fundamental quantities.[123]

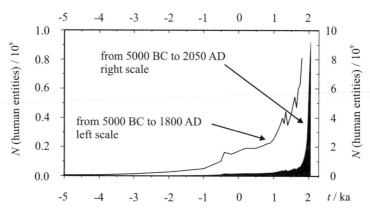

Figure 1 *Exponential growth of human population*

Human population according to the lower estimate of the U.S. Census Bureau from 5000 BC to 1850 AD (left curve), with projection to 2050 AD (right curve, which has been shaded for clarity). [a]

Some population milestones: 10 ka BC: 10^6; 6 ka BC: 10^7; 500 a BC: 10^8; 1835 AD: 10^9; 1950 AD: 2.5×10^9; 1999 AD: 6.0×10^9; 2009 AD: 6.8×10^9.

Chemistry and society

Goods and services must be produced and distributed to sustain society. This requires resources of energy and material and unavoidably affects the environment. International scientific and technical expertise is the basis for making well-founded decisions regarding energy, water, food, consumer products, industrial processes, and so forth. Here we present an outline of the structure of the international scientific organizations that manage these tasks.

A hundred years after the start of the industrial revolution the number of known chemical substances had become very large and it was no longer possible for any individual to command the entire field of chemistry. Many different systems of nomenclature were in use: chaos reigned. Therefore, an international chemical conference was organized in Karlsruhe, Germany, in 1860. The objective was to reach agreement on the theory of organic chemistry. That, in turn, required standardization of nomenclature and fundamental constants. Although the 140 chemists who attended did not reach consensus on any of the issues, committees were appointed to make recommendations, which were eventually published.

During the following years, several national chemical societies were founded,[b] and they began formulating national nomenclatures. In order to coordinate these attempts, an international conference on chemistry was held in Paris in 1889 during the World Exhibition, and here the first International Commission of Chemical Nomenclature was formed. After these preliminary steps, a series of meetings on scientific chemistry, *pure chemistry*, was organized, and in 1911 the International Association of Chemical Societies was formed.

---

[a] 5 ka BC ≈ 7 ka BP. Capital letters denote the zero of timescales: BC = before Christ, BP = before present, AD = Anno Domini (i.e., the Year of the Lord). For geological timescales, see Section 1.3.

[b] For example, The American Chemical Society 1876 and the Danish Chemical Society 1879.

During the period up to World War I, several congresses on industrial chemistry, *applied chemistry*, were held. The first was confined to sugar refining, agriculture, foodstuffs, and fermentation, but the scope of the following congresses was extended to include metallurgy, mines and explosives, electrochemistry, effluents, and chemistry applied to medicine, toxicology, pharmacy, and hygiene.

This was the background for forming the International Union of Pure and Applied Chemistry, IUPAC, in 1919 after World War I.[14,21,233] Today (2012) the union has 58 national members (i.e., chemical societies or national academies), and the work is carried out by seven divisions covering each of the major subjects of chemistry: Physical and Biophysical Chemistry, Inorganic Chemistry, Organic and Biomolecular Chemistry, Polymer Chemistry, Analytical Chemistry, Chemistry and the Environment, and Chemistry and Human Health. Eight committees have special duties including nomenclature and symbols. The coordination of the CHEMRAWN conferences will be discussed later.[a]

IUPAC is one of 31 scientific unions (2012) that are members of the International Council for Science, ICSU.[b,229] ICSU coordinates inter union projects, an example being SCOPE,[c] the Scientific Committee on Problems of the Environment. Other examples include programs aimed at collecting data on a global scale,[12,62] such as the International Polar Year, IPY,[d] and the International Geosphere-Biosphere Programme, IGBP.[e] ICSU has the important role of being the scientific advisor of the United Nations Educational, Scientific and Cultural Organization, UNESCO.

As a final illustration, in December 2008, the 63rd General Assembly of the United Nations adopted a resolution proclaiming 2011 as the International Year of Chemistry (IYC 2011). Its implementation was left to the associations: IUPAC, the chemical societies, and the national committees for chemistry.

---

[a] CHEMRAWN = Chemical Research Applied to World Needs.

[b] ICSU, founded in 1931, is the continuation of the International Research Council from 1919. Its original name was International Council of Scientific Unions, whose acronym has been preserved.

[c] See footnote *a*, Section 7.1.

[d] http://www.ipy.org (2012).

[e] http://www.igbp.net (2012).

The Earth

Since ancient times, humanity has wanted to understand the Earth. This desire has driven the development of modern physics, chemistry, geology, and biology.

Section 1.1 gives a brief survey of the present understanding of the formation of the Earth. This knowledge is based on observations of electromagnetic radiation coming from space, as well as direct observations of the Sun, the Earth, the Moon, a few planets, comets, and various meteorites found on the surface of the Earth.

In the remaining part of Chapter 1 we discuss classical results dealing with the form of the Earth, including its surface and internal structure, how its tremendous age was determined, and the important role that life has played in the development of our present environment.

1.1 Origin of the Earth

A study of environmental chemistry would not be complete without a description of the origin of the Earth and its relation to the rest of the universe. This section describes the materials of which the Earth is made, and how and why these elements are distributed among the various spheres of the Earth.

a. Big Bang

A number of observations support the theory that the present universe had an explosive start more than 10^{10} a ago. First, it was discovered that all galaxies are receding: the astronomer Hubble (1929) compiled the available data to show that the universe is expanding, implying a starting point for the expansion at about 20 Ga BP. The "Big Bang" would have given off a great deal of light as blackbody radiation, and as the universe expanded, the temperature of this radiation field would have decreased, with a simultaneous increase in the wavelength of the light.[a] The universe is still bathed in this light today, as discovered accidentally by Penzias and Wilson (1964).[102] During the testing of one of the first communication satellites, noise was detected in the microwave region of the spectrum that came from all directions in space. Subsequent research established that the temperature of the cosmic microwave radiation field depends on the direction, the mean value being 2.728 K.[b] The temperature of the

---

[a] See Section 10.1b.
[b] This value is the constant term (the monopole) of an expansion of the temperature profile onto spherical harmonic functions.

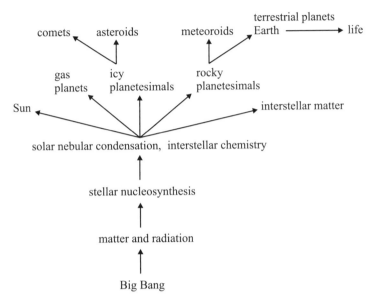

The development of the Universe, Earth, and life
Time proceeds upward.

young universe has been estimated by calculating backward from current conditions. At less than 10^{-4} s after the beginning, the temperature of the radiation field/plasma of matter-antimatter pairs would have been about 10^{12} K. About 10^6 a later, the temperature of the universe would have been cold enough for atoms to form. A summary of the major steps in the succeeding development is given in Figure 1.1.

The Big Bang model predicts values for the temperature and density of the universe as a function of time. When this information is combined with rates of formation of atomic nuclei from elementary particles, models show that the temperature and density of the universe between 3 and 20 minutes of age would have allowed synthesis of nuclei with atomic mass less than 5 Da. The distribution of elements is controlled by the ratio of photons to baryons, and the model predicts mole fractions of $x(^1\text{H}) \approx 0.92$, $x(^4\text{He}) \approx 0.08$, and smaller amounts of ^2H, ^3He, and ^3Li.

The emission and absorption of light are characteristic of specific atoms and molecules, and the compositions of celestial bodies such as stars and nebulae[a] have been determined using spectroscopy. The analysis shows that the universe does indeed contain the distribution of elements predicted by the Big Bang nucleosynthesis model, providing further verification of the theory.

Interstellar space

Table 1.1 lists more than 100 molecular entities found in interstellar space and identified using microwave spectroscopy. The kinds of entities observed are explained as resulting from the abundances of the elements (Figure 1.2) and their electronegativities (Table 1.2). Among the binary entities, one finds one fluoride, a few carbides,

---

[a] Lat. *nebula* $\hat{=}$ cloud, mist.

Table 1.1 Molecular entities identified in interstellar space

Two-atomic species

H_2, CH, CH^+, CO, CO^+, CN, CS, PC, C_2, SiC, SiN, SiO, SiS, NH, NO, NS, PN, SO, SO^+, $HO^•$, HCl, NaCl, KCl, AlF, AlCl

Three-atomic species

CH_2, COS, HCN, HCO, HCO^+, HCS^+, HOC^+, $HC≡C^•$, C_2O, C_2S, C_3, *cyclo*-SiC_2, H_2O, H_2S, HNO, NH_2, N_2H^+, N_2O, SO_2, NaCN, NaOH, MgCN, MgNC

Four-atomic species

H_2CO, H_2CS, HNCO, HNCS, $HOCO^+$, H_2CN, $HCNH^+$, CH_2D^+, HC≡CH, HCCN, *cyclo*-$HC_3^•$, $HC_3^•$, NC−C≡$C^•$, C_3O, C_3S, H_3O^+, NH_3

Five-atomic species

CH_4, HCOOH, $H_2C=NH$, $H_2N−CN$, CH_2CN, $H_2C=CO$, C_3H_2, *cyclo*-C_3H_2, HC≡C−CN, HC≡C−NC, HC≡C−C≡$C^•$, C_5, SiH_4, SiC_4

Six-atomic species

$H_2C=CH_2$, CH_3CN, CH_3NC, CH_3OH, CH_3SH, $HC≡CCNH^+$, HC≡C−CHO, $HCOCH_2$, $HCONH_2$, HC≡C−C≡CH

Species with seven or more atoms

HC≡C−C≡C−C≡$C^•$, $CH_2=CHCN$, HC≡C−CH_3, HC≡C−C≡C−CN, CH_3CHO, CH_3NH_2; $CH_3C≡C−CN$, $HC(O)OCH_3$; $CH_3−C≡C−C≡CH$, CH_3CH_2CN, $(CH_3)_2O$, CH_3CH_2OH, HC≡C−C≡C−C≡C−CN; $CH_3C≡C−C≡C−CN$, $(CH_3)_2CO$; HC≡C−C≡C−C≡C−C≡C−CN; HC≡C−C≡C−C≡C−C≡C−C≡C−CN

Atoms and cations of metals, such as Na, K, Ca, Ti, and Fe, are not included, and more than 30 isotopologues[a] of H, N, O, Si, and S are not listed. The elements of binary compounds are ordered according to IUPAC recommendations,[124a] for example, NS instead of SN, as previously used for binary sulfur-nitrogen compounds. The list (from 2005) is growing continuously.

Table 1.2 Electronegativities of the elements listed in Table 1.1												
K	Na	Ca	Mg	Al	Si	P	H	C	S	N	Cl	O	F
0.8	0.9	1.0	1.3	1.6	1.9	2.2	2.2	2.6	2.6	3.0	3.2	3.4	4.0

Pauling electronegativities, cf. inside front cover.

nitrides, and sulfides, even more chlorides, and several hydrides and oxides. Larger binary inorganic species such as $AlCl_3$ have not been observed. In contrast, the number of entities with carbon chains illustrates the remarkable ability of carbon to catenate.

It is worth noting that the terms *electropositive* and *electronegative*, referring to elements, had already been used by Berzelius in 1817.[6a] He ordered the 52 elements known at that time into a series with hydrogen in the center, in a sequence almost identical to that of the elements according to Pauling electronegativities, shown inside the front cover. Thus, these concepts are basic to an experimental chemist's

[a] See footnote *c*, Section 3.2c.

understanding of the elements. However, it is Pauling's definition (1932)[74] that stands: electronegativity is the ability of an element in a molecule to attract electrons to itself. Evidently, electronegativity is not a property of an isolated atom. Tables such as the one on the inside of the front cover are quite useful when sorting out properties of chemical species.

b. The solar system

The heavier elements in the periodic table are formed in stars and supernovae through stellar nucleosynthesis. Gravitational collapse of interstellar gas and dust during the formation of a star leads to the high densities and temperatures needed for the nuclear-synthetic reactions that give them their energy. Stars do not live forever, and supernova explosions distribute their mass into the interstellar medium; this is the so-called secondary material of the universe.

The solar system began as a spinning disk of primary and secondary material, with the protosun at its center. The pioneering isotope chemist Harold Urey wrote in 1952 that the primary mechanism fractionating material during the formation of a solar system is the separation of dust from gas. A young star ejects large amounts of ions and particles called the solar wind, which blows gas away from the center of the disk. The outward pressure is counteracted by the gravitational attraction of the star. Different types of material respond differently to these opposing forces, depending on each material's charge, mass, and collision cross section. Because of the temperature gradient inward to the star, the composition of the dust varies from containing only refractory[a] material near the center to containing semivolatile material farther out in the disk.

Before continuing the discussion of the partitioning of material in the solar nebula, we look at the (high-temperature) geochemical classification of the elements.

Abundances and classification of the elements

Analysis of spectroscopic data has provided information about the distribution of elements in the Sun (see Figure 1.2).[149] Data of this type have been critical for developing theories of nucleosynthesis in stars. A closer inspection of the figure shows an odd-even variation: elements with an even atomic number Z are generally more abundant than the adjacent odd-Z elements. This fact has been explained in terms of a stable nuclear structure (Harkins, 1917 and 1931; Goeppert-Mayer, 1948). Another feature of stellar nucleosynthesis is that nuclei whose mass number (sum of protons and neutrons) is a multiple of 4 are typically very stable, and these elements are the main components of the Earth.

Ninety-one elements occur naturally on the Earth, namely, the elements from $_1$H to $_{92}$U, minus $_{43}$Tc and $_{61}$Pm, plus $_{94}$Pu, which has been found in uranium ores from southern California.[b] By mass, half of the mantle is ^{16}O and one-fourth ^{28}Si. In all,

---

[a] The term *refractory* refers to the ability of a chemical species to retain its structure and physical strength at high temperature.

[b] The nomenclature of indices around an element symbol is given in Appendix A1.

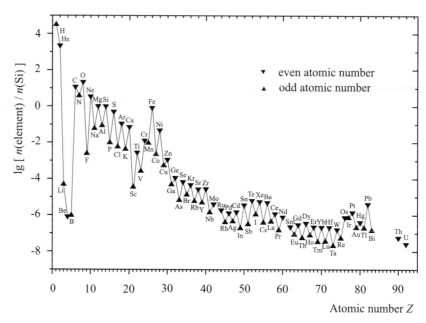

Solar abundance of the elements as a function of atomic number[149]
The logarithm of the abundance of the elements relative to that of Si as found in the
Sun. Meteorites provide abundance data for nonvolatile elements as well; their
abundances differ by at most a factor of 3 from those of the Sun.

the four nuclides ^{16}O, ^{24}Mg, ^{28}Si, and ^{56}Fe make up 85 % of the mass of Earth. A
few other nuclides with mass number divisible by four, ^{12}C, ^{32}S, ^{40}Ca, and ^{48}Ti are
also key components. If we add ^{23}Na, ^{27}Al, and ^{39}K to the list, we have accounted for
99 % of the mass of the mantle of the Earth (see Section 1.2).

While analyzing an earthquake, Mohorovičić (1909) saw that the velocity of seismic waves changed with depth (seismic discontinuities; see Section 1.2b). This observation eventually led to an increased understanding of the chemical structure of the
interior of the planet: the change from a light silica-alumina (sial) rich crust to a
silica-magnesia (sima) rich mantle resting on a dense nickel-iron (nife) core.

Such observations gave rise to the geochemical classification[168] of the elements,
shown in Figure 1.3 (Goldschmidt, 1937).[c] The original discussion was concerned
with partition equilibria of the elements among liquid iron (rich in free electrons),
liquid sulfides (of semimetallic nature), and fused silicates (ionic liquid). When
taken in the context of a cooling protoplanet, the classification gains substantially in
chemical significance. Elements soluble in molten iron are called siderophiles; those
with an affinity for molten sulfides, chalcophiles; and those prefering liquid silicates,
lithophiles.[d]

Figure 1.3 shows the geochemical classification of periods 2 through 6 of the
periodic table. Classes I through V are distinguished in the following discussion. In
general, classes I through IV comprise the metals (the electropositive elements) and

---

[c] Goldschmidt's discussion[168] included atmophilic elements (Table 3.18) and biophilic elements
   (Table 3.26).
[d] Gk. λιθος $\hat{=}$ stone; σιδηρο- $\hat{=}$ iron-; χαλκ $=$ ore; φιλο $-\hat{=}$ loving-; γενναω $\hat{=}$ form.

	1	2	3	4	5	6	7	8	9	10	11	12	13	14	15	16	17
2	Li	Be											B	C	N	O	F
3	Na	Mg											Al	Si	P	S	Cl
4	K	Ca	Sc	Ti	V	Cr	Mn	Fe	Co	Ni	Cu	Zn	Ga	Ge	As	Se	Br
5	Rb	Sr	Y	Zr	Nb	Mo	Tc	Ru	Rh	Pd	Ag	Cd	In	Sn	Sb	Te	I
6	Cs	Ba	La	Hf	Ta	W	Re	Os	Ir	Pt	Au	Hg	Tl	Pb	Bi		

I   II   III   IV   V

Figure 1.3   *Geochemical classification of selected elements*

  I Lithophiles; volatile metals; moderately refractory oxides.

 II Lithophiles; refractory oxides; the group includes Th and U; the transition metals V, Cr, and Mn are moderately oxyphilic elements.

III Siderophiles; the group includes C and P. Note that Tc has no stable isotopes.

IV Chalcophilic and volatile elements; the group includes S and Se.

 V Nonmetals.

class V the nonmetals (the electronegative elements). The elements of classes I and II are lithophiles; those of class I form moderately refractory oxides [$\theta_{fus}(Na_2O) = 1132\ °C$, $\theta_{fus}(K_2O) > 740\ °C$],[a] whereas the oxides of class II are refractory with high melting points [$\theta_{fus}(MgO) = 2826\ °C$; $\theta_{fus}(Al_2O_3) = 2054\ °C$]. The siderophilic elements (class III + C + P) are those that have an affinity for liquid iron [$\theta_{fus}(Fe) = 1538\ °C$, $\theta_{vap}(Fe) = 2861°C$]; at lower temperatures, they form sulfides. Class IV comprises the chalcophiles: that is, the elements that have an affinity for the elements of group 16.[b] Some of elements of class IV are quite volatile [$\theta_{vap}(Hg) = 357\ °C$]. The limits between the groups are not sharp; for example, the sulfides of the coinage metals (Cu, Ag, and Au) are somewhat soluble in molten class IV sulfides, and in Nature these elements are found as sulfides and in the pure state.

Compositions of the planets

The compositions of the planets are functions of chemical and physical mechanisms. The inner planets are almost completely composed of refractory compounds and siderophilic elements, whereas the outer planets are made of gas around a rocky core. The three planets Venus, Earth, and Mars were formed by similar processes and together are known as the terrestrial planets; some data are given in Table 1.3. Rocks from the Moon brought back by the Apollo and Luna missions show a mineral composition not found on Earth; the Moon is believed to have been formed by a bolide collision with the Earth. In contrast to the gas giants Jupiter and Saturn,[c] the terrestrial planets are not massive enough to have attracted a significant atmosphere

---

[a] θ_{fus} denotes the temperature of fusion (solid $\rightarrow$ liquid); θ_{vap} denotes the temperature of vaporization (liquid $\rightarrow$ gas).[126]

[b] The chalcogens are the elements that form ores, that is, group 16.

[c] Jupiter: mass $= 1.899 \times 10^6$ Yg, density $= 1.33$ g cm^{-3}.
   Saturn: mass $= 0.569 \times 10^6$ Yg, density $= 0.70$ g cm^{-3}.

Table 1.3 Some properties of the terrestrial planets[129]			
	Venus ♀	Earth ⊕	Mars ♂
Mass / Yg [1]	4869.0	5957.4	641.9
Radius / km [1]	6051.9	6366.7	3397
Gravity / m s^{-2}	8.87	9.80	1.62
Density / g cm^{-3}	5.24	5.51	3.94
Escape velocity / km s^{-1}	10.4	11.2	2.37
Distance to Sun / au [2]	0.723	1	1.524
Albedo [3]	0.65	0.30	0.15
T_s / K [4]	730	287	218
p_s / bar [5]	90	1.013	0.007
Atmospheric composition as mole fraction:			
CO_2	0.964	380×10^{-6}	0.9532
N_2	0.034	0.7808	0.027
O_2	69×10^{-6}	0.2095	0.0013
H_2O	0.001	0 to 0.03	300×10^{-6}
Ar	4×10^{-6}	0.0093	0.016
Ne	–	18×10^{-6}	3×10^{-6}
CO	20×10^{-6}	1×10^{-6}	0.0007

[1] See the discussion at Table 1.5.
[2] The astronomical unit: 1 au = 149.6 Gm.
[3] See Section 10.1c.
[4] Mean surface temperature.
[5] Surface pressure.

of primary material, gas trapped by the gravitational potential of the planet. Instead, their atmospheres are secondary, resulting from the outgassing of the material from which the planet is made. One indication that Earth has a secondary atmosphere is that the relative abundance of He in the atmosphere is 10^{14} less than in the Sun or the universe as a whole.

Material originating with the formation of the solar nebula can be found in a special class of meteorites called carbonaceous chondrites. These meteorites are made of grains of material that are believed to have condensed directly from gaseous material in the solar nebula. This condensation likely took place within 50 ka of the formation of the disk, about 4.6 Ga BP. The relative abundances of many of the elements in these meteorites are about the same as the abundances found in the Sun, (see Figure 1.2).

c. The Earth

Condensation of the solar nebula
Urey's description of processes in the solar nebula aims to account for the chemical compositions of the planets and their densities based on the assumption that the initial

nebula was homogeneous and in thermodynamic equilibrium. After the separation of dust from gas, matter in the inner nebula started to condense as it cooled to about 1500–1300 °C (a total pressure of 10^{-3} bar is assumed in the model).[172] Under these conditions, refractory metal oxides such as Al_2O_3 (corundum) and CaO were formed, and at lower temperatures minerals such as $Ca_2Al_2SiO_7$ (melitite), $MgAl_2O_4$ (spinel), and $CaTiO_3$ (perovskite) were produced.[a] At 1200 °C, metallic Fe-Ni alloy was condensed together with small amounts of the siderophilic elements C, P, and Co. Still assuming equilibrium between the liquid and solid phases and the gases of the nebula, $MgFeSiO_4$ (olivines) and $CaMgSi_2O_6$ (pyroxenes) separated out at 1100 °C. Next, alkali metal vapors reacted at 700 °C with the calcium aluminosilicates produced at higher temperature to give alkali aluminosilicates (feldspars). Mixtures of such minerals, formed at high temperature, are called *igneous rocks* (see Section 3.1c). At lower temperature, Fe(s) reacted with H_2S(g) around 430 °C to give FeS (troilite), which is very common on the Moon, Mars, and the meteorites, and with H_2O(g) around 200 °C to give iron oxides. At still lower temperature, water reacted with pyroxenes to give amphiboles; H_2O(s) condensed at temperatures below -50 °C.

Oceans

Based on the D/H ratio (see Section 1.3) in cometary and meteoric water, less than 10 % of Earth's water is derived from icy comets. Noble gases have been used to place even tighter constraints on cometary input of volatiles to Earth. According to current understanding (2007), most of Earth's water was acquired during planetary formation and then escaped during mantle degassing after the event that created the Moon.

Seawater is an ionic solution whose present composition is well known (see Table 3.13). However, the question of how it was formed is still debated.[193] The table shows that the ionic strength is rather high ($I \approx 0.7$ M) and the acidity slightly basic (pH $\approx$ 8.1); further, the buffer value of the hydrogencarbonate–carbonate system is close to a minimum (see Figure 5.5).

Over geological timescales, material partitioned among the solid, aqueous, and gas phases:

$$\text{igneous rocks} + \text{volatiles} \; \leftrightharpoons \; \text{seawater} + \text{sediments} + \text{air} \qquad (1.1)$$
$$0.6 \text{ kg} \qquad\qquad \approx 1 \text{ kg} \qquad 1 \text{ liter} \qquad \approx 0.6 \text{ kg} \qquad 3 \text{ liter}$$

The igneous rocks include kaolinite, orthoclase, and calcite; volatiles H_2O and HCl; and sediments, micas (see Table 3.4). It is convincingly argued using the phase rule (see Section 2.3b) that this system acts as a pH-stat rather than as a buffer solution.[206] The seawater of this equation has been treated as an open physicochemical system (see Section 2.2b).[116]

Atmosphere

The early Earth was a violent place. Volcanic activity or outgassing of the planet was continuous. One of the main components of this gas was water, giving the planet a steamy atmosphere. In addition, the young planet was continually bombarded by

---

[a] The minerals mentioned here are discussed in detail in Section 3.1.

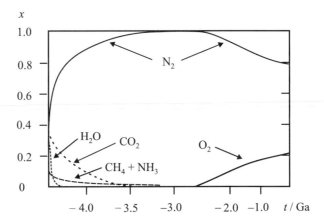

Figure 1.4   *Evolution of the atmosphere*
Mole fraction x as a function of time t before present. The atmosphere was formed
after the Moon-forming impact, which took place 4.527 ± 0.010 Ga BP. [150, 156, 219]

material from space, dust, ice, and rocks. With time, the planet cooled and water
condensed, forming the oceans. Carbon dioxide dissolved in the oceans, eventually
forming sedimentary carbonate rocks. The evolution of Earth's atmosphere with time
is shown in Figure 1.4. The loss of water vapor and carbon dioxide led to a decrease
of atmospheric pressure at the surface by roughly a factor of 100. The main gas still
remaining in the atmosphere was nitrogen, and it has stayed this way to the present day.

Table 1.3 summarizes some of the key parameters of the atmospheres of the
terrestrial planets. It is interesting to note that although they share a common origin,
they did not evolve in the same manner. Because Venus lacked an ocean, Venusian
CO_2 was left in the atmosphere. As seen in the table, the atmosphere of Venus is
composed mainly of CO_2. The amount of nitrogen in the atmosphere of Venus (a mole
fraction of 0.03 times an atmospheric pressure at the surface of 91 bar) is roughly
equal to the amount of nitrogen present in Earth's atmosphere. This is in contrast
to the relatively small amount of nitrogen found in the atmosphere of Mars. The
explanation is that the Martian nitrogen has escaped to space. Mars is smaller than
Earth and Venus, and its gravitational potential cannot prevent the thermal escape of
gas to space (Jeans' escape). The Boltzmann distribution of kinetic energies in gas at
the top of the atmosphere implies that a small fraction can achieve escape velocity.
The isotopic enrichment of ^{15}N in the $^{14}N_2$ remaining in the atmosphere of Mars is
strong evidence for this explanation, because the heavier isotope will escape from the
atmosphere more slowly. The ratio of gravitational binding energy to thermal kinetic
energy is higher for a heavier isotope.

Current theories say that the Earth has never been so cold that the oceans froze
completely, nor so warm that they vaporized completely. One question is why the
surface temperature of Earth was not colder in the first 2 Ga after the oceans formed,
given that the radiant excitance of the young Sun would have been 25 to 30 % lower
than today.[a] One theory is that there was an enhanced greenhouse effect warming the

---

[a] See Section 10.1.

Table 1.4 Characteristics of organic compounds found in carbonaceous meteorites[111]			
Class	w	Number	Chain length
Amino acids	0.105	74	C_2–C_7
Aliphatic hydrocarbons	>0.060	140	C_1–$C_{\leq 23}$
Aromatic hydrocarbons	2.5–5.0×10^{-2}	87	C_5–C_{20}
Carboxylic acids	>0.525	20	C_2–C_{12}
Dicarboxylic acids	>0.052	17	C_2–C_9
Hydroxy acids	0.026	7	C_2–C_6
Purines and pyrimidines	0.002	5	na
Basic N-heterocycles	0.012	32	na
Amines	0.013	10	C_1–C_4
Amides	9.6–12.2×10^{-2}	2	na
Alcohols	0.019	8	C_1–C_4
Polyols[154]	0.10	19	C_3–C_6
Oxo compounds	0.047	9	C_1–C_5

Relative mass fraction, w, of organic species whose total mass fraction in a meteorite amounts to $\approx 0.57 \times 10^{-3}$.[a] "Number" indicates the number of compounds identified. na = not applicable.

surface, caused by the conversion of volcanic sulfur to COS, a powerful greenhouse gas, in a weakly reducing atmosphere.[212] By 3.5 Ga BP, the partial pressure of carbon dioxide had dropped. As carbon dioxide dissolved in the ocean and calcium was weathered from the crust, massive deposits of calcium carbonate (limestone) were formed. The amount of carbon dioxide found in carbonate rocks corresponds to a partial pressure of about 60 bar.

Life formed atmospheric oxygen by photosynthesis, the overall process shown later in Equation 1.2. In the early atmosphere, there could not have been very much free oxygen, as it would have reacted with iron(II) in the oceans. The simultaneous existence of oxidized iron and reduced uranium in rocks from about 2.2 Ga BP indicates that atmospheric oxygen levels at that time must have been between 10^{-12} and 10^{-3} of the present atmospheric level. Further, the appearance of mass-independent fractionations in sulfur isotopes shows that earlier than 2.5 Ga BP, oxygen levels were below 10^{-5} of the present atmospheric level.[212]

Life

It is not known how life started on Earth. However, the organic building blocks of life are everywhere, even in space (Table 1.4). Although the formula $CH_2O(biota)$[b] accounts for 98.7 % of the mass of living biological material,[c] more than two dozen elements are essential for life as we know it. Life was found on Earth as early as 3.8 Ga BP, as evidenced by microbes fossilized in stromatolites.

[a] "Organic species" are chemical species that have some carbon-hydrogen bonds.
[b] Carbohydrates in biota are denoted as a state of aggregation: $CH_2O(biota)$.
[c] See Table 3.26.

Some of the chemical species listed in Table 1.1 may have been precursors for biomolecules. In fact, more than 35 amino acids can be synthesized by passing an electric discharge through a reducing gas mixture containing ammonia, methane, and water (the Miller-Urey experiment, 1953), and half of those match the relative abundance of amino acids in carbonaceous meteorites (see Table 1.4). However, it is uncertain whether the early atmosphere was reducing enough to preserve these species for an extended time.

Sustained life was not possible on the surface of the young planet because of energetic ultraviolet radiation from the Sun, active volcanoes, and bombardment by meteors. Eventually, the temperature decreased and the environment stabilized. The earliest forms of life are thought to have obtained their energy from reduction of various chemical species found in the environment. However, by at least 2.7 Ga BP, certain bacteria had developed a mechanism for harvesting the energy of the Sun through photosynthesis, turning carbon dioxide and water into carbohydrate and free oxygen:

$$CO_2(aq) + H_2O(l) \xrightarrow{h\nu} CH_2O(biota) + O_2(g) \tag{1.2}$$

This process requires low-frequency visible light. Life was possible in the ocean because water shielded aquatic organisms from solar UV radiation. After a time the concentration of oxygen in the atmosphere began to increase and eventually it became high enough that a layer of ozone (trioxygen) in the stratosphere could be formed by the photolysis of dioxygen (at wavelengths less than 200 nm):

$$O_2(g) \xrightarrow{h\nu} 2O(g) \tag{1.3}$$

$$O(g) + O_2(g) \xrightarrow{M} O_3(g) \tag{1.4}$$

The level of ozone in the atmosphere was increasing during the Precambrian, and at 0.4 Ga BP, the present atmospheric levels of oxygen and ozone were reached. Ozone absorbs damaging UV radiation, greatly increasing the opportunities for life on land. The formation of the ozone layer and its attenuation of the solar spectrum are discussed in greater detail in Chapter 4.

1.2 Structure of the Earth

a. Classical measurements

Shape and volume

The spherical shape of the Earth has been known since antiquity, as the oldest known "measurement of degrees" (i.e., determination of the shape of the Earth) was made by Eratosthenes (276–194 BC).

A measurement of degrees requires a determination of the distance and angle between two vertical lines located at different places on the Earth's surface. A well in the town of Syene (now Aswan in Egypt) had a unique property. It was located on

Table 1.5 Physical properties of Earth, part 1	
Equatorial radius[1]	$r_{equator} = 6378.14$ km
Distance from center to pole[1]	$r_{pole} = 6356.76$ km
Mean polar compression	$(r_{equator} - r_{pole})/r_{equator} \approx 1/298$
Average radius[2]	$r_\oplus = 6367$ km
Mass[3]	$m_\oplus = 5957$ Yg
Mean density[4]	$\rho_\oplus = 5.51$ g/ml

[1] Determined by Helmert (1886).
[2] Based on the nautical mile; see text.
[3] Determined as explained in the text using $r_\oplus$.
[4] Determined using $r_\oplus$ and $m_\oplus$.

the Tropic of Cancer, and therefore, at midday on the summer solstice, the Sun would shine directly to the bottom of the well without casting a shadow. At just this time, Eratosthenes measured the length of the shadow of an obelisk located in Alexandria, due north of Syene. The height of the obelisk and the length of its shadow were used to determine the angle between the obelisk and the well shaft. He calculated the arc to be 1/50 of the Earth's circumference. The distance between the two points was paced by the military at 5,000 stadiums (a stadium is ca. 157 m), and thus Eratosthenes determined that the Earth's radius was approximately 6.25 Mm.[a]

A more accurate measurement of degrees using the method of triangulation was carried out in northern France by Picard (1669). He found the length of a degree of latitude to be 111.2 km, corresponding to a distance between the equator and the pole of 10.009 Mm. At the very end of the 18th century, Napoleon's nomenclature committee set the length of the metre with the intention that a quarter of a great circle would be exactly 10 Mm, but this definition was soon replaced by a better standard of length, the "normal metre." During the last half of the 19th century, meticulous measurements were carried out by the geodesist Helmert, whose values of the equatorial radius $r_{equator}$ and distance from center to pole r_{pole} are still in use today (Table 1.5).

As seen, the Earth is not a perfect sphere. Instead, as a consequence of rotational centrifugal force, it is slightly farther from the center of the Earth to the equator than it is from the center to the poles. Therefore, the length of the arc of a degree of latitude is not constant. However, if we use the average length of an arc minute (i.e., a nautical mile = 1852 m) and assume that the Earth is spherical, its radius would be $r_\oplus = 6366.71$ km, the value we use in this text.

Mass

The mass of the Earth can be determined from Newton's second law (1687): that is, the force F of the gravitational field on a body is proportional to its mass m and the acceleration of gravity g:

$$F = mg \tag{1.5}$$

[a] 0.157 km $\times 5000 \times 50 / (2\pi)$.

Table 1.6 Physical properties of Earth, part 2			
	Area / Mm²	Mass / Yg	Volume / Mm³
Planet Earth	509	5957	1081
Continents	149	–	–
Hydrosphere[1]	360 (oceans)	1.66	1.62
Atmosphere[2]	509	5.2×10^{-3}	41
Biota[3]	–	7×10^{-7}	–
Humanity[4]	–	3×10^{-10}	3×10^{-10}

[1] See Table 3.12.
[2] Defined using the altitude 80 km, which includes 99.99 % of the mass of the atmosphere.
[3] Measured as mass of carbon.
[4] Assumes 6.6×10^9 humans with a mass of 50 kg each and a density of 1 g/ml.

On this basis, the law of mutual gravitation says that the attractive force F between masses m_1 and m_2 separated by a distance r is given by

$$F = G\frac{m_1 m_2}{r^2} \tag{1.6}$$

Here $G = 6.674 \times 10^{-11}$ m³ kg⁻¹ s⁻², known as the Newton constant of gravitation,[126] was determined by Cavendish (1798) using his invention, the torsional balance. Applying these expressions, the values of G, $r_\oplus$, and the standard acceleration of gravity $g_n = 9.807$ m s⁻², one arrives at the mass of the Earth $m_\oplus = 5957.4$ Yg and the mean density $\rho_\oplus = 5.511$ g/ml.

Some of Earth's characteristics are collected in Tables 1.5, 1.6, and 1.7.

Surface

Table 1.6 shows that most of the surface of the Earth (71 %) is covered by the oceans. Ocean currents transport tremendous amounts of energy and have a profound impact on the physical conditions we experience on the surface of the planet. One example is the heat transported by the Gulf Stream, which gives northern Europe a mild climate

Table 1.7 Physical properties of Earth, part 3		
	Average / km	Maximum / km
Land surface (height)	0.88	8.85
Oceans (depth)	3.79	11.04
Atmosphere (height)	7.46[1]	2000

[1] The characteristic height used for the atmosphere is the "scale height" (see Equation 2.81). It is the altitude over which the atmosphere's density has fallen by $1/e \approx 37$ %. The top of the atmosphere is often defined as the altitude at which escape to space becomes significant. This is the base of the exosphere.

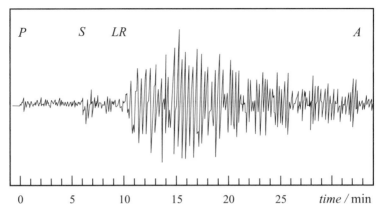

Figure 1.5 *Internal structure of the Earth, part 1*
Seismogram of an earthquake in Iceland recorded on May 6, 1912, 22° from the epicenter.[72] At *P* the primary wave starts, at *S* the secondary wave, and at the point *LR* the Lowe-Rayleigh wave begins. *A* marks the start of the aftershocks.

far different from that of its latitudinal companions Labrador and Kamchatka.[a] It is also interesting to compare the masses of the various components. As discussed later, minor components such as biomass or humanity have a large effect on the environment.

b. Internal structure of the Earth

The density of the rocks found at the Earth's surface are typically in the range of 2.7 to 3.3 g/ml, and the density of the ocean is 1.02 g/ml. Because the average density of the Earth is 5.5 g/ml, it has been known for centuries that the inner density must be significantly greater than that near the surface. A more detailed description of the interior structure of the Earth is mainly due to the analysis of seismic data. Seismology[b] is the science concerned with measurement and interpretation of the vibrations of the Earth. In Figure 1.5 we see a seismogram of a powerful earthquake that took place in Iceland in 1912, recorded quite some distance away.

Seismograms show several distinct regions that can be interpreted in terms of the Earth's structure. The origin of an earthquake is known as the epicenter, which typically lies 15 to 30 km under the surface. The rapidly shifting crust produces longitudinal and transverse waves that propagate through the planet. The longitudinal waves result from pressure changes and are called *P*-waves, whereas the transverse waves result from shearing (sideways) motions are called *S*-waves. Shear waves cannot be transmitted through liquids, and they move more slowly than pressure waves in a given solid. In contrast, *P*-waves propagate quickly through the Earth's interior with a speed dependent in part on the density of the material through which they are passing and on its state of aggregation. The speed of the *P*-wave increases

---

[a] See footnote *b*, p. 122.
[b] Gk. σεισμος $\hat{=}$ tremor; λογος $\hat{=}$ reckoning.

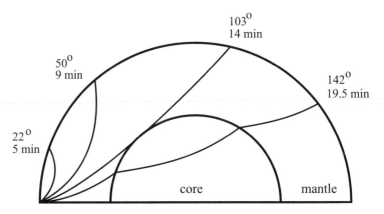

Figure 1.6 *Internal structure of the Earth, part 2*[81]
Longitudinal waves in the inner Earth. Chosen observation points and delay times
for a powerful earthquake's *P*-wave.

from 5.6 km/s at the surface to 13.5 km/s at a depth of 2.9 Mm, after which the speed
abruptly decreases to about 8 km/s. The slower *S*-wave does not reach depths greater
than 2.9 Mm. As seen in Figure 1.5, the *S*-wave reaches the point of observation later
than the *P*-wave. The main earthquake, denoted by the letters *LR* for Lowe-Rayleigh
wave, is a different type of elastic vibration that propagates along the surface and
therefore is observed later. The *P*- and *S*-waves propagate in volume (decreasing as
d^3, where d is a distance), whereas the *LR*-wave moves on the surface (decreasing as
d^2). This means that the surface wave's intensity decays more slowly with distance
than the two other types of waves. The main earthquake is caused by the surface
wave, and its amplitude is used to describe the earthquake's strength using Richter's
logarithmic scale (1932).[81] Today, large earthquakes are measured by the amount of
energy released using the so-called moment magnitude scale.

The remarkable change in the velocity of *P*- and *S*-waves observed at 2.9 Mm is
the clearest example of a seismic discontinuity. The fact that shearing *S*-waves do
not propagate below this depth demonstrates that there is a phase transition from
solid to liquid. In Figure 1.6 we see a cross-sectional diagram of the Earth derived
from seismic measurements. The main conclusion is that the Earth is composed of a
relatively heavy core surrounded by a lighter mantle.

At the beginning of the 20th century, a seismic discontinuity was observed by
Mohorovičić (1909) that has come to be known as *Moho*. Moho lies at a depth of 30
to 40 km under the continents and is known to reach approximately 700 km under
folded mountain chains; it occurs at a depth of only 6 to 7 km under the ocean floor.
Moho is the division between crust and mantle and arises from an abrupt change in
chemical composition. These early results are marked on the right side of Figure 1.7,
together with the individual region's characteristic elements.

Knowledge of the structure of the mantle and the crust is now very detailed. In part
this is due to the development of the discrete Fourier transform, allowing frequency
analysis of seismograms. It is also due to the introduction of scanning methods
(seismic tomography) whereby one can transform reflected sound waves into three-
dimensional images. Specific applications of these techniques include monitoring

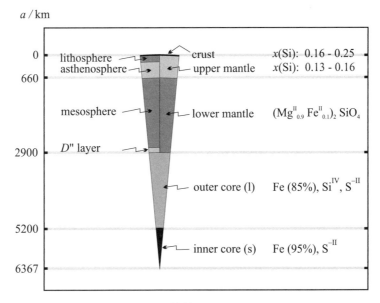

a / km

Figure 1.7 *Internal structure of the Earth, part 3*[16, 81]

The figure represents a cut through the Earth corresponding to a length at the surface of approximately 1330 km. The quantity a marks the distance from the Earth's surface to each characteristic seismic discontinuity, and $x(Si)$ is the average silicon mole fraction of the rocks in the crust and the upper mantle. Interpretations of the early seismic measurements are marked to the right of the cut. As measurements and interpretations were refined, it became possible to divide the mantle into the lithosphere, the asthenosphere, the mesosphere, and the D'' layer as shown on the left side.

nuclear weapons testing and searching for oil and gas fields, as well as for valuable deposits of ores and minerals.

Modern understanding of the geological structure of the Earth is as follows. The top layer is called the crust, a hard but relatively brittle shell. It includes the continents and the ocean floor, and it extends downward from the surface to Moho. The upper parts of the continents are granitic (density $\rho \approx 2.7$ g ml^{-1}; see Table 3.1), whereas the lower basaltic parts are somewhat denser ($\rho \approx 3.0$ g ml^{-1}; see Table 3.1) and have the same chemical composition as the ocean floor. The crust rests on the lithosphere, which extends to about 300 km; its lower parts are composed of peridotite ($\rho \approx 3.3$ g ml^{-1}; see Table 3.1).

The lithosphere rests on the asthenosphere, which is also made of the rock peridotite. At these depths, however, temperature has increased to about 1600 K, close to peridotite's melting point, and so its viscosity is relatively low. At greater depths, the pressure increases enough that rocks become solid once again. This phase transition occurs at the 660-km discontinuity in the lower mantle that marks the beginning of the mesosphere. The average composition of the mantle is olivinic, that is, resembling the mineral olivine, $(Mg,Fe)_2SiO_4$.

The seismic D'' layer marks the transition to the core. The core's inner region transmits S-waves, but at a low speed. Thus, one can conclude that it is a distinct

a	Name	ρ	w	φ	T	p	v_P
	Table 1.8 Internal structure of the Earth, part 4[16]						
0	Surface					10^{-4}	
	Crust	2.7–3.0	5	19	0.3–1.0		5–6
40	Moho					1	
	Upper mantle	3.3–4.3	186	276	1.1–2.5		>8.2
660						26	
	Lower mantle	4.3–5.5	481	544			<13.5
2900	D″ layer				45	135	13.5
	Outer core	10.0–12.3	309	154			8.5–11
5200					62	334	
	Inner core	13.3–13.6	19	8			≈12
6367	Center					370	

The shadowed areas correspond to the shaded lines of Figure 1.7.

a / km	Distance from the Earth's surface to a seismic discontinuity
ρ / g ml^{-1}	Density
w	Mass fraction
φ	Volume fraction
T / kK	Temperature
p / GPa	Pressure; 1 GPa = 10 kbar
v_P / km s^{-1}	Speed of the seismic P wave

region, close to the melting point. The main component of the core appears to be iron, but interpretation of seismic data and computer modeling requires a certain content of elements of lower atomic mass. It turns out that nickel is not required, as was previously thought.

Key physical data concerning Earth's interior are collected in Table 1.8. It is seen that pressure as well as temperature increase with depth. The increase in pressure occurs because the planet is essentially a plastic body. A monotonic increase of pressure p with depth a is observed:

$$-\frac{\Delta p}{a} \approx 0.3 \, \frac{\text{kbar}}{\text{km}}, \qquad a < 100 \, \text{km} \qquad (1.7)$$

This equation only takes hydrostatic pressure into account (see Euler's equation, Equation 2.74). The actual pressure may be much higher if water, carbon dioxide, or other gases are present.

The temperature of the Earth changes with depth and location. Under the mid-oceanic ridge at depths less than 12 to 15 km, it follows the empirical equation

$$-\frac{\Delta T}{a} \approx 100 \, \frac{\text{K}}{\text{km}}, \qquad a < 15 \, \text{km} \qquad (1.8)$$

Under the ocean floor in geologically stable areas (e.g., along the Atlantic Ocean), the increase is approximately 30 K/km ($a < 80$ km), whereas under continental bedrock, it is approximately 13 K/km ($a < 80$ km). This is associated with an outward flux density of energy of 80 mW m^{-2} at the midoceanic ridge, which falls to 40 mW m^{-2} for the Precambrian parts of the continents. Overall, the outward energy flux density at the surface due to heat coming from the core is about 60 mW m^{-2}. To put this number into perspective, the flux density of energy from the Sun, averaged over the entire surface, is 240 W m^{-2} (see Equation 10.19). The source of the planet's own heat is radioactive decay.

1.3 Geological periods and dating

a. Geological periods

At the end of the 18th century, sufficient observations had been made to conclude that the upper parts of the continental crusts are largely stratified. Geologists and paleontologists sought to characterize these layers according to their mineral composition and the content of fossils, thereby determining the age of the individual layers. For example, a geological formation was found in England in which a layer of red sandstone occurred both above and below the economically important coal layer. A key observation was that the fossils occurring in the two sandstones are not the same.

Analysis of the geological data shows that there are three distinct layers (strata) of the crust. The lowest "primary" stratum is the oldest and most primitive, because it does not contain sedimentary rocks or fossils. The "secondary" stratum, located immediately above the primary, has evidently been processed by the Earth's biogeo-chemical system and contains limestone, sandstone, slate, and a wealth of fossils of now-extinct species. The topmost "tertiary" stratum is the youngest developmentally, because the rocks are relatively fresh: this layer contains chalk, along with marble and sandstone, and fossils from species of plants and animals that may still be found today.

Table 1.9 Geological periods, part 1[232]			
t/Ma	Δt /Ma	Era	Eon
0			
	66	Cenozoic ("recent life")	
66			
	185	Mesozoic ("middle life")	Phanerozoic
251			
	291	Paleozoic ("ancient life")	
542			
	1960	Proterozoic	
2500			Precambrian
	2070	Archean	
4570			

By the end of the 19th century, the secondary stratum was known to have formed in the three ages, the Paleozoic era, the Mesozoic era, and the Cenozoic era[a] (Table 1.9). The name *Tertiary period* was reserved for the time period during which the topmost

---

[a] Gk. παλαι- $\hat{=}$ old-; ζωον $\hat{=}$ animal; μεσο $\hat{=}$ between; καινο- $\hat{=}$ new-.

t/Ma	Δt/Ma	Period	Some events or data
0	66		
66		Tertiary	Grasses; mangroves
			Alpine orogenesis
	80	Cretaceous[a]	
146			
	54	Jurassic	Mammals; birds
200			Oldest ocean floor; modern corals
	51	Triassic	Dinosaurs
251			
	48	Permian	Flowers; insect pollination
299			
	60	Carboniferous[b]	Coniferous trees; evolution of lignin
359			
	57	Devonian	Insects; vertebrates on land
416			Caledonian orogenesis (Scotland)
	28	Silurian	Land plants with spores
444			Atmosphere: $x(O_2) \approx 0.2$, $x(CO_2) \approx 0.01$
	44	Ordovician	First corals; invertebrates on land
488			
	54	Cambrian	Trilobites; vertebrates
542			
		Precambrian	

Table 1.10 Geological periods, part 2

The Cenozoic era.[232]

The names Cambria (Wales), Ordovices, Silures, Devon, Perm, Trias, and Jura are names of the areas where rocks from the time period in question were first found.

layer of the Cenozoic was formed. Researchers were able to further characterize the layers as shown in Tables 1.10 and 1.11. The fossils found in the lowest layer from the Paleozoic, the Cambrian, do not occur in the layer below, the Precambrian, which is therefore considered to be an older one.

Determinations of the age of the Earth were very uncertain until the beginning of the 20th century. As late as 1897, the famous English physicist Thomson (Lord Kelvin) calculated that the age of Earth was between 20 and 40 Ma, based on estimates of the rate of cooling and the amount of heat received from the Sun. Twenty years later Barrell (1917) succeeded in establishing the tremendous length of Earth's prehistory by studying natural radioactive decay. In the middle of the century, Holmes (1960) used direct methods to show that the age of the Earth is approximately 4.55 Ga. Recent improvements in sampling and detection techniques have led to the figure 4.5662 Ga as the age of the oldest rock.[139] The basis for direct dating methods is discussed in the next section.

[a] Lat. *creta* $\hat{=}$ chalk.
[b] Lat. *carbon* $\hat{=}$ coal; *fero* $\hat{=}$ carry.

t/Ma	Δt/Ma	Period	Examples of events
Table 1.11 Geological periods, part 3			
0			
		Holocene	
0.01			
	1.80	Pleistocene	
1.81			Humans
	3.52	Pliocene	Fossils of apes
5.33			
	17.70	Miocene	Grasses
23.03			
	10.9	Oligocene	Cats, dogs, pigs
33.9			
	21.9	Eocene	Hoofed mammals
55.8			
	9.7	Paleocene	Primates
65.5			

The Tertiary period.[232]
Pleistocene (in Northern Europe and North America: ice ages and interglacial periods) and Holocene (present time), approximately 2 Ma, are parts of the Quaternary period.

b. Radioactive dating

The oldest known volcanic rocks, with ages just below 4.56 Ga, are found in western Greenland, Canada, and western Australia. The age of these rocks was determined by measuring (using mass spectrometry) the concentration of the first and last elements in a naturally occurring radioactive decay series. Only elements whose half-lives are longer than the age of the minerals in which they are found are useful for dating. The relevant series are listed in Table 1.12, but the last two series are very rare. The half-life $t_{1/2}$ is given in parentheses after the mother element, and the broken arrow $\cdots\rightarrow$ denotes a series of short-lived intermediate species.

In some important cases a radiogenic end product is identical to a nonradiogenic isotope occurring in a given sample. This occurs in decay series 3 and 4 of Table 1.12 as discussed next.

Table 1.12 Naturally occurring radioactive decay series				
1.	^{238}U (4.47 Ga) $\cdots\rightarrow$ ^{206}Pb	6.	^{87}Rb (48.8 Ga) $\rightarrow$ ^{87}Sr	
2.	^{235}U (0.70 Ga) $\cdots\rightarrow$ ^{207}Pb	7.	^{147}Sm (106 Ga) $\rightarrow$ ^{143}Nd	
3.	^{232}Th (14.0 Ga) $\cdots\rightarrow$ ^{208}Pb	8.	^{176}Lu (35.7 Ga) $\rightarrow$ ^{176}Hf	
4.	^{40}K (11.9 Ga) $\rightarrow$ ^{40}Ar (11 % of ^{40}K)	9.	^{187}Re (43.0 Ga) $\rightarrow$ ^{187}Os	
5.	^{40}K (1.47 Ga) $\rightarrow$ ^{40}Ca (89 % of ^{40}K)			

The symbol $\cdots\rightarrow$ denotes a series of intermediate species. The branching rule for ^{40}K is explained in footnote c on p. 25.

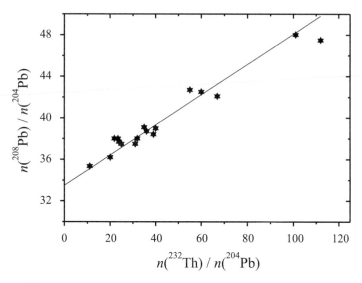

Figure 1.8 *An isochron*[28]

The figure shows the thorium-lead isochron (defined in the text) for a rock from the Granite Mountains in Wyoming, USA. The slope of the isochron is 0.148, which means that the age of the rock is 2.79 Ga.

Four stable isotopes of lead occur naturally, ^{204}Pb, ^{206}Pb, ^{207}Pb, and ^{208}Pb. All ^{204}Pb originates with the solar system itself, as it is not radiogenic. The other three are also formed by radioactive decay, and their present concentrations depend on the amount of the mother and daughter elements in the rock when they were formed, along with the amount of time that has passed. The Cañon Diablo meteorite is an iron meteorite that contains the mineral troilite, FeS, as do most iron meteorites. The interesting thing about this particular meteorite is that its troilite crystals contain the lowest level of radiogenic lead that has ever been observed. It is therefore taken to represent the original distribution of lead isotopes found on the Earth, on the Moon, and in iron meteorites. The following ratios have been determined:[a]

$$n(^{208}\text{Pb}) : n(^{207}\text{Pb}) : n(^{206}\text{Pb}) : n(^{204}\text{Pb}) \approx 29 : 10 : 9 : 1 \qquad (1.9)$$

and with these values as references, it has been possible to infer the ages of the oldest rocks, both from the Earth and from the Moon. The age of the samples ranges between 4.56 and 4.58 Ga.

The Th-Pb dating of a sample of granite from Wyoming, USA, is a good example of the application of radiometric dating. Th, U, and Pb are relatively easily washed out as rock weathers, but crystals of zircon, $ZrSiO_4$, are almost indestructible. Some Th and U are and trapped in zircon, and this is the basis for the analysis. It is standard to record all three decay series ending in Pb and compare the results. Figure 1.8 shows some of the different mineral samples from the *same* rock, analyzed for their content of ^{232}Th, ^{208}Pb, and ^{204}Pb. The samples are all of the same age, but the Th content of the original crystals was different and lowest in the sample with the smallest

[a] Actual experimental values: 29.476 : 10.294 : 9.307 : 1.000.

observed Th-Pb ratio. The amount of thorium formed up to time t (for example, up to the present day) is given by

$$n_t(^{232}\text{Th}) = n_0(^{232}\text{Th}) \times e^{-kt} \qquad (1.10)$$

where $t = 0$ denotes the time at which the crystal was formed. The rate constant k for radioactive decay can be calculated from the relation $k \times t_{1/2} = \ln 2$ and the half-life given in Table 1.12; use Equation 2.50 with $\mu = 0$ and concentration c replaced by amount n. The amount of ^{208}Pb that is found depends on radiogenic lead, that is, on the original amount of thorium, as well as on the original amount of lead:

$$\begin{aligned} n_t(^{208}\text{Pb}) &= n_0(^{232}\text{Th}) \times (1 - e^{-kt}) + n_0(^{208}\text{Pb}) \\ &= n_t(^{232}\text{Th}) \times (e^{kt} - 1) + n_0(^{208}\text{Pb}) \end{aligned} \qquad (1.11)$$

and therefore the slope at any given time depends on the age of the rock:

$$\frac{dn_t(^{208}\text{Pb})}{dn_t(^{232}\text{Th})} = \frac{d(n_t(^{208}\text{Pb})/n(^{204}\text{Pb}))}{d(n_t(^{232}\text{Th})/n(^{204}\text{Pb}))} = e^{kt} - 1 \qquad (1.12)$$

The line shown in Figure 1.8 is called an isochron.[a] The slope of the line can be used to determine the age of the rock,[b] whereas its intercept with the ordinate axis gives the isotope ratio at the time the minerals crystallized, $n_0(^{208}\text{Pb})/n_0(^{204}\text{Pb})$, and thereby became isolated from the surrounding rock.

The oldest sedimentary rocks, from about 3.86 Ga BP have been found at Isua in West Greenland near the edge of the inland ice in a belt approximately 40 km long and 1 to 3 km wide. The rocks are composed of typical erosion products, formed by the action of a CO_2- and H_2O-containing atmosphere on the original volcanic rocks, of which there are still traces.[38] The samples contain calcite, $CaCO_3$, chalcedony, SiO_2 (fibrous microcrystalline quartz, deposited from water solutions), and clays. In addition, the occurrence of magnetite, Fe_3O_4, orthopyroxene, $(\text{Mg}, \text{Fe}^{II})SiO_3$, and other minerals with a certain amount of Fe^{II} provides clear evidence that the atmosphere at that time was reducing. The extremely low lead isotope ratios $n(^{206}\text{Pb})/n(^{204}\text{Pb})$ and $n(^{207}\text{Pb})/n(^{204}\text{Pb})$ in the sediment's galena crystals PbS, the lowest that have been observed in any terrestrial material, made the dating possible.

^{40}K decays into ^{40}Ca and ^{40}Ar. Because calcium is widespread in silicates, it is not suitable for age determinations. In contrast, the noble gas argon, not otherwise occurring in rocks, is very suitable provided that none of the radiogenic Ar has been lost to the atmosphere.

Natural potassium contains 1.17 % ^{40}K. As it decays, 11 % of the product is ^{40}Ar, and the rest Ca.[c] The age of a mineral may be determined through the ratio $n(^{40}\text{Ar})/n(^{40}\text{K})$ using an equation analogous to Equation 1.12. At low temperature, Ar atoms cannot escape from the site of the parent K ion in the mineral's crystal lattice, because the two atoms have nearly the same radius. At higher temperatures,

---

[a] Gk. χρόνος $\hat{=}$ time.
[b] From Figure 1.8 and Equation 1.12 : $t = 14.0$ Ga $\times \ln 1.148 / \ln 2 = 2.79$ Ga.
[c] Use of Equation 2.50 (with $\mu = 0$), eq 2.56 and the half-lives from Table 1.12 yields:
  $k_{\text{K}\rightarrow\text{Ar}} / (k_{\text{K}\rightarrow\text{Ar}} + k_{\text{K}\rightarrow\text{Ca}}) = 0.11$ and $k_{\text{K}\rightarrow\text{Ca}} / (k_{\text{K}\rightarrow\text{Ar}} + k_{\text{K}\rightarrow\text{Ca}}) = 0.89$.

argon diffuses away and age determinations become inaccurate, but on the other hand, one may gain the temperature history of the sample as follows: About 93.26 % of naturally occurring potassium is ^{39}K. A known fraction of this can be transformed into ^{39}Ar using neutron activation. The amount of $n(^{39}$Ar$)$ is proportional to $n(^{40}$K$)$, and determination of the ratio $n(^{40}$Ar$)/n(^{39}$Ar$)$ gives the same result as before. However, now it is straightforward to obtain the result by heating small samples directly into the mass spectrometer. Measurements made on very thin slices of a mineral grain can be used to give a temperature profile of the sample.

The dating of continental rocks with the Rb-Sr and Sm-Nd methods shows that there have been several continent-forming periods in Earth's history, with the first possibly forming from 3.0 to 2.5 Ga BP. The oldest supercontinent[a] known with certainty is Rodinia. It formed between 1.3 and 1.0 Ga BP and fragmented at 750–600 Ma BP into pieces called Gondwana and Laurentia. These two recombined from 450 to 320 Ma BP into a final supercontinent called Pangea, which broke up into the parts of the world that we know today around 160 Ma BP.[16] The latter facts have been established mainly using the K-Ar and ^{40}Ar-^{39}Ar methods, together with the study of the remanent magnetism of the ocean floor (see Section 1.4a).

Example: Radioactive decay and lung cancer

The noble gas radon is formed as a decay product in the radioactive series that starts with ^{238}U. Radon can escape from the rocks in which it is formed. It is quickly transformed via several radioactive species, including an alpha-emitting lead isotope, into the stable product ^{206}Pb:

$$^{238}\text{U}\,(4.47\,\text{Ga}) \quad \cdots \rightarrow \quad ^{226}\text{Ra}\,(1.59\,\text{ka}) \quad \rightarrow$$
$$^{222}\text{Rn}\,(3.83\,\text{d}) \quad \cdots \rightarrow \quad ^{210}\text{Pb}\,(22\,\text{a}) \quad \cdots \rightarrow \quad ^{206}\text{Pb} \qquad (1.13)$$

Thus, there is a risk that residents of houses built on ground containing uranium will inhale radioactive radon and lead (in house dust) with consequent health effects. An increase in the level of background radiation is also found in buildings containing granite (e.g., churches from the Early Middle Ages) because of their uranium content. In the beginning of the 21st century, 20,000 U.S. citizens died each year of radon-related lung cancer (2005), the most common cause of lung cancer after tobacco smoke.

c. Isotopic fractionation

Before 1960, physicists and chemists used slightly different definitions of atomic mass, both based on the mass of an oxygen atom from water. However, studies of groundwater showed that its isotopic composition varied depending on the source, giving this mass scale an uncertain foundation.[157] The adaptation of a new scale

---

[a] A supercontinent is a large continent composed of several or all of the existing continents.

in which the $^{12}_{6}C$ isotope is assigned the mass 12 Da overcame the problem.[a] The choice of carbon-12 was motivated by its utility for mass spectrometry.[192] Presently (2008), 2,850 nuclides are known, and mass spectrometry is an indispensable tool in environmental studies.

Isotope effects are largest for light elements and for reactions that are carried out at low temperature. Therefore, isotopic fractionation is especially pronounced for the nonradiogenic isotopes that are most widespread in rocks and biota: hydrogen, carbon, nitrogen, oxygen, and sulfur. Studies of the distributions of stable isotopes are the foundation of our present understanding of the chronological development of climate and biota. Similarly, isotopic studies have been the basis for our understanding of such important parts of geochemistry as sedimentary rocks and magmatic rocks that have been subject to hydrothermal reactions. In addition, isotope effects are useful in the study of atmospheric chemistry and transport.

In order to make a quantitative determination of isotopic effects, the ratio, r, between the concentrations of the individual isotopes in the sample is determined experimentally. In the case of ^{18}O, one would measure $r = n(^{18}O)/n(^{16}O)$. The deviation of this ratio from that of an accepted reference sample is defined as

$$\delta(^{18}O) \quad = \quad \frac{r_{\text{sample}} - r_{\text{reference}}}{r_{\text{reference}}} \tag{1.14}$$

The δ is positive if the sample is enriched in the heavy isotope relative to the reference material, and negative if it is depleted.

One example is the determination of the relative concentration of oxygen isotopes in rocks. BrF_5 can be used to oxidize oxygen from SiO_2 to the free element, which is converted to CO_2 using graphite.[b] The oxygen isotopes are then determined by mass spectrometry of the CO_2.

Hydrogen and oxygen

The isotopes in water are ^1H, ^{16}O, ^{17}O, ^2H (D), and ^{18}O (see Table 3.17). Most work is concerned with the determination of $\delta(^{18}O)$ and $\delta(^2H)$, for which the symbol $\delta(D)$ is normally used. The internationally accepted reference material used to calibrate mass spectrometers is "Vienna Standard Mean Ocean Water,"[c] for both $\delta(^{18}O)$ and $\delta(D)$.

Meteoric water is water originating from precipitation. Its isotopic composition reflects the temperature at which water liquefied or solidified in the atmosphere. Such temperature-dependent isotopic fractionation is especially pronounced for hydrogen. As an example, the value of $\delta(D)$ in meteoric water varies over the North American continent from $-30\,‰$ in Louisiana (ca. $30°$ N) to $-170\,‰$ in Alaska (ca. $65°$ N).

Table 1.13 gives some of the intervals of $\delta(^{18}O)$ values for different classes of materials. The broad range of values for meteoric water is noted. There is an empirical

---

[a] See Appendix A1.4.

[b] For example, the reaction with the feldspar albite at 450 °C is
$$NaAlSi_3O_8 + 8\,BrF_5 \rightarrow NaF + AlF_3 + 3\,SiF_4 + 4\,O_2 + 8\,BrF_3.$$

[c] In specialist literature, the acronym is VSMOW; despite the name, it contains no salt.

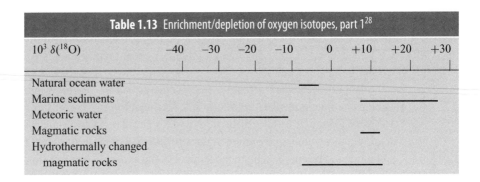

Table 1.13 Enrichment/depletion of oxygen isotopes, part 1[28]

$10^3 \, \delta(^{18}O)$	−40	−30	−20	−10	0	+10	+20	+30
Natural ocean water						▬		
Marine sediments							▬▬▬▬	
Meteoric water		▬▬▬▬▬▬▬						
Magmatic rocks						▬		
Hydrothermally changed magmatic rocks						▬▬▬▬		

relationship between meteoric water's isotopic enrichment and the temperature at which it condensed:[192]

$$\delta(^{18}O) = (\theta/^\circ C - 20)/1490 \tag{1.15}$$

Loosely stated: the warmer the climate, the larger the value of $\delta(^{18}O)$ in the precipitation. By using both $\delta(D)$ and $\delta(^{18}O)$ it is possible to obtain both the temperature of evaporation and condensation. Isotope ratios are used to determine the Earth's temperature record by analyzing annual precipitation in cores taken from the Greenland and Antarctic glaciers (Chapter 10).

Trends in the enrichment and depletion of oxygen isotopes in rocks are given in Table 1.13. Normal $\delta(^{18}O)$ values are +7 to +12 ‰ for granites and +5 to +7 ‰ for basalts. Higher values indicate "contamination" by sedimentary rocks, which generally have higher values of $\delta(^{18}O)$.

Important hydrothermal chemistry takes place under circumstances where water is above the critical temperature, 374°C. Consider the chemical equilibrium between quartz, water, and silicic acid:

$$SiO_2(s) + 2\,H_2O(critical) \;\rightleftarrows\; H_4SiO_4(aq) \tag{1.16}$$

where orthosilicic acid is dissolved in the supercritical water. Figure 1.9 shows the distribution of ^{18}O between quartz and water as a function of temperature. The value of $\delta(^{18}O)$, measured relative to the reactant water, becomes negative at temperatures higher than about $810\,^\circ C$.

Carbon

Carbon has two stable isotopes, ^{12}C ($x \approx 0.9889$) and ^{13}C ($x \approx 0.0011$), and the value of $\delta(^{13}C)$ in a sample is defined by the ratio $r = n(^{13}C)/n(^{12}C)$ via Equation 1.14. An early reference material for carbon isotopes was calcite, $CaCO_3$, from Cretaceous belemnite fossils found in the bank of the Pee Dee River in South Carolina, USA.[a] For practical reasons this was changed in 1995 to Vienna Pee Dee Belemnite, a calcium carbonate.[b]

[a] In specialist literature, the acronym is PDB, Pee Dee Belemnite.
[b] In specialist literature, the acronym is VPDB.

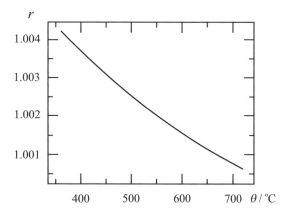

Figure 1.9 *Enrichment/depletion of oxygen isotopes, part 2*
Hydrothermal oxygen isotopic fractionation between quartz and water above its
critical temperature.
$r = [n(^{18}O)/n(^{16}O)]_{quartz}/[n(^{18}O)/n(^{16}O)]_{water}$, cf. eq 1.16.

The process of photosynthesis in bacteria and plants fractionates carbon isotopes,
in part because $^{12}CO_2$ diffuses more quickly through cell walls than $^{13}CO_2$ and
partly because of the greater reaction rate of ^{12}C in photosynthesis. The isotopic
fractionation of photosynthesis results in residual carbon dioxide gas that is enriched
in $^{13}CO_2$ and plant carbohydrates that are depleted in $^{13}CO_2$. Photosynthesis gives
rise to a difference of $\delta(^{13}C) \approx -28$ ‰. One example of the use of isotopic analysis
is to determine what fraction of atmospheric carbon dioxide is taken up by the
fractionating process of photosynthesis as opposed to the virtually nonfractionating
process of solvation in ocean water. The discovery of organic material with very low
$\delta(^{13}C)$ values in 3.8 Ga BP rocks is evidence of the activity of the first prokaryotic cells.

Sulfur

Sulfur has four stable isotopes, two of which are quite common, ^{32}S ($x \approx 0.9502$) and
^{34}S ($x \approx 0.0421$).[120] The value of $\delta(^{34}S)$ is defined through the ratio $n(^{34}S)/n(^{32}S)$ as
shown previously, the reference material being troilite, FeS, from the Cañon Diablo
meteorite.[a] Table 1.14 shows the ranges of $\delta(^{34}S)$ values found in different classes of
materials.

There is some variation of $\delta(^{34}S)$ in acidic volcanic rocks (granites), but the
variation in sulfide- and sulfate-containing sediments arising from anaerobic sulfate-
reducing bacteria is especially pronounced. These bacteria convert the sulfate naturally found in the oceans into sulfide:

$$2\,CH_2O(biota) + SO_4^{2-}(aq) \;\rightarrow\; H_2S(aq) + 2\,HCO_3^-(aq) \qquad (1.17)$$

The sulfide reacts with iron(II) in the sediment to form pyrite, FeS_2, which is a
disulfide of Fe^{II}. Bacteria react more quickly with the lighter isotope, leading to an

---

[a] In specialist literature, the acronym is CDT. In 1997, the troilite scale was replaced by the "Vienna-
Cañon-Diablo-Troilite" scale, VCDT, which uses a standard silver sulfide, Ag_2S, as reference.

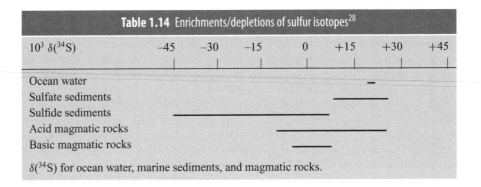

Table 1.14 Enrichments/depletions of sulfur isotopes[28]

$10^3\ \delta(^{34}S)$	−45	−30	−15	0	+15	+30	+45
Ocean water					−		
Sulfate sediments						———	
Sulfide sediments		—————————————					
Acid magmatic rocks				—————			
Basic magmatic rocks				—			

$\delta(^{34}S)$ for ocean water, marine sediments, and magmatic rocks.

enrichment of ^{32}S in pyrite, whereas the residual sulfate contains relatively more ^{34}S. These relationships are reflected in Table 1.14, which in particular shows the observable result of bacterial metabolism in the past (sulfate and sulfide sediments). Such isotopic separation is ubiquitous under anaerobic conditions in mud in brackish water.

The data of Table 1.15 show the competition as a function of time between the processes of bacterial production of isotopically heavy sulfate and the weathering of rocks, which produces isotopically light sulfate. The larger values of $\delta(^{34}S)$ in the Ordovician, Devonian, Carboniferous, and Tertiary periods relative to modern values are due to the past activity of life (Section 1.4).

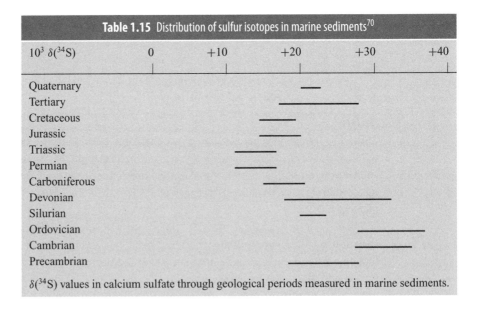

Table 1.15 Distribution of sulfur isotopes in marine sediments[70]

$10^3\ \delta(^{34}S)$	0	+10	+20	+30	+40
Quaternary			—		
Tertiary			———		
Cretaceous			—		
Jurassic			—		
Triassic		—			
Permian		——			
Carboniferous		—			
Devonian			———		
Silurian			—		
Ordovician				———	
Cambrian				———	
Precambrian			———		

$\delta(^{34}S)$ values in calcium sulfate through geological periods measured in marine sediments.

Example: Use of mass spectrometry

Lead poisoning disturbs the victim's judgment and causes death.

During the British Franklin Expedition of 1845 to 1848, 129 men sought the Northwest Passage, which Franklin hoped would allow boats to sail from the Atlantic to the Pacific. Mysteriously, the entire crew perished. In the 1980s, some of the bodies, which had been preserved in the Arctic cold, were recovered and brought to research laboratories for examination.[221] Using atomic absorption spectroscopy, it was found that they all contained lead in toxic quantities. Further, mass spectrometry showed that the ratio $n(^{208}Pb)/n(^{204}Pb)$ to $n(^{206}Pb)/n(^{204}Pb)$ in the bodies was the same as in the solder used to seal the seams of the cans that stored the food for the expedition. This isotopic ratio turned out to be unique and was not seen in contemporary bones of Inuit or of North American reindeer, or in cans from the second half of the 19th century. It was concluded that the Franklin expedition's fate was sealed by lead from their canned food.[186]

The utility of isotopic analysis in environmental science is obvious.

1.4 Features of the Earth's development

Modern science arose in the 17th century with the work of the philosopher Francis Bacon, the astronomer Johannes Kepler, and the physicist Galileo Galilei. At that time, maps of the Earth were sufficiently detailed that Bacon (1620) could observe that the east coast of South America resembled the shape of the west coast of Africa, such that the two continents could be put together like the pieces of a puzzle (Figure 1.10). In the middle of the 19th century, the geological similarities (von Humboldt, 1845) and related fossils (Sneider-Pellegrini, 1858) of the coasts were described. Explanations ranged from the Biblical flood theory and the theory that the moon was taken to form the Atlantic basin, to the theory that the Earth is expanding, creating space for an increase of the oceans.

a. Plate tectonics

The idea that the continents were once (merely 160 Ma BP) united but have since split was proposed by Wegener (1912).[16] The most important evidence is the observation that the continents and the ocean floor have fundamentally different compositions. Geological and geophysical measurements since the 1960s have given Wegener's theory a solid experimental foundation. Today we have a mature science called *plate tectonics* that describes the dynamics of the Earth's crust.[16]

The Earth's current status is as follows. Its surface area does not change with time. The lithosphere, composed of six large, seven small, and many very small plates, rests on the viscous asthenosphere. At some locations, the plates move away from one another at a rate of 2 to 10 cm per year. At the same time, material pours up from the Earth's interior, resulting in new ocean floor. At other places, the plates collide with one another, burying surface and/or building mountains. Finally, the plates may move past one another without changing the composition of the surface. Practically all volcanic, seismic, and tectonic activity takes place at the borders between the plates, which themselves are geologically stable. The largest plates are shown in Figure 1.10, which also shows the most important construction and destruction zones.

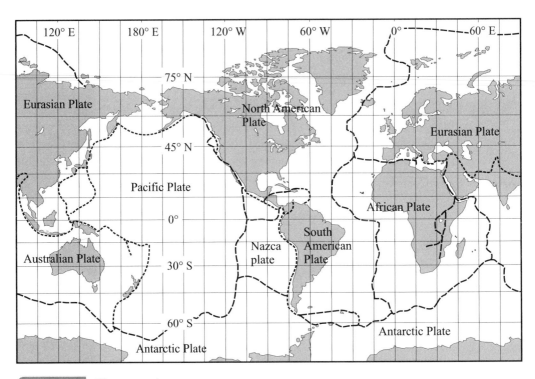

Plate tectonics

The map shows the largest tectonic plates. The convergence zones are marked with dashed lines; they comprise zones of subduction (e.g., west of South America) and collision (e.g., the Himalaya Mountains). Divergent zones are marked with larger dashed lines; they include construction zones with spreading faults (e.g., the mid-Atlantic ridge) and sideways faults (e.g., the border between the Pacific Ocean plate, which is displaced to the west, and the Antarctic plate).

The Wilson cycle

Figure 1.11 details the main stages of the so-called Wilson cycle (1966), the most comprehensive of the Earth's geochemical cycles (see Chapter 7). At the beginning of the process, all land surface was located in a supercontinent. This was torn apart because of a combination of mechanical stress and physical weakness similar to that found today in the East African rift valley. Basaltic magmas flowed into the weak areas and divided the plates (see Figure 1.11A). This process is active today under the Red Sea, and in a more developed form along the mid-Atlantic ridge. Here the sea floor slowly moves away from the fault, and only minor volcanic activity occurs along the margins of the ocean; sedimentary debris collects on the sea floor as shown in Figure 1.11B. Eventually a margin fractures, and the plate slides under the continental plate, disappearing into the depths, where it melts. This process is taking place under the Pacific Ocean east of Japan where the Pacific plate slides under the Eurasian plate. Such subduction is associated with violent volcanic activity (Figures 1.11C and D). Deep ocean trenches are also associated with these zones, reaching depths of about 10 km under the ocean surface. In addition, the tectonic cycle involves the collision

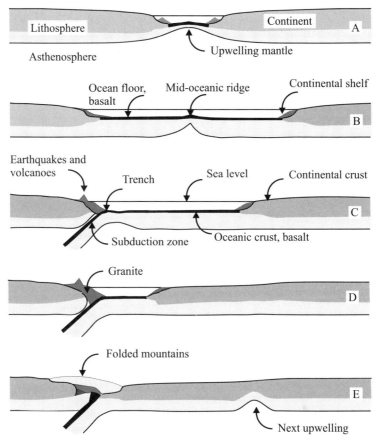

Figure 1.11 *The Wilson cycle*[28]
The events begin at A, and the cycle is completed at E.

of continents, resulting in the formation of mountain chains (e.g., the Ural Mountains in the Permian era, and the Himalayas today; see Figure 1.11E). Recrystallization of rocks in the subduction zones gives granite.

Magnetism

The so-called anomalous magnetization found in bands parallel to the midocean ridges is an important indication of sea floor spreading. The Earth generates a magnetic flux density, with a magnitude of 30 to 60 μT at the surface,[a] but for reasons that are not completely understood, the direction of this field changes every 0.1 to 1 Ma. As the ocean floor spreads, extruded basaltic magmas cool and solidify. While the material is plastic, naturally occurring magnetic crystals of magnetite, Fe_3O_4, align with the Earth's magnetic field, and this orientation is frozen in place when the magma cools. The direction and magnitude of the residual magnetism of the ocean floor has been measured using ships and airplanes with sensitive magnetometers and in samples

---

[a] The magnetic induction; SI-unit tesla; an old unit is gauss, 1 G = 10^{-4} T.

taken from the bottom. As the ocean floor spreads outward from the center, the history of Earth's magnetic field is recorded in the residual magnetization, which is displayed symmetrically on each side of the ridge.

Curie temperature

The most important minerals that cause rock magnetism are magnetite, Fe_3O_4; hematite, Fe_2O_3; and ilmenite, $FeTiO_3$. Above a certain temperature, the Curie temperature, the three minerals are simple paramagnets, that is, their unpaired electrons are unordered. When cooled below the Curie temperature (500 °C to 600 °C) these minerals become ferromagnetic, that is, a phase transition takes place, the electron spins align in domains, and the material becomes magnetic.

b. Chemistry mediated by water and biota

Sediments found in Isua, West Greenland, show that the oceans existed 3.8 Ga BP, and that atmospheric carbon dioxide had already reacted with the water to form hydrogencarbonate.[a] This in turn reacted with silicates from volcanic rocks to form carbonates and hydrated quartz in aqueous reactions such as the following:

$$CO_2(g) + H_2O(l) \rightarrow H^+(aq) + HCO_3^-(aq) \tag{1.18a}$$

$$CaSiO_3(s) + HCO_3^-(aq) + H^+(aq) + H_2O(l)$$
$$\rightarrow CaCO_3(s) + H_4SiO_4(aq) \tag{1.18b}$$

$$H_4SiO_4(aq) \rightarrow SiO_2 \cdot aq(s) + H_2O(l) \tag{1.18c}$$

Water and carbon dioxide very probably originated from the Earth's outer mantle and crust and escaped as the planet formed and cooled. Under conditions of 30 kbar pressure and 1.9 kK temperature, minerals such as pyroxenes[b] dissolve up to 25 % H_2O and 5 % CO_2, and these are also the main components of modern volcanic gases. Part of this CO_2 is formed through metathesis in igneous rocks:

$$CaCO_3(s) + SiO_2(s) \rightarrow CaSiO_3(s) + CO_2(g) \tag{1.19}$$

If all of the carbon dioxide currently bound in deposited carbonates were in gaseous form, as is thought to have been the case 4.5 Ga BP, the CO_2 pressure in the atmosphere would have been about 60 bar. This is not so different from the composition of the atmosphere of the planet Venus, which is 96 % CO_2 at a total pressure of 91 bar (see Table 1.3).

When the Earth was young, solar irradiation (insolation[c]) was weaker than it is today, and the increased CO_2 content resulted in a greenhouse effect that kept the Earth's surface temperature above the freezing point of water. At later times the silicate/carbonate equilibrium described earlier removed CO_2, with further CO_2

---

[a] See Figure 1.4.
[b] See Figure 3.2.
[c] The insolation formula, Equation 10.29, is discussed in Section 10.2a.

reductions occurring because photosynthetic microorganisms increased the level of oxygen in the atmosphere.

The released oxygen partially oxidized dissolved Fe^{II} in the oceans, which precipitated as the *banded iron formations* found today.[a] They have a ratio of $n(Fe^{II})/\{n(Fe^{II}) + n(Fe^{III})\}$ in the range of 0.3 to 0.6, which is the approximate composition of magnetite, Fe_3O_4. Isua contains the oldest deposits of Fe_3O_4 from 3.8 Ga BP. More than 1.5 Ga later, large amounts of banded iron were deposited at several places, such as in the Lake Superior region, USA.

The continuing photosynthetic production of oxygen is reflected in immense layers of pure iron(III) oxide. For example, these can be seen in Provence, France, where the town of Roussillon is located on top of a 100-m-thick cement of red Fe_2O_3 (as the mineral ochre). Beneath the iron(III) oxide is a layer of green iron(II) oxide. It was not possible for oxygen to accumulate in the atmosphere before the dissolved iron(II) in the ocean had been oxidized. It is interesting to note that once this was completed, atmospheric oxygen slowly increased until it reached the present level, just below that required for wood to burn spontaneously.

c. Global chemistry of life

The Earth's collective mass of organic material increased more during the Carboniferous age than in any other geological period. The swamps of the time were well suited for storing huge peat deposits, which, via diagenesis,[b] were transformed into the present-day deposits of coal and oil. Large quantities of carbon dioxide were removed from the atmosphere and sea and transformed into biota by photosynthesis:

$$CO_2(aq) + H_2O(l) \overset{h\nu}{\rightarrow} CH_2O(biota) + O_2(g) \tag{1.20}$$

Although this process consumes carbon dioxide, the fossil records show that neither the amount of carbon dioxide in the atmosphere nor the concentration of hydrogencarbonate in the ocean changed significantly. This means that there must have been a significant simultaneous input of material from the surrounding inorganic world.

Reactions such as Equation 1.19 release CO_2 from the upper mantle. Other sources are weathering of the ubiquitous carbonate minerals calcite, $CaCO_3$; dolomite, $(Ca,Mg)CO_3$; and magnesite, $MgCO_3$:

$$CO_2(aq) + MCO_3(s) + H_2O(l) \rightarrow M^{2+}(aq) + 2HCO_3^-(aq) \tag{1.21}$$

where M stands for Ca or Mg.

Hydrogencarbonate is the actual reactant in the buildup of biota:

$$HCO_3^-(aq) + H_2O(l) \overset{h\nu}{\rightarrow} CH_2O(biota) + O_2(g) + HO^-(aq) \tag{1.22}$$

Some calcium may precipitate as biogenic $CaCO_3$ via the reverse of reaction 1.21, but the precipitation of gypsum, $CaSO_4 \cdot 2H_2O(s)$, is also an important process:

$$Ca^{2+}(aq) + SO_4^{2-}(aq) + 2H_2O(l) \rightarrow CaSO_4 \cdot 2H_2O(s) \tag{1.23}$$

---

[a] In specialist literature, the acronym is BIF.
[b] See Section 3.1.

Magnesium ions react with quartz, SiO_2, and water to give silicates, here represented as a metasilicate:

$$Mg^{2+}(aq) + SiO_2(s) + H_2O(l) \rightarrow MgSiO_3(s) + 2H^+(aq) \quad (1.24)$$

Isotopic studies show that the sea's sulfate concentration was (and is) constant. Accordingly, the system must have had a source of sulfur, and the obvious source is pyrite, FeS_2, which is found in large quantities in igneous rocks and as a hydrothermal mineral:

$$2\,FeS_2(s) + \frac{15}{2}O_2(g) + H_2O(l)$$
$$\rightarrow 2\,Fe^{3+}(aq) + 4\,SO_4^{2-}(aq) + 2\,H^+(aq) \quad (1.25)$$

This reaction simultaneously removed most of the excess oxygen produced by photosynthesis, and it was completed by the hydrolysis and diagenesis of iron(III) to hematite:

$$2\,Fe^{3+}(aq) + 3\,H_2O(l) \rightarrow Fe_2O_3(s) + 6\,H^+(aq) \quad (1.26)$$

This is in accord with the fact that hematite is rare in igneous rocks but common in sediments and their metamorphosed products (see Chapter 3).

To sum, the global process of photosynthesis after the Precambrian is approximately:

$$calcite + magnesite + quartz + pyrite + water$$
$$\rightarrow biota + oxygen + silicates + gypsum + iron(III)\ oxide \quad (1.27)$$

in order to account for the significant quantities of matter and energy required to achieve the reaction

$$carbon\ dioxide + water \xrightarrow{h\nu} biota + oxygen \quad (1.28)$$

The qualitative nature of Equation 1.27 is obvious because the equation is composed of independent reactions. However, each process illustrates important chemistry. For example, the photosynthetic reaction, Equation 1.22, produces base, whereas *all* oxidations of nonmetals using oxygen in aqueous environments yield acids (Lavoisier, 1776).

2 Environmental dynamics

2.1 Introduction

The environment is in a constant state of change. In order to describe this adequately, this chapter reviews *environmental dynamics*, the unification of fluid dynamics, chemical thermodynamics, and chemical kinetics.

An important goal of environmental chemistry is to model Nature, including predicting the fate of a given chemical species, whether of natural or anthropogenic origin. The purpose of this chapter is to establish the nomenclature used in the book and to present fundamental concepts and expressions. It is particularly important to derive the explicit time dependence of the relevant physical quantities. In separate sections of this introduction, we first describe and define a few basic concepts, and then introduce important concepts for the time dependence of concentration, field, and transport.

a. Basic concepts

System and phase

The concepts of system, surroundings, and phase used in thermodynamics are the cornerstones of any environmental investigation. A *system* is the part of the world that has been chosen as the subject of study; it may contain matter, in which case it also contains energy. Everything else is said to belong to the *surroundings*. The separating boundary may be real or imaginary, but must be defined before the study begins. Depending on the properties of the boundary, one can define three types of system. An *open system* is one that can exchange substance and energy (work, heat, radiation) with the surroundings. A *closed system* cannot exchange substance with the surroundings, and an *isolated system* can exchange neither substance nor energy with the surroundings. Except in the last case, the surroundings enter the physical description, because they must have certain properties as sources or sinks for the exchanged physical quantities.[a]

All systems considered in this book are *macroscopic systems*, that is, systems for which bulk thermodynamic properties apply. The lower limit is somewhere around 10^{10} (small) molecular entities, depending on the properties considered. As an example, the number of entities of a liquid must be sufficiently large to make a surface.

---

[a] In some applications of fluid mechanics, *systems* are defined differently; for example, the system called *control volume* may have properties that are here considered as belonging to the surroundings.[69]

The *state of a system* is specified by the values of all measurable physical quantities, including the chemical composition. The system is said to be in a state of equilibrium when *all* of its physical quantities are independent of time.[a]

A system whose intensive properties[b] are uniform is called a *homogeneous system* or a *phase*. Any system consists at least of one phase, and systems with more than one phase are called *heterogeneous systems*. A phase that consists of one chemical species is a *pure substance* and can be characterized by an empirical formula. A phase that contains more than one experimentally discernible chemical species is called a *mixture*.

The number of independent intensive properties for any system in equilibrium is determined by the *phase rule* (see Equation 2.178). When applied to a phase, it asserts that the state is given by just two intensive properties; with no chemical reactions, they are temperature and pressure (see Equation 2.6).

Fluid dynamics

Fluid dynamics is the science of the motion of fluids (i.e., gases and liquids). It deals with the kinematics and dynamics of fluid matter as well as the transfer of mass, amount of substance, energy, and entropy. The systems studied are normally not in a state of equilibrium in either the mechanical or the thermodynamic sense. The methods and results of fluid dynamics find their application in the description and calculation of currents in the atmosphere, the oceans, gas wells, groundwater, and more viscous fluids such as muds and lava.

The major physical laws in this area are the continuity equation (Equation 2.1) and the equation of motion (Equation 2.3). They are discussed in some detail later, but the following survey introduces their scope of application.

Consider a system that is a volume element of a fluid. One use of the continuity equation is to equate the rate of concentration change r_B of a chemical species B with the time derivative of the concentration due to chemical reactions within the volume element plus the changes due to flows of B across the borders of the element:

$$r_B = \frac{\partial c_B}{\partial t} + \boldsymbol{div} \, \boldsymbol{J}_B \qquad (2.1)$$

Here, c_B is the concentration (amount density), $\partial c_B/\partial t$ is its time derivative, and "flows" are quantitatively described by the flux density $\boldsymbol{J}_B = c_B \, \boldsymbol{u}$, where $\boldsymbol{u}$ is the velocity, and the divergence operator enables a calculation of the difference between input and output of B in the element. If the system is closed and homogeneous, then Equation 2.1 reduces to the usual expression for the rate of reaction in a vessel:

$$r_B = \frac{d c_B}{d t} \qquad (2.2)$$

because the rate is now independent of spatial coordinates.

---

[a] Note the discussion of a steady state of an open system, Section 2.2c.

[b] Examples: extensive properties: mass, volume, amount of substance, energy, and entropy; intensive properties: pressure, temperature, chemical potential, and the ratio between any two extensive quantities.

For an infinitesimal volume of a fluid, Newton's second law is Euler's equation. It may be written in the form

$$\frac{d\boldsymbol{J}_m}{dt} = -\,\boldsymbol{grad}\ p \tag{2.3}$$

where the time derivative of $\boldsymbol{J}_m = \rho\boldsymbol{u}$ ($\boldsymbol{J}_m$ is the flux density with respect to mass, ρ is the mass density) is equal to the gradient of pressure (i.e., a pressure difference per length, such as winds in atmospheric systems).

Chemical thermodynamics

Thermodynamic quantities are physical quantities, such as temperature, pressure, chemical potential, volume, amount of substance, energy, and entropy, that do not depend on the rate at which something happens.[67] However, the standard machinery of thermodynamics is well suited for application to properties that depend on time, and we shall see how time dependence is included in the description of environmental processes (see Section 2.3a).

For an infinitesimal change in the state of a system, one may write

$$dU = TdS - pdV + \sum_B \mu_B dn_B \tag{2.4}$$

No experiments are known that have contradicted this equation, the fundamental equation for change of state, which is considered to be an axiom. The equation connects the changes of the extensive quantities internal energy U, entropy S, volume V, and amount of substance n_B with the intensive quantities: temperature T, pressure p, and chemical potential μ_B. Thermodynamics lays down procedures that transform Equation 2.4 into equivalent forms where the roles of extensive and intensive quantities have been interchanged.[137] Most useful for our applications are the infinitesimal change of the Gibbs energy dG:

$$dG = -SdT + Vdp + \sum_B \mu_B dn_B \tag{2.5}$$

and the Gibbs-Duhem equation

$$0 = -SdT + Vdp - \sum_B n_B d\mu_B \tag{2.6}$$

Most environmental phenomena have their foundation in one of these three equations.

Chemical kinetics

Chemical kinetics addresses the rates of chemical reactions and is important for understanding how quickly systems approach equilibrium and the interplay between transport and reaction. As we shall see from Equation 2.161, the rate of entropy change is fundamental to the applications: examples from atmospheric chemistry include the fate of polluted air and reactions initiated by the energy of sunlight.

The rate of concentration change across an interface between two phases is a complicated subject. Among the examples to be discussed are (1) the solid-aqueous solution interfaces in soils; this chemistry includes the ion-exchange properties of clay minerals and humus, and (2) the ocean-atmosphere interface where an important part of the global carbon dioxide cycle takes place.

b. Time dependence of concentration

Measures of mass concentration and amount-of-substance concentration of a phase are given in Table A1.7.

The rate of concentration change

The change of the concentration of a species B as a function of time is a key parameter in environmental chemistry. Basically, we have the definition

$$c_B(t) = \frac{n_B(t)}{V(t)} \tag{2.7}$$

for *concentration* and

$$r_B = \frac{dc_B}{dt} \equiv \dot{c}_B \tag{2.8}$$

for the *rate of concentration change* of the species B in a phase. The dot indicates the derivative with respect to time.[a] We stress that r_B may also be defined using the partial pressure (p_B), number concentration (C_B), surface concentration (Γ_B), and so forth, with analogous expressions.

Logarithmic differentiation of Equation 2.7 shows that[b]

$$d \ln c_B = d \ln n_B - d \ln V \tag{2.9}$$

leading to

$$r_B = \frac{\dot{n}_B}{V} - c_B \frac{d \ln V}{dt} \tag{2.10}$$

The second term on the right-hand side of Equation 2.10 corrects for the change in concentration brought about by a change in volume.

The definition of concentration as a scalar field is given in Section 2.1c.

Example

Assume a closed phase (i.e., a homogeneous system) where the amount of species B is constant and the volume is changing: for example, $\dot{V} = \beta V_0$, where β is a constant and V_0 is the volume at $t = 0$. Then $V(t) = V_0 (1 + \beta t)$, and it follows from Equation 2.9 or directly from Equation 2.7 that

$$c_B(t) = \frac{n_B}{V_0(1 + \beta t)} \tag{2.11}$$

---

[a] The dot does *not* indicate a partial derivative with respect to time. Further, we shall not use a prime to indicate a derivative with respect to the argument of a function of one variable.

[b] First take the logarithm, then differentiate, and finally, use the definition: $d \ln x = x^{-1} dx$.

The extent of a chemical reaction

The law of conservation of mass states that the increase in mass of the products of a chemical reaction is equal to the decrease in mass of the reactants. The law was used by Lavoisier (1789) but was not proved because he considered it to be an axiom, handed down from the ancient philosophers.[a] Experimental verification was performed by Landolt (1909), who spent 17 years carrying out the most meticulous experiments ever made in classical chemistry.

Consider a closed phase. The only way the amount of substance of the phase can change is by chemical reaction (or nuclear decay). The stoichiometric[b] laws for chemical reactions were formulated around 1800, and some years later Berzelius (1814) created modern notation. Let the balanced reaction equation be written in the generic form

$$cC + dD + \cdots = \cdots + yY + zZ \qquad (2.12)$$

Here, uppercase (roman) letters designate chemical species and lowercase (italic) letters stoichiometric coefficients. We use the letters C, D, $\cdots$ for reactants and $\cdots$, Y, Z for products. Basically this stoichiometric equation expresses ratios between amounts of substances:

$$\frac{n_C}{c} = \frac{n_D}{d} = \cdots = \frac{n_Y}{y} = \frac{n_Z}{z} \qquad (2.13)$$

All classical quantitative analysis is based on Equation 2.13.[c]

In order to apply Equation 2.13 to physicochemical problems we proceed by defining *stoichiometric numbers v* by means of the stoichiometric coefficients of the reaction equation:

$$v_C = -c; \; v_D = -d; \; \cdots; \; v_Y = y; \; v_Z = z \qquad (2.14)$$

They are negative for reactants and positive for products. The reaction equation may now be written in a very general form:

$$\sum_B v_B B = 0 \qquad (2.15)$$

where B is a generic symbol for a molecular species. However, the following use of the v's is more important: changes in the amounts of the chemical species entering Equation 2.13 are constrained, and using the sign convention of Equation 2.15, we have the relation

$$\frac{dn_C}{v_C} = \frac{dn_D}{v_D} = \cdots = \frac{dn_Y}{v_Y} = \frac{dn_Z}{v_Z} = d\xi \qquad (2.16)$$

which defines the differential of the *extent of reaction*, ξ. The formal definition is

$$d\xi \overset{\text{def}}{=} \frac{dn_B}{v_B} \qquad (2.17)$$

---

[a] In 55 BC, Lucretius wrote in his book *De Rerum Natura* (On the Nature of Things): *nil posse creari de nihilo* (nothing can be created from nothing). Since then, the law has been considered to be an axiom.
[b] From stoichiometry: Gk. $\sigma\tau o\iota\chi\varepsilon o\nu \hat{=}$ element; $\mu\varepsilon\tau\rho o\nu$ = measurement.
[c] See Equation 3.10d and *Example 2* in Section 5.3b.

for all B. Note that the extent is defined only with respect to a particular reaction equation. Integration of Equation 2.17 gives:

$$\Delta n_B = v_B \Delta \xi \tag{2.18}$$

The change of the extent $\Delta \xi = 1$ mol gives precision to the phrase "1 mol has been converted according to Equation 2.12."

The rate of a chemical reaction

Consider a closed system in which chemical reaction 2.12 is taking place and for which the stoichiometric numbers are time independent. One defines the *rate of conversion, $\dot{\xi}$*:

$$\dot{\xi} = \frac{d\xi}{dt} \tag{2.19}$$

Note that *this quantity requires knowledge of the balanced reaction equation* that makes the definition eq 2.17 possible.

The progress of the reaction as a whole may be characterized using the *rate of the chemical reaction v*:

$$v = \frac{\dot{\xi}}{V} \tag{2.20}$$

where V is the volume of the system. The introduction of the volume means that the quantity v describes concentrations, and for a general chemical species B, one has[a]

$$v = \frac{1}{v_B} \frac{d[B]}{dt} \tag{2.21}$$

For some reactions the rate of reaction v can be expressed by an equation of the form

$$v = k[C]^\gamma [D]^\delta \tag{2.22}$$

where k, γ, and δ are independent of time and concentration. The coefficient k of Equation 2.22 is called the *rate constant* or *rate coefficient*,[b] and the exponents γ and δ are called the partial orders of the reaction (see Equation 2.12). The sum $\gamma + \delta$ is the overall order or simply the *order of reaction*. There is not necessarily a relation between the stoichiometric coefficients c and d and the partial reaction orders γ and δ.

A reaction need not have an order. For example, reactions where a substrate S is catalyzed by enzymes frequently obey the Michaelis-Menten equation

$$v = \frac{V[S]}{K + [S]} \tag{2.23}$$

where V and K are "kinetic parameters."

---

[a] [B] is a generic expression for concentration. One uses the label v_c when referring to amount concentration and v_C when referring to number concentrations. See Table A1.7, note g.
[b] *Rate constant* is the preferred term; see ref. 126.

To conclude: determination of the rate of concentration change r_B as given by Equation 2.8 requires no assumption of a particular reaction equation. But when such a stoichiometric equation is known, then r_B and the rate of reaction υ are related:

$$r_B = \nu_B \upsilon \tag{2.24}$$

The equilibrium constant

An early attempt to establish a direct connection between reaction rates (chemical kinetics) and equilibrium constants (chemical thermodynamics) followed this argument (Guldberg and Waage, 1865): Consider the reaction

$$c\mathrm{C} + d\mathrm{D} \;\rightleftharpoons\; y\mathrm{Y} + z\mathrm{Z} \tag{2.25}$$

and suppose that the rate of the forward reaction and the reverse reaction are given by

$$\upsilon_f = k_f[\mathrm{C}]^c[\mathrm{D}]^d \quad \text{and} \quad \upsilon_r = k_r[\mathrm{Y}]^y[\mathrm{Z}]^z \tag{2.26}$$

respectively. During reaction the rate $\upsilon = \upsilon_f - \upsilon_r$ approaches zero, and at equilibrium the ratio

$$\frac{[\mathrm{Y}]^y [\mathrm{Z}]^z}{[C]^c [D]^d} = \frac{k_f}{k_r} = K \tag{2.27}$$

is equal to a constant K, the equilibrium constant.[71]

This derivation is applicable to elementary processes (Section 2.4) and is often valid in gas-phase kinetics. As seen, the derivation uses rates expressed by Equation 2.26, but rate expressions cannot in general be derived from stoichiometric equations. A reaction such as Equation 2.25 is generally composed of many elementary processes, and the condition for equilibrium must be that all of them are in equilibrium. Although the concentrations of intermediates will disappear from the final expression, the equilibrium constant depends on the rates of the elementary processes. The thermodynamic connection between the equilibrium Equation 2.25 and the equilibrium constant (Equation 2.27) is discussed in Section 2.3, and modern results unifying kinetics and thermodynamics are discussed in Section 2.4c.

c. Field

Consider a fluid in motion. This system is described here using a fixed coordinate system, given by the mutually orthogonal basis vectors $\boldsymbol{i}, \boldsymbol{j}$, and $\boldsymbol{k}$. At any time t a fluid element at the point $\boldsymbol{r} = x\,\boldsymbol{i} + y\,\boldsymbol{j} + z\,\boldsymbol{k}$ will have a velocity $\boldsymbol{u}(\boldsymbol{r}, t) \equiv \boldsymbol{u}(x, y, z, t)$ given by

$$\boldsymbol{u}(\boldsymbol{r}, t) = u_x(\boldsymbol{r}, t)\,\boldsymbol{i} + u_y(\boldsymbol{r}, t)\,\boldsymbol{j} + u_z(\boldsymbol{r}, t)\,\boldsymbol{k} \tag{2.28}$$

A function that is defined in this way is called a *vector field*. Similarly, a *scalar field* is defined as a function of the variables x, y, z, t that gives the value of a scalar quantity such as $p(\boldsymbol{r}, t)$, $\rho(\boldsymbol{r}, t)$, or $u_x(\boldsymbol{r}, t)$. Over time, new fluid elements arrive at position $\boldsymbol{r}$. The functions give the values of the physical quantities at this point in space at a given time; in other words, the functions belong to points fixed in space and not to moving elements of the fluid.

The state of motion of a system that consists of one phase is completely determined by five functions of the spatial coordinates x, y, and z and of the time t. These are the three components of the distribution of velocity in the fluid, $u(x, y, z, t)$, which is a vector field, and two thermodynamic functions, the pressure, $p(x, y, z, t)$, and the mass density (mass concentration), $\rho(x, y, z, t)$, which are scalar fields.[a] Functions describing normal, well-behaved fluids must be smooth; that is, they are continuous functions with continuous first partial derivatives.[b] In this case standard calculus identifies the velocity of a moving infinitesimal volume element as the derivative of its position vector with respect to time:

$$u(r, t) = \dot{r} = \dot{x}i + \dot{y}j + \dot{z}k \tag{2.29}$$

We shall return to the subject of chemical thermodynamics in Section 2.3, but we will need the general definition of *concentration*: the concentration, $c_X(r, t)$, of any extensive property X of a fluid system is a scalar function of r and t defined as the ratio between $X(r, t)$ and the volume, $V(r, t)$. Different types of concentration may have specific names and symbols; for example, the mass concentration is known as the mass density, $\rho(r, t)$.

d. Transport

Transport is a central phenomenon in environmental sciences. For example, two apparently distinct types of system are encountered. One is a river in which the solvent (water) flows in a certain direction and transports solutes and dispersed matter at a certain rate. The other is a lake with a solvent (water) of constant mass (i.e., unchanging in time), containing a certain amount of solute (e.g., a pollutant) whose fate in time is of interest; it may appear or disappear by migration across the boundaries of the system (brooks and drains; interaction with the bottom) or by chemical reaction within the aqueous phase.

The quantitative description of such events uses the physical quantities *flux density* and *flux*.

The *flux density*, $J_X(r, t)$, of an extensive property X of a fluid system is a vector field, defined as the product of the concentration of that property and its velocity:

$$J_X(r, t) = c_X(r, t)\, u(r, t) \tag{2.30}$$

Important examples are the mass flux density, $J_m = \rho(r, t)\, u(r, t)$, and the amount flux density, $J_B = c_B(r, t)\, u(r, t)$.

The *flux* is a scalar to be formally defined in Equation 2.32. It is a measure of the amount of an extensive property that is carried in and out through the borders of an open system per time. An example is the mass flux,

$$\dot{m} = \frac{dm}{dt} \tag{2.31}$$

---

[a] This statement is a straightforward application of the phase rule, Equation 2.178.
[b] Some natural events cannot be described in this way, such as the shock waves around lightning.

This is not a field because it is not a function of spatial coordinates. Under certain circumstances the flux may be used to characterize large systems, such as lakes or the atmosphere (see Section 2.2).

It is noted that the dimension of flux density is flux divided by area (see Table A1.5).

2.2 Fluid dynamics

From the age of the Vikings through the industrial revolution, water power was the most important source of mechanical energy in Europe. Surprisingly, the first correct results in the field of fluid dynamics were not obtained until the 18th century (Bernoulli, $\approx$ 1740; L. Euler, $\approx$ 1760; J. Euler, $\approx$ 1780), ultimately leading to equations of motion and laws of conservation. Later, in the last part of the 19th century, the problem of transforming laboratory results into real-world applications was overcome for ships, turbines, and tubes (Froude, $\approx$ 1870; Reynolds, $\approx$ 1880). In the 20th century, many remarkable hydrodynamic results were embodied in the technology of aircraft and large vessels.[69] Finally, but most important for the environmental sciences, the principles of fluid dynamics were applied to the physics of the atmosphere and hydrosphere (hydrodynamics), which is basic to a proper description of the chemistry of these domains.

The main task in fluid dynamics is to set up and solve the equations of motion. Among the subjects to be studied, therefore, are the treatment of the force fields and the laws of conservation of mass and energy.

a. Basic properties of the flux density

The purpose of this section is to give a reader who is not familiar with vector calculus a feeling for the behavior of flux density, and descriptions of how calculations may be performed.[a]

The flux

Figure 2.1 shows the flux density $J_X(r, t)$ of an extensive property X passing through an open system[b] with volume V and boundary surface area A. The infinitesimal area $d\sigma$ is represented as the area element vector $d\sigma$, which is the product of the area $d\sigma$ and a unit vector normal to the area with its direction pointing away from the volume. The volume element is denoted $d\tau$.

The *flux* of an extensive property X of the system is defined by the expression

$$\dot{X}(t) = \oint_A J_X(r, t) \cdot d\sigma(r, t) \tag{2.32}$$

---

[a] We use the symbols **grad**, **div**, and **rot** for the vector differential operators rather than those involving the nabla operator (∇, $\nabla\cdot$, and $\nabla\times$, respectively) because the former set gives physical insight regarding the property being described.

[b] The flux density must be a differentiable function within and on the boundary surface of the system.

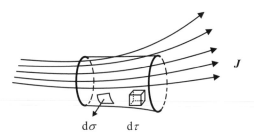

$$d\sigma \qquad d\tau$$

Figure 2.1   *A flux density **J** drawn at a fixed time*
***J** passes the volume $V = \iiint d\tau$ with the total boundary surface $A = \oint d\sigma$.*

The right-hand side of this equation is a surface integral of the scalar product of the flux density **J** and the surface element d**σ**, taken over the closed surface A.[a] It includes the total amount of X entering and leaving the system during a time interval. The mass flux of the system (Equation 2.31) is an example of Equation 2.32, which will be discussed in detail in the next section.[b]

The divergence

The value of the integral Equation 2.32 may be calculated by means of Gauss's theorem (the divergence theorem):

$$\oint_A \boldsymbol{J} \cdot \mathrm{d}\boldsymbol{\sigma} = \iiint_V \boldsymbol{div\, J}\, \mathrm{d}\tau \tag{2.33}$$

Here, the right-hand side is the volume integral of the scalar function ***div J***, and the volume element is $d\tau = dx\,dy\,dz$. The divergence of the vector field **J** is defined as the scalar field

$$\boldsymbol{div\, J} \overset{\mathrm{def}}{=} \frac{\partial J_x}{\partial x} + \frac{\partial J_y}{\partial y} + \frac{\partial J_z}{\partial z} \tag{2.34}$$

(in Cartesian coordinates). The essential point is that given the explicit expressions for the flux density, the integral is straightforward to calculate. Note that the equations written here deal only with spatial coordinates, not temporal.

The gradient

In some cases it may happen that the flux density is a gradient field; that is, it is a vector field that can be created from a differentiable scalar field φ in the following way:

$$\boldsymbol{J}(\boldsymbol{r}, t) = \boldsymbol{grad}\; \varphi(\boldsymbol{r}, t) \tag{2.35}$$

---

[a] Note: We use the integral symbol with a circle to indicate integration over a closed surface or (as in Equation 2.40) a closed curve.

[b] The term *flow rate of X* (not to be used in this book) is the vector quantity $(\mathrm{d}X/\mathrm{d}t)\boldsymbol{e}$, where $\boldsymbol{e}$ is a unit vector in the direction of the flow.[126]

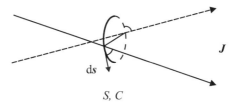

Figure 2.2   *An example of the rotation of a vector field*
The flux density J, drawn at a fixed time, is represented by two skew vectors. Such a pair defines a helical system, here a right-hand screw. The area S has the circumference C. The line element ds is a tangent to C.
With the chosen direction of ds, the circulation Γ is positive because the dot product $J \cdot ds$ is positive along C.

The gradient operator is the vector differential operator

$$\boldsymbol{grad}\ \varphi \overset{\text{def}}{=} \frac{\partial \varphi}{\partial x}\boldsymbol{i} + \frac{\partial \varphi}{\partial y}\boldsymbol{j} + \frac{\partial \varphi}{\partial z}\boldsymbol{k} \tag{2.36}$$

which converts a scalar field into a vector field. Here are some examples of Equation 2.35 found in daily life: the gradient points in the direction of maximum increase of the scalar function, for example, toward the top of a hill (height gradient) or toward high pressure on a weather map (pressure gradient).

The line integral of a vector field J over a curve C is defined as the integral of the scalar product of J with the line element dr:

$$\int_C \boldsymbol{J} \cdot d\boldsymbol{r} = \int_C (J_x\, dx + J_y\, dy + J_z\, dz) \tag{2.37}$$

The right side is simply a procedure (in Cartesian coordinates) for evaluating the line integral. Line integrals of gradient fields are particularly important because they are independent of the path of integration: Assume that the path C has the end points a and b. Then

$$\int_a^b \boldsymbol{grad}\ \varphi \cdot d\boldsymbol{r} = \varphi(b) - \varphi(a) \tag{2.38}$$

(this is called the gradient theorem) and, as a special – but important – case,

$$\oint_C \boldsymbol{grad}\ \varphi \cdot d\boldsymbol{r} = 0 \tag{2.39}$$

when C is a closed path.

The rotor or the curl

In some cases the flux density may be twisted or curled[a] (Figure 2.2). In order to discuss this phenomenon, we define the circulation Γ as the line integral of the scalar

---

[a] The rotor (German) or the curl (English) in vector integral calculus are identical constructions. Since 2007, the internationally accepted name has been *rotor* for the concept and **rot** for the operator.[126]

product of the flux density J where the line element ds taken is over the closed curve C:

$$\boldsymbol{\Gamma} = \oint_C \boldsymbol{J} \cdot ds \tag{2.40}$$

It is easy to think of Γ as the quantity X per length of curl per time (see Figure 2.2). Γ may be calculated by means of Stokes's theorem (the *rot* or *curl* theorem):

$$\oint_C \boldsymbol{J} \cdot ds = \iint_S \boldsymbol{rot J} \cdot d\sigma \tag{2.41}$$

The right-hand side is the surface integral of the scalar product between the vector field *rot J* and the surface element $d\sigma$ taken over a surface S, bounded by the closed curve C. The equation may be interpreted in two ways: First, the circulation gives a measure of the angular motion: Γ increases with increasing J and with increasing cosine of the angle between J and s. Second, the right-hand side is a procedure for calculating the circulation; this requires knowledge[a] of the vector differential operator *rot* and of the flux density J.

Concluding remarks

The equations of motion of a fluid phase are discussed later in Section 2.2d. However, it should be pointed out that the significance of the mass flux density is a restatement of Newton's second law: the mass flux density is the linear momentum of the fluid, and its time derivative is the force per volume of the fluid.

This section concludes with two useful identities to be used in the following sections. First, it may be verified by direct calculation that the divergence of the product of a scalar field and a vector field, $\rho(\boldsymbol{r}, t)\, \boldsymbol{u}(\boldsymbol{r}, t)$, say, is a scalar field given by

$$\boldsymbol{div}\,\rho\,\boldsymbol{u} = \boldsymbol{u}\cdot\boldsymbol{grad}\,\rho + \rho\,\boldsymbol{div}\,\boldsymbol{u} \tag{2.42}$$

As a special case, let the vector field be a constant one, $\boldsymbol{c}$. Application of the divergence theorem (Equation 2.33) gives

$$\boldsymbol{c}\cdot\oint_A \rho\,d\sigma = \boldsymbol{c}\cdot\iiint_V \boldsymbol{grad}\,\rho\,d\tau \tag{2.43}$$

where $d\tau$ is the volume element. Because this applies to all scalar fields, we find for the special case of pressure, $p(\boldsymbol{r}, t)$, the important expression

$$\oint_A p\,d\sigma = \iiint_V \boldsymbol{grad}\,p\,d\tau \tag{2.44}$$

which is a vector field.

---

[a] We have not included the explicit expression of the *rot* operator, because it is not needed in this book. The discussion of atmospheric vortex flows is not made quantitative, and the Bernoulli equation is confined to cases where use of *rot* is unnecessary.

b. Open physicochemical systems

In this section we model a flow system frequently used in environmental modeling. It is a special case of the continuity equation, to be discussed in Section 2.2c (see also Equation 2.115). The model may represent a lake, a river, or part of the atmosphere.

Assume an open, one-phase system consisting of a solution of nonreacting chemical species. The mass m of the system is given as a function of the time t:

$$m(t) = \sum_{C} m_C(t) \tag{2.45}$$

where the sum over all C includes the solvent and all dissolved species. The mass that flows through the system is mainly that of the solvent, but small amounts of other species contribute because they are carried through by advection.[a] The system is assumed to be isotropic (without internal structure), meaning that the mass of the system is a scalar quantity but not a scalar field.

A particular species B may have several sources and sinks. Note, however, the peculiar interaction between an open system and its surroundings: when the system acts as a sink of some quantity, then the surroundings are the source and vice versa. This is what is meant by an open system: the surroundings play an important role in the physics and chemistry of the system, but they do not appear explicitly.

Next, we discuss the behavior of the solvent and the solute.

The solvent

The solvent, the chemical species A, flows through the system. Suppose that the flux is

$$\dot{m}_A = \frac{d\,m_A}{d\,t} = 0 \tag{2.46}$$

implying that m_A is constant in time; this is tantamount to having a nonvanishing mass flux

$$\dot{m}_{A,in} = -\dot{m}_{A,out} \neq 0 \tag{2.47}$$

because a flow was assumed. Here "in" signifies all solvent that enters the system from the surroundings, and "out," all that leaves the system. The ratio

$$\left| \frac{m_A}{\dot{m}_{A,in}} \right| = \tau_A \tag{2.48}$$

is a characteristic property called the *lifetime* of solvent A in the system. It is noted that τ_A behaves qualitatively correctly in the limit because $\tau_A \to \infty$ for $\dot{m}_{A,in} \to 0$.

---

[a] *Advection*: Solutes or dispersed matter are carried with the solvent; they may have the velocity of the solvent (see Section 2.2f).

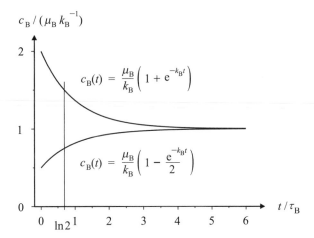

$c_B / (\mu_B k_B^{-1})$

$$c_B(t) = \frac{\mu_B}{k_B}\left(1 + e^{-k_B t}\right)$$

$$c_B(t) = \frac{\mu_B}{k_B}\left(1 - \frac{e^{-k_B t}}{2}\right)$$

Figure 2.3 *Examples of solutions, Equation 2.50*
Upper curve: The initial condition is $c_B(0) = (2\mu_B)/k_B$.
Lower curve: The initial condition is $c_B(0) = \mu_B/(2k_B)$.

The solute

Assume that a dissolved or dispersed species B with the concentration $c_B(t)$ is carried through the system by advection.[a] Assume that the rate of increase of c_B due to sources outside the system is constant in time: $\dot{c}_B = \mu_B$ with $\mu_B > 0$. Assume also that the rate of decrease of c_B, removed through the boundaries or by chemical reactions, is proportional to the actual concentration $c_B(t)$: $\dot{c}_B = -k_B c_B$ with $k_B > 0$, $k_B = \Sigma_i\, k_{B,i}$. The actual change of concentration in time $\dot{c}_B$ is equal to the difference between the rate of gain μ_B and the rate of loss $k_B\, c_B(t)$. Overall, one has

$$\dot{c}_B = \mu_B - k_B\, c_B(t) \tag{2.49a}$$

For convenience, the designation B will be omitted in the following, and we have

$$\dot{c} = \mu - kc \quad \text{with} \quad k = \sum_i k_i > 0 \tag{2.49b}$$

This is a differential equation that, when brought to the form

$$d\ln[\mu - kc] = -k dt \quad \text{with} \quad \mu \neq kc \tag{2.50}$$

can be solved to give

$$c(t) = \frac{\mu}{k} - \left(\frac{\mu}{k} - c(0)\right) e^{-kt} \tag{2.51}$$

This function is depicted in Figure 2.3 for two initial values of c_B. For large values of the time t, the limit of the concentration of the solute is

$$\lim_{t \to \infty} c(t) = \frac{\mu}{k} \tag{2.52}$$

[a] See Section 2.2f.

A characteristic time, the *lifetime* τ, is defined through

$$\tau \overset{\text{def}}{=} \frac{1}{k} = \left(\sum_i k_i \right)^{-1} \tag{2.53}$$

The lifetime is the mean time for which a chemical species stays in the system before being removed. The reciprocal lifetime is also given by Equations 2.49 and 2.50:

$$-\frac{\dot{c}}{c} = -\frac{\mathrm{d}\ln(\mu - kc)}{\mathrm{d}t} = k = \frac{1}{\tau} \tag{2.54}$$

(see Equation 2.48). Another characteristic time is the *half-life* $t_{1/2}$, which is defined as the time at which the concentration has reached a value that is the arithmetic mean of its initial and final values:

$$c(t) = \frac{c(0) + c(\infty)}{2} \quad \Rightarrow \quad t_{1/2} = \frac{\ln 2}{k} = \tau \ln 2 \tag{2.55}$$

We conclude by noting that the additivity of the constants $k_{B,i}$ leads to the following relationships between lifetimes and constants:

$$k_B = \sum_i^N k_{B,i} \quad \text{and} \quad \frac{1}{\tau_B} = \sum_i^N \frac{1}{\tau_{B,i}} \tag{2.56}$$

For an example, see Equation 7.1.

c. Continuity equations

General expressions

Consider an open system comprising a single fluid phase. Its volume V is simply connected[a] and bounded by the surface A. Let $c_B(\boldsymbol{r}, t)$ be the concentration of B. The total amount of B is then given by

$$n_B(t) = \iiint_V c_B(\boldsymbol{r}, t)\, \mathrm{d}\tau \tag{2.57}$$

where $\mathrm{d}\tau$ is the volume element. The derivative with respect to time is given by

$$\dot{n}_B = \iiint_V \left(\frac{\partial c_B}{\partial t} \right)_{x,y,z} \mathrm{d}\tau \tag{2.58}$$

where the variables $\boldsymbol{r}$ and t have been omitted for brevity. There are only two ways in which the amount of B can change. One is that B may enter or leave the system through its boundaries. This is accounted for by a term containing the integral in Equations 2.32 and 2.33. The other is that B may be created or annihilated in situ (through chemical reactions): Let $r_B(\boldsymbol{r}, t)$ be the time rate of the amount of B that is

---

[a] In plain terms, a domain is simply connected if it has no holes.

created ($r_B > 0$) or annihilated ($r_B < 0$) in a volume element.[a] Then the net gain of B is

$$\dot{n}_B = \iiint\limits_V r_B \, d_\tau - \oiint\limits_A J_B \cdot d\sigma = \iiint\limits_V (r_B - div J_B) \, d\tau \quad (2.59)$$

where $J_B = c_B \, u$ (see Appendix A1.5). Equating 2.58 and 2.59 gives the general *continuity equation*

$$\frac{\partial c_B}{\partial t} = - div J_B + r_B \quad (2.60)$$

We now discuss some important properties and applications of this equation.

Eulerian and Lagrangian interpretations

Let the flux density in Equation 2.60 be the product $J_B = c_B \, u$. Then the divergence term can be expanded as follows:

$$r_B = \frac{\partial c_B}{\partial t} + u \cdot grad \, c_B + c_B \, div \, u \quad (2.61a)$$

$$= \frac{dc_B}{dt} + c_B \, div \, u \quad (2.61b)$$

$$\overset{\text{def}}{=} \dot{c}_B + c_B \, div \, u \quad (2.61c)$$

using mathematical identities as explained later.

Equation 2.61a follows from Equation 2.42, and the notation $\dot{c}_B$ in Equation 2.61c means the total derivative given in Equation 2.61b by definition. The total differential dc_B is given as the sum[b]

$$dc_B = \left(\frac{\partial c_B}{\partial t}\right)_{x,y,z} dt + \left(\frac{\partial c_B}{\partial x}\right)_{t,y,z} dx + \left(\frac{\partial c_B}{\partial y}\right)_{t,z,x} dy + \left(\frac{\partial c_B}{\partial z}\right)_{t,x,y} dz \quad (2.62)$$

Dividing[c] by dt and noting the relation $u = dr/dt$ leads from the first two terms of the right-hand side of Equation 2.61a to the term dc_B/dt of Equation 2.61b.

1. The first two terms of the right-hand side of Equation 2.61a describe the change of the scalar function $c_B(r, t)$ identified with the field point r at time t:

$$\frac{\partial c_B}{\partial t} + u \cdot grad \, c_B = - c_B \, div \, u + r_B \quad (2.63)$$

This method of describing the fluid motion is called the *Eulerian method*.

2. The term dc_B/dt of Equation 2.61b does not give the change of density of B at a fixed spatial point. Instead, it gives the change of density of B for a fluid element that moves in space:

$$\dot{c}_B = - c_B \, div \, u + r_B \quad (2.64)$$

---

[a] The dimension of r is concentration per time.
[b] Equation 2.62 is known as the *fundamental lemma of differential calculus*.[44]
[c] The expression after division by dt is known as the *chain rule* of composite functions.

This method of describing the fluid motion is called the *Lagrangian method*.

See the discussion of Equations 2.28 and 2.29 and that of Equations 2.114 and 2.115.

Steady fields; flux of entropy

A vector field or a scalar field that is independent of time is called a *steady field*. In the case of a system described by the scalar field c_B (see Equations 2.61 and 2.62), this situation is equivalent to the statement

$$\frac{\partial c_B}{\partial t} = 0 \tag{2.65}$$

and the system is said to be in a *steady state*.

Characterization of the steady state of an open system: An infinitesimal change of entropy in an open system is the sum of two terms[46]

$$dS = d_i S + d_e S \tag{2.66}$$

where the index "i" refers to the internal contribution due to irreversibility, and "e" to the reversible exchange of entropy through the boundary surface. A steady state is characterized by the flux of entropy $\dot{S}$ being zero with $\dot{S}_i > 0$ and $\dot{S}_e = -\dot{S}_i$.[a]

Concentration

We will give three examples of applications of the continuity equations:

1. Density, that is mass concentration

The continuity equation for mass is[b]

$$\frac{\partial \rho}{\partial t} + \boldsymbol{div} \, \boldsymbol{J}_m = 0 \tag{2.67}$$

because mass has no sources or sinks within a system, i.e., $r_m = 0$. Furthermore, if the fluid is assumed to be incompressible with ρ = constant, $d\rho/dt = 0$, then $\boldsymbol{div} \, \boldsymbol{u} = 0$, imposing restrictions on the possible velocity fields.[c]

In the discussion of Equation 2.46 it was assumed that the divergence of the flux density vanishes, i.e., $\boldsymbol{div} \, \boldsymbol{J}_m = 0$. Accordingly,

$$\frac{dm}{dt} = 0 \tag{2.68}$$

and the flow of the solvent (typically water) is steady.

2. Amount concentration; number concentration

The general continuity equation for the amount of substance of species B in an open system is given by Equation 2.60. Use of Lagrangian variables gives the appearance

---

[a] See Equation 2.151.
[b] Substitute in Equations 2.61 c_B with concentration of mass $c_m = \rho$.
[c] This $\boldsymbol{u}$-field is called a solenoidal field.[44,52]

shown in Equation 2.64. Assuming a closed system, *div **u*** = 0, Equation 2.8 emerges as expected:

$$\dot{c}_B = r_B \tag{2.69}$$

Note, however, that this system is *not* in a steady state, and the amount of substance (but not the mass!) may increase or decrease.

Equation 2.69 applies analogously to the number concentration C_B (see Appendix 1.7).

3. Surface adsorption

The derivation of the Langmuir adsorption isotherm is a simple application of the continuity equation 2.60. The system is a surface A, of which a fraction θ is covered by chemical species B, and the fraction $1 - \theta$ is bare. Species B is desorbed (i.e., dissociated) from the surface with the flux density $\boldsymbol{J}_{des}$, and the resulting flux is $A\,\theta\,\boldsymbol{J}_{des}$. Simultaneously the surface is bombarded with the species B with the flux density $\boldsymbol{J}_{ads}$, and every collision is adsorbed when space allows such that the adsorbed flux is $A(1 - \theta)\,\boldsymbol{J}_{ads}$. At equilibrium,

$$\theta A \boldsymbol{J}_{des} = (1 - \theta) A \boldsymbol{J}_{ads} \tag{2.70}$$

leading to an expression for the surface coverage θ,

$$\theta = \frac{\boldsymbol{J}_{ads}}{\boldsymbol{J}_{des} + \boldsymbol{J}_{ads}} \tag{2.71}$$

which is the most general form of the Langmuir adsorption isotherm. Langmuir[65] assumed the desorption flux to be kinetically of zeroth order, $AJ_{des} = k_{des}$, and the adsorption flux to be of first order in [B], $AJ_{ads} = k_{ads}$ [B], and arrived at the surface coverage

$$\theta = \frac{k_{ads}\,[B]}{k_{des} + k_{ads}\,[B]} = \frac{K\,[B]}{1 + K\,[B]} \tag{2.72}$$

where $K = k_{ads}/k_{des}$. This equation is further discussed in Section 6.3. Note the identical mathematical form of Equation 2.48 with $\dot{c}_B = 0 : \mu = kc$, and Equation 2.70: $\theta\,k_{des} = (1 - \theta)\,k_{ads}$ [B].

d. Equations of motion

Consider a closed system consisting of a single fluid phase in a steady state. It is assumed to be simply connected with a volume V and bounded by the surface A. The pressure and mass density are given by the scalar functions $p(\boldsymbol{r}, t)$ and $\rho(\boldsymbol{r}, t)$, respectively. The force that acts on the system is equal to the integral

$$-\oint_A p\,\mathrm{d}\boldsymbol{\sigma} \tag{2.73}$$

of the pressure p over the closed surface A. The minus sign is due to the convention that the surface element vector $d\sigma$ is positive when pointing away from the volume. Equation 2.73 may be expressed as a volume integral as demonstrated in Equation 2.44. This shows that the volume element $d\tau$ is subject to the force $-d\tau$ $\textbf{grad}\,p$ from its surroundings, giving a force of $-\textbf{grad}\,p$ per volume element.

According to Newton's second law,[a, 50] this force is equal to the product of the density ρ (mass per volume) and the acceleration $d\textbf{u}/dt$

$$\rho \frac{d\textbf{u}}{dt} = -\textbf{grad}\,p \tag{2.74}$$

which is *Euler's equation of motion* of a fluid (L. Euler, 1755). Because the mass flux density is given as $\textbf{J}_m = \rho\,\textbf{u}$, Euler's equation may be written $\dot{\textbf{J}}_m = -\textbf{grad}\,p$.

In Equation 2.74, Euler's equation is expressed using Lagrangian variables, but it may equally well be expressed in Eulerian variables:

$$\rho \left(\frac{\partial u}{\partial t} + (\textbf{u} \cdot \textbf{grad})\,\textbf{u} \right) = -\textbf{grad}\,p \tag{2.75}$$

The derivation of this expression is straightforward, using the same line of reasoning that connected Equation 2.61b with Equation 2.61a.

The physical interpretation is as follows: $d\textbf{u}/dt$ is the rate of change of the velocity of a volume element that moves in space, but it must be expressed through quantities that belong to fixed points in space. In order to do so, consider the change $d\textbf{u}$ of the velocity of the volume element in the small interval of time dt. It consists of two terms: (1) the change of velocity at the given space-fixed point $\textbf{r}$ at time dt, and (2) the difference between the velocities at the same time t at two points very close to $\textbf{r}$, whose spacing $d\textbf{r}$ is the distance that the volume element travels in the time dt.

Including other possible accelerations $\textbf{a}$ gives the general expression

$$\dot{\textbf{J}}_m = -\textbf{grad}\,p + \rho\,\textbf{a} \tag{2.76}$$

which is *the equation of motion of a fluid*. Because this equation is Newton's second law applied to fluids, it is worthwhile pointing out that the mass flux density $\textbf{J}_m = \rho\,\textbf{u}$ is the linear momentum of the fluid, and its time derivative is the force per volume of the fluid.

We shall discuss this equation more generally as the Navier-Stokes equation (Equation 2.101), which includes viscosity as a deceleration of the fluid.

---

[a] Newton's second law for a particle with the mass m is $F = m\,d\textbf{u}/dt$, where $\textbf{F}$ is the acting force and $d\textbf{u}/dt$ the imposed change in the velocity; $\textbf{u} = d\textbf{r}/dt$.
Case 1. $\textbf{F}\,dt = d(m\textbf{u}) \equiv d\textbf{p}$: The impulse $\textbf{F}\,dt$ is equal to the change of linear momentum $\textbf{p}$.
Case 2. $\textbf{F} \cdot d\textbf{r} = m\,\textbf{u} \cdot d\textbf{u} = d(1/2\ m\textbf{u}^2) \equiv dE_k$: The mechanical work $\textbf{F} \cdot d\textbf{r}$ is equal to the change of kinetic energy E_k. If the sole change of potential energy is $-\textbf{F} \cdot d\textbf{r} \equiv dE_p$, then the total change of energy is $dE \equiv d(E_p + E_k) = 0$, and the force field $\textbf{F}$ is a conservative one.

e. Applications of the equations of motion

The hydrostatic equation

Small particles suspended in a fluid phase exhibit special behavior. The particles have a gravitational potential energy depending on their height, giving rise to a Boltzmann distribution at equilibrium (L. Boltzmann, 1877) characterized by a scale height. The scale height depends on the balance between gravitational settling and diffusional mixing (Brownian motion; R. Brown, 1827) driven by thermal energy.

The equations of motion for a fluid at rest in a gravitational field, characterized by acceleration g, are given by Equation 2.76 with $J_m = 0 \Rightarrow \dot{J}_m = 0$:

$$\mathbf{grad}\, p = \rho \mathbf{g} \tag{2.77}$$

Introducing a Cartesian coordinate system x, y, h with the positive h-axis pointing antiparallel to the direction of g leads to the three equations

$$\frac{\partial p}{\partial x} = \frac{\partial p}{\partial y} = 0 \quad \text{and} \quad \frac{\partial p}{\partial h} = -\rho g \tag{2.78}$$

It follows that the partial derivative of p with respect to h may be replaced by the ordinary derivative $dp/dh = -\rho g$. This equation is called the *hydrostatic equation*. It will be shown in Appendix 4 that for most fluids a necessary condition for mechanical rest is that entropy increases with altitude.

Example: The barometric law

The estimation of the pressure profile of the atmosphere is a simple application of Equation 2.78. Assume a perfect gas mixture, that is, one obeying the equation of state $pV = RT \sum_B n_B$. This is equivalent to the relation

$$\rho = p \frac{M_{air}}{RT} \tag{2.79}$$

where M_{air} is the average molar mass of the atmosphere.[a] Then Equation 2.78 may be converted into the expression

$$d \ln p = -\frac{M_{air}g}{RT}\, dh \tag{2.80}$$

which is the barometric law.

Integration assuming a constant atmospheric temperature[b] of $T = 255$ K and using the value of M_{air}, the recommended (CODATA[126]) values for the standard acceleration of gravity g_n, and the molar gas constant R yields[c]

$$p(z) = p(0)\, e^{-h/h^{\circ}} \tag{2.81}$$

---

[a] See Equation 3.17 and Table 3.18.
[b] See Equation 10.19.
[c] $M_{air} = 28.9668$ g mol^{-1}; $g_n = 9.80665$ m s^{-2}; $R = 8.314472$ J K^{-1} mol^{-1}.

The constant $h° = (R\ 255\ \text{K}) / (M_{\text{air}}\ g_n) \approx 7.46$ km is called the scale height of the atmosphere. Pressure is predicted to fall to $1/e$ of its value at a given height when the altitude is increased by $h°$.

Geostrophic flow

This section considers six marked subjects.

1. Geocentric coordinates

The validity of Newton's second law and the equation of motion of a fluid (Equation 2.76), presupposes an inertial reference frame. This is a frame in which the inertial principle (Newton's first law) is valid; an example is the heliocentric (Copernican) reference frame used for calculations of the trajectories of planets and comets. However, it is intuitive to describe motions on the Earth in a frame fixed to the Earth itself. This is the geocentric frame to which we refer positions, velocities, and accelerations as they are observed here. The geocentric frame is not an inertial frame, and its motion relative to the heliocentric frame includes a translational velocity of ≈ 30 km s^{-1} with a centripetal acceleration[a] of 6 mm s^{-2} and a rotation around the Earth's N-S axis with the angular frequency[b]

$$\omega_\oplus = \frac{2\pi}{86400\ \text{s}}\frac{366.242}{365.242} \approx 7.292 \times 10^{-5}\ \text{s}^{-1} \tag{2.82}$$

The direction of the vector $\omega_\oplus$ is from S to N.

2. Coriolis acceleration; centrifugal acceleration

The fact that the geocentric frame is not an inertial frame means that a certain observed acceleration a_{obs} is different from the acceleration a that would have been measured in the heliocentric frame.[c] One goal of classical mechanics was to reformulate Newton's second law to describe directly observed velocities and accelerations.

A detailed analysis of the relations between the two frames of reference is outside the scope of this book, but the main result is[d]

$$a = a_{\text{obs}} - 2\,\omega_\oplus \times u_{\text{obs}} - \omega_\oplus \times (\omega_\oplus \times r_{\text{obs}}) \tag{2.83}$$

Here u_{obs} is the observed velocity of an object at position r_{obs} that is the position vector perpendicular to and directed away from the axis of rotation. The second term is the Coriolis acceleration (G.-G. Coriolis, 1828), and the third term is the centrifugal acceleration. Both "forces" exist only as a consequence of the selection of Earth-fixed coordinates; they do not fulfill Newton's third law (the law of reciprocal actions).

[a] See Figure 2.4. The Sun-Earth distance is ≈ 150 Gm; see footnote a in Section 10.1c.
[b] One revolution of the Earth around the Sun takes 366.242 d when referred to the heliocentric frame; 1 d = 86400 s.
[c] The heliocentric frame is an inertial frame with the Sun at the origin.
[d] A nonconstant angular frequency requires an additional term in Equation 2.83.

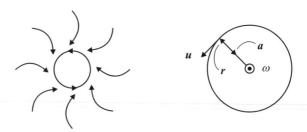

Figure 2.4   *A cyclone in the Northern Hemisphere*
Left panel: The cyclone as a geostrophic wind; the arrows represent flows.
Right panel: A circular motion with constant angular frequency, $\boldsymbol{\omega}$. The velocity is
$\boldsymbol{u} = \boldsymbol{\omega} \times \boldsymbol{r}$ and the centripetal acceleration is $\boldsymbol{a} = \boldsymbol{\omega} \times \boldsymbol{u}$.

3. Discussion of Equation 2.83

a. The second term of Equation 2.83 is the Coriolis acceleration. Evaluation of the vector product shows that the Coriolis acceleration, $\boldsymbol{a}_{\text{Coriolis}}$, is always perpendicular to the velocity. At latitudes[a] $\theta > 0$, a horizontal speed u_{h} gives rise to $a_{\text{Coriolis}} = 2\,\omega\,u_{\text{h}}\,\sin\theta$ pointing to the right; at $\theta < 0$, it is pointing to the left. In a similar way, a vertical speed u_{v} gives $a_{\text{Coriolis}} = 2\omega\,u_{\text{v}}\,\cos\theta$.

b. The third term of Equation 2.83 is the centrifugal acceleration due to the daily revolution of the Earth. As an example, consider an object at rest on the equator, $r_{\text{obs}} = r_{\text{E}} = 6378$ km, and with the acceleration of gravity equal to 9.81 m/s. Then $a/(\text{m s}^{-2}) = 9.81 - 0.03$, so this term is of practical significance for launching satellites, submarine navigation, and so forth, but not for daily life.[b] At a latitude θ, $r_{\text{obs}} \approx r_{\text{e}} \cos\theta$, so this acceleration is absent at the poles.

4. Water currents

a. The Great Belt ($56°$ N) is a Danish strait approximately 30 km wide. A northgoing current of about 1 m/s (≈ 1.9 knot) gives rise to a water level that is 30 cm higher on the eastern side.

a. In the Southern Hemisphere, rivers erode the left bank more quickly than the right.

5. Cyclones in the Northern Hemisphere

Horizontal winds are driven by forces arising from gradients of pressure. The ensuing velocity creates a Coriolis force that in the Northern Hemisphere deflects winds to the right. At regional distances, the resultant motion is the counterclockwise circulation of air around a low pressure (Figure 2.4).[c] This is called a cyclone; cyclones cannot form on the equator.

6. Geostrophic flow

Consider a cyclone in the Northern Hemisphere (see Figure 2.4) for which we shall apply the equation of motion (Equation 2.76), using the local coordinate system

---

[a] $\theta > 0$ for N latitudes; $\theta < 0$ for S latitudes.
[b] If $\omega_{\oplus}$ was ≈ 17 times greater than the actual value, then $a \approx 0$ and an object would be suspended.
[c] An anticyclone is a clockwise (in the Northern Hemisphere) circulation of air around a high pressure.

(Equation 2.83). The path of the air is approximately circular, and an element of air travels with constant speed. Accordingly there must be a centripetal force, directed toward the center, and it must be the result of a pressure gradient and the Coriolis force. Such *a flow whose velocity is perpendicular to the pressure gradient and the Coriolis force (of opposite direction) is called a geostrophic flow*. Friction between wind and the surface slows air, reducing the Coriolis force, which depends on velocity. The result is that low-altitude air has a greater velocity component along the direction of the pressure gradient, and low and high winds are not parallel.

Consider a cyclone with its center at a northern latitude θ where air with density ρ travels in a circular motion with radius r and the constant speed u. The pressure gradient that is necessary for maintaining this motion, $-\mathrm{d}p/\mathrm{d}r$, may be found from Equation 2.76, using colinear forces, taken to be positive away from the center:

$$-\rho \frac{u^2}{r} = -\frac{\partial p}{\partial r} + 2\,\rho\,u\,\omega \sin\theta \tag{2.84}$$

A numerical example: $\theta = 60°$ N; $r = 300$ km; $u = 8$ m/s (fresh breeze, ≈ 20 knot), $\rho = 1.25$ g/l, to give $-\rho u^2/r = -0.267$ Pa/m; $2\,\rho\,u\,\omega \sin\theta = 1.263$ Pa/m; and the gradient $-\partial p/\partial r = 1.53$ Pa/m $= 15.3$ mbar/km. It is noted that the resulting force, the left-hand side of Equation 2.84, is small when compared to the driving forces.

Bernoulli's equation

The gradient field is part of the equation of motion (Euler's equation) for a fluid in a state of steady flow without eddies or curls:

$$\boldsymbol{grad}\left(gz + \frac{p}{\rho} + \frac{u^2}{2}\right) = 0 \tag{2.85}$$

This is a special case of the general Bernoulli equation 2.93 (D. Bernoulli, 1738). The applications used in this book stem from Equation 2.85. The structure of this remarkable equation is clearly seen after integration:

$$\rho g + \frac{p}{z} + \rho \frac{u^2}{2z} = \frac{force}{volume} = \frac{pressure}{length} \tag{2.86}$$

Thus, the force that causes a fluid to move is generally a *sum* of three terms due to gravity, pressure, and velocity, respectively. The equation often appears in the form

$$z + \frac{p}{\rho g} + \frac{u^2}{2g} = h_{\mathrm{h}} \tag{2.87}$$

where all terms have the physical dimension of length. This fact is emphasized by the symbolic expression:

$$h_z + h_p + h_u = h_{\mathrm{h}} \tag{2.88}$$

The terms are called *heads*, and from left we have the elevation head, the pressure head, and the velocity head, respectively. The total head is called the hydraulic head.

The heads are scalar fields, and the practical goal is to measure or calculate their gradients, which are dimensionless numbers. For applications to fluids (gas, oil, water) that move in underground sediments and rocks, Bernoulli's equation is often written

$$P = zg\rho + p + 1/2\,\rho u^2 \tag{2.89}$$

The term P has dimensions of pressure, that is, force divided by area.[a] In the soil the flow of water is slow enough that the velocity head can be ignored. If u in Equation 2.89 is set to zero and the equation multiplied by dV, an equation is obtained that describes potential energy. Making the substitution $PdV = d\Phi_h$, one obtains

$$d\Phi_h = (zg\rho + p)\,dV \tag{2.90}$$

In soil science, these quantities are called potentials: $z\,\rho\,g\,dV$ is the elevation potential, $p\,dV$ is the pressure potential, and the sum $d\Phi_h$ is called the hydraulic potential (see Section 6.2).

We conclude by giving a heuristic derivation of Bernoulli's equation from Equation 2.75 in order to provide some insight as to the nature of the approximations. The following equation is a known vector identity:

$$(\boldsymbol{u}\cdot\boldsymbol{grad})\boldsymbol{u} = \boldsymbol{grad}\left(\frac{u^2}{2}\right) - \boldsymbol{u}\times\boldsymbol{rot\,u} \tag{2.91}$$

which is inserted into Equation 2.75. The effect of gravity

$$\boldsymbol{a} = -\boldsymbol{grad}\,(gz) \tag{2.92}$$

is added on the right-hand side, and the general Bernoulli equation is obtained:

$$\left(\frac{\partial \boldsymbol{u}}{\partial t}\right)_r - \boldsymbol{u}\times\boldsymbol{rot\,u} = -\boldsymbol{grad}\left(gz + \frac{u^2}{2} + \frac{p}{\rho}\right) \tag{2.93}$$

It follows that a steady $\boldsymbol{u}$-field (no acceleration) and no curl (turbulence-free flow) causes the left-hand side to vanish, giving Equation 2.85.

Poiseuille's equation

The dynamic viscosity η is a measure of how much a fluid phase resists flow: for example, the frictional drag met by a solid body moving through a gas or liquid. In order to understand the size and nature of this resistance, we consider the special case of a fluid confined between two extended and parallel plates separated by a distance a (Figure 2.5). The lower plate is kept fixed while the upper plate is moved with the constant velocity $\boldsymbol{u}$. Experiments show (Newton, 1687) that in order to maintain the motion one must exert a constant force $\boldsymbol{F}$ parallel to the moving plate.

---

[a] For future estimates and comparison with literature, using the standard acceleration of gravity, $g_n = 9.80665$ m s^{-2}, and the density of liquid water at 10 °C, $\rho = 999.73$ kg m^{-3}, gives $\boldsymbol{grad}\,p = 1$ bar/m corresponding to $\boldsymbol{grad}\,h_p = 10.200$.

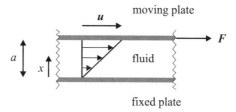

Figure 2.5 *A viscous flow between a fixed plate and a moving plate*
(The so-called couette flow.) The distance between the parallel plates is a, and the moving plate is dragged with velocity u. The molecules very close to the surface of a plate move with its velocity. If the flow is laminar,[a] then the layer at the position x moves with the velocity $x\,u/a$.

Table 2.1 Viscosity of water						
θ / °C	−5	0	10	20	30	100
η / cP	2.15	1.79	1.31	1.002	0.80	0.28

The dynamic viscosity of water as a function of temperature. Relation between units: cP = mPa s = g s^{-1} m^{-1}.

Table 2.2 Viscosities of selected fluids				
θ / °C		0	20	40
Substance	Unit			
Air (g)	η / μPa s	17.08	18.10	19.04
Methane (g)	η / μPa s	10.26	10.88	11.45
Hexane (l)	η / cP	0.401	0.326	0.271
Octane (l)	η / cP	0.706	0.542	0.433
Light fuel oil (l)	η / cP		110	
Lubricating oil (l)	η / cP		660	

The dynamic viscosities η of substances that may occur in soil as a function of temperature. Relation between units: cP = mPa s = g s^{-1} m^{-1}.

This force is proportional to the velocity u, the area A of the plate and the inverse of the distance a:

$$F = \eta \frac{A}{a} u \qquad (2.94)$$

where the material constant η is called the dynamic viscosity.[b] This parameter depends on the nature of the fluid and its temperature but is independent of pressure. As can be seen from Equation 2.94, the SI-unit of viscosity is Pa s. The unit mPa s is so widespread that it has been given the special name centipoise, cP = 10^{-3} kg s^{-1} m^{-1}. Tables 2.1 and 2.2 show the viscosities of some different fluids.

---

[a] Examples: A laminar flow is the "smooth" flow of viscous syrup onto a pancake; the counterpart, a turbulent flow, is "irregular," for example, the splashing of water from a faucet into the sink.
Lat. *lamina* $\hat{=}$ thin plate; Lat. *turbulentus* $\hat{=}$ disturbance.
[b] The quantity *kinematic viscosity*, η/ρ, is not used in this book.

It is complicated to derive an explicit expression for mass flux density in systems with an appreciable viscosity (see Equation 2.101). However, it is not difficult to see how the physical quantities must enter such an expression:

$$\boldsymbol{J}_m \propto -\rho \frac{r^2}{\eta} \kappa \, \boldsymbol{grad}\, p \tag{2.95}$$

It suffices to consider the velocity: it must be proportional to the pressure gradient, $-\, \boldsymbol{grad}\, p$, and inversely proportional to the viscosity η. Dimensional analysis[a] shows that $\boldsymbol{J}_m$ must be proportional to the area, r^2, of the tube; the number κ accounts for the exact form of the tube. For example, Poiseuille found (1844) that for a tube with a circular cross section, $\kappa = \pi/8$. Equation 2.95 with this value of κ is called Poiseuille's equation.

Example. Determination of relative viscosities. Insertion of $\boldsymbol{J}_m = \rho\, \boldsymbol{u}$ in Equation 2.95, where $\boldsymbol{u}$ is the velocity of a fluid in a vertical tube whose ends are open to the atmosphere,[b] gives $\eta/t \propto -\, r^2 \kappa\, \boldsymbol{grad}\, p$. Thus, the viscosity is proportional to the time t it takes for the top surface of a fluid to pass a given length of a certain tube. An unknown viscosity η may be determined relative to a known viscosity η_{ref} using the relationship $\eta = \eta_{\text{ref}}\, t/t_{\text{ref}}$.

Darcy's equation

The estimation of gas, oil, and groundwater flows in porous media (e.g., soils) is an important application of Poiseuille's equation. The velocity factor of the flux density, $\boldsymbol{J}_m = \rho\, \boldsymbol{u}$ of Equation 2.95, can be estimated empirically using Darcy's equation (H. Darcy, 1856). This equation is the empirical relationship

$$\boldsymbol{u}_{\text{D}} = -\mu_{\text{D}}\, \boldsymbol{grad}\, h_{\text{h}} \tag{2.96}$$

The constant μ_{D} is called the hydraulic conductivity (see Table 6.3). The constant depends on the medium as well as the fluid and is normally assumed to be known; it can be found in tables.

Alternatively, one may use an expression for the velocity factor of Equation 2.95 directly,

$$u_{\text{D}} = -\frac{A_{\text{D}}}{\eta}\, \boldsymbol{grad}\, p \tag{2.97}$$

where the empirical constant A_{D} has dimension of area. The original unit, called the Darcy, is still in use. In SI units,

$$1\, \text{Darcy} = \frac{\mu m^2}{1.01325} \approx 1\, \mu m^2 \tag{2.98}$$

The denominator is due to the conversion from atm to bar. A_{D} is called the (Darcy) permeability of the system (rock or soil); it is nominally independent of the kind of fluid phase.

---

[a] Substitute SI-units into the equation to see that m^2 is missing; thus, just the factor r^2 is needed to complete the relation.

[b] For crude measurements or very viscous fluids, a vessel with a draincock will suffice.

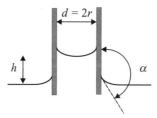

Figure 2.6 *Capillary tube*
A capillary tube with inner diameter d in water, enlarged.

Laplace's equation

Experiments show that flows through narrow tubes are subject to significant friction with the walls. This means that forces at the liquid-solid interface are particularly important in porous materials such as rocks and soils. Figure 2.6 shows an enlarged picture of an experiment in which one end of a narrow glass tube (a capillary tube) is placed in a fluid, with the other end open to the atmosphere. The contact angle between the liquid and the tube, shown on the drawing, is indicated by α. For water and glass, this angle is a little less than $180°$; the surface of water is concave, and the water is drawn up into the tube, as if there had been a reduction in the pressure over the water. Some liquids have very small contact angles; for example, mercury has a contact angle of less than $45°$. This leads to a convex surface inside the tube that is lower than the surface of the liquid on the outside, as if there were a larger pressure in the tube. An expression for the apparent pressure difference over the liquid in the capillary tube has been given by Young (1805) and Laplace (1806):[a]

$$\Delta p = \rho g \Delta h = \frac{2\gamma}{R} = \frac{2\gamma \cos\alpha}{r} \tag{2.99}$$

Here, the material constant γ is the surface tension, which is defined later. R is the radius of curvature of the meniscus, and r is the radius of the tube. Equation 2.99 plays an important role in soil chemistry (Section 6.2b). We anticipate the discussion by noting that if we define the level of the free water surface as the zero point, the capillary pressure of water (see Figure 2.6) is negative, $p = -4\gamma/d$, and $p > 0$ below the free surface.

Surface tension

The surface tension, γ, describes the forces at the interface between a liquid phase and another phase. It is defined as the Gibbs energy required to increase the surface area of the liquid system by an amount dA_s, divided by this area:

$$\gamma \stackrel{\text{def}}{=} \left(\frac{\partial G}{\partial A_s}\right)_{p,T,n} \tag{2.100}$$

Table 2.3 shows some values for the surface tension of water as a function of temperature. The surface tension depends on the temperature and on the chemical composition of the two phases (see Section 9.4c).

---

[a] Use Equation 2.76 with $\boldsymbol{J}_m = 0$: $\Delta p/\Delta h = \rho g$.

Table 2.3 Surface tension of water[129]							
$\theta / °C$	-5	0	10	20	30	100	374
$\gamma / (\text{mN m}^{-1})$	76.4	75.6	74.2	72.86	71.2	58.9	0

The critical temperature of water is 374 °C. A relation between units: $\text{N m}^{-1} = \text{J m}^{-2}$.

Concluding remarks

The preceding section has described some basic applications of the equations of motion, and we now briefly mention three additional concepts: the Navier-Stokes equations, characteristic numbers of turbulent flow and transport, and eddy diffusion.

1. The Navier-Stokes equations

Random collisions of particles in fluid systems give rise to the dissipation of energy and a never-decreasing entropy. Such irreversible features of the motion cannot be neglected, because they give rise to several observable macroscopic phenomena. As an example, the general Euler equation (Equation 2.76) must be modified to the Navier-Stokes equations (given here for incompressible fluids), which take viscosity into account:

$$\boldsymbol{j}_m = -\boldsymbol{grad}\, p + \rho \boldsymbol{a} + \eta \Delta \boldsymbol{u} \qquad (2.101)$$

Here η is the dynamic viscosity and Δ a differential operator.[a] This accounts for the fact that the equation of motion for a real fluid must include friction. It is relatively easy to set up the equations embodied in Equation 2.101, but they have only been solved in a few simple cases (e.g., Equations 2.94 and 2.95).[b] The equations are nonlinear partial differential equations for which no analytical solutions are known for flow past solid objects, for example as a sphere, a cube, a building, or an airplane.

2. Turbulent flow

The flow of a fluid in a pipe may be laminar or turbulent, or a combination of the two.[c] Reynolds (1883) was the first to distinguish between these two classes of flow. He found that when a fluid of density ρ and viscosity η runs through a pipe of diameter d, then, for increasing velocity u, the motion shifts from laminar to turbulent at the value of $u \approx 2000\, \eta/(\rho\, d)$. Experiments show this to be a general property of fluids, and one defines the *Reynolds number Re* by the quantity

$$Re = \frac{\rho\, u\, l}{\eta} \qquad (2.102)$$

of dimension 1. Here l is a characteristic length of the system, for example, a pipe or an immersed body.[d] One may think of the Reynolds number as the ratio between

---

[a] See footnote *a* on page 67 in this chapter.
[b] It can be shown that Equation 2.101 implies Equations 2.94 and 2.95[69]; η enters the equations correctly.
[c] See footnote *a* on page 61.
[d] Note a possible factor of 2: For a pipe, l may be chosen as the diameter (technical literature) or the radius (physical literature).

inertial forces and viscous forces $(\rho\, u^2\, d^2)/(\eta\, u\, d)$ or as the ratio between kinetic energy and energy of friction $(\rho\, u^2\, d^3)/(\eta\, u\, d^2)$.

An important idea in fluid mechanics is the concept of similitude: that is, measurements made on a model system in the laboratory can be used to describe the behavior of a similar, larger system. A well-known example is the use of wind tunnels to test properties of airplanes or bridges. The Reynolds number is one of more than 20 named "transport characteristic numbers" of dimension 1 that are in use in similitude experiments.[69,126]

3. Eddy diffusion
The average mixing generated by eddy diffusion processes can be parameterized as a simple diffusion constant (see Equation 2.110),

$$J_{eddy} = -D_{eddy}\ \boldsymbol{grad}\ c \tag{2.103}$$

This type of parameterization is used to model mixing between air in the planetary boundary layer and the free troposphere, between the upper troposphere and the stratosphere, in groundwater, and in ocean currents.

f. Applications of the continuity equation

Advection and retention
Generally, solutes and dispersed matter are carried with a moving solvent; this phenomenon is called *advection*. *Retention* is the word used when the solute[a] is delayed relative to the solvent.

In order to study some basic properties of advection, consider the continuity equation 2.63 for a nonreacting species B traveling in the x direction with a constant velocity, that is, $\boldsymbol{div}\ \boldsymbol{u} = 0$. We also assume $r_B = 0$ and have

$$\frac{\partial c_B}{\partial t} = -u\frac{\partial c_B}{\partial x} \tag{2.104}$$

meaning that the distribution of concentration is traveling with speed u along the x-axis. Assume that the distribution of the concentration of B is $f(x)$ at a certain time t_0 and that the amount of substance B is given by the integral

$$c_B(x, t_0) = f(x) \quad \text{and} \quad n_B(t_0) = \int_{-\infty}^{+\infty} f(x)\,dx \tag{2.105}$$

Figure 2.7 shows the profile $f_1(x)$ that was at position $f(x)$ at an earlier time $t_0 - \Delta t_1$, and the profile $f(x)$ that arrives at position $f_1(x)$ at a later time $t_0 + \Delta t_1$:

$$f_1(x) = f(x - u\Delta t_1) \quad \Leftrightarrow \quad f_1(x + u\Delta t_1) = f(x) \tag{2.106}$$

---

[a] In this section solutes include suspended matter.

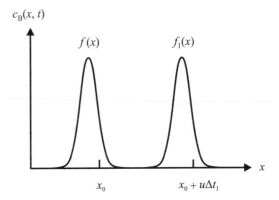

$c_B(x, t)$

$f(x)$

$f_1(x)$

x

x_0

$x_0 + u\Delta t_1$

Figure 2.7 *A moving concentration profile*
A concentration profile described by $f(x)$ is traveling to the right with a speed u.

Retention of a dissolved species is observed in several systems in the environment, for example, in groundwater. The phenomenon is described quantitatively by the retention ratio R_f, defined as the ratio

$$R_f = \frac{u(\text{solute})}{u(\text{solvent})} \tag{2.107}$$

where u denotes speed. Typically, retention is caused by the large surface area of soil combined with an affinity by the solute for the surface. Further discussion of this equation has two facets. One is the use of Equation 2.107 as the point of departure for quantitative analytical chromatography;[a] the other is a thorough investigation of adsorption as a surface phenomenon (see Section 6.3), the chemistry of soils.

Diffusion and dispersion

Early in the 19th century it was observed that a solute spontaneously migrates from a region of higher concentration to a region of lower concentration (Berthollet, 1803). Thomas Graham's experiments significantly improved understanding of diffusion in gases (1833) and liquids (1851). Independently, Fick presented (1855) the law of diffusion (see Equation 2.109), which was of immediate use in solving the problem of heat conduction being studied by Fourier (1811). The form of the equation for heat conduction is the same as for diffusion (see Equation 2.110).

Fick's first law is an empirical relation that Nernst[71] aptly explained as follows: Consider a diffusion tube, a long, narrow tube filled with a viscous gel, the solvent. The amount of substance dn of solute B that passes a cross-section A of the tube in the time dt when moving from position x with concentration c to position $x + dx$ with concentration $c - dc$ is proportional to

$$dn_B = -D_B A \frac{dc_B}{dx} dt \tag{2.108}$$

[a] For example, Equation 20–1 of ref. 29 or Chapter 26 of ref. 88.

Table 2.4	Diffusion coefficients of chemical species in air and water		
Species	$D/\mathrm{m^2\ s^{-1}}$	Species	$D/\mathrm{m^2\ s^{-1}}$
CH_4	10.6×10^{-6}	Sucrose	0.52×10^{-9}
Ar	14.8×10^{-6}	Glucose	0.67×10^{-9}
CO_2	16.0×10^{-6}	Alanine	0.91×10^{-9}
H_2O	24.2×10^{-6}	Ethylene glycol	1.16×10^{-9}
He	58.0×10^{-6}	Ethanol	1.24×10^{-9}
H_2	62.7×10^{-6}	Acetone	1.28×10^{-9}

Left columns: In atmospheric air, $p = 1.0$ bar, $\theta = 25.0\ °C$
Right columns: In aqueous solution, $\theta = 25.0\ °C$.

The material-dependent constant D_B is the diffusion coefficient. It follows from the discussion of the continuity equation (see Section 2.2c) that the general expression for the flux density of diffusion according to Fick is given by

$$J_B = -D_B\,\boldsymbol{grad}\,c_B \tag{2.109}$$

Some values of diffusion coefficients are given in Table 2.4. Application of Equation 2.109 to the continuity equation 2.60, with $r_B = 0$, gives[a]

$$\frac{\partial c_B}{\partial t} = -\boldsymbol{div}\,J_B = D_B\,\boldsymbol{div}\,\boldsymbol{grad}\,c_B = D_B\,\Delta c_B \tag{2.110}$$

to be solved for $c_B(r, t)$. One finds[51]

$$c(r, t) = \frac{n}{(4\pi Dt)^{3/2}}e^{\frac{-r^2}{4Dt}} \tag{2.111}$$

However, the important features of this equation can be derived from the one-dimensional case,

$$c(x, t) = \frac{n}{\sqrt{4\pi Dt}}\,e^{\frac{-x^2}{4Dt}} \tag{2.112}$$

where the dimension of $c(x, t)$ is amount per length. Consider a diffusion tube. In the middle, $x = 0$; a sample of a solute is introduced as a narrow zone, $x = \pm 1$, at time $t = 0$. The initial condition $c(x, 0)$ of the solute is given by the profile

$$c(x, 0) = 0 \quad \text{for} \quad -\infty < x < -1 \tag{2.113a}$$

$$c(x, 0) = c_0 \quad \text{for} \quad -1 \leq x \leq +1 \tag{2.113b}$$

$$c(x, 0) = 0 \quad \text{for} \quad +1 < x < +\infty \tag{2.113c}$$

The development of the system in time and space is shown in Figure 2.8. Note that the peak concentration decreases exponentially as a function of time, and also that the area of the curve in each slice is constant because it is assumed that the amount of solute is constant in time.

[a] The explicit form of the differential operator is
$$\Delta \overset{\mathrm{def}}{=} \boldsymbol{div}\,\boldsymbol{grad} = \frac{\partial^2}{\partial x^2} + \frac{\partial^2}{\partial y^2} + \frac{\partial^2}{\partial z^2}.$$

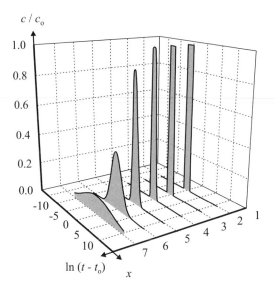

Figure 2.8 *Solutions of the one-dimensional diffusion equation*
Six different instances of Equation 2.112 with the boundary conditions Equations
2.113. The calculations were restricted to the interval $-11 < x < +11$.

Examples from atmospheric chemistry

In environmental chemistry one is often interested in the interplay between transport
and chemical reaction. For example, the rate of a chemical reaction in the troposphere
may compete with transport to the stratosphere, or the rate of transport of a pollutant
within a region may compete with the rate of chemical breakdown.

Simple descriptions use the continuity equation 2.61, and, depending on the approximation, one of Equation 2.61c or 2.61a will be the point of departure.

Puff models

A puff is a moving, infinitesimal parcel of air. A puff model is concerned with the
composition of a puff that is followed through time. A puff model uses Equation
2.61c,

$$\dot{c}_{B} = -c_{B}\, \boldsymbol{div}\, \boldsymbol{u} + r_{B} \qquad (2.114)$$

which refers to an infinitesimal volume of a phase. Of course, the volume (i.e., the
physicochemical system) is chosen to be as large as possible, consistent with the
system being homogeneous. The basic assumption of the puff approximation is that
the divergence vanishes, $\boldsymbol{div}\, \boldsymbol{u} = 0$. This reduces the chemistry to reactions within a
confined space (see Equation 2.69), changing the problem of transport into a question
of obtaining the trajectory of the puff by some other means.

The quantity r_{B} contains all contributions to the rate of concentration change of
B. It may be given explicitly as the sum of contributions of B created ($r_{B,c} > 0$) or
annihilated ($r_{B,a} < 0$) by chemical processes within the system. In some cases terms
describing emission ($r_{B,e}(t) > 0$) or deposition ($r_{B,a}(t) < 0$) can be added.[39] However

this approach may introduce an error as it assumes the system is homogeneous, whereas these terms are of the type $c_B \, \mathbf{div} \, \mathbf{u}$.

Box models

A box is a spatially fixed, infinitesimal volume through which air flows. A box model uses the continuity equation in the form of Equation 2.61a:

$$\frac{\partial c_B}{\partial t} + \mathbf{u} \cdot \mathbf{grad} \, c_B \;=\; -c_B \, \mathbf{div} \, \mathbf{u} + r_B \qquad (2.115)$$

A simple approximation is to remove the gradient of the concentration, that is, set $\mathbf{grad} \, c_B = 0$. The model of an open physicochemical system, discussed in Section 2.2b, makes use of this approximation and sets $r_B = 0$:

$$\frac{\mathrm{d}c_B(t)}{\mathrm{d}t} \;=\; -c_B(t) \, \mathbf{div} \, \mathbf{u} \qquad (2.116)$$

More sophisticated applications include multibox models where the divergence terms describe the exchange of B between the boxes.

2.3 Chemical thermodynamics

Classic thermodynamics was developed early in the 19th century in order to explain the behavior of steam engines, including their fuel efficiency (Carnot, 1824). The equivalence of heat and work was established almost a quarter of a century later (Mayer, Colding, Joule; 1842). Soon after, the first version of the second law of thermodynamics was presented (Clausius, 1850), followed by the thermodynamic scale of temperature (Thomson [Kelvin], 1854) and entropy (Clausius, 1865). In the following decades, the macroscopic description of irreversible processes was brought to the level of development that we shall use here.[a] Chemical thermodynamics was founded by J. W. Gibbs through the years 1873–1876, and the subject was made generally accessible through the classic textbook by Lewis and Randall (1923).[56] Since then major progress – of interest for environmental chemistry – has been made in understanding complex formation (Chapter 5) and chemical kinetics (Section 2.4; Chapter 4).

Environmental issues are of growing importance to modern life, and thermodynamic principles are basic to their description. It is a surprising fact that the past few decades have witnessed a flourishing of the scientific methods and tools that were developed from 1850 to 1950. As an example, data used to determine solubility or the distribution of chemical species between phases are a prerequisite for reliable predictions of the fate of a pollutant in Nature, and such calculations can now be made on the basis of readily accessible tables of physicochemical constants.

The purpose of Section 2.3 is not to derive thermodynamic expressions, but to make thermodynamic concepts accessible to environmental chemists. First we consider

---

[a] We shall not make use of the concepts of irreversible thermodynamics[46] as laid down by L. Onsager (1931) and I. Prigogine (1947).

thermodynamics and chemical reactions, including absence of thermal equilibrium in a fluid phase. Section 2.3b, describes the partitioning of chemical species between two phases. The chemistry of interfaces is discussed in Sections 6.3 and 8.9.

a. Basic concepts

The fundamental equations

Two of the fundamental equations of thermodynamics, Equations 2.4 and 2.5, were presented in the introduction together with the Gibbs-Duhem equation, Equation 2.6. When applied to a single phase α, Equations 2.5 and 2.6 are[a]

$$dG_m^\alpha = -S_m^\alpha dT^\alpha + V_m^\alpha dp^\alpha + \sum_B \mu_B^\alpha dx_B^\alpha \qquad (2.117)$$

$$0 = -S_m^\alpha dT^\alpha + V_m^\alpha dp^\alpha - \sum_B x_B{}^\alpha d\mu_B^\alpha \qquad (2.118)$$

The purpose of this section is to discuss the significance of the terms of Equations 2.117 and 2.118. We start with a partial molar quantity, the partial molar volume V_m. Next we go into detail with the chemical potential μ_B, the choice of standard state, and the connection of μ_B to activity, measures of concentration, and the standard equilibrium constant.

Partial molar quantities

The formal definitions of a partial molar quantity are as follows. An intensive quantity X_B of species B corresponding to an extensive quantity X of a phase with more than one chemical species is defined by the relation

$$X_B \overset{\text{def}}{=} \left(\frac{\partial X}{\partial n_B} \right)_{p,T,n'} \qquad (2.119)$$

X_B is called the partial molar quantity, and the prime on the symbol n after the partial derivative means that all n_C with $C \neq B$ are kept constant. X_B measures the change of X when a small amount of B is added to a large amount of the phase. Generally, if the independent variables of a phase are p, T, and n, then the differential of X is given by

$$dX = \left(\frac{\partial X}{\partial p} \right)_{T,n} dp + \left(\frac{\partial X}{\partial T} \right)_{p,n} dT + \sum_B X_B dn_B \qquad (2.120)$$

where the n after the partial derivatives means all n_B.

Euler's theorem for homogeneous functions[44] states that given the third term of Equation 2.120, $dX = \sum_B X_B dn_B$, then

$$X = \sum_B X_B n_B \qquad (2.121)$$

---

[a] In this section the phase α may be gas (g), liquid (l), or solid (s).

exists with the derivative

$$dX = \sum_B X_B \, dn_B + \sum_B n_B \, dX_B \tag{2.122}$$

When this equation is subtracted from Equation 2.120, one obtains

$$0 = \left(\frac{\partial X}{\partial p}\right)_{T,n} dp + \left(\frac{\partial X}{\partial T}\right)_{p,n} dT - \sum_B n_B \, dX_B \tag{2.123}$$

which is the Gibbs-Duhem equation (Gibbs, 1875; Duhem, 1886). Note that these mathematical identities do not require that $dp = 0$ or that $dT = 0$.

Division of Equation 2.121 by $\Sigma_B \, n_B$ gives

$$X_m = \sum_B X_B x_B \tag{2.124}$$

With the further condition that $dp = 0$ and $dT = 0$, Equation 2.123 is reduced to

$$0 = \sum_B x_B \, dX_B \qquad (p, \; T \text{ constant}) \tag{2.125}$$

Finally, differentiation of Equation 2.124 and use of Equation 2.123 leads to

$$dX_m = \sum_B X_B dx_B \qquad (p, \; T \text{ constant}) \tag{2.126}$$

The expressions presented here in a general form are used repeatedly in the application of thermodynamics to environmental problems: Equations 2.124 and 2.126 are the basis for measurement of partial molar quantities.[a] When X_m is equal to G_m, the molar Gibbs energy, Equation 2.125 is of fundamental importance in the study of phase equilibria (Section 2.3b), and the fundamental equation (see Equation 2.117) is the basis for determinations of equilibrium constants. We shall discuss how Equation 2.117 can be standardized for practical use.

Measurement of partial molar volume

In order to illustrate Equations 2.119–2.126, we discuss the determination of partial molar volumes based on measurements of the densities of mixtures of methanol (1) and tetrachloromethane (2).[204] Figure 2.9 graphically shows the data converted to molar volume V_m as a function of mole fraction x_i, $i = 1, 2$. We use Equation 2.124,

$$V_m = V_1 x_1 + V_2 x_2 = (V_2 - V_1) x_2 + V_1 \tag{2.127}$$

whose derivative dV_m is

$$dV_m = V_1 \, dx_1 + V_2 \, dx_2 = (V_2 - V_1) \, dx_2 \tag{2.128}$$

because $x_1 + x_2 = 1$. The slope α of the tangent at any point of curve V_m is

$$\alpha \stackrel{\text{def}}{=} \frac{dV_m}{dx_2} = V_2 - V_1 \tag{2.129}$$

---

[a] Examples include partial molar energies, volumes, conductivities, and radiant excitances.

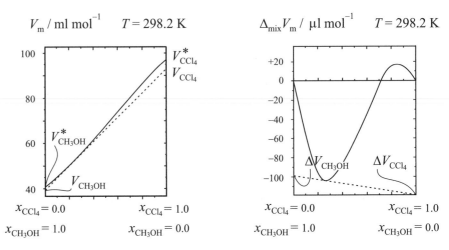

Figure 2.9 *Partial molar volume of methanol-tetrachloromethane mixtures*
Left panel: Full curve: V_{m} (the curvature is exaggerated); broken line: the tangent of $V_{\mathrm{m}}(x_2 \approx 0.26)$.
Right panel: Full curve: $\Delta_{\mathrm{mix}}V_{\mathrm{m}}$; broken line: the tangent of $\Delta_{\mathrm{mix}}V_{\mathrm{m}}(x_2 \approx 0.26)$.
The drawings are based on data from ref. 204.

The intercepts of the tangent with the vertical axes are the partial molar volumes,

$$V_1(x_2) = V_{\mathrm{m}}(x_2) - \alpha x_2 \tag{2.130}$$

and

$$V_2(x_1) = V_{\mathrm{m}}(x_1) + \alpha x_1 \tag{2.131}$$

Figure 2.9 (left-hand side) shows V_{m}, the tangent at $x_2 \approx 0.26$, and the partial molar volumes for this value of x_2.

Two or more components are said to give an ideal mixture if the attraction of the components to one another is the same as their attraction to themselves.[a] Its molar volume V_{m}^* may be represented by a straight line that connects the molar volumes of the pure components, V_1^* and V_2^*,

$$V_{\mathrm{m}}^* = V_1^* x_1 + V_2^* x_2 = (V_2^* - V_1^*)x_2 + V_1^* \tag{2.132}$$

where the last term is expressed in terms of the variable x_2. The "volume of mixing" $\Delta_{\mathrm{mix}}V_{\mathrm{m}}$ is defined as the difference

$$\Delta_{\mathrm{mix}}V_{\mathrm{m}} \overset{\mathrm{def}}{=} V_{\mathrm{m}} - V_{\mathrm{m}}^* = \left((V_2 - V_1) - (V_2^* - V_1^*)\right)x_2 + (V_1 - V_1^*) \tag{2.133}$$

which is of the same form as Equation 2.127, leading to analogous conclusions as shown in Figure 2.9 (right-hand side).

---

[a] Generally, an ideal mixture is a mixture whose extensive properties are the sum of those of each component times its mole fraction.

Some comments on Figure 2.9:

1. In the drawing, the curvature of V_m is exaggerated; only the endpoints and the point at $x_2 \approx 0.71$ are correctly depicted. Therefore, in many cases, the introduction of the mixing volume is an experimental advantage.
2. $\Delta_{mix} V_m$ represents the deviation of the molar volume of the mixture from that of an ideal one. In connection with the thermodynamic functions such deviations from ideality are termed *excess functions* (see Equation 2.147 and Section 2.3b).
3. Solutions rich in methanol mix with volume contraction, whereas solutions with less than 30 % methanol experience volume expansion. This phenomenon lacks an explanation in terms of more fundamental quantum chemistry.
4. The physical quantity volume has a natural reference point, namely, zero volume. In contrast, energies have no natural reference points, and only differences in energy can be determined experimentally. This gives additional significance to the procedure of Equation 2.133. In order to communicate physical data accurately, the international scientific community[a] has agreed on arbitrary reference points for partial molar energies and the associated concentrations.

Measurement of chemical potential

The chemical potential of a species B in phase α is defined as the partial molar Gibbs energy

$$\mu_B^\alpha = -\left(\frac{\partial G^\alpha x}{\partial n_B}\right)_{T^\alpha, p^\alpha, n^i} \tag{2.134}$$

(see Equation 2.119). In this section we discuss how chemical potentials are measured and how their reference points, so-called standard states, are defined. Three different cases are considered: (1) B is a gas, (2) B is a pure phase in the liquid or solid state, and (3) B is a solute in solution.

1. A gas phase

For a pure chemical species B in the gas phase at constant temperature, the Gibbs-Duhem equation 2.118 is

$$d\mu_B^g = V_B^g \, dp_B^g \qquad (T \text{ constant}) \tag{2.135}$$

and the goal is to integrate this differential equation.[b] Suppose that B is a perfect gas.[c] Then,

$$d\mu_B^g = V_{m,B}^g dp_B^g = \frac{RT}{p_B^g} dp_B^g = RT \, d \ln p_B^g \qquad (T \text{ constant}) \tag{2.136}$$

---

[a] The international scientific unions IUPAC, IUPAP (physics), and IUB (biochemistry) and the International Organization for Standardization, ISO.

[b] See Equation 2.142.

[c] A perfect gas is a mixture that obeys the equation of state: $pV = RT \sum_B n_B$.

and after integration,

$$\mu_B^g(T, p_B) - \mu_B^{go}(T) \; = \; RT \ln \frac{p_B}{p^\circ} \qquad (T \text{ constant}) \qquad (2.137)$$

with two constants of integration, μ_B^{go} and p°. They are marked with a $^\circ$ representing "standard": the standard chemical potential, $\mu_B^{go}(T)$, is the chemical potential of B for a specified standard pressure, p°.[a]

The result, Equation 2.137, states that a change in chemical potential is given by $R\,T$ times the logarithm of the ratio between two partial pressures.

Before considering the definition of standard state for gases, consider the fundamental properties of a real gas using van der Waals' qualitative equation of state (1873),

$$\left(p + \frac{a}{V_m^2} \right) (V_m - b) \; = \; RT \qquad (2.138)$$

Here V_m is the molar volume and the empirical parameters a and b correct for intermolecular attractive forces and the eigenvolume of the molecules, respectively. Such interactions can be ignored in perfect gases, and it is an experimental fact that at low pressure and high temperature, the limit $a = 0$ and $b = 0$ exists for all real gases as shown in the following example.

Example

The van der Waals constants for water are[129] $a = 5.54$ bar M^{-2} and $b = 0.030$ M^{-1}. Using Equation 2.138:

At $p = 1$ bar, $T = 373.15$ K: $V_{m, \text{perfect gas}} = 31.03$ l/mol, $V_{m, \text{water}} = 30.38$ l/mol.

At $p = 0.1$ bar, $T = 373.15$ K: $V_{m, \text{perfect gas}} = 310.3$ l/mol, $V_{m, \text{water}} = 310.1$ l/mol.

At $p = 1$ bar, $T = 1573.15$ K: $V_{m, \text{perfect gas}} = 130.8$ l/mol, $V_{m, \text{water}} = 130.8$ l/mol.

This illustrates that a gas approaches the perfect state at low pressures and high temperatures.

Returning to the definition of standard state for gases, the internationally accepted definition of the standard chemical potential is the limit

$$\mu_B^{go}(T) \; = \; \lim_{p \to 0} \left[\mu_B^{go}(T, p, y_B, \ldots) - RT \ln \frac{y_B p}{p^\circ} \right] \qquad (T \text{ constant}) \qquad (2.139)$$

where y_B is the mole fraction and $p_B = y_B p$ the partial pressure of a gas B. The most common choice of standard pressure is $p^\circ = 0.1$ MPa $= 1$ bar.[b] The standard chemical potential of a gas is the chemical potential at the standard pressure, but with the physical properties of a perfect gas. Note that Equation 2.139 is valid for any component of a gas mixture.

For historical reasons, deviations from ideality are described by the fugacity, $\tilde{p}_B$, in place of p_B, and Equation 2.139 shows that at sufficiently low pressures the fugacity is equal to the partial pressure.

---

[a] The symbol $^\circ$ or $^\ominus$ (the plimsoll) is used to indicate standard.
[b] Most data published before 1982 used the value $p^\circ = 1$ atm $= 101325$ Pa. The term *normal*, for example, used in "normal boiling temperature," continues to refer to this standard.[162]

2. A liquid or solid phase

In this case the species B is either a pure substance or a component of a mixture. It can be shown that for ideal mixtures the chemical potential is given by a relation similar to Equation 2.137:

$$\mu_B^\alpha(T, p^\circ) = \mu_B^{\alpha\circ}(T) + RT \ln x_B^\alpha \qquad (p, T \text{ constant}) \qquad (2.140)$$

with $\alpha = l$ or s. The standard state is the state of pure B in the phase α at the standard pressure p°, and the chemical potential is defined as

$$\mu_B^{\alpha\circ}(T) = \mu_B^{*\alpha}(T, p^\circ) \qquad (2.141)$$

where $*$ signifies the pure species. Thus, a change in the chemical potential of B is expressed in terms of the mole fraction x_B. Several examples of the application of relation 2.140 are given later in Table 2.5.

Suppose that the value of the pressure p changes from p_1 to p_2 at constant T. Then, from the Gibbs-Duhem equation, the change in chemical potential is given by

$$\mu_B^\alpha(T, p_2) = \mu_B^\alpha(T, p_1) = \int_{p_1}^{p_2} V_B^* \, dp \qquad (2.142)$$

where V_B^* is the molar volume of B (see Equation 2.135).

It is a matter of convenience whether a liquid or solid mixture is called a mixture or a solution. In the latter case, the solvent is the component in surplus and it is treated as discussed earlier.[a]

3. A solute

For a solute B in a liquid or solid solution, the standard state is often taken as the ideal dilute behavior of the solute. This is a hypothetical state where the solute B is at standard concentration c° and standard pressure p° but behaves as an infinitely dilute solution. The defining equation is[126]

$$\mu_B^{\alpha\circ}(T) = \left[\mu_B^\alpha(T, p^\circ, c_B, \ldots) - RT \ln \frac{c_B}{c^\circ} \right]^\infty \qquad (2.143)$$

Molality, m, may be used in place of concentration. The standard molality $m^\circ = 1 \text{ mol kg}^{-1}$ and the standard concentration $c^\circ = 1 \text{ mol dm}^{-3} = 1 \text{ M}$ are universally accepted.

Activity

In order to discuss nonideal mixtures, the activity a_B^α was introduced by G. N. Lewis (1907) through the exponential form of Equations 2.137 and 2.140:[b]

$$a_B^\alpha = \exp\left(\frac{\mu_B^\alpha - \mu_B^{\alpha\circ}}{RT} \right) \qquad (2.144)$$

---

[a] Frequently, mole fraction x_A is used for the solvent A and amount concentration c_B or molality m_B for the solute B.

[b] In specialist literature a_B^α is called relative activity. Quantities defined by $\lambda_B = \exp(\mu_B/RT)$ are called absolute activities; they are not used in this book.

From Equation 2.137 it is clear that the activity includes a term defining the standard state. Deviation from ideal behavior is measured by the activity coefficient, $\gamma_B{}^\alpha$, which is defined by

$$a_{x,B}^\alpha = \gamma_{x,B}^\alpha x_B^\alpha, \quad a_{c,B}^\alpha = \frac{\gamma_{c,B}^\alpha c_B^\alpha}{c^\circ} \quad \text{or} \quad a_{m,B}^\alpha = \frac{\gamma_{m,B}^\alpha m_B^\alpha}{m^\circ} \quad (2.145)$$

depending on the choice of standard state. The activity coefficient is always equal to 1 in an ideal mixture. In nonideal mixtures it may be more or less than 1, but it approaches unity with increasing dilution.

In terms of mole fraction, the chemical potential is

$$\begin{aligned}
\mu_B^\alpha &= \mu_B^{\alpha o} + RT \ln a_B^\alpha \\
&= \mu_B^{\alpha o} + RT \ln x_B^\alpha + RT \ln \gamma_{x,B}^\alpha \\
&= \mu_B^{\alpha o} + RT \ln x_B^\alpha + \mu_{x,B}^{E\alpha} \quad (2.146)
\end{aligned}$$

with the definition

$$\mu_{x,B}^{\alpha E} = RT \ln \gamma_{x,B}^\alpha \quad (2.147)$$

which is called the excess chemical potential. Analogous definitions apply for other choices of standard state. Application of the excess chemical potential to problems of an environmental nature is discussed in Section 2.3b.

Choices of standard state
Equation 2.139 implies the expressions

$$\lim_{p \to 0} \left(\ln \frac{p^\circ}{y_B p} \right) = 0 \quad \text{and} \quad \lim_{p_B \to 0} \left(\frac{p^\circ}{p_B} \right) = 1 \quad (2.148)$$

with $y_B p = p_B$. They clearly show that the standard state of a gas is the standard pressure, except with properties as if the pressure were zero.

Similarly, the standard state of a solute, given by Equation 2.143 and the middle part of Equation 2.145, may be formulated as

$$\lim_{c_B \to 0} \left(\frac{a_{c,B} c^\circ}{c_B} \right) = 1 \quad (2.149)$$

where for simplicity the designation of phase α has been omitted. This choice of standard state indicates that the activity $a_{c,B}$ of a solute B (e.g., an ion) approaches its concentration as the composition of the solution approaches that of pure solvent.

Other standard states in general use include the following:

1. In work on ionic equilibria in aqueous solution, it is convenient to define the standard state with respect to a salt solution instead of the pure solvent. Common media include 0.5 M NaCl, 1.0 M NaClO$_4$, 3.0 M NaClO$_4$, and synthetic seawater. Such standard states have the advantage that electrode measurements are better defined, and one may use concentrations with the unit "mole per liter solution" directly in the equilibrium constants.[159]

2. The standard state of Equation 2.149 implies a hydron concentration of $[H^+] = 1$ M. A standard state that is close to physiological conditions may be chosen such that $[H^+] = 10^{-7}$ M:

$$\mu_{H^+}^{o\prime} = \mu_{H^+}^{o} - 7RT \ln 10 \qquad (2.150)$$

The addition of a prime is the accepted form indicating the physiological standard state. The size of the difference $7\,RT \ln 10 \approx 39.956$ kJ mol^{-1} at $\theta = 25\,^{\circ}$C between the physiological standard state and the thermodynamic standard state may change the sign of the affinity of a chemical reaction.[a]

The affinity of a chemical reaction

Until the middle of the 19th century, *reactivity* and *affinity* were loosely used as phenomenological descriptions of a chemical reaction reflecting both the equilibrium constant and the reaction rate. However, the study of irreversible thermodynamics changed these concepts, because it was realized that chemical reactions can be quantified using the rate of entropy production[b] (see Equation 2.162, later).

The theme of this section is the entropy differential dS of Equation 2.4. The entropy changes in any system can be divided into two components,[c]

$$dS = d_e S + d_i S \qquad (2.151)$$

where $d_e S$ is due to interactions with the surroundings (e = exterior) and $d_i S$ is due to changes within the system (i = interior) (see Equation 2.66). The latter is never negative; it is only zero when the system undergoes reversible changes and positive if irreversible processes take place. For isolated systems the relation $dS = d_i S \geq 0$ is equivalent to the classical statement that "entropy can never decrease," and simultaneously, it is a tool for the experimental detection of the presence of irreversible processes. Similar criteria exist for other systems. For example, for closed systems at constant pressure and temperature, the Gibbs energy decreases when irreversible changes occur and remains constant otherwise. From Equation 2.117:

$$dG_m = \sum_B \mu_B dx_B \leq 0 \qquad (2.152)$$

Returning to Equation 2.4, the first task is to modify the last term using the extent of reaction (Equation 2.17) to give the result:

$$dU = TdS - pdV + d\xi \sum_B \nu_B \mu_B \qquad (2.153)$$

In order to do so, the reaction equation 2.15 for a reaction "r" must be known:

$$\sum_B \nu_B B = 0 \qquad (2.154)$$

---

[a] See Equation 2.156. The reaction NADH(aq) + H$^+$(aq) → NAD$^+$(aq) + H$_2$(g) at 37 $^{\circ}$C has $A^{\circ} = 21.8$ kJ/mol and $A^{\circ\prime} = -19.7$ kJ/mol. NAD is nicotinamide adenine dinucleotide, and the redox reaction is part of the respiratory processes.

[b] Gk. $\varepsilon\nu$-$\tau\rho\omega\pi\eta \,\hat{=}\,$ that which changes itself.

[c] See textbooks on thermodynamics, such as refs. 4, 56, or 78.

This relation is the reason why several amount parameters can be reduced to one, the extent (see Equation 2.17).

It is customary to define the *affinity* of the a chemical reaction as

$$A \stackrel{\mathrm{def}}{=} - \sum_B \nu_B \, \mu_B \tag{2.155}$$

(Th. De Donder, 1928). We note the relationships[a]

$$A = - \sum_B \nu_B \, \mu_B = - \left(\frac{\partial G}{\partial \xi} \right)_{T,p} = - \Delta_r G \tag{2.156}$$

where the symbol $\Delta_r = \partial / \partial \xi$, the derivative with respect to the extent of reaction r, is much used in applications of thermodynamic tables (see later discussion).

The entropy differential, Equation 2.153, can be written

$$\mathrm{d}S = \frac{\mathrm{d}U + p\mathrm{d}V}{T} + \frac{A}{T} \, \mathrm{d}\xi \tag{2.157}$$

in terms of affinity. Application of the first law of thermodynamics, $\mathrm{d}U = \mathrm{d}Q - p\mathrm{d}V$, gives rise to the expression

$$\mathrm{d}S = \frac{\mathrm{d}Q}{T} + \frac{A}{T} \, \mathrm{d}\xi \tag{2.158}$$

where each of the three terms is a total differential. We now restrict the discussion to a closed system, because then $\mathrm{d}Q$ is simply heat transfer. Nonetheless, despite this simplifying assumption, general features will emerge at the end. The entropy differential of Equation 2.158 is composed of two terms, which parallels the split form of Equation 2.151. One term is due to interactions with the surroundings, and the entropy change is given by the heat exchange,

$$\mathrm{d}_e S = \frac{\mathrm{d}Q}{T} \tag{2.159}$$

The other term describes entropy production due to an irreversible chemical reaction within the system:

$$\mathrm{d}_i S = \frac{A}{T} \, \mathrm{d}\xi > 0 \tag{2.160}$$

For a system in equilibrium, $A = 0$. If, for example, the reaction consists of the transfer of species B from phase α to phase β, then the equilibrium condition becomes $\mu_B^\alpha = \mu_B^\beta$. These relations show why the chemical potential is fundamental to the thermodynamics of equilibria.

The reactivity of a chemical reaction

It follows from Equation 2.160 that the rate of entropy change is given by the expression

$$\frac{\mathrm{d}_i S}{\mathrm{d}t} = \frac{A}{T} \dot{\xi} > 0 \tag{2.161}$$

---

[a] Use Equations 2.134 and 2.17.

where $\dot{\xi} = \mathrm{d}\xi/\mathrm{d}t$ is the rate of conversion according to a reaction scheme (see Equation 2.154). Thus, the affinity and the rate of conversion always have the same sign, and entropy production ceases when either of the factors A or $\dot{\xi}$ is zero. The term *reactivity* is used for the rate of entropy change. In many cases a vanishing reactivity is due to the lack of a suitable reaction mechanism, which overrides favorable thermodynamics.[a]

Equation 2.161 is easily extended to the case of several simultaneous reactions:

$$\frac{\mathrm{d_i}S}{\mathrm{d}t} = T^{-1} \sum_i A_i \dot{\xi}_i > 0 \tag{2.162}$$

Consider two simultaneous reactions, 1 and 2. It may happen that the reactivities differ in sign,

$$A_1 \dot{\xi}_1 < 0 \quad \text{and} \quad A_2 \dot{\xi}_2 > 0 \tag{2.163a}$$

which is possible provided that the sum

$$A_1 \dot{\xi}_1 + A_2 \dot{\xi}_2 > 0 \tag{2.163b}$$

is positive. Thus, the coupling of reactions allows one reaction to progress against its affinity if this is more than compensated by the simultaneous consumption of reactivity by other reactions.[78] For example, this kind of concerted reaction accounts for the entropy production of life processes taking place without destroying life itself.

The equilibrium constant

When a reaction r governed by Equation 2.154 has reached equilibrium, the affinity $A = 0$. An important consequence of this statement is obtained by entering Equation 2.144 into Equation 2.156:[b]

$$A = -\sum_B v_B \left(\mu_B^\circ + RT \ln a_B \right) = 0 \qquad (T \text{ constant}) \tag{2.164}$$

leading to

$$A^\circ = RT \ln \left(\prod_B a_B^{v_B} \right) = RT \ln K^\circ \qquad (T \text{ constant}) \tag{2.165a}$$

or its equivalent

$$K^\circ = \exp \left(\frac{-\sum\limits_B v_B \mu_B^\circ}{RT} \right) = \exp \left(\frac{A^\circ}{RT} \right) \qquad (T \text{ constant}) \tag{2.165b}$$

Here we have made use of the two definitions,

$$A^\circ = -\sum_B v_B \mu_B^\circ \qquad \text{the } \textit{standard affinity} \tag{2.166}$$

---

[a] An example: Thermodynamic equilibrium of the atmosphere and the hydrosphere would result in a dilute nitric acid solution.

[b] Note the relation $\ln(\prod_i x_i^{y_i}) = \sum_i (y_i \ln x_i)$.

and

$$K^\circ = \prod_B a_B^{\nu_B} \qquad \text{the } standard\ equilibrium\ constant \qquad (2.167)$$

Equations 2.165 constitute the thermodynamic connection between the affinity and the standard equilibrium constant of a given chemical reaction.

It is important to note that the standard equilibrium constant K° is a physical quantity of dimension 1 (one). This is a convention that requires specification of the standard state.

Generally used equilibrium constants usually differ from standard equilibrium constants because the former generally have a physical dimension different from 1. Consider an equilibrium constant K_c measured using concentrations. It is connected to the standard equilibrium constant as seen by substituting the middle expression of Equation 2.165 into Equation 2.167,[a]

$$K^\circ = \left(\prod_B (c_B \gamma_{c,B})^{\nu_B} \right) \left(\prod_B (c^\circ)^{-\nu_B} \right) \overset{\text{def}}{=} K_c \left(\prod_B (c^\circ)^{-\nu_B} \right) \quad (2.168)$$

Accordingly, the equilibrium constant K_c refers to a definite chemical reaction, and the concentrations are measured using a standard concentration as the unit.

Some remarks on the standard affinity:

1. The standard reaction Gibbs energy $\Delta_r G^\circ$ is related to the standard affinity A° by $A^\circ = -\Delta_r G^\circ$.
2. By use of the relation $\Delta G = \Delta H - T\Delta S$, the standard affinity may be expressed using the other thermodynamic functions:

$$A^\circ = - \left(\frac{\partial H^\circ}{\partial \xi} \right)_{p,T} + T \left(\frac{\partial S^\circ}{\partial \xi} \right)_{p,T} \qquad (2.169)$$

and

$$A^\circ = - \Delta_r H^\circ + T \Delta_r S^\circ = RT \ln K^\circ \qquad (2.170)$$

which will be used on several occasions later in the book. A useful approximation for chemical reactions in condensed phases is that entropy changes are small; thus:

$$A^\circ \approx - \Delta_r H^\circ \qquad (2.171)$$

from which the rate of entropy production follows:

$$\frac{d_i S^\circ}{dt} = \frac{A^\circ}{T} \dot{\xi} \approx - \frac{1}{T} \frac{dQ}{dt} \qquad (p,\ T\ \text{constant}) \qquad (2.172)$$

In this case the statement $A_1^\circ \dot{\xi}_1 < 0$ means that the reaction is endothermic, which can only take place in the presence of an exothermic reaction having $A_2^\circ \dot{\xi}_2 > -A_1^\circ \dot{\xi}_1$.

---

[a] A common standard is $c^\circ = 1\ \text{mol}\ l^{-1} = 1\ \text{M}$.

Temperature dependence of the equilibrium constant

The temperature dependence of the equilibrium constant is given by the van't Hoff equation (1884):

$$\frac{d \ln K}{dT} = \frac{\Delta_r H}{RT^2} \tag{2.173}$$

where

$$\Delta_r H = \sum_B v_B H_B = \left(\frac{\partial H(T)}{\partial \xi} \right)_{T,p} \tag{2.174}$$

noting that the enthalpy H is a function of T (see Equation 2.156) and that the equation is only valid when referring to a definite reaction equation r (see Equation 2.154).

The van't Hoff equation may be integrated:

$$\ln \frac{K_2}{K_1} = - \int_{T_1}^{T_2} \frac{\Delta_r H(T)}{R} \, d\left(\frac{1}{T} \right) \approx \frac{\Delta_r H}{R} \left(\frac{T_2 - T_1}{T_1 T_2} \right) \tag{2.175}$$

where the last step requires that the reaction enthalpy be approximately constant over the considered interval of temperature, which is frequently the case. We will make use of this equation at several places, such as in Chapters 5 and 8.

b. Phase equilibria

Phase equilibria are central to environmental applications. One example is the chemistry of soil, which involves the mobility of a chemical species via exchange among the gas phase, aqueous solutions, and surface adsorption. Another example is the partitioning of species between the gas and condensed phases of the atmosphere.

We now discuss the phase rule and apply it to the description of the thermodynamic behavior of mixtures. The discussion includes Henry's law and Nernst's distribution law, which are frequently used to evaluate environmental problems.

The phase rule

To a large extent, the thermodynamics of a heterogeneous system can be described using modifications of the equations applicable to a single phase. The independent intensive variables that specify a state are the *degrees of freedom* of the system; they are denoted $\mathscr{F}$. The phase rule concerns determination of $\mathscr{F}$ for an arbitrary number of phases and chemical reactions. Recall the Gibbs-Duhem equation (Equation 2.118) for a single phase in thermal, hydrostatic, and diffusive equilibrium. The terms in the sum Σ_B are called *components* and their number denoted $\mathscr{C}$. Because the number of intensive variables is $\mathscr{C} + 2$ and Equation 2.118 furnishes one constraint, the number of independent variables is $\mathscr{F} = \mathscr{C} + 1$.

In a system with $\mathscr{P}$ phases in mutual equilibrium, there are $\mathscr{P} - 1$ additional equations imposing $\mathscr{P} - 1$ restrictions on the same intensive variables. Thus,

$$\mathscr{F} = (\mathscr{C} + 1) - (\mathscr{P} - 1) = \mathscr{C} + 2 - \mathscr{P} \tag{2.176}$$

which is called the *phase rule* (Gibbs, 1876).

Finally, we shall include the possibility of chemical reaction. At equilibrium, each independent chemical reaction must fulfill Equation 2.156,

$$\sum_{B} \nu_B \mu_B = 0 \qquad (2.177)$$

and each such constraint will reduce $\mathscr{F}$ by 1. With $\mathscr{R}$ independent chemical reactions, the number of degrees of freedom is

$$\mathscr{F} = \mathscr{C} + 2 - \mathscr{P} - \mathscr{R} \qquad (2.178)$$

which is the general form of the phase rule.

1. Clapeyron's equation

Consider two phases α and β of a pure species in equilibrium. From Equation 2.178 with $\mathscr{C} = 1$, $\mathscr{P} = 2$, $\mathscr{R} = 0$, one obtains $\mathscr{F} = 1$. If the temperature T is chosen as the independent variable, then the equilibrium pressure p becomes the dependent variable and is called the *vapor pressure*. On the other hand, if the pressure p is taken as the independent variable, then the equilibrium temperature T becomes the dependent variable and is called the *melting point*, T_m (solid $\rightarrow$ liquid) or the *boiling point*, T_b (liquid $\rightarrow$ gas).

The Gibbs-Duhem equation relates intensive quantities. For each phase $\gamma = \alpha$ or β, Equation 2.118 takes the form

$$S_m^{*\gamma} dT - V_m^{*\gamma} dp + d\mu^* = 0 \qquad (2.179)$$

where $*$ refers to the pure species. Eliminating $d\mu^*$ from the two equations and rearranging, using T as the independent variable, yields

$$\frac{dp}{dT} = \frac{S_m^{*\beta} - S_m^{*\alpha}}{V_m^{*\beta} - V_m^{*\alpha}} \overset{\text{def}}{=} \frac{\Delta_\alpha^\beta S_m^*}{\Delta_\alpha^\beta V_m^*} \qquad (2.180)$$

Because at equilibrium $\Delta_\alpha^\beta G_m^* = 0$, one uses the relation $\Delta_\alpha^\beta H_m^* = T\Delta_\alpha^\beta S_m^*$. Substituting into Equation 2.180 gives an alternative and very useful form:

$$\frac{dp}{dT} = \frac{\Delta_\alpha^\beta H_m^*}{T\Delta_\alpha^\beta V_m^*} \qquad (2.181)$$

which is also exact. Either of Equations 2.180 or 2.181 is the *Clapeyron equation* (1834).

2. Clausius-Clapeyron equation

Assume that the phase α is the condensed phase, $\alpha = cd$, and its molar volume V_m^{*cd} is negligible compared with V_m^{*g} of the gas phase, $\beta = g$. In addition, assume that the gas is perfect, $pV_m^{*g} = RT$; then Equation 2.181 can be written

$$d\ln p \approx \frac{\Delta_{vap} H_m^*}{R} d\left(\frac{1}{T}\right) \qquad (2.182)$$

where $\Delta_{vap} H_m^* = \Delta_l^g H_m^*$. This equation may be used to estimate the vapor pressure $p(T)$ from the enthalpy of evaporation and a known value of the vapor pressure at a certain temperature, $p_0 = p(T_0)$. This less general equation is known as the Clausius-Clapeyron equation (R. Clausius, 1850).

3. Trouton's and Kistiakowsky's relations

Estimation of the vapor pressures of xenobiotic organic compounds is an important issue in environmental chemistry. One tool is the following set of expressions. It has been observed that for most nonpolar organic species, the entropy of evaporation is

$$\Delta_{vap}S(T_b) = \frac{\Delta_{vap}H(T_b)}{T_b} \approx 85\,\mathrm{J\,K^{-1}\,mol^{-1}} \tag{2.183}$$

at 1 bar.[4] This relation by Trouton (1884) was later confirmed by Kistiakowsky (1923), who derived the relation[85]

$$\Delta_{vap}S(T_b) = R\{4.40 + \ln(T_b/K)\} \tag{2.184}$$

at 1 bar, using thermodynamic arguments.

4. Conditions for equilibrium

Consider n species distributed between two phases in equilibrium. From Equation 2.178 with $\mathscr{C} = n$, $\mathscr{P} = 2$, $\mathscr{R} = 0$, one obtains $\mathscr{F} = n$. The case of $n = 1$ was governed by the Clapeyron equation.

For $n > 1$, we choose an arbitrary chemical species, called B; the phases are labeled α and β. The reaction equation and the condition for equilibrium with respect to transport are

$$B^\alpha \rightleftharpoons B^\beta \qquad \mu_B^\alpha = \mu_B^\beta \tag{2.185}$$

respectively. It must be emphasized that the chemical potential difference $(\mu_B^\alpha - \mu_B^\beta)$ is defined (that is to say, measurable) only when $T^\alpha = T^\beta$. In more detail, the fact that the chemical potential must be the same in both phases may be written

$$\mu_B^{\alpha o} + RT \ln a_B^\alpha = \mu_B^{\beta o} + RT \ln a_B^\beta \tag{2.186}$$

where a_B^α and a_B^β are relative activities referring to the chemical potential of species B in the standard states $\mu_B^{o\,\alpha}$ and $\mu_B^{o\,\beta}$ in the two phases, respectively. We shall make frequent use of Equation 2.186 expressed in terms of the standard equilibrium constant:

$$\frac{a_B^\alpha}{a_B^\beta} = \exp\left(\frac{-(\mu_B^{\alpha o} - \mu_B^{\beta o})}{RT}\right) = K_B^{\alpha\beta o} \tag{2.187}$$

A useful special case arises when the standard states of B in the two phases are chosen to be identical. Then $\mu_B^{\alpha o} = \mu_B^{\beta o}$, leading to

$$a_B^\alpha = a_B^\beta \quad \Rightarrow \quad \gamma_{x,B}^\alpha x_B^\alpha = \gamma_{x,B}^\beta x_B^\beta \tag{2.188}$$

The temperature dependence of the equilibrium constant, $K_B^{\alpha\beta}$, is given by

$$d \ln K_B^{\alpha\beta} = -\frac{\Delta_\alpha^\beta H_B}{R} d\left(\frac{1}{T}\right) \tag{2.189}$$

at constant p (see Equation 2.173). For most applications, this equation is integrated by assuming that the transition enthalpy, $\Delta_\alpha^\beta H_B$, is independent of temperature.

An ideal binary mixture

We now discuss in greater detail the case in which two chemical species A and B are distributed between two phases, namely, a liquid phase, $\alpha = 1$, and a gas phase, $\beta = g$. From the phase rule (Equation 2.178), with $\mathscr{C} = 2$, $\mathscr{P} = 2$, $\mathscr{R} = 0$, one obtains $\mathscr{F} = 2$. It is common to use the mole fraction x_A of A as one of the independent variables and either the temperature T or the pressure p as the other.

1. General properties

Consider the case of an ideal liquid mixture and a perfect gas mixture. For each of the species A and B, its activity in a given phase is equal to its mole fraction in that phase. Switching to the generic symbol B, the activity in the liquid phase is equal to x_B, and the activity in the gas phase is equal to y_B, which is proportional to the partial pressure, p_B:

$$a_B^l = x_B \qquad a_B^g = y_B = p_B/p^\circ \tag{2.190}$$

For $x_B = 1$, p_B is simply the vapor pressure of pure B, $p_B = p_B^*$. It follows from Equation 2.187 that

$$p_B = x_B \, p_B^* \tag{2.191}$$

because the standard equilibrium constant is $K_B^{\lg \circ} = p_B^*/p_B^\circ$ or $K_B^{\lg} = p_B^*$.

For example, we will consider a system that is close to being ideal: a mixture of hexane (C_6H_{14}, denoted A) and octane (C_8H_{18}, denoted B).

1. The left panel of Figure 2.10
The temperature is kept constant, and the remaining independent variable is the composition, $x_A = 1 - x_B$. Equation 2.191 shows that the partial pressure of each compound is proportional to its vapor pressure in the pure state and that the sum $p = p_A + p_B$ varies linearly with x_A. The dotted line shows the total pressure as a function of the composition $y_A = 1 - y_B$ of the gas phase:

$$y_A = \frac{x_A p_A^*}{x_A(p_A^* - p_B^*) + p_B^*} \tag{2.192}$$

which follows from Equation 2.191.
2. The right panel of Figure 2.10
Pressure is kept constant, and the remaining independent variable is the composition, $x_A = 1 - x_B$. First we shall discuss the operation of a normal distillation column and then the shape of the boiling curve.

From Equation 2.191, one has

$$\frac{y_A}{y_B} = \frac{x_A \, p_A^*}{x_B \, p_B^*} \overset{\text{def}}{=} K \frac{x_A}{x_B} \tag{2.193}$$

but note that the equilibrium constant $K = p_A^*/p_B^*$ is dependent on temperature.

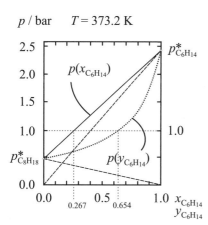 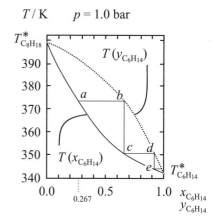

Figure 2.10   *Mixtures of hexane and octane at equilibrium*

Left panel: Vapor pressure as a function of the mole fractions, x of hexane (l) (solid line) and y of hexane (g) (dotted curve) at $T = 373.2$ K ($\theta = 100\,°C$).
$p^*_{C_6H_{14}} = 2.448$ bar and $p^*_{C_8H_{18}} = 0.472$ bar.
Example: Fixed T and p. A certain mixture that at equilibrium boils at $100\,°C$, 1 bar, has the composition $x_{C_6H_{14}} = 0.267$ and $y_{C_6H_{14}} = 0.654$.

Right panel: Boiling point curves at $p = 1.0$ bar. The function $T(x_{C_6H_{14}})$ is the bubble point curve (solid curve), and the function $T(y_{C_6H_{14}})$ is the dew point curve (dotted curve).
$T^*_{C_6H_{14}} = 341.9$ K and $T^*_{C_8H_{18}} = 398.8$ K.
Example: With A = hexane and B = octane; $K = p^*_A/p^*_B = 5.186$. Using Equation 2.194 and the composition at the point a, $x_A = 0.267$, one can calculate the composition at point e, $x_A = 0.907$.

In a distillation column at constant pressure, the temperature drops gradually toward the top. At a point n in the column (the nth plate) the equilibrium condition is $(y_A/y_B)_n = K(x_A/x_B)_n$. Slightly above this point, liquid at the point $n + 1$ has the composition $(x_A)_{n+1} = (y_A)_n$. Using this condition, one obtains $(y_A/y_B)_n = K(x_A/x_B)_{n-1}$. If K is approximately independent of temperature, then

$$\left(\frac{y_A}{y_B}\right)_n \approx K^n \left(\frac{x_A}{x_B}\right)_0 \tag{2.194}$$

The ideality of the mixture of hexane and octane means that the intermolecular forces are unchanged when the two components are mixed; accordingly, K is constant.

However, neither p^*_A nor p^*_B is independent of temperature but varies according to the Clapeyron equation. This is the reason for the curved shape of the boiling curve.

2. Raoult's law

A rearrangement of Equation 2.191 gives

$$p_A = x_A p^*_A = (1 - x_B)p^*_A \tag{2.195}$$

which links the vapor pressure, p_A, of a solvent, A, to the mole fraction, x_B, of the solute, B:

$$\frac{p_A^* - p_A}{p_A^*} = x_B \qquad (2.196)$$

This is Raoult's law (1888): the relative decrease in vapor pressure of the solvent is equal to the mole fraction of the solute. It is applicable to all compositions of an ideal binary mixture.

3. The chemical potential

Suppose that a small amount δn_B of liquid substance B is transferred from a pure state to a large excess of the mixture. The change in Gibbs energy of the total system is then $(\mu_B^l - \mu_B^{*l})\delta n$. The change of the partial molar Gibbs energy is (see Equation 2.140):

$$\mu_B^l - \mu_B^{*l} = RT \ln x_B \qquad (2.197)$$

This is always a negative quantity: dilution is spontaneous, and the system is able to deliver work to the surroundings.

Real binary mixtures

Figure 2.11 shows two examples of nonideal behavior, which are to be compared with the left panel of Figure 2.10. The vapor pressure of a mixture of acetone and chloroform (left panel) is less than expected for an ideal mixture, and the vapor pressure of a mixture of acetone and carbon disulfide (right panel) is greater than expected. In the discussion of the thermodynamics of such mixtures, we shall define and use the term *lyophilic* for solutes of the left-panel type, whereas the right-panel-type solutes will be classified as *lyophobic*.[a]

1. Raoult's law

Although Equation 2.195 applies to any ideal mixture, for nonideal solutions it is only applicable to solvent mole fractions in the range $x_A > 0.8$–0.9, as was known by Raoult. Because of this, the form shown in Equation 2.196 is more useful. An explanation for this apparent ideality is that the structure of the solvent (caused by intermolecular forces) is undisturbed by small amounts of solute molecules. Note that it is a basic thermodynamic principle that the activity of a solvent, here its vapor pressure, follows Raoult's law.

2. Henry's law

Henry's law (1803) describes properties of the solute, B, using Equation 2.187:

$$a_B^\beta = a_B^\alpha \, K_B^{\alpha\beta\circ} \qquad (2.198)$$

---

[a] The generic term is lyo- (Gk. λυοφλικος $\hat{=}$ belonging to the solution; φοβος $\hat{=}$ fright). The special terms *hydro-* and *lipo-* refer to water and lipids, respectively. Quite often a hydrophilic species (soluble in water) is lipophobic, and conversely, a lipophilic species (soluble in chloroform) is hydrophobic.

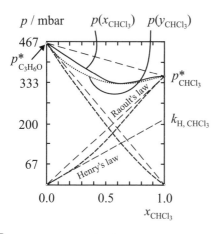
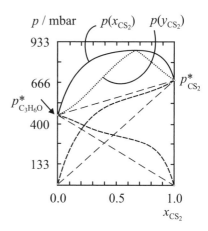

Figure 2.11 *Vapor pressure for binary mixtures at 35.2 °C* [a]
Left panel: Acetone ($p^* = 466$ mbar) + chloroform ($p^* = 395$ mbar).[136]
The Henry's law constant of chloroform in acetone is $k_{H,CHCl_3} = 215$ mbar
$\mu^E_{CHCl_3} < 0$: $CHCl_3$ is "acetonophilic."
Right panel: Acetone ($p^* = 466$ mbar) + carbon disulfide ($p^* = 695$ mbar).[136]
The Henry's law constant of carbon disulfide in acetone is $k_{H,CS_2} = 2.78$ bar
$\mu^E_{CS_2} > 0$: CS_2 is "acetonophobic."
Superscript [E] denotes excess; see the following section.

The activity, $a^\beta_B = \tilde{p}_B/p^\circ$, of a chemical species B in the gas phase, $\beta = $ g, is the
fugacity, $\tilde{p}_B$ divided by the standard pressure, p°. In the present example, however,
the gas phase is treated as a perfect gas mixture, and we set $a^\beta_B = p_B/p^\circ$. There
are two options for the standard state of the liquid phase, $\alpha = l$, on the right side of
Equation 2.198. The first is to define the standard state as that of pure species B and
describe the deviation from ideality using the activity coefficient, γ_B:

$$p_B = x_B \, \gamma_B \, p^*_B \tag{2.199}$$

The second is to set the activity equal to the mole fraction, $a^\alpha_B = x_B$, and then define
the standard state using an infinitely dilute solution. This leads to Henry's law:

$$p_B = x_B \, k_{H,B} \tag{2.200}$$

where the constant $k_{H,B}$ is the *Henry's law constant* and is given the generic symbol
k_H.[b]

The practical determination of $k_{H,B}$ reflects the choice of standard state:

$$k_{H,B} = \lim_{x_B \to 0} \frac{p_B}{x_B} \tag{2.201}$$

---

[a] Original data: A. E. Makovietsky (1908).
[b] The Henry's law constant was originally considered to be just a constant of proportionality,[123] and use
of a lowercase letter for this equilibrium constant has been maintained.[126]

Thus, when a series of experimental values of p_B/x_B is plotted against x_B, $k_{H,B}$ is found as the extrapolated value at $x_B = 0$.

3. The excess chemical potential

The activity coefficient γ_B plays an important role in environmental chemistry. We shall determine its value γ_B^∞ in the range of mole fractions x_B where Henry's law is valid. Equate Equation 2.199 with Equation 2.200 and divide by x_B. The limit of infinite dilution ($x_B \to 0$) gives

$$\gamma_B{}^\infty = \frac{k_{H,B}}{p_B^*} \tag{2.202}$$

Substituting this into Equation 2.146 and rearranging according to Equation 2.197 gives

$$\mu_B^l - \mu_B^{l*} = RT \ln x_B + \mu_B^{E\infty} \tag{2.203}$$

with $\mu_B^{E\infty} = RT \ln \gamma_B^\infty$. A negative excess chemical potential means that the work that the system may perform on the surroundings is increased relative to the ideal case, Equation 2.197. This is equivalent to $\gamma_B^\infty < 1$, but for $\mu_B^{E\infty} > -RT \ln x_B$, the molar Gibbs energy changes sign. Most xenobiotic substances have very large values of γ_B^∞ in water, and values of the order of $10^6 - 10^9$ are not uncommon.

Although this discussion has been about the activity coefficient of a solute in the Henry's law range of concentrations, we shall make the following general definition: a solute whose activity coefficient is less than 1 is lyophilic, and a solute whose activity coefficient is greater than 1 is lyophobic.

Example: Excess chemical potential

Using data from Figure 2.11, Equation 2.202, and Equation 2.147, one finds

$$\gamma_{CHCl_3}^\infty = 215/395 = 0.544 \quad \text{and} \quad \mu_{CHCl_3}^{E\infty} = -1.51 \frac{kJ}{mol}$$

$$\gamma_{CS_2}^\infty = 2780/695 = 4.00 \quad \text{and} \quad \mu_{CS_2}^{E\infty} = +3.55 \frac{kJ}{mol}$$

with $RT = 2564$ J/mol.

Example: Standard states

Consider Figure 2.11 and the following two systems of solutes, B, in the solvent acetone: [h]

    a. $B = CHCl_3$ $x_B = 0.280$ $p_B = 67$ mbar
    b. $B = CS_2$ $x_B = 0.125$ $p_B = 267$ mbar

Application of Equation 2.199 in the form $p_B = a_B\, p_B^0 = x_B\, \gamma_B\, p_B^0$ gives the activity of B in the liquid phase based on five different standard states as shown in Table 2.5.

Table 2.5 Example: Various standard states of two liquid mixtures

Standard state	System a	System b
1. Pure B(l)		
$p_B^o = p_B^*$	$a_{CHCl_3}^l = 67/395 = 0.170$	$a_{CS_2}^l = 267/695 = 0.384$
2. B(g) at 1 bar		
$p_B^o = 1$ bar	$a_{CHCl_3}^l = 0.067$	$a_{CS_2}^l = 0.267$
3. B(g) at 1 mbar		
$p_B^o = 1$ mbar	$a_{CHCl_3}^l = 67$	$a_{CS_2}^l = 267$

Standard state	Activity	Activity coefficient
4. $\gamma_B = 1$ in pure B(l)		
System a	$a_{CHCl_3}^l = 0.170$	$\gamma_{CHCl_3}^l = 0.170/0.280 = 0.607$
System b	$a_{CS_2}^l = 0.384$	$\gamma_{CS_2}^l = 0.384/0.125 = 3.07$
5. $\gamma_B = 1$ in an infinitely dilute solution of B in acetone		
System a	$a_{CHCl_3}^l = 67/215 = 0.312$	$\gamma_{CHCl_3}^l = 0.312/0.280 = 1.11$
System b	$a_{CS_2}^l = 267/2780 = 0.096$	$\gamma_{CS_2}^l = 0.096/0.125 = 0.768$

4. Calculation of Henry's law constants

The basic thermodynamic expression (Equation 2.198) has been discussed using mole fraction as the measure of concentration and assuming a perfect gas phase. We have arrived at Henry's law:

$$p_B = x_B \gamma_B^\infty p_B^* \tag{2.204}$$

which is valid for a certain range of concentrations called the Henry's law range. In this range, additional expressions of Henry's law are in use,

$$p_B = c_B k_{H,B}' \quad \text{and} \quad p_B = m_B k_{H,B}'' \tag{2.205}$$

depending on the choice of standard state.[a]

For a perfect gas, $p_B = c_B RT$; therefore, the number $k_{H,B}/(RT)$ expresses the ratio between concentration of B in the gas phase and the liquid phase.

A rather large collection of experimental values for the three entries of Equation 2.202 are available in the literature.[243] It is therefore of interest to express the amount concentration constants of Equation 2.204 and Equation 2.205 in terms of those of Equation 2.202. Let V_m be the molar volume[b] of the liquid solution; then

$$k_{H,B,\text{ concentration}} = \gamma_B^\infty p_B^* V_m = k_{H,B,\text{ mole fraction}} V_m \tag{2.206}$$

---

[a] *1.* In the remaining part of this book $k_{H,B}''$ is referenced to Henry's law on a concentration basis.

2. For example $k_{H,B}$ using a molality basis is used in the NIST databases, http://webbook.nist.gov/chemistry.

[b] See Table A1.7, note h: $x_B = c_B V_m$.

Because the determinations of p_B^* and V_m are straightforward, the problem is to obtain a value for γ_B^∞. However, in the Henry's law range, Equation 2.188 gives directly

$$\gamma_B^\infty x_B = 1 \qquad (2.207)$$

because of the choice of standard state. Further, in many cases it is a good approximation to set the molar volume of the solution equal to the molar volume of the pure solvent, $V_m \approx V_{m,A}$.

Example: The Henry's law constant of 1,1,1-trichloroethane in water

CH_3CCl_3 is denoted B, and the water phase, w. At 25 °C:

$p_B^* = 0.128$ bar; the solubility water is : $\rho_B^w = 0.073$ g/dl; $M_B = 133.4$ g/mol; $V_{m,w} = 18.02$ ml/mol.

Then:

$$x_B^{\;w} = 9.85 \times 10^{-5}; \; \gamma_B^\infty = 1/x_B^{\;w} = 1.02 \times 10^4; \; k_{H,B,\text{ mole fraction}} = 1.30 \text{ kbar};$$

$$k_{H,B,\text{ concentration}} = 23.4 \text{ l bar mol}^{-1}; \; k_{H,B,\text{ concentration}}/(RT) = 0.943.$$

Nernst's distribution law

Consider a solute, B, distributed between two solvents, A (liquid phase α) and D (liquid phase δ). At equilibrium, Equation 2.187 applies:

$$\frac{a_B^\alpha}{a_B^\delta} = K_B^{o\alpha\delta} \qquad (2.208)$$

For many applications, the conditions of equilibrium may be written

$$\frac{\gamma_{x,B}^\alpha \, x_B^\alpha}{\gamma_{x,B}^\delta \, x_B^\delta} = K_{x,B}^{\alpha\delta} \quad \text{or} \quad \frac{\gamma_{c,B}^\alpha \, c_B^\alpha}{\gamma_{c,B}^\delta \, c_B^\delta} = K_{c,B}^{\alpha\delta} \quad \text{or} \quad \frac{\gamma_{mB}^\alpha \, m_B^\alpha}{\gamma_{x,B}^\delta \, m_B^\delta} = K_{m,B}^{\alpha\delta} \qquad (2.209)$$

depending on the choice of standard state.[a] The three formulas are collectively called Nernst's distribution law (W. Nernst, 1891) and the constants partitioning ratios or distribution constants.

Calculating partition ratios from known data is an important objective in environmental science. As they stand, the three expressions of Equation 2.209 are independent, but a careful choice of standard states makes a relation between the first two possible. For this purpose, the partitioning ratio is defined as the ratio between the amount concentration of the solute in the two phases (the middle of the formulas

---

[a] The constants of Equation 2.209 are not necessarily standard equilibrium constants because the standard state may be chosen differently in the two phases.

of Equation 2.209 with $\gamma_{C,B}^{\alpha} = \gamma_{C,B}^{\delta} = 1$) By choosing identical standard states (Equation 2.188) and introducing molar volumes, one obtains

$$K_{c,B}^{\alpha\delta} \overset{def}{=} \frac{c_B^{\alpha}}{c_B^{\delta}} = \frac{\gamma_B^{\infty\delta} \, V_m^{\delta}}{\gamma_B^{\infty\alpha} \, V_m^{\alpha}} \tag{2.210}$$

where the constant activity coefficients determined in the Henry's law range, have been used.

In environmental work, Equation 2.210 is frequently used to describe the distribution of a xenobiotic species between two particular solutes: octanol saturated with water (phase $\alpha = o$) and water saturated with octanol (phase $\delta = w$). The amphiphilic[a] nature of octanol makes it suitable as a reference phase, mimicking biophases. Given the value of the octanol-water partition ratio of a species B, $K_{c,B}^{wo}$, a biological partition ratio, $K_{c,B}^{wb}$, may be estimated using the product $K_{c,B}^{bw} = f^b K_{c,B}^{wo}$, where f^b is an empirical factor that depends on the nature of the biological material.

Example: Octanol-water

Consider the partition ratio of 1,1,1-trichloroethane (T) between octanol saturated with water (phase o) and water saturated with octanol (phase w). Data at 25 °C:

$V_{m,C_8H_{17}OH} = 0.157\,l/mol$; $\rho_{C_8H_{17}OH} = 0.828\,g/ml$ (for reference, not used here);

$V_m^o = 0.120\,l/mol$; $\rho^o = 0.887\,g/ml$; $x_{H_2O}^o = 0.21$; $\gamma_T^{\infty o} = 5.19$;

$V_m^w = 0.018\,l/mol$; $\rho^w = 1.00\,g/ml$; $x_{C_8H_{17}OH}^w = 8\times10^{-5}$; $\gamma_w^{\infty T} = 1.02\times10^4$.

Then:

$$K_{c,T}^{ow} = \frac{c_T^o}{c_T^w} = \frac{1.02 \times 10^{-4} \times 0.018}{5.19 \times 0.120} = 295$$

Assume an average value of $f^b = 0.05$ for b = "biota." After some time fish in a lake with a given c_T^w will have $c_T^b = 1.5c_T^w$.

2.4 Chemical kinetics

The present section is particularly concerned with ideas and background material for application to atmospheric chemistry, Chapter 4.

The concentration changes of chemical species of macroscopic systems were described in Section 2.1b. An important conceptual tool was the stoichiometric equation 2.12, and the stoichiometric numbers may be large numbers or fractions. In the present section we discuss reactions on a molecular level. Here the reaction equations

---

[a] Gk. $\alpha\mu\phi\iota$- $\hat{=}$ double, to a certain extent octanol dissolves both hydrophilic and lipophilic species.

deal with molecular entities, and the stoichiometric numbers are typically the integers 1, 2, or 3.

a. Basic concepts

Definitions

An *elementary reaction* is a reaction that occurs in a single step. The number of reactant entities that are involved in each event is called the *molecularity*. A unimolecular reaction involves just one reactant entity and has a first-order rate law. A bimolecular reaction involves two reactant entities and has a second-order rate law. A termolecular reaction involves three reactant entities and has a third-order rate law. It is normal to define elementary reactions as unidirectional, written with a simple arrow from left to right.

A reaction that involves more than one elementary reaction is said to occur by a composite mechanism. The combination of elementary reactions, that is, the net reaction, is not an elementary reaction and has no molecularity. The *rate of the chemical reaction* (see Equation 2.20) that is based on the net reaction is just called *the reaction rate*.

Example

$$
\begin{array}{rcl}
A & \rightarrow & B + C \\
B + C & \rightarrow & D + E \\
\hline
A & = & D + E
\end{array}
$$

Unimolecular elementary reaction
Bimolecular elementary reaction

Net reaction

The last expression is a stoichiometric equation.

Note

A chemical reaction with a first-order (second-order) rate expression need not be a unimolecular (bimolecular) reaction. Thus, a reaction that is of first order with respect to the reactant A and of zero order with respect to all other reactants may correspond to various elementary reaction schemes, such as $A \rightarrow Z$, $A \rightarrow Z + Z$, $A + A \rightarrow Z$, $A + C \rightarrow Z$, $A + A + C \rightarrow Z$. However, a unimolecular (bimolecular) elementary reaction will always lead to a first-order (second-order) rate expression.

b. Description of elementary reactions

Unimolecular reactions

Elementary reaction: $A \rightarrow Y + Z$
Stoichiometric reaction and reaction rate:

$$
A = Y + Z \qquad \upsilon = -\frac{d[A]}{dt} = k\,[A] \qquad (2.211)
$$

The most important unimolecular processes (in the atmosphere) are photolysis and pyrolysis:[a] a molecule is broken apart with light or heat, respectively.

Example: Photolysis
A prerequisite for photolysis is that the energy of the photon be at least as great as the bond dissociation energy. However, even if a molecule absorbs light with sufficient energy, there is no guarantee that it will dissociate. The metastable photoexcited molecule can be relaxed in several ways – for example, by isomerization, collisional quenching, ionization, or luminescence – in addition to dissociation. The quantum yield ϕ is defined as the number of photolytic events divided by the number of photons absorbed. For a photochemical reaction, it is the rate of conversion divided by the rate of photon absorption:

$$\varphi = |\dot{\xi}/\dot{\eta}_\gamma| \tag{2.212}$$

Typically, photolysis breaks apart a stable molecule with a *closed shell* of electrons (all electrons spin paired in chemical bonds; the compound is diamagnetic) and creates two *radicals*, species with unpaired electrons. An example is

$$H_2O_2 + h\nu \rightarrow HO^\bullet + HO^\bullet \tag{2.213}$$

(see Equation 4.18). Radicals are unstable and reactive relative to most closed-shell molecules and drive the photochemistry of the atmosphere.

Bimolecular reactions
Elementary reactions:

$$A + C \rightarrow Y + Z \qquad \upsilon = -\frac{d[A]}{dt} = k\,[A][C] \tag{2.214}$$

$$A + A \rightarrow Y + Z \qquad \upsilon = -\frac{1}{2}\frac{d[A]}{dt} = k\,[A]^2 \tag{2.215}$$

Bimolecular reactions propagate chain reactions in the atmosphere. A bond is broken and another is formed. An example is

$$CH_4 + HO^\bullet \rightarrow CH_3^\bullet + H_2O \tag{2.216}$$

(see Equation 4.19). In many bimolecular reactions, a closed-shell species reacting with a radical results in a new closed-shell species and a different radical.

Termolecular reactions
Elementary reaction:

$$A + C + D \rightarrow Y + Z \qquad \upsilon = -\frac{d[A]}{dt} = k\,[A][C][D] \tag{2.217}$$

Termolecular reactions form a bond between two species; here the entity Z may be the entity D in a different quantum state. They are of particular interest because they

---

[a] Lat. *pyra* $\hat{=}$ bonfire (pyre) Gk. λυσι $\hat{=}$ disintegrate.

terminate chain reactions: two radicals combine to form a closed-shell species, for example,

$$HO^{\bullet} + HO^{\bullet} + N_2 \quad \rightarrow \quad H_2O_2 + N_2^* \qquad (2.218)$$

(see Equation 4.3). The energy of the new bond must be removed from the nascent molecule because it is sufficient to break the molecule apart. The quenching of the excess energy is accomplished through collision with another molecule. In the example, atmospheric nitrogen carries kinetic energy away from hydrogen peroxide.

c. Temperature dependence of reaction rates

It is found experimentally for many reactions that the temperature dependence of the reaction rate may be described by a two-parameter equation called the Arrhenius equation (1889)[188]:

$$k = A \exp\left(-\frac{E_a}{RT}\right) \qquad \text{constant } p \qquad (2.219)$$

According to this equation, the rate constant k is the product of a constant A called the preexponential factor and an exponential term containing the constant E_a, which is called the activation energy.

Frequently, the constants of Equation 2.219 are independent of temperature, and they can be determined from a plot of $\ln k$ ($= \ln A - E_a/RT$) versus $1/T$. This is a straight line with the slope $-E_a/R$ and intersection with the $\ln k$-axis equal to $\ln A$. In some cases the temperature dependence is non-Arrhenius, meaning that the plot is nonlinear. In such cases one defines the activation energy as

$$E_a(T) \stackrel{\text{def}}{=} RT^2 \left(\frac{\partial \ln k}{\partial T}\right)_p = -R \left(\frac{\partial \ln k}{\partial (1/T)}\right)_p \qquad (2.220)$$

that is, E_a at the temperature of interest is equal to $-R$ times the slope at that temperature. The preexponential factor may also depend on the temperature; this is usually expressed by the relation

$$k(T) = A' \left(\frac{T}{T_0}\right)^n \exp(-E_a'/RT) \qquad \text{constant } p \qquad (2.221)$$

where A' and E_a' are temperature independent parameters and T_0 is a reference temperature allowing a unit less power.

Discussion of the Arrhenius equation

We now discuss the Arrhenius equation (Equation 2.219) in general terms.

It took more than 60 years (ca. 1850–ca. 1910) from the first kinetic measurements until an Arrhenius-type equation was fully accepted in theoretical descriptions. In the period up to 1935, a statistical mechanical formulation of transition state theory for chemical reactions was created that, since then, has been applied to a wide variety of problems. The developments in the past 60 years include studies of ultrafast reactions (femtosecond chemistry) and calculations of the reaction kinetics of large systems.

Current understanding of the ideas behind this complex of experiments, theory, and calculations will be outlined using as an example a general reaction that is bimolecular in both directions:

$$A + BC \overset{k_f}{\underset{k_r}{\rightleftharpoons}} AB + C \tag{2.222}$$

The reaction is assumed to be exothermic with the rate constants k_f and k_r for the forward and reverse reactions, respectively. According to Equation 2.27, the equilibrium constant is related to the rate constants: $K = k_f/k_r$. Substituting this into the van't Hoff equation (Equation 2.173) leads to

$$\frac{d \ln k_f}{dT} - \frac{d \ln k_r}{dT} = \frac{\Delta E}{RT^2} \tag{2.223}$$

where ΔE is the potential energy difference between products and reactants. It was suggested by van't Hoff that Equation 2.223 may be split into two,

$$\frac{d \ln k_f}{dT} = \frac{E_f}{RT^2} + Q_f \quad \text{and} \quad \frac{d \ln k_r}{dT} = \frac{E_r}{RT^2} + Q_r \tag{2.224}$$

where $\Delta E = E_f - E_r$. Later it was found experimentally that the constants Q are close to zero, so that Equation 2.219 is obtained after integration.

The interpretation of the two parameters E_a (1) and A (2) of the Arrhenius equation is as follows. Consider the reactants, namely, the two molecular entities A and BC. To react, they must be brought together to form a cluster, the entity A-B-C, which decomposes to yield the product entities AB and C. As long as the reactants are far from each other, the potential energy of the system is unchanged. But as A approaches BC, the nuclei of both reactants are forced apart somewhat, and the potential energy increases. This continues until the entity A-B-C is formed at the maximum of potential energy. The entity A-B-C is called an activated complex or transition state, and we have the general picture

$$\underset{\text{reactants}}{A + BC} \rightleftharpoons \underset{\substack{\text{activated} \\ \text{complex}}}{A - B - C} \rightleftharpoons \underset{\text{products}}{AB + C} \tag{2.225}$$

The change of potential energy during the reaction is shown in Figure 2.12. The abscissa is the reaction coordinate q, a generalized coordinate describing the changes of interatomic distances and bond angles that lead from reactant entities to product entities.

1. Reaction takes place if the kinetic energy of the reactants is sufficient to overcome the potential energy barrier, E_f. Only a fraction, dependent on temperature and determined by a Boltzmann distribution of velocities, fulfill this requirement, which is given by the exponential part of the Arrhenius equation. The figure shows qualitatively that an endothermic reaction (the reverse reaction) is likely to be very slow except at high temperatures because E_r includes ΔE.

2. Suppose that an entity of the shape A-B-C is not formed due to the collision geometry. Then, some encounters with otherwise sufficient kinetic energy will not lead to reaction. The preexponential factor describes the probability that an encounter leads to reaction.

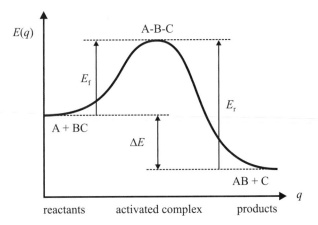

Figure 2.12 *Energy profile of an exothermic bimolecular reaction*
Potential energy $E(q)$ versus reaction coordinate q for the reaction, Equation 2.222.

Concluding remarks

Attempts to understand the connections among chemical thermodynamics, chemical kinetics, and the mechanism of chemical reactions have been a main theme of theoretical chemistry for more than a century. In many cases the point of departure is an Arrhenius-type equation for the rate constant k, and numerous papers have been published that connect the constants A and E_a to other physicochemical properties or calculate them using quantum chemistry. We shall mention just two examples of ubiquitous reactions: (1) proton transfer and (2) electron transfer.

1. Brønsted and Pedersen's (1924)[148] discovery of free energy relations for proton transfer reactions opened the area of physical organic chemistry concerned with *linear free-energy relationships* (LFER).[63] The original Brønsted relation was based on the relation between the rate constant k of the base-catalyzed decomposition of nitramide[a] as a function of the pK values of a series of conjugate acids:

$$\ln(k/k_{\text{ref}}) = \beta \ln(K_a^o/K_{\text{ref}}^o) \qquad (2.226)$$

where β is a constant. The relation describes a linear relationship between rate constants and differences in reaction affinities[b] and was the first experimentally established connection between chemical kinetics and chemical thermodynamics. The constant β has been shown to include terms describing the structural reorganization that accompanies proton transfer.[49]

2. R. Marcus was awarded the Nobel Prize in Chemistry in 1992 in recognition of his work on the rate of the electron transfer between an electron donor (a reductor) and an electron acceptor (an oxidator). The Marcus theory of electron transfer reactions builds on the Arrhenius equation to include interaction with the solvent: the preexponential factor depends on the coupling between the initial and final states, and the activation energy depends on the reorganization energy of the system over the

[a] $NH_2NO_2 \rightarrow N_2O + H_2O$.
[b] One may write $\ln k = \alpha + \beta' \Delta A^\circ$, with $\alpha = \ln k_{\text{ref}}$, $\beta' = \beta/RT$ and $\Delta A^\circ = A^\circ - A_{\text{ref}}^\circ$, cf. affinity, Equation 2.170, and pK, Equation 5.8.

course of the reaction plus the standard affinity of the process. A surprising feature of the theory is that the rate constant k as function of the standard affinity A° has a maximum, meaning that for small values of A°, an increase of A° causes k to increase, but at a certain point further increase of A° causes k to decrease. This prediction was controversial when it was proposed (1956), but later it was observed experimentally (J. Miller, 1984)[4,49] using a series of closely related reactants. Now the phenomenon is a well-established fact.[59]

3

The Spheres

It seems "natural" to describe the Earth in terms of spherical shells, given the shape of the planet and its layered structure. The *atmosphere* is a good example of such a shell.[a] This chapter begins by introducing some basic ideas and continues with a more thorough discussion, including the chemical structure and composition of the spheres, along with some key features of the physics of each sphere and their interactions.

From a chemical perspective, the Earth has a distinct central region, called the *barysphere*, on which the *lithosphere*, the *hydrosphere*, and the *atmosphere* rest. The term *sphere* is used in a more abstract sense than the simple meaning of "ball"; for example, the hydrosphere, comprising all states of aggregation of water, is irregular in form and extent. In the context of the environment, it is convenient to discuss separately the *pedosphere*, which consists of the material in which plants grow, and to delimit the *biosphere* as follows later.[b]

Suess (1875) was the first to attempt to unite the fields of geophysics and biology. His contribution was to recognize and characterize the biosphere as the region that gives rise to life. The concept was refined by Vernadsky (1925) to mean the top part of the lithosphere, part of the hydrosphere, and the *troposphere*,[c] the lowest layer of the atmosphere. According to his definition, the biosphere is the part of the Earth that sustains life, but not including *biota*, defined as living and dead organisms. Today, *biosphere* is understood to mean all organic material *and* that part of the inorganic surroundings in which life is found. It has been proposed that the word *ecosphere*[d] be used with this meaning, in contrast to *biosphere*, which would then be used to mean all living and dead organisms and their degradation products; however, these definitions have not been widely accepted. The part of the Earth that is not the biosphere is called the *geosphere*.[e]

The thermodynamic state of the Earth can be described conveniently by observing that the planet is essentially a closed physicochemical system. Solar radiation provides energy with relatively low entropy to the system, see Table 10.1. The system in turn gives off energy with higher entropy as radiation to space. Energy from the Sun heats the atmosphere, the upper part of the oceans, and the lithosphere, and with a few rare exceptions it sustains life. Each of these subsystems produces entropy. The lithosphere also receives energy from the core via conduction and convection;

---

[a] Gk. $\alpha\tau\mu\acute{o}\varsigma \triangleq$ vapor; $\sigma\varphi\alpha\iota\rho\alpha \triangleq$ ball.
[b] Gk. $\beta\alpha\rho\acute{v}\varsigma \triangleq$ heavy; $\lambda\iota\theta o\varsigma \triangleq$ stone; $v\delta\omega\rho \triangleq$ water; $\pi\acute{\epsilon}\delta ov \triangleq$ soil; $\beta\iota\omega\sigma \triangleq$ life.
[c] Gk. $\tau\rho\acute{o}\pi o\varsigma \triangleq$ turning.
[d] Gk. $o\iota\kappa o\varsigma \triangleq$ house, meaning household of Nature.
[e] Gk. $\gamma\eta \triangleq$ Earth.

the main source of this energy is radioactive decay. Here, the important radioactive elements are uranium, thorium, and potassium, and the energy of their decay results in volcanic activity and drives the circulation of material in the Earth's mantle as described in Section 1.4. This circulation in turn moves the continental plates and is responsible for the burial of old rocks and the formation of new. Circulation in the mantle also gives rise to the Earth's magnetic field, which together with the atmosphere protects life on Earth from the solar wind and particles from outside the solar system.

In this chapter, we have chosen to present the spheres from the inside out, from the lithosphere to the hydrosphere, atmosphere, and biosphere. This parallels the formation of the planet and the masses of the spheres listed in Table 1.6. The composition of the hydrosphere follows from that of the lithosphere, the atmosphere follows from the lithosphere and hydrosphere, and life occurs under the preconditions of these spheres. In contrast, we have chosen to present in later chapters the atmosphere (Chapter 4), hydrosphere (Chapter 5), and pedosphere (Chapter 6) in an order given by chemistry. The photochemistry of the atmosphere occurs largely in the gas phase, with the introduction of some basic heterogeneous reactions. The liquid phase introduces new phenomena, and the concepts introduced in the chapter on the chemistry of aqueous solutions are then used in discussing the soil, a mixture of gas, liquid, and solid.

3.1 The lithosphere

The lithosphere is composed of *rocks*. Rocks are heterogeneous mixtures of *minerals*. Minerals in turn are naturally occurring pure substances, that is, well-defined chemical species whose composition can be described using an empirical formula. Rocks are characterized by their mineral composition and the geological processes by which they were formed. At the temperatures and pressures found in the lithosphere, virtually all minerals are solid, and the great majority crystalline.[a] As heterogeneous mixtures, rocks melt over a range of temperatures. For typical rocks (at relatively low pressure), the melting point will stretch over a range of 100 to 200 K, at temperatures of 1100 to 1600 K. This is the basis of the plasticity of the outer asthenosphere (see Figure 1.7).

Some chemical properties of rocks are directly connected to their average chemical composition, defined by the mole fraction of the various elements in the mixture. The dominant rocks are peridotite (asthenosphere), basalt (ocean floor), and granite (continents) (Table 3.1).

Oxygen is found as O^{-II} in the rocks, that is, in a negative oxidation state, whereas the other elements given in Table 3.1 are always in positive oxidation states. This is in accord with the relative electronegativities as seen on the inside front cover. Rocks having a silicon mole fraction $x(Si)$ that is larger than $1/5$ are called *acidic*, those with $1/5 > x(Si) > 1/6$ *intermediate*, those with $1/6 > x(Si) > 1/7$ *basic*, and those with $x(Si) < 1/7$ *ultrabasic*. This terminology refers to silicon dioxide's ability to act as a

---

[a] Gases are not minerals. Mineral oil is considered a mineral; liquid water is not.

	Peridotite[a] $\rho \approx 3.3$ g/ml	Basalt[b] $\rho \approx 3.0$ g/ml	Granite[b] $\rho \approx 2.7$ g/ml
Table 3.1 Chemical composition of the lithosphere			
Element			
O	0.578	0.596	0.615
Si	0.153	0.167	0.227
Al	0.017	0.064	0.064
Fe^{III}	0.007	0.011	0.003
Fe^{II}	0.029	0.024	0.007
Mg	0.179	0.049	0.008
Ca	0.013	0.035	0.013
Na	0.003	0.025	0.026
K	0.001	0.007	0.013
Ti	0.003	0.010	0.001
H	0.016	0.012	0.023

The average composition (mole fraction of elements) of certain rocks with approximate density.[34]

Lewis acid,[c] which is important for chemical reactions in molten silicates. Similarly, the physical properties of rocks depend on the minerals of which they are composed. For example, several minerals containing Fe^{III} are ferromagnetic, and the direction of their magnetic moments reflect the magnetic field at the time they solidified.

The aluminium content of the crust is high, but it falls in the asthenosphere. The asthenosphere has a high content of iron relative to the crust, which among other things results in its higher density. In addition, the amounts of group 1 and 2 metals in rocks have characteristic variations that serve, together with the amount of silicon, as the basis of a more detailed classification of volcanic rocks (see Tables 3.8 and 3.9, later).

a. Abundance of the elements

The subject of geochemistry considers the distribution of the elements on Earth. On the basis of the chemical analysis of more than 5,000 carefully chosen samples of rocks, the geochemist Clark (1924) was able to construct a table of the incidence of the elements in the lithosphere (Table 3.2). Table 3.3 shows these data converted into mole fractions and volume fractions. It can be seen that about 60 % of the atoms in the lithosphere are oxygen and 20 % are silicon, and that just 10 elements make up more than 99.7 % of the total. It should also be noted that no less than 90 % of the volume of rocks is oxygen, and that on the basis of volume, just four elements make up more than 97 % of the total.

---

[a] After a major component, the green olivine mineral peridot.
[b] Lat. *basaltes* $\hat{=}$ hard stone; *granum* $\hat{=}$ grain.
[c] Lewis acid-base definition: an acid is an electron pair acceptor and a base an electron pair donor.

Table 3.2	Abundance of the elements of the crust, part 1				
		Mass fraction			Mass fraction
1	O	0.4550	40	Sm	7.0×10^{-6}
2	Si	0.2720	41	Gd	6.1×10^{-6}
3	Al	83.0×10^{-3}	42	Dy	3.5×10^{-6}
4	Fe	62.0×10^{-3}	43	Er	3.5×10^{-6}
5	Ca	46.6×10^{-3}	44	Yb	3.1×10^{-6}
6	Mg	27.6×10^{-3}	45	Hf	2.8×10^{-6}
7	Na	22.7×10^{-3}	46	Cs	2.6×10^{-6}
8	K	18.4×10^{-3}	47	Br	2.5×10^{-6}
9	Ti	6.32×10^{-3}	48	U	2.3×10^{-6}
10	H	1.52×10^{-3}	49	Sn	2.1×10^{-6}
11	P	1.12×10^{-3}	50	Eu	2.1×10^{-6}
12	Mn	1.06×10^{-3}	51	Be	2.0×10^{-6}
13	F	544×10^{-6}	52	As	1.8×10^{-6}
14	Ba	390×10^{-6}	53	Ta	1.7×10^{-6}
15	Sr	384×10^{-6}	54	Ge	1.5×10^{-6}
16	S	340×10^{-6}	55	Ho	1.3×10^{-6}
17	C	180×10^{-6}	56	Mo	1.2×10^{-6}
18	Zr	162×10^{-6}	57	W	1.2×10^{-6}
19	V	135×10^{-6}	58	Tb	1.2×10^{-6}
20	Cl	126×10^{-6}	59	Lu	0.8×10^{-6}
21	Cr	122×10^{-6}	60	Tl	0.7×10^{-6}
22	Ni	99×10^{-6}	61	Tm	0.5×10^{-6}
23	Rb	78×10^{-6}	62	I	0.46×10^{-6}
24	Zn	76×10^{-6}	63	In	0.24×10^{-6}
25	Cu	68×10^{-6}	64	Sb	0.20×10^{-6}
26	Ce	66×10^{-6}	65	Cd	0.16×10^{-6}
27	Nd	40×10^{-6}	66	Ag	0.08×10^{-6}
28	La	35×10^{-6}	67	Hg	0.08×10^{-6}
29	Y	31×10^{-6}	68	Se	0.05×10^{-6}
30	Co	29×10^{-6}	69	Pd	0.02×10^{-6}
31	Sc	25×10^{-6}	70	Pt	0.01×10^{-6}
32	Nb	20×10^{-6}	71	Bi	8.0×10^{-9}
33	N	19×10^{-6}	72	Os	5.0×10^{-9}
34	Ga	19×10^{-6}	73	Au	4.0×10^{-9}
35	Li	18×10^{-6}	74	Ir	1.0×10^{-9}
36	Pb	13×10^{-6}	75	Te	1.0×10^{-9}
37	Pr	9.1×10^{-6}	76	Re	0.7×10^{-9}
38	B	9.0×10^{-6}	77	Ru	0.1×10^{-9}
39	Th	8.1×10^{-6}	78	Rh	0.1×10^{-9}

Arranged according to decreasing mass fraction. The total mass of the crust is estimated to be 23.6 Yg.[129]

Table 3.3	Abundance of the elements of the crust, part 2			
		Mole fraction	Mass fraction	Volume fraction
1	O	0.5944	0.4550	0.9098
2	Si	0.2024	0.2720	0.0056
3	Al	64.3×10^{-3}	0.0830	0.0056
4	H	31.5×10^{-3}	0.0015	
5	Ca	24.3×10^{-3}	0.0466	0.0213
6	Mg	23.8×10^{-3}	0.0276	0.0085
7	Fe	23.2×10^{-3}	0.0620	0.0083
8	Na	20.6×10^{-3}	0.0227	0.0191
9	K	9.84×10^{-3}	0.0184	0.0212
10	Ti	2.76×10^{-3}	0.0063	0.0006
11	P	750×10^{-6}	0.0011	
12	F	600×10^{-6}	0.0005	
13	Mn	400×10^{-6}	0.0011	
14	C	310×10^{-6}	0.0002	
15	S	220×10^{-6}	0.0003	
16	Sr	92×10^{-6}	0.0004	
17	Cl	74×10^{-6}	0.0001	
18	Ba	60×10^{-6}	0.0004	
19	V	56×10^{-6}	0.0001	
20	Li	54×10^{-6}		
21	Cr	49×10^{-6}		
22	Zr	37×10^{-6}		
23	Ni	35×10^{-6}		
24	N	28×10^{-6}		
25	Zn	24×10^{-6}		
26	Cu	22×10^{-6}		
27	Rb	19×10^{-6}		
28	B	17×10^{-6}		
29	Sc	12×10^{-6}		
30	Co	10×10^{-6}		

The data are derived from Table 3.2. Mole fractions are arranged according to decreasing value. For comparison, mass fractions as given in Table 3.2 are shown. The volume fractions have been calculated using Shannon's ion radii.[37]

b. The rock-forming minerals

There is a small group of minerals that make up most rocks, and together they account for more than 96 % of the mass of the crust. These so-called rock-forming minerals are summarized in Table 3.4. The remaining minerals, many of which are economically important, are labeled "rare minerals" in the table. It can be seen that silicates and quartz make up 91 % of the mass of the crust, iron oxides and titanate about 4 %, the carbonates barely 2 %, and the balance some 3 %.[a]

[a] Formulas are given in Table 3.6.

Table 3.4 Mineral composition of the crust[79]		
Mineral[a]	w	φ
Plagioclases	0.40	0.43
Orthoclases	0.18	0.20
Olivines, pyroxenes, amphiboles	0.16	0.12
Quartz	0.13	0.15
Hematite, magnetite, ilmenite	0.04	0.02
Micas	0.03	0.03
Calcite	0.015	
Clay minerals, kaolinite	0.010	
Dolomite, magnesite	0.002	
Rare minerals	0.034	

Mass fraction w; volume fraction φ.
Silicates: feldspars (plagioclases, orthoclases), olivines, pyroxenes, amphiboles, micas, clay minerals, kaolinite. Iron oxides: hematite, magnetite. Iron titanate: ilmenite. Carbonates: calcite, dolomite, magnesite. Rare minerals include approximately 3200 minerals and 1600 structural types (1996).

Rocks can be classified into three major groups according to the chemical processes by which they were formed: igneous rocks, sedimentary rocks, and metamorphic rocks. A crude chemical distinction is that igneous rocks crystallized from salt melts, sedimentary rocks are precipitates from aqueous suspensions that have been subject to heat and pressure, and metamorphic rocks have been formed through the partial melting of igneous, sedimentary, or metamorphic rocks. The chemistry of these processes is discussed in more detail later. Table 3.5 shows an estimate of the volume of the three groups and the most common subclasses. Note that sedimentary rocks, which are common near the surface, covering two-thirds of the continents and coastal areas, make up less than 8 % of the volume of the crust.

Table 3.5 Composition of the rocks of the crust, volume fraction φ [15]		
Class of rock	φ	$\Sigma\,\varphi$
Igneous rocks		0.647
Granites	0.214	
Basalts	0.427	
Other	0.006	
Sedimentary rocks		0.079
Shales	0.065	
Sandstones	0.009	
Limestones	0.005	
Metamorphic rocks		0.274

Igneous rocks are classified according to chemical composition (Section 3.1c). Sedimentary rocks are classified according to size and type of grains (Section 3.1d). Classification of metamorphic rocks is discussed in Section 3.1e.

	Table 3.6 Some important minerals[19]	
Group	Name	Chemical formula
Olivine	Forsterite	$Mg_2[SiO_4]$
Olivine	Fayalite	$Fe_2[SiO_4]$
Pyroxene	Enstatite	$Mg_2[Si_2O_6]$
Pyroxene	Diopside[2]	$Ca(Mg,Fe^{II})[Si_2O_6]$
Pyroxene	Hypersthene	$(Mg,Fe^{II})_2[Si_2O_6]$
Amphibole	Hornblende[3]	$Ca_2(Mg,Fe^{II})_5[Si_6Al_2O_{22}](OH)_2$
Feldspar[1]	Albite	$Na[AlSi_3O_8]$
Feldspar[1]	Anorthite	$Ca[Al_2Si_2O_8]$
Feldspar[1]	Orthoclase	$K[AlSi_3O_8]$
Nepheline[1]	Nepheline	$Na_3K[Al_4Si_4O_{16}]$
Leucite[1]	Leucite	$K[AlSi_2O_6]$
Mica	Muscovite (light mica)	$K_2Al_4[Si_6Al_2O_{20}](OH,F)_4$
Mica	Biotite (dark mica)[4]	$K_2Fe_6[Si_6Al_2O_{20}](OH)_2F_2$
Clay mineral	Kaolinite	$Al_4[Si_4O_{10}](OH)_8$
Oxide	Quartz	SiO_2
Oxide	Corundum	Al_2O_3
Oxide	Hematite	Fe_2O_3
Oxide	Magnetite	Fe_3O_4
Oxide	Ilmenite	$Fe^{II}Ti^{IV}O_3$
Carbonate	Calcite	$CaCO_3$
Carbonate	Magnesite	$MgCO_3$
Carbonate	Dolomite	$CaMg(CO_3)_2$
Phosphate	Apatite	$Ca_5(OH,F,Cl)(PO_4)_3$

Square brackets denote Si and Al with tetrahedral coordination, for example, $[Si_6Al_2O_{22}]$ in hornblende. Isomorphic substitution is marked by soft brackets, for example, (Mg, Fe^{II}) in diopside.

[1] Tectosilicates include the feldspars and nepheline, and the analcime group, which includes leucite.

[2] Augite is a common pyroxene-aluminosilicate:
$(Ca,Na)(Mg,Fe^{II},Mn,Fe^{III},Al,Ti)[(Si,Al)_2O_6]$.

Minerals with the given names occur with compositions according to the formulas:

[3] $(Na,K)_{0-1}Ca_2(Mg,Fe^{II},Fe^{III},Al)_5[Si_{6-7}Al_{2-1}O_{22}](OH,F)_2$.

[4] $K_2(Mg,Fe^{II})_{6-4}(Fe^{III},Al,Ti)_{0-2}[Si_{6-5}Al_{2-3}O_{20}](OH,F)_4$.

Table 3.6 provides the chemical formulas of the most important minerals of the groups, listed in Table 3.4. Because silicates dominate the lithosphere, not only by mass but also by structural richness, a discussion of their structure and chemistry is indispensable in an environmental context.

Silicates are built using silicon(IV) as the central ion coordinated to four oxygen($-$II) ligands, which are placed at the corners of a regular tetrahedron. One type of diversity arises because oxygen has two binding modes: single-bonded (O) or bridging (μ-O), with bridging index of 2; that is, two silicon atoms are linked by each bridging ligand. Another type of versatility occurs because tetrahedrally coordinated Si^{IV} can be exchanged with tetrahedrally coordinated Al^{III}, with the difference in charge being compensated by the addition of cations from groups 1 or 2 in the periodic table.

We shall now show how these elements interact in the structural chemistry of the silicates.

Monomeric and oligomeric silicates

The monomeric silicates are orthosilicates, that is, salts of orthosilicic acid, H_4SiO_4; a mineral example is olivine, $(Mg,Fe^{II})_2 [SiO_4]$,[a] a structure of independent, isolated $[SiO_4]^{4-}$ units linked by divalent cations in octahedral coordination. This is the origin of the generic geological name *nesosilicate*,[b] which is used for all orthosilicate minerals. However, one may equally well describe the crystal structure as a hexagonal closest packed structure of oxygen atoms, with the silicon atoms occurring in tetrahedral sites and the divalent metals in octahedral.

In olivine, magnesium and iron can be exchanged freely,

$$0 \leq \frac{n(Fe^{II})}{n(Fe^{II}) + n(Mg^{II})} \leq 1 \tag{3.1}$$

without affecting the crystal structure. This is an example of isomorphic substitution,[c] and olivine minerals span a continuous series of minerals from forsterite, $Mg_2[SiO_4]$, to fayalite, $Fe_2[SiO_4]$. From a physicochemical point of view, an olivine may be considered as a solid solution of the two metal ions in the silicate: it resembles a binary mixture whose components are miscible at all ratios and whose phase diagram is simple, analogous to that depicted later in the upper part of Figure 3.2 (fayalite melts at 1205 °C, forsterite at 1890 °C). Mg^{II} and Fe^{II} (and Mn^{II}) have virtually identical ionic radii, and even the smaller ions Zn^{II} and the larger Ca^{II} and Pb^{II} are observed in olivine, resulting in a great variety of minerals that share a common crystal structure.

Hydrothermal processes and weathering can easily change olivine. For example, the clay mineral talc can be formed via chrysotile (a commercially important asbestos):

$$3\,Mg_2SiO_4 + 4\,H_2O + SiO_2 \; \rightarrow \; 2\,Mg_3Si_2O_5(OH)_4 \tag{3.2}$$

$$2\,Mg_3Si_2O_5(OH)_4 + 3\,CO_2$$
$$\rightarrow \; Mg_3Si_4O_{10}(OH)_2 + 3\,MgCO_3 + 3\,H_2O \tag{3.3}$$

Another example of the weathering of olivine is presented later in Table 3.11.

---

[a] In the mineral formulas of Section 3.1, we follow geological/geochemical custom and place square brackets around the silicate anion. Note that metal ions placed in this part of the formula *always* occur with tetrahedral coordination.

[b] Gk. $\nu\eta\sigma\iota \; \hat{=}$ island.

[c] Gk. $\iota\sigma\sigma\varsigma \; \hat{=}$ identical; $\mu\sigma\rho\varphi\eta \; \hat{=}$ shape.

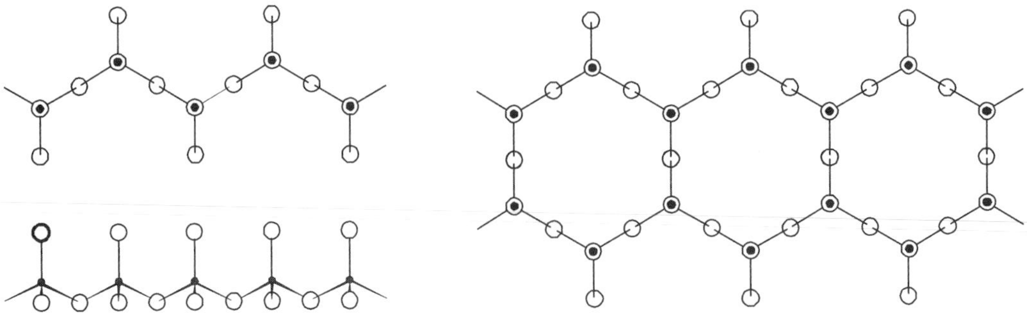

A   B

Figure 3.1   *Structure of pyroxenes and amphiboles*
Panel A. A pyroxene anion viewed in two directions perpendicular to the fiber axis.
The empirical formula is $[Si_2O_4(\mu\text{-}O)_2{}^{4-}]_n$ for a single-strand chain polymer.
Panel B. An amphibole anion viewed perpendicular to the fiber axis. The projection
shown at the lower left applies to the amphiboles as well. The empirical formula is
$[Si_4O_6(\mu\text{-}O)_5{}^{6-}]_n$ for a double-strand chain polymer.
In both cases the silicon atoms are located in a plane; silicon is tetrahedrally
coordinated with three oxygen atoms located on one side of the silicon plane and
one on the other side.

Metasilicic acids, $(H_2SiO_3)_n$, are polymers that are formally derived from orthosilicic acid by abstraction of water. For small n, their salts are called oligosilicates,
an example being *cyclo*-hexametasilicate,[a] $[SiO_2(\mu\text{-}O)^{2-}]_6$. These are rare minerals
such as beryl, $Al_2Be_3[Si_6O_{18}]$, the main source of beryllium, and the gem emerald,
which is beryl colored green by a few percent of chromium(III).

Chain silicates

We begin the discussion of polymer silicates by considering one-dimensional polymers. This group consists of single-strand chain polymers, called pyroxenes, and
double-strand chain polymers, called amphiboles. The generic geological name is
inosilicate, meaning chain silicate.[b] The molecular structure of pyroxenes gives rise
to a clear macroscopic fibrous structure as seen in asbestos.

The common silicate anion in pyroxenes is $[SiO_2(\mu\text{-}O)^{2-}]_n$, *catena*-poly-
[(dioxidosilicate-μ-oxido) (2−)], of which an example is shown in Figure 3.1, panel
A. The $[SiO_3{}^{2-}]$ units can be located differently relative to one another, giving rise
to distinct structures. As an example, the repetitive unit of the particular pyroxene
shown in the figure is dimeric, $[(\mu\text{-}O)_{1/2}O_2SiOSiO_2(\mu\text{-}O)_{1/2}{}^{4-}]_n$. In the top view,
every other tetrahedron is oriented toward the top of the figure, with the rest point-
ing downward. In other pyroxenes, all of the tetrahedra could be pointing in the

[a] *cyclo*-hexa[(dioxidosilicate-μ-oxido)(2−)].
[b] Gk. $\iota\nu\alpha \,\hat{=}\,$ fiber.

same direction, or two could be pointing up for every one down, and so forth. The cations are bound to the oxygen between parallel silicate chains in 6- or 8-coordinate sites.

Water, hydroxide, and fluoride are not found in pyroxenes.[a] The name arises because they were once considered to be impurities in lavas, but actually they are formed at rather high temperatures in volcanoes. This explains the lack of water and hydroxide (see the Bowen series, Table 3.10).

The formula of augite (marked with 2 in Table 3.6), a very common pyroxene mineral, shows a characteristic feature: tetrahedrally coordinated Si^{IV} may to some extent be replaced by tetrahedrally coordinated Al^{III}. The charge difference is compensated by the substitution of, for example, Mg^{II} with Fe^{III} or Al^{III} at one of the octahedral positions outside the polymeric anion, or by the substitution of Ca^{II} for Na^{I} in an 8-coordinate site. This type of substitution occurs over wide ranges of compositions up to the point where the crystal substitution is so substantial that the substance is classified as a new mineral. The isomorphic substitutions that were described for olivine are also observed here. These various possibilities result in an enormous variety of minerals.

The common silicate anion in amphiboles is $[Si_4O_6(\mu\text{-}O)_5{}^{6-}]_n$ (see Figure 3.1, panel B).[b] These minerals display a macroscopic fibrous structure similar to that seen in the pyroxenes. Mg^{II} and Ca^{II} are found in 8-coordinate sites as in the pyroxenes, but in contrast hydroxide (or its substitution analogue fluoride) always occurs and is necessary for the stability of the crystal matrix. The chemistry of the amphiboles is similar to that of the pyroxenes, but substitution is more widespread, making it more difficult to express the composition of different mineral types as simple formulas. The name reflects the fact that many amphiboles are virtually indistinguishable to the naked eye (Haüy, 1784).

Amphiboles form in cooling magma, to be discussed in the next section, and some of them occur only as metamorphic rocks. For example, the amphibole tremolite, $Ca_2Mg_5[Si_8O_{22}](OH)_2$, is formed by the metamorphosis of dolomite and quartz in the presence of water:

$$5\,CaMg(CO_3)_2 \;+\; 8\,SiO_2 \;+\; H_2O$$
$$\rightarrow \quad Ca_2Mg_5[Si_8O_{22}](OH)_2 \;+\; 3\,CaCO_3 \;+\; 7\,CO_2 \qquad (3.4)$$

At higher temperatures, diopside can be formed from calcite, tremolite, and quartz, and forsterite from dolomite and quartz. Amphiboles are hydrolyzed as shown by some of the examples shown later in Table 3.11.

Sheet silicates

Two-dimensional polymers make up a second class of polymer silicates. Here, the silicate anions have a layered structure that in many cases gives rise to a characteristic

---

[a] Gk. $\pi\upsilon\rho-\,\hat{=}$ fire; $\xi\acute{\varepsilon}\nu o\varsigma\,\hat{=}$ stranger.
[b] Gk. $\alpha\mu\upsilon\iota\beta o\lambda o\varsigma\,\hat{=}$ ambiguous.

sheetlike macroscopic structure. Some minerals from this class can be divided into flakes using a knife. The generic geological name is *phyllosilicate*, meaning sheet silicate.[a]

The general formula of the phyllosilicate anion is $[Si_2O_2(\mu\text{-O})_3{}^{2-}]_n$, and an example is shown later in Figure 6.4. Mica and most of the proper clay minerals, that is, most soil minerals, belong to the phyllosilicate class, and their structure and chemistry are discussed in more detail in Chapter 6, the pedosphere.

Framework silicates

A third class of polymer silicate comprises three-dimensional polymers formally derived from quartz. The generic geological name is *tectosilicate*, meaning framework silicate.[b] Tectosilicates include plagioclases, orthoclases, and quartz, and thus they are common and ubiquitous minerals.

The molecular structure of the tectosilicates may be envisaged using quartz as the point of departure. Quartz, SiO_2, exists in several isomers, all of which have the general composition poly[(silicon-di{μ-oxido})(0)], $[Si(\mu\text{-O})_2]_n$. This means that each SiO_4 tetrahedron shares all four ligands with other Si atoms, which themselves are central atoms in similar coordination tetrahedra, resulting in the three-dimensional structure. For convenience, we shall use a tetrameric unit, $[Si_4O_8]$, as the basis of the following discussion. The feldspar albite is obtained when Si^{IV} in quartz is replaced with Al^{III}. In order to maintain charge neutrality, the resulting charge imbalance is compensated by Na^{I}. The feldspar anorthite is obtained in a similar way by replacing two Si^{IV} with two Al^{III} and by compensating the charge imbalance with Ca^{II}. Thus, we have arrived at the two minerals albite, $Na[AlSi_3O_8]$, and anorthite, $Ca[Al_2Si_2O_8]$. They form an isomorphic series (fully miscible solid solutions) over a broad range of temperature and are known collectively as plagioclases, denoted $(Na,Ca)[Al(Al,Si)Si_2O_8]$. Albite and orthoclase, $K[AlSi_3O_8]$, are likewise completely miscible, forming the isomorphic series known as alkali feldspars. Unordered structures can arise at lower temperatures when the species are not fully miscible. In general, plagioclases may only contain minor amounts of K^{I}, and likewise, alkali feldspars may only contain minor amounts of Ca^{II}. Examples of the hydrolysis of feldspars are given later in Table 3.11.

The phase diagram for the three-component system albite(Ab)-anorthite(An)-water(W) is shown in Figure 3.2. In contrast to simple binary phase diagrams, it can be seen that the role of water is simply to change the melting point, which is extremely sensitive to the water vapor pressure. At $p_W = 1$ kbar and $1100\,°C$, there will only be a single solid phase, and a melt ($\theta > 1230\,°C$) will contain $w_{An} = 0.5$. For the same system, subject to $p_w = 5$ kbar (Figure 3.2b), the melt will be in equilibrium with crystals with $w_{An} = 0.9$. When the water pressure is increased further, the system becomes completely molten. Relationships of this kind are central to the formation of minerals.

[a] Gk. φύλλο $\hat{=}$ leaf.
[b] Gk. τεκτον $\hat{=}$ timberwork.

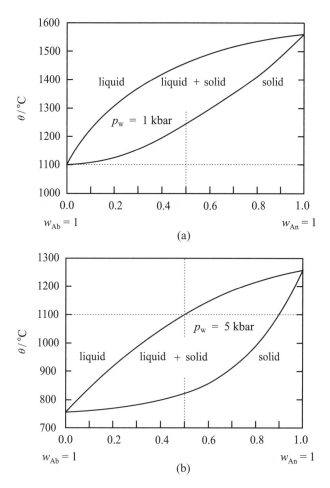

Figure 3.2 *Phase diagram for albite(Ab)-anorthite(An)-water(w)*
The abscissa indicates the mass fraction w of anorthite in the mixture.

The remaining minerals in Table 3.6 are oxides and species with discrete anions. Their physical and chemical properties are detailed in most textbooks on general or inorganic chemistry.

c. Igneous rocks

Magma is a fluid mixture of molten silicates, water, and other gases that is formed at high temperature within the mantle. The rocks that crystallize during cooling are called igneous rocks.[a] As ascending magma cools, successive types of material will crystalize, and igneous rocks can be characterized by two quantities: the mineral composition and the texture. Equilibrium between the liquid and solid phases is controlled by chemical potentials. First, that is, at high temperature, basic rocks of

---

[a] Lat. *ignis* $\hat{=}$ fire.

Mineral	Nephelinite (ijolite)	Basalt (gabbro)	Andesite (diorite)	Rhyolite (granite)
Quartz	–	–	12.37	32.87
Corundum	–	–	–	1.02
Orthoclase	3.16	6.53	9.60	25.44
Albite	–	24.66	29.44	30.07
Anorthite	7.39	26.62	26.02	4.76
Leucite	13.57	–	–	–
Nepheline	21.95	–	–	–
Diopside	32.36	14.02	4.84	–
Hypersthene	–	15.20	9.49	1.34
Olivine	2.32	1.50	–	–
Magnetite	7.95	5.49	4.74	2.14
Ilmenite	5.05	3.49	1.65	0.54
Apatite	2.51	0.82	0.50	0.17
Calcite	1.37	0.26	0.11	0.17

Table 3.7 Mineral composition of some igneous rocks[28]

Average mineral composition of volcanic rocks; mass fractions in percent. Names of equivalent plutonic rocks are given in parentheses.

olivine type precipitate, and it follows that the content of silicic acid in the remaining melt increases. Therefore, the rocks that precipitate subsequently become progressively more acidic. Texture refers to the size, shape, and distribution of the mineral grains and is controlled by the cooling rate. Based on texture, one can define two major types of rock, plutonic and volcanic. A plutonic rock is a coarse-grained (>3 mm) rock in which the individual crystals can be distinguished with the naked eye. A volcanic rock is fine grained (<1 mm) and most of the crystals cannot be distinguished with the eye. Volcanic rocks may also contain glasses. Tables 3.7 to 3.9 illustrate the variation of mineral composition, chemical composition, and texture for some major rock types.

We consider first the process that forms basalt at the midoceanic ridges. Typical conditions at a depth of 15 km are 1300 °C and 5 kbar of hydrostatic pressure. The magma is basaltic (cf. composition shown in Tables 3.7 and 3.8) and has a relatively low viscosity, approximately 500 P to 1000 P (see Section 2.2e). In addition to the minerals and elements shown in the tables, magma contains supercritical H_2O, CO_2, and SO_2. The resulting increase in pressure means that the temperature required to precipitate crystalline material is reduced relative to the melting point of the pure materials. Magma at the midoceanic ridge is transported to the surface rather quickly, and the rapid cooling results in relatively small crystals. In these zones, the magmatic rock basalt is formed instead of large-grained gabbro, which requires a long cooling time. At lower temperatures, the presence of water during precipitation can lead to the formation of large crystals. For example, pegmatites, which are rocks formed from a melt containing water, contain grains more than 20 mm in length. Water allows rarer elements to be concentrated in the pegmatites, and it is not uncommon to find rare minerals and gem stones in pegmatic veins.

Element	Nephelinite (ijolite)	Basalt (gabbro)	Andesite (diorite)	Rhyolite (granite)
O	0.572	0.595	0.607	0.618
Si	0.143	0.173	0.200	0.243
Al	0.060	0.065	0.069	0.052
Fe^{III}	0.015	0.010	0.009	0.004
Fe^{II}	0.018	0.021	0.012	0.003
Mg	0.034	0.035	0.017	0.002
Ca	0.045	0.038	0.025	0.007
Na	0.033	0.020	0.023	0.037
K	0.016	0.005	0.007	0.030
Ti^{IV}	0.007	0.005	0.002	0.001
$H(+)^{I}$	0.038	0.022	0.019	0.040
$H(-)^{I}$	0.017	0.010	0.000	0.011
Mn^{II}	0.001	0.001	0.001	0.000
P^{V}	0.003	0.001	0.000	0.000
C^{IV}	0.003	0.001	0.000	0.001

Table 3.8 Elemental composition of some igneous rocks

Average elemental composition (mole fraction) of the rocks of Table 3.7.
[I] Hydrogen from water driven off above (+) and below (−) 105 °C, respectively.

Element	Nephelinite (ijolite)	Basalt (gabbro)	Andesite (diorite)	Rhyolite (granite)
x_{Si}	0.143	0.173	0.200	0.243
$x_{Na} + x_{K}$	0.049	0.025	0.030	0.067
$x_{Mg} + x_{Ca}$	0.079	0.073	0.044	0.009

Table 3.9 Composition of some igneous rocks

Some characteristic mole fractions of elements in the rocks of Table 3.7.

Magma is formed under the subduction zone at a depth of approximately 150 km when the subducted ocean floor reaches a temperature of around 1300 °C. Magma rises in this region, but the distance to the surface is much farther, and the longer timescales this implies allow the material to fractionate as it cools. Plagioclase minerals precipitate, changing the composition of the upwelling material, resulting in so-called andesitic magma.[a] The fractionation continues, resulting finally in magma with a granitic composition (see Tables 3.7 and 3.8). Magnesium-rich olivine, followed by iron-rich olivine, precipitates at the same time as the plagioclase minerals (anorthite precipitates first; see Figure 3.2). Overall, this process means that the remaining molten material is depleted in calcium and magnesium and enriched in silicon (Table 3.9). During the entire upwelling process, some of the materials will come to rest on the underside of the cooler tectonic plate and be sequestered from

[a] Named after the Andes Mountains.

Composition of melt	Solid phase 1, heterogeneous series		Solid phase 2, homogeneous series	Approximate temperature
Gabbro	Olivine		Anorthite	1200 °C
Diorite	Pyroxene		Albite	
Granite	Amphibole		Oligoclase	
Granite		Dark mica		
Granite		Light mica		
Granite		Quartz		
Granite		Pegmatite minerals		600 °C
		Pneumatolytic minerals		
		Hydrothermal minerals		200 °C

Table 3.10 Bowen's reaction series

The two series, marked 1 and 2, are discussed in the text. Only a single solid phase is formed at temperatures where the micas precipitate.

further reaction, while the remaining material will sink into the warmer magma where it will dissolve and react with the melt.

The precipitation series discussed in the preceding paragraph was first described by Bowen (1928), and "Bowen's reaction series" is shown in Table 3.10. Traditionally, a distinction is made between two types of condensed phase. One is a discontinuous series, labeled "solid phase 1, heterogeneous series" in the table. In this series, the various mineral groups precipitate in a discontinuous fashion because they are not miscible with one another. However, as described in the section on minerals, each such mineral group may comprise an isomorphic crystallization series. The second type of condensed phase, labeled "solid phase 2, homogeneous series," is composed of plagioclase minerals and alkali feldspars. They constitute a so-called continuous series because the minerals are miscible to a certain extent.

d. Sedimentary rocks

Air, water, and ice carry suspensions of particles that deposit at the surface. This sediment can become buried, and if the lithospheric pressure becomes great enough, it will be transformed into sedimentary rock. The term *diagenesis* is used to describe the chemical, physical, and biological changes that take place during the formation of sedimentary rocks from sediment.

Several processes generate the suspended particles that form sediments. Igneous rocks are weathered at the surface by acid hydrolysis and oxidation involving carbon dioxide and oxygen, respectively. Weathering results in the production of dissolved ions and the formation of insoluble salts. In addition to these chemical processes, weathering includes the physical breaking of rocks into smaller particles by wind, water, and ice. Weathering produces sand and dust that can be transported over long distances in the atmosphere. For example, it is not uncommon that dust from the Sahara Desert is blown to Brazil or Scandinavia. It is clear that a particle's size and density will affect how it is transported in the atmosphere and hydrosphere.

	Table 3.11 Weathering of some rock-forming minerals[53]		
	Weathering reaction	$A_{tot}°$	$A_{Si}°$
a.	$Fe_2SiO_4 + \frac{1}{2}O_2 \xrightarrow{aq} Fe_2O_3 + SiO_2$	220.5	220.5
b.	$Mg_2SiO_4 + 4H^+ \rightarrow 2Mg^{2+} + 2H_2O + SiO_2$	184.1	184.1
c.	$MgSiO_3 + 2H^+ \rightarrow Mg^{2+} + H_2O + SiO_2$	87.5	87.5
d.	$CaMgSi_2O_6 + 4H^+ \rightarrow Ca^{2+} + Mg^{2+} + 2H_2O + 2SiO_2$	159.5	79.7
e.	$Mg_7Si_8O_{22}(OH)_2 + 14H^+ \rightarrow 7Mg^{2+} + 8H_2O + 8SiO_2$	574.2	71.8
f.	$Ca_2Mg_5Si_8O_{22}(OH)_2 + 14H^+ \rightarrow 2Ca^{2+} + 5Mg^{2+} + 8H_2O + 8SiO_2$	515.6	64.5
g.	$CaAl_2Si_2O_8 + 2H^+ + H_2O \rightarrow Ca^{2+} + Al_2Si_2O_5(OH)_4$	100.0	50.0
h.	$2NaAlSi_3O_8 + 2H^+ + H_2O \rightarrow 2Na^+ + Al_2Si_2O_5(OH)_4 + 4SiO_2$	96.7	16.1
i.	$2KAlSi_3O_8 + 2H^+ + H_2O \rightarrow 2K^+ + Al_2Si_2O_5(OH)_4 + 4SiO_2$	72.4	12.1
j.	$2KAl_3Si_3O_{10}(OH)_2 + 2H^+ + 3H_2O \rightarrow 2K^+ + 3Al_2Si_2O_5(OH)_4$	72.4	12.1

The affinity $A_{tot}°$ of a reaction is $-\Delta_r G°$; unit: kJ mol^{-1}; $A_{Si}°$ is the affinity for an extent of weathered Si equal to 1 mol; unit: kJ mol^{-1}. The minerals are: (a) fayalite (olivine) → hematite + quartz; (b) forsterite (olivine); (c) clinoenstatite (pyroxene); (d) diopside (pyroxene); (e) anthophyllite (amphibole); (f) tremolite (amphibole); (g) anorthite (Ca-feldspar) → kaolinite; (h) albite (Na-feldspar); (i) microline (K-feldspar); (j) muscovite (mica). The difference in affinity between reactions a and b is mainly due to the oxidation in reaction a.

The ease with which igneous rocks weather has been determined empirically and is shown in Table 3.11 with the most vulnerable listed first. Note that weathering taking place on geological timescales is governed largely by differences in Gibbs energy.

Diagenesis occurs over time as sediment becomes buried. The increased pressure squeezes water out of porous sediment, whose grains are subsequently held together by minerals such as gypsum, anhydrite, calcite, dolomite, quartz, iron oxides, sulfides, or bitumen that act as cement. Diagenesis occurs at pressures up to about 8 kbar and temperatures up to about 300 °C. At higher pressures and temperatures, the process changes into metamorphism.

Sedimentary rocks are divided into three main categories (see Table 3.5). The first categories comprise the clastic sedimentary rocks, which are composed of discrete fragments of other rocks. From small to large grain size, clastic sedimentary rocks are made up of clay (diameter $d < 0.004$ mm), silt ($0.004 < d < 0.06$ mm), sand ($0.06 < d < 2$ mm), and gravel ($d > 2$ mm); see Table 6.3. Shales and sandstones have silt- and sand-sized grains, respectively. The last category, limestone, consists mainly of skeletal fragments ($CaCO_3$) of marine organisms such as corals, mollusks, and foraminifera, or of calcite ($CaCO_3$) and dolomite ($CaCO_3 \cdot MgCO_3$), formed by evaporation of seawater.

The upper, weathered part of the solid surface is called the regolith. The regolith is the basic material that is turned into soil by biological activity. Soil genesis and the chemical reactions taking place in soil are described in Chapter 6. Biological activity gives rise to carbon dioxide concentrations in soil that can be 200 times larger than found in the atmosphere. To illustrate the effect, consider that groundwater at 5 °C may have a CO_2 concentration of 4.5 mM, see Section 5.1c. Under such conditions, the primary mechanism weathering plagioclase minerals is acid hydrolysis:

$$CaAl_2Si_2O_8 + 2CO_2 + 3H_2O$$
$$\rightarrow Al_2Si_2O_5(OH)_4 + Ca^{2+} + 2HCO_3^- \quad (3.5)$$

$$2NaAlSi_3O_8 + 2CO_2 + 11H_2O$$
$$\rightarrow Al_2Si_2O_5(OH)_4 + 2Na^+ + 2HCO_3^- + 4H_4SiO_4 \quad (3.6)$$

Orthosilicic acid polymerizes (forming metasilicic acids), becoming quartzite, and calcite can precipitate at lower CO_2 partial pressures, becoming marble. The series shown in Table 3.11, which otherwise is the same as Bowen's reaction series, is not changed by the presence of carbon dioxide.

e. Metamorphic rocks

Metamorphic rocks are formed from other rocks by structural and/or chemical changes. Metamorphism can occur through a partial recrystallization in which some of the solid phase is present during the entire process. This is in contrast to igneous rocks, which form via cooling of liquid magma. Depending on the rock, metamorphism occurs at relatively low temperatures, from 250 to 600 °C. It is not always straightforward to distinguish between sedimentary rocks that are formed by diagenesis and metamorphic rocks formed at low temperature.

Metamorphic rocks are divided into two groups depending on their origin. Regional metamorphic rocks are formed in large areas in mountain collision zones. Some important rocks of this type are quartzite, schists, marble, amphibolite, and various gneisses. An example of metamorphism is provided by the minerals in the garnet group. They are common, characteristic of regional metamorphism, and produced by reactions such as the following:

$$\underset{\text{olivine}}{Mg_2SiO_4} + \underset{\text{anorthite}}{CaAl_2Si_2O_8} \rightarrow \underset{\text{garnet}}{Mg_2CaAl_2Si_3O_{12}} \quad (3.7)$$

The reaction is promoted by high pressure as it is associated with a 5/6 reduction in volume.

Contact metamorphic rocks are formed locally around intrusive and extrusive igneous rocks. The conversion is driven by heat, and the original material is to a certain degree mixed with the intrusive magma. Contact metamorphism of clay-containing sediments gives rise to hornfels of different types. An example of a relatively simple reaction that can occur in contact metamorphism is[15,47]

$$\underset{\text{calcite}}{CaCO_3} + \underset{\text{quartz}}{SiO_2} \rightarrow \underset{\text{wollastonite}}{CaSiO_3} + \underset{\text{carbon dioxide}}{CO_2} \quad (3.8)$$

Wollastonite is often found where granite and limestone meet. This reaction is endothermic, so equilibrium shifts to the right as temperature increases and pressure decreases. At the surface pressure of 1 bar the reaction occurs at 400 °C, whereas at a pressure of 6 kbar, corresponding to a depth of 20 km, a calcite-quartz mixture is stable at 800 °C. The presence of wollastonite means that the temperature has been at least 400 °C, and if the CO_2 pressure is known, the temperature of metamorphosis can be determined even more precisely. It appears from the discussion of this equation and Equation 3.7 that the phase rule (Equation 2.178) and phase diagrams (see, for example, Figures 3.2) are indispensable tools for analyses of metamorphic rocks and their reactions.

Facies are assemblages of rocks. Specifically, a metamorphic facies is a group of rocks that has been formed under the same conditions of pressure and temperature. Contact metamorphic rocks are mainly formed at low pressure and regional metamorphic rocks at high pressure, resulting in two series of facies whose members can be characterized as a function of temperature.

3.2 The hydrosphere

All water on the Earth is collectively called the hydrosphere. Water exists as a solid, a liquid, and a gas under the conditions of temperature and pressure found on the surface (Table 3.12). However, many geological processes take place in the presence of supercritical water, as its critical point is rather low ($\theta_c = 374$ °C and $p_c = 220.6$ bar) compared with the conditions found in the lithosphere. Although several structures of solid water different from ice are known to occur at $p > 6$ kbar and $\theta > 0$ °C, only traditional ice is important in normal environmental contexts.

Table 3.12 The distribution of the hydrosphere

	Area, A / Mm^2	Mass, m / Zg	Mass fraction	Mass flux, $\dot{m}$ / Zg a^{-1}
Oceans	360	1625	0.978	
Groundwater		9.1	0.005	
Cryosphere		29	0.017	
Surface waters		0.28	10^{-4}	
Atmosphere		5.2		
Water in atmosphere		0.013–0.014	10^{-5}	
Flow from continents				0.032–0.040
Precipitation = evaporation				0.45–0.49

Note: The assessment of the magnitudes of the masses and the mass fluxes of the hydrosphere varies in the literature.

A survey of the general chemistry of water is given in ref. 22.

Water circulates quickly between the ocean, the atmosphere, and the surface as depicted in Figure 3.3. As seen, the atmosphere contains approximately 13 Eg water, and water enters the atmosphere, and is removed, at a rate of 490 Eg/a, giving a lifetime of about 10 days.[a] Using these figures, the annual rainfall, averaged over the Earth's surface, is 96 cm.[b]

[a] From Equation 2.48: 13 Eg / (490 Eg/a) = 0.027 a ≈ 9.7 d.
[b] Assuming a density of 1 g/ml: 490×10^{12} m^3 / (509 × 10^{12} m^2) = 0.96 m.

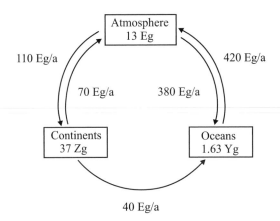

Figure 3.3   *The hydrological cycle*
Mass fluxes. The assessment of the fluxes varies $\pm 10\,\%$ in the literature.
Using $\Delta_{vap}H \approx 2.5$ kJ/g (see Table 3.16) gives the heat turnover ≈ 1 YJ/a.[a]

a. Chemical composition of natural waters

Natural waters contain dissolved inorganic salts and organic material. The composition of the inorganic salts in seawater (Table 3.13), is quite different from that found dissolved in freshwater (Table 3.14). One also finds various amounts of a brown-yellow organic material, plainly called "dissolved organic matter." Analyses show that it is made of polysaccharides, proteins, lignins, and fulvic acids, but the exact composition varies with the location. In lakes with freshwater, its origin is "soil organic matter," as is also the case for coastal seawater.[b] In the oceans, organic material is dilute and its origin cannot be ascertained.

Table 3.13 shows the concentrations of some of the chemical species that are found in seawater. Even though it is possible to find virtually all elements of the periodic table in the sea, the majority of inorganic material is due to just a few species. Although the salinity of seawater can vary significantly, the relative fractions of the dissolved ions hardly change. The planet's freshest seawater is found near freshwater sources (e.g., the mouths of rivers or near the Arctic ice cap). In contrast, water is saltier than average in places with large evaporation and little circulation. For example, the mass fraction of salt is found to be 0.007 in the surface water of the Baltic Sea, rising to 0.010 near the Danish Straits. This number is 0.035 to 0.037 around the equator in both the Pacific and Atlantic Oceans and 0.039 in the Eastern part of the Mediterranean. In contrast, the value is 0.15 in the Dead Sea, between Jordan and Israel. The acidity (pH) is 8.1 ± 0.2 in the world's oceans, but because the pH scale is logarithmic, this variation spans a factor of 4 in hydron concentration.[c]

[a] $420 \times 10^{18} \times 2500$ (g a^{-1} J g^{-1}) $\approx 10^{24}$ J a^{-1} ≈ 33 PW. See footnote p. *153*: such simple estimates frequently vary 10–20 %.
[b] In specialist literature, the acronyms DOM and SOM, respectively, may be seen.
[c] The different usages in the past of the term *acidity* are discussed in Section 5.1b, alkalinity. In this book the concept is used in a qualitative sense: the acidity increases (decreases) when pH decreases (increases).

	$10^3\,w$	$m\,/\,\mathrm{mol\,kg^{-1}}$	x_{solute}	$c\,/\,\mathrm{mmol\,l^{-1}}$
Table 3.13 The hydrosphere, part 1				
Na^+	10.760	0.46803	0.41816	479.9
Mg^{2+}	1.294	0.05324	0.04757	54.9
Ca^{2+}	0.412	0.01028	0.00918	10.5
K^+	0.399	0.01021	0.00912	10.5
Sr^{2+}	0.0079	0.00009	0.00008	0.1
Cl^-	19.350	0.54579	0.48764	559.3
SO_4^{2-}	2.712	0.02823	0.02522	28.9
HCO_3^-	0.145	0.00238	0.00212	2.4
Br^-	0.067	0.00084	0.00075	0.8
H_4SiO_4	0.0029	0.00003	0.00003	–
H_3BO_3	0.0046	0.00007	0.00007	0.1
F^-	0.0013	0.00007	0.00006	0.1

The average concentration of the most common chemical species in seawater.[92] Note that the concentrations of dissolved oxygen and nitrogen have not been included because they are highly variable, depending on salinity, temperature, and biological activity. w, mass fraction; m, molality; x_{solute}, relative mole fraction of dissolved species; c, amount concentration at 20 °C where the mass density is 1.02476 g ml^{-1}.
Characteristic quantities: $w_{\mathrm{solute}} = 35.16 \times 10^{-3}$; $w_{\mathrm{salinity}} = 34.99 \times 10^{-3}$; $w_{\mathrm{Cl}} = 19.35 \times 10^{-3}$; $w_{\mathrm{chlorinity}} = 19.37 \times 10^{-3}$; ionic strength,[a] $I = 0.6972$ mol kg^{-1}; total molality, $m_{\mathrm{tot}} = 1.1193$ mol kg^{-1}. The content of dissolved inorganic carbon is given as the alkalinity of HCO_3^-; see Example 3 of Section 5.2c.

There is great variation in the kinds of inorganic salts found in different freshwaters, for example, from a mountain lake, farmland groundwater, or a forest stream. Table 3.14 shows a representative example of river water. When evaporated, the major component remaining is calcium carbonate, but local conditions vary. For example, the crystals found at the shores of Great Salt Lake, Utah, are mainly sodium chloride.

The concentration of salts in freshwater is typically 10^{-3} of that found in seawater, and the chemical composition is different. The ions found in river water are largely due to crustal weathering, especially of secondary sedimentary rocks. The acidity of freshwater is typically less than pH 7 because of dissolved carbon dioxide, and the dominant ions are calcium and hydrogencarbonate. It is clear that seawater is not simply evaporated river water. The reason is linked to the origin and evolution of the different waters, and the evolution of continental rocks.

b. Analytical characteristics of environmental waters

This section will discuss some characteristic analytical properties of natural and waste water, and some tools used to analyze water in the environment. The discussion deals

---

[a] See Equations 6.18 and 5.56.

Table 3.14	The hydrosphere, part 2		
	$10^6\,w$	m / mmol kg^{-1}	$10^3\,x_{solute}$
Na$^+$	5.15	0.224	0.115
Mg^{2+}	3.35	0.138	0.071
Ca^{2+}	13.4	0.334	0.172
K$^+$	1.3	0.033	0.017
Sr^{2+}	0.03	0.000	0.000
Cl$^-$	5.75	0.162	0.083
SO$_4^{2-}$	8.25	0.086	0.044
HCO$_3^-$	52.0	0.852	0.439
Br$^-$	0.02	0.000	0.000
H$_4$SiO$_4$	10.4	0.108	0.056
H$_3$BO$_3$	0.01	0.000	0.000
F$^-$	0.1	0.005	0.003

The average concentration of the most common chemical species in river water. Note that the content of dissolved oxygen and nitrogen are not included because they are highly variable, depending on salinity, temperature, and biology.

w, mass fraction; m, molality; x_{solute}, relative mole fraction of dissolved species.

Ionic strength,[a] $I = 1.75$ mmol kg^{-1}; total molality, $m_{tot} = 1.942$ mmol kg^{-1}; total mass fraction, $w_{tot} = 99.8$ mg kg^{-1}.

with principles rather than explicit analytical procedures and spectroscopic methods, which may be found in refs. 29 and 88.

Chlorinity

The ratio between the concentrations of the main ionic constituents of seawater is reasonably constant, so variations in the content of inorganic salts may be characterized by a single parameter, the chlorinity or the salinity.

The chlorinity $w_{chlorinity}$ of a sample of water is the mass fraction of chloride and bromide, where bromide is measured as the mass of the equivalent amount of chloride.

Consider, for example, the seawater in Table 3.13. An argentimetric determination[b] will show that the content of Cl and Br is equivalent to 58.965 g/l of Ag.[c] This in turn is equivalent to 58.965 g/l × $M(Cl)/M(Ag)$ = 19.380 g/l of "Cl" as required by the definition. The ratio $M(Cl)/M(Ag) = 0.32867$ is called a *gravimetric factor*, referring to the analytical chemical procedure used in in the first half of the 20th century. Note however, that it is maintained[d] that the gravimetric factor is 0.3285233, giving the chlorinity as 19.371; this is the number listed in Table 3.13.

[a] See Equations 6.18 and 5.56.
[b] Argentimetric methods include classical titrations according to Mohr and Volhard as well as more recent electrometric titrations.
[c] Atomic weights (2009) are:[215] $M(Ag) = 107.8682$ g/mol; $M(Cl) = 35.453$ g/mol; $M(Br) = 79.904$ g/mol.
[d] American Meteorological Society, September 2006.[220] Note in particular that the factor has too many significant digits and is not in accordance with tables of atomic weights.[215]

Salinity

The so-called absolute salinity, w_{solute}, of a sample of water is the mass fraction of dissolved salts. However, it is customary to define the salinity, $w_{salinity}$, as a secondary quantity derived from the chlorinity by multiplying the chlorinity by the factor 1.80655. Applying this method to the seawater of Table 3.13 gives a value of $w_{salinity}$ that is only slightly lower than the absolute salinity. The salinity may be determined by comparing the conductivity of a sample (of known dilution) with that of a standard KCl solution.[29]

Hardness

The hardness w_{CaO} of a sample of water is the mass fraction of Ca and Mg, measured as CaO. In the determination Mg must be converted into the equivalent amount of CaO.

Water with w_{CaO} in the range 0 to 50×10^{-6} is called *soft*; if w_{CaO} is in the range 100×10^{-6} to 200×10^{-6}, it is called *hard.*[a]

A distinction has been made between temporary hardness (where the anion is carbonate) and permanent hardness (where the anion is sulfate) since the days of steam engines. Temporary hardness may be removed by boiling

$$\text{Ca}^{2+} + 2\,\text{HCO}_3^- \quad \rightarrow \quad \text{CaCO}_3(\text{s}) + \text{CO}_2(\text{g}) + \text{H}_2\text{O} \tag{3.9}$$

or by adding slaked lime, Ca(OH)_2, which also precipitates the calcium ions (see Equation 9.63). $\text{CaSO}_4(\text{aq})$ cannot be removed in this way but may be precipitated as $\text{CaCO}_3(\text{s})$ using soda, $\text{Na}_2\text{CO}_3(\text{aq})$, which also removes $\text{MgCl}_2(\text{aq})$ as basic magnesium carbonates of varying composition (see Equations 9.64).

Biological Oxygen Demand

The *biological oxygen demand*[b] w_{BOD} is a measure of the amount of oxygen required to bio-oxidize all organic material present in a sample of water. It is used as an indication of water quality and marks the degree of pollution.

The w_{BOD} of water is determined as a mass ratio, namely, the mass of oxygen consumed in a biomediated oxidation of the organic substance in a sample, divided by the mass of the sample. The number w_{BOD} is sometimes given in units of parts per million (Appendix A1.4).

Biological oxygen demand is determined using an old biological method (1865) that seeks to copy Nature in a reproducible way. A small amount of bacteria and ciliates (which eat bacteria) from activated sludge is added to the sample. It is then stored in the dark (to prevent photosynthesis) at a temperature of 20 °C for 5 days,[c] and the amount of oxygen that has been consumed is measured. For a result to be accepted, the remaining mass fraction of oxygen must be larger than 2×10^{-6}. Considering that

---

[a] In technical literature one uses the unit *German degree*, °dH: $w_{CaO} = 1°\text{dH} = 10 \times 10^{-6} = 10$ ppm $\hat{=}$ $[\text{Ca}^{2+}] = 0.18$ mM.

[b] Note that we follow the ISO-IUPAC-IUPAP recommendation of denoting a physical quantity by an italicized letter with a subscript rather than using an acronym directly; for example, we write w_{BOD} rather than BOD.

[c] The incubation period may be seven days, indicated by w_{BOD7}.

freshwater has $w(O_2) \approx 9 \times 10^{-6}$ (see Table 3.24), dilution is normally required before the test procedure can be started.

Example. A clean river will have $w_{BOD} \approx 1 \times 10^{-6}$, whereas waste water having $w_{BOD} > 10 \times 10^{-6}$ is considered very polluted. (Note: International convention deprecates use of the unit ppm = 10^{-6}.[126])

Determination of dissolved oxygen

The determination of w_{BOD} requires two determinations of oxygen in aqueous solution, one before and one after biological action. The Winkler method (1888) is reliable and is still in use today. It is used directly and to calibrate modern oxygen-sensitive electrodes, as follows:

Freshly prepared $Mn(OH)_2$ (from $MnSO_4(aq)$ and base, reaction 3.10a) is added to the sample, where it reacts quantitatively with the dissolved oxygen according to reaction 3.10b. The next step is standard iodometry: a surplus of KI(aq) reduces manganese(IV), (reaction 3.10c), and the resulting iodide, I(0), is titrated with a standard solution of thiosulfate (reaction 3.10d).

$$Mn^{2+}(aq) + 2\,HO^-(aq) \rightarrow Mn(OH)_2 \cdot aq(s) \tag{3.10a}$$

$$Mn(OH)_2 \cdot aq(s) + \tfrac{1}{2}O_2(aq) \rightarrow MnO_2 \cdot aq(s) + H_2O(l) \tag{3.10b}$$

$$MnO_2 \cdot aq(s) + 3\,I^-(aq) + 4H^+(aq) \rightarrow$$
$$Mn^{2+}(aq) + I_3^-(aq) + 2\,H_2O(l) \tag{3.10c}$$

$$I_3^-(aq) + 2\,S_2O_3^{2-}(aq) \rightarrow S_4O_6^{2-}(aq) + 3\,I^-(aq) \tag{3.10d}$$

The overall reaction is

$$\tfrac{1}{2}O_2 + 2\,S_2O_3^{2-} + 2\,H^+ \rightarrow S_4O_6^{2-} + H_2O \tag{3.10e}$$

showing that oxygen is effectively determined by the amount of thiosulfate used[a] and that the exact forms of the intermediate manganese species are unimportant.

Chemical Oxygen Demand

Chemical oxygen demand w_{COD} is a measure of the amount of oxygen required to oxidize all the organic substance present in a sample. It is used for solid samples as well as aqueous solutions. In the latter case, a calibrated determination of the chemical oxygen demand can be used as a substitute for a determination of the biological oxygen demand.

The w_{COD} of a sample is a mass ratio, namely, the mass of oxygen consumed in a chemical oxidation of the organic substance in a sample, divided by the mass of the sample. Obviously, this number may be larger than 1 for a solid sample.

The chemical oxygen demand depends on the sample and the mode of oxidation. For example, direct combustion in oxygen oxidizes C, N, and S to their oxides,

---

[a] Equations 3.10d and 2.13 give the basic analytical relation: $n_{\frac{1}{2}O_2} = n_{2S_2O_3^{2-}}$ or $n_{O_2} = \tfrac{1}{4}n_{S_2O_3^{2-}}$.

Table 3.15 Vapor pressure and density of water[129]					
θ / °C	p_{aq}^* / hPa	ρ_{aq} / g ml^{-1}	θ / °C	p_{aq}^* / hPa	ρ_{aq} / g ml^{-1}
-30	0.380	0.922	$+20$	23.38	0.99823
-10	2.599	0.920	$+25$	31.67	0.99707
-5	4.018	0.918	$+30$	42.46	0.99567
0	6.111	0.99987 (l)	$+100$	1013.25	0.95838
$+4$	8.136	0.99997	$+200$	15536	0.8628
$+10$	12.28	0.99973	$+374$	220640	

The asterisk in p_{aq}^* refers to pure water. The critical temperature of water is $\theta_c = 373.99$ °C. *Note:* 1013.25 hPa = 1.01325 bar = 1 atm.

but the result is strongly dependent on temperature. The most common method is standard dichromate titration in acidic solution ($Cr_2O_7^{2-} + 14\,H^+ + 6\,e^- \rightarrow 2\,Cr^{3+} + 7\,H_2O$) followed by back titration of surplus dichromate with Fe^{II} ($\rightarrow Fe^{III}$). This gives $C \rightarrow CO_2$, $S^{-II} \rightarrow S^0$, but note that N^{-III} is not oxidized. Therefore, in order to compare w_{COD} with w_{BOD}, nitrogen must be determined independently, for example, using Kjeldahl's titration method (1883): organic N^{-III} is converted into NH_4^+ by boiling the sample in concentrated sulfuric acid with a mercury catalyst (HgO); after the addition of a surplus of sodium hydroxide, ammonia is boiled off, collected, and determined using standard acid-base titration.

Note that determinations of chemical and biological oxygen demands use different analytical methods, and for certain types of samples w_{COD} can only replace w_{BOD} after an intercalibration of the methods.

Example

For some applications, dissolved organic matter in municipal waste water may be assumed to have the composition $C_{18}H_{19}O_9N$ (see Table 1.28). For this species $w_{COD} = 1.588$, or 1.424, for oxidations with and without nitrification ($N^{-III} \rightarrow N^V$), respectively.

c. Physicochemical properties of water

Note: Some properties of water are given in connection with the discussion of hydrosphere-atmosphere equilibria in Section 3.3b.

Density and vapor pressure

One of the remarkable properties of water is that its density has a maximum at 4 °C (Table 3.15). A consequence is that deep freshwater lakes can become stratified with warm, relatively light water lying on top of dense cold water. The upper layer is called the epilimnion. It is exposed to the atmosphere, is mixed by the effects of wind and waves, and typically has a higher pH and amount of dissolved oxygen

θ / °C	$\Delta_{vap} H$ / kJ mol^{-1}	θ / °C	$\Delta_{vap} H$ / kJ mol^{-1}
0	51.059 (s)	120	39.684
0	45.054 (l)	200	34.962
25	43.990	300	25.300
40	43.350	360	12.966
100	40.657	374	2.066

Table 3.16 Water: Enthalpy of vaporization

than the layer below, called the hypolimnion. The temperature changes rapidly in the so-called thermocline, the thin layer separating the epilimnion and the hypolimnion. In many tropical lakes, this temperature profile is permanent. In temperate areas, spring warming and fall cooling act to make the density equal throughout the water column, allowing vertical mixing by the action of wind and waves. In the winter, temperate lakes are also divided into layers with ice on top and light water near 0 °C immediately below, followed by denser 4 °C water.

The temperature of ocean water decreases rapidly to about 5 °C at a depth of around 1 km and thereafter more slowly, reaching a minimum of approximately 2 °C.

Enthalpy of vaporization and heat capacity

The enthalpy of vaporization of water at different temperatures is given in Table 3.16. Based on the figures given here and in Table 3.12, one may estimate the amount of heat exchange, mediated by water, between the atmosphere and the surface of the Earth to be about 37 PW.[a]

Among the notable deviations of water relative to other liquids is its unusually large heat capacity C_p and the correspondingly large specific heat capacity c_p at constant pressure. At 15 °C, 1 bar, one finds

$$C_p = 75.4030 \, \text{J K}^{-1} \, \text{mol}^{-1}; \quad c_p = 4.1855 \, \text{J K}^{-1} \, \text{g}^{-1} \qquad (3.11)$$

This value varies less than 1 % in the temperature interval 0–100 °C. The magnitude of the heat capacity is important to life in its present form. For example, the Gulf Stream transports heat from the tropical Atlantic to the western coast of northern Europe, resulting in mild, cloudy winters. Assume that the stream is 100 km wide and 1.0 km deep and flows at 1.0 km/h, and that the total temperature drop is about 12 K. Accordingly, it delivers a power of about 1.4 PW to Europe.[b] For a comparison, the world production of energy from coal is approximately 1.1 TW (2005).[c]

[a] $0.47 \times 10^{21} \frac{\text{g}}{\text{a}} 45 \times 10^3 \frac{\text{J}}{\text{mol}} \times \frac{1}{18.02} \frac{\text{mol}}{\text{g}} \times \frac{1}{31.6 \times 10^6} \frac{\text{a}}{\text{s}} = 37 \times 10^{15} \, \text{W}$

[b] Assume that the Gulf Stream is an open physicochemical system of constant mass. Then

$$100 \times 1.0 \times 1.0 \times 10^9 \frac{\text{m}^3}{\text{h}} \times 1.03 \times 10^6 \frac{\text{g}}{\text{m}^3} \times 4.19 \frac{\text{J}}{\text{gK}} \times 12 \, \text{K} \times \frac{1}{3600} \frac{\text{h}}{\text{s}} = 1.4 \, \text{PW}.$$

[c] Total of worldwide mined coal (2005) is *ca.* 5.3 Pg.[222] The most efficient power plant, Drax, in England, produced (2005) 6.57 J g^{-1} (energy per mass of coal).

Table 3.17 Physical properties of isotopologues of water					
Entity	$^1H_2{}^{16}O$	$^1H_2{}^{17}O$	$^1H_2{}^{18}O$	$^1H^2H^{16}O$	$^2H_2{}^{16}O$
x^1	0.99731	0.00038	0.00200	0.00031	–
θ_{fus} / °C	0.00				3.82
θ_{vap} / °C	100.00		100.15		101.42
ρ_{max} / g ml^{-1} ^{2}	0.99995		1.11249		1.1060
$\theta_{\rho_{max}}$ / °C	3.984		4.211		11.185
$\Delta_{fus} H$ / kJ mol^{-1}	5.98		6.03		
$p_{25\,°C}$ / kPa^3	3.165				2.724
$pK_w{}^4$	13.9991				14.82

Note: The values of all quantities, except the vapor pressure at 20 °C, are measured at "normal" pressure, that is, 1 atm.[a]
[1] Mole fraction, calculated from the isotopic distribution in VSMOW (see Section 1.3c): $n(^2H) / n(^1H) = 0.15576 \times 10^{-3}$;
$n(^{18}O) / n(^{16}O) = 2.0052 \times 10^{-3}$; $n(^{17}O) / n(^{16}O) = 0.3799 \times 10^{-3}$.
[2] Maximum mass density, occuring at the temperatures noted in the next row.
[3] Vapor pressure.
[4] Ion product at 25 °C.

Colligative properties of seawater

Important colligative properties of water relevant in environmental contexts include freezing point depression and osmotic pressure. From Table 3.13:

1. The cryoscopic constant for water is 1.86 K kg mol^{-1}, giving a freezing point of about −2 °C, as observed.
2. The osmotic pressure, $\Pi = c\,R\,T$, is around 27 bar at 20 °C.

Isotopes and isotope effect

The average natural isotopic distribution of the elements of water is given in Table 3.17. The ^{18}O and D isotopologues differ from the parent species in virtually every way, including vapor pressure, standard electrode potential, vibrational frequencies, chemical reaction rate, and rate of diffusion.[b] Some examples are given in the table. The macroscopic properties of HDO are not known because this species cannot be isolated, because of rapid exchange of hydrogen and deuterium in water.

The effect on the rate or equilibrium constant of two reactions that differ only in the isotopic composition of one (or more) of their otherwise chemically identical components is called a kinetic or thermodynamic isotope effect. Fundamentally, the two effects have a common root in the relative difference of mass between the

[a] By convention, "normal boiling points," and so forth, refer to a pressure of 1 atm.[155] The "standard-state" pressure is 1 bar.
[b] The term *isotopomer* is a contraction of *isotopic isomer*, that is, molecular entities that have the same number of each isotope atom but differing in their position, such as acetaldehyde CH_2DCHO and CH_3CDO. *Isotopologue* molecular entities differ only in the number of isotopic substitutions, for example, H_2O, HDO, D_2O.

isotopes in question. Therefore, the greatest isotope effects are found in the deuterium-hydrogen system, a frequent object of study in experimental and theoretical chemical kinetics.

The kinetic deuterium isotope effect is defined as the ratio k_H/k_D between the rate constants of a reaction carried out with hydrogen and deuterium isotopologues. The major contributions to this ratio are quantum mechanical quantities such as vibrational frequencies and tunneling effects,[a] which are both related to the atomic mass. It is understandable, therefore, that isotope effects are most pronounced for elements with low atomic numbers and for reactions carried out at low temperatures.

3.3 The atmosphere

The atmosphere is a thin layer of gases held in place by Earth's gravitation. In addition, it contains colloidal suspensions of liquid and solid particles known as aerosols (see Table 6.8). The pressure and density of the atmosphere decrease with altitude. The mole fractions of the major components of the atmosphere, with the exception of water vapor, are constant up to an altitude of 100 km, above which they begin to separate according to molecular weight. It is useful to define the top of the atmosphere as the surface where gravity and kinetic energy balance, and beyond which escape to space becomes likely. The region beyond is called the exosphere; here gases have a net velocity outward. Escape velocity can be achieved thermally, or after photolysis when part or all of the photon energy in excess of the energy of the broken bond is released as kinetic energy to the molecule's fragments. The base of the exosphere varies depending on temperature, the solar wind, and other factors, but it is normally 500 to 1,000 km above the surface. The region just below the exosphere is called the thermosphere, and it extends down to an altitude of about 100 km.

The physical structure of the atmosphere up to the thermosphere is shown in Figure 3.4. The troposphere is mixed rapidly by weather. In contrast, the stratosphere is quite stable, due to a temperature inversion. Between the different spheres are pauses, for example, the troposphere and the stratopause.

a. Chemical composition

The atmosphere is a gas mixture of virtually constant chemical composition up to about 100 km; 99.999 % of the atmosphere's mass is below this altitude.[b] Table 3.18 shows the mole fractions of the eight gases that constitute the *standard atmosphere*; to this must be added a varying content of water (gas and aerosols) and other gases (Tables 3.19 and 3.20). The content of carbon dioxide and many other trace gases is not constant (see Figure 7.3).

The sources of the gases of Table 3.18 have been discussed previously, except for methane, which is produced by anaerobic processes (see Section 3.4b).

---

[a] In addition, symmetry-dependent isotope effects have been observed in systems such as O_3.
[b] The mass of the atmosphere is $m_{atm} = 5.136 \times 10^{18}$ kg = 5.136 Zg.

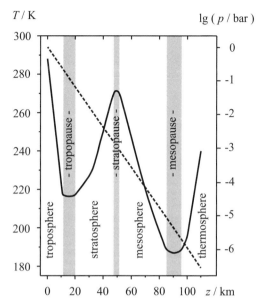

Figure 3.4 *Atmospheric temperature and pressure as functions of altitude*
Variation of temperature (full line, left scale) and pressure (dotted line, right scale)
of the atmosphere at $\pm 40°$.[129] The lapse rate (negative temperature gradient; see
Equation 2.11) of the troposphere is $\Gamma = -dT/dz = 6.6$ K / km. The tropopause (tp)
has $dT/dz \approx 0$, $\theta_{tp} \approx -56$ °C, $z_{tp} \approx 8$ km (at the poles), $z_{tp} \approx 17$ km (at the
equator). The stratosphere is an inversion layer, $dT/dz > 0$. The stratopause and
mesopause are also approximate isotherms, each covering a total of $\Delta T \approx 2.5$ K, see
Figure 2.10.

Table 3.18 The standard atmosphere			
Entity	Mole fraction	Entity	Mole fraction
N_2	0.78084 ± 4	Ne	18.18×10^{-6}
O_2	0.20946 ± 6	He	5.24×10^{-6}
Ar	0.00934 ± 1	CH_4	2.0×10^{-6}
CO_2	0.000380	Kr	1.14×10^{-6}

Composition (mole fraction) of the dry standard atmosphere (1996).[129] The mole fraction
of CO_2 is set[a] to $y(CO_2) = 380 \times 10^{-6}$, and the molar mass of this atmosphere is $M_{air} =$
28.9668 g mol^{-1}. With this figure and the known mass, the amount of substance of the
atmosphere is $n_{atm} = 177.3$ Emol.

Trace gases of the dry troposphere are listed in Table 3.19. Here Xe is the only
abiotic gas. All of the reduced gases have biological sources. The oxidized species
are produced from them through photochemical chain reactions. A number of gases
such as CO and CH_3Cl are produced when biomass burns.

Atmospheric chemistry is driven by the production of primary radicals by sunlight.
They in turn react with almost all species. As seen from the tables of trace gases,

---

[a] In October 2010, the value was $y(CO_2) = 388 \times 10^{-6}$.

Table 3.19 Natural trace gases in the dry troposphere[121]

Entity	Mole fraction	Entity	Mole fraction
H_2	0.55×10^{-6}	NH_3	$<3 \times 10^{-9}$
N_2O	0.32×10^{-6}	CH_3Cl	0.61×10^{-9}
CO	0.125×10^{-6}	COS	0.50×10^{-9}
Xe	0.087×10^{-6}	NO_x	$<0.30 \times 10^{-9}$
O_3	$<0.1 \times 10^{-6}$	H_2S	$<0.10 \times 10^{-9}$
C_nH_m	$<0.1 \times 10^{-6}$	SO_2	$<0.09 \times 10^{-9}$
HNO_3	$<4 \times 10^{-9}$	$(CH_3)_2S$	$<0.06 \times 10^{-9}$
H_2O_2	$<3 \times 10^{-9}$	CH_3I	2×10^{-12}
terpenes	$<3 \times 10^{-9}$		

C_nH_m marks aliphatic hydrocarbons; most of these gases also have significant anthropogenic sources.

most are hydrophobic. But when oxidized in the atmosphere, they can be converted to hydrophilic condensed phase species, which then are removed by precipitation. This is the basic mechanism of the atmosphere's purifying effect in Nature. As discussed in Chapter 4, the most important modes of reaction are photooxidation (mainly initiated by hydroxyl radicals) and photolysis.

Particles play an important role in the physics and chemistry of the atmosphere. Atmospheric aerosols include cloud droplets and dust, smoke, and photochemically generated smog. High particle concentrations are associated with increased mortality from heart disease and cancer.[158] Particles act as catalysts in the stratosphere, converting inert chlorine reservoirs into active species that lead to ozone depletion. Further, particles scatter light and thus have a cooling effect on climate. The chemical composition of a particle determines its ability to act as a nucleus for cloud formation. Atmospheric particles, both natural and anthropogenic, have a large impact on climate, and this is an area of intense research activity; the topic is discussed in Chapter 4.1c.

b. Hydrosphere-atmosphere equilibria

Measures of humidity of air

The humidity of air is a general term referring to air's content of water vapor, that is, $H_2O(g)$. The content may be defined in several ways:[a]

1. The mole fraction $y_{aq} = n_{aq} / \Sigma_B n_B = p_{aq} / \Sigma_B p_B$, where the last equation assumes perfect gases.

2. The relative humidity, $\pi = p_{aq}/p_{aq}^*$. The relative humidity is the ratio between the partial pressure of water $p_{aq}(T)$ and the vapor pressure of water $p_{aq}^*(T)$ at the same temperature (see Table 3.15).[b]

[a] In the following the indices, "aq" denotes the species H_2O and "air" the dry gas mixture of Table 3.18 with the molar mass M_{air}. The sum over chemical species B runs over all species and includes "aq."

[b] Example: The relative humidity of air, π_{max}, over a 2 M aqueous solution of sucrose ($\rho = 1.255$ g/ml, $M = 342$ g/mol) is 0.94. This is also the value for the seawater in Table 3.13.

Table 3.20 Anthropogenic trace gases in the dry troposphere[121]			
C_6H_6 (<1000)[1]	CF_2Cl_2 (300)	$CFCl_3$ (178)	CH_3CCl_3 (157)
CCl_4 (121)	CF_4 (69)	CHF_2Cl (59)	$C_2F_3Cl_3$ (<40)
CH_2Cl_2 (30)	$(CH_2Cl_2)_2$ (26)	CH_3Br (22)	$CHCl_3$ (16)
$(CF_2Cl)_2$ (14)	C_2H_5Cl (12)	$ClHC=CCl_2$ (8)	C_2Cl_5F (4)
C_2F_6 (4)	CF_3Cl (3)	$CHFCl_2$ (1.6)	CF_2ClBr (1.2)

(mole fraction) / 10^{-12}
[1] Includes derivatives of benzene.

3. The specific humidity is the mass fraction, $w_{aq} = m_{aq} / m_{total}$.
4. The humidity ratio, $\eta = m_{aq} / m_{air}$.

There is no general relationship between y_{aq} and π, but w and η are related:

$$w = \frac{\eta}{1+\eta} \quad \text{and} \quad \eta = \frac{w}{1-w} \tag{3.12}$$

Table 3.22 shows examples of the use of this equation.

Table 3.21 gives as an example the composition of the *standard atmosphere* when saturated with water at 20 °C.

Solubility of gases in water

The water solubilities of the species B(g), c_B, at a partial pressure of 1 bar are given in Table 3.22 for selected temperatures. The entries of the table are the equilibrium constants K,

$$c_B = Kp_B \tag{3.13}$$

Here c_B is the stoichiometric concentration of the species B in the solution, that is, this c_B includes all species derived from reactions between B and water.

Henry's law for real binary mixtures was discussed in Chapter 2 (see Equation 2.198 and in particular Equations 2.204–2.205). Note that the expression

$$c_B = k_{H,B}^{-1} p_B \tag{2.205}$$

is concerned with uncharged species B(l). Accordingly, for species like N_2 (for example, O_2, CO, NO, CH_4), the relation $K = k_{H,B}^{-1}$ holds true, but for species like CO_2 (such as H_2S, SO_2, NH_3), this simple relation does not apply. As an example, for CO_2, Equation 2.205 is concerned with CO_2(aq) and H_2CO_3(aq), whereas Equation 3.13 further includes HCO_3^- and CO_3^{2-} (see Equations 5.46 and 5.55).

Table 3.21 An atmosphere with relative humidity $\pi = 1$ at 20 °C				
N_2	O_2	Ar	H_2O	CO_2
0.7630	0.2047	0.0091	0.0229	0.0003

The composition (mole fraction) is calculated using Tables 3.18 and 3.15.

Table 3.22 Solubilities of gases in water							
θ / °C	$N_2{}^1$	CO	O_2	NO	CO_2	H_2S	SO_2
0	1.037	1.55	2.142	3.23	75.04	204.7	3.52×10^3
10	0.815	1.22	1.656	2.49	51.99	148.1	2.50×10^3
20	0.673^2	1.00	1.338	2.03	37.86	111.4	1.74×10^3
30	0.572	0.85	1.107	1.70	28.19	85.1	1.20×10^3

Solubilities $c_B = K p_B$. The entries of the table are K / mM bar^{-1}. The sum of the partial pressure of a gas and the vapor pressure of water is 1.000 bar.

[1] The figures in this column refer to a gas mixture of N_2 and Ar with $y(N_2) = 0.9882$ and $y(Ar) = 0.0118$.

[2] At a partial pressure of 1.0000 atm, the mole fraction of N_2 in water is $x(N_2) = 1.274 \times 10^{-5}$. Using $M_{aq} = 18.016$ g/mol, the molality is 0.6822 mol/kg. Then using 1 atm = 101325 Pa, $p_{aq}{}^*$/hpa = 2338 Pa, $\rho_{aq} = 0.99823$ kg/l (at 20 °C, cf. Table 1.15), and $y(N_2) = 0.9882$, one obtains the value $c(N_2) = 673$ mmol/l.

As an example of the use of Table 3.22, the mass concentration of oxygen γ in water at 10 °C in equilibrium with the atmosphere, where the partial pressure of oxygen is 0.21 bar, is:

$$\gamma = 0.21 \, \text{bar} \times 1.656 \frac{\text{mmol}}{1 \, \text{bar}} \times 32 \frac{\text{mg}}{\text{mmol}} = 11 \frac{\text{mg}}{\text{l}} \qquad (3.14)$$

Such calculations lead to the entries of Tables 3.23 and 3.24.

Table 3.25 shows the decreasing solubility of air as a function of increasing salinity.

Table 3.23 Hydrosphere-atmosphere, part 1					
$10^3 \, w_{salinity}$	0	10	20	30	35
c / mmol l^{-1}	1.01	0.94	0.89	0.83	0.81

Solubility, c, of water-saturated air at 1 bar in air-saturated water at 10 °C as a function of the salinity, $w_{salinity}$ (see Section 3.2b).
The mole fraction of oxygen, $x(O_2)$, in the dissolved air is approximately 0.33.
Note: Water with $c_{air} = 1.0$ mM contains ≈ 10 mg / l of dioxygen.

c. The physics of the atmosphere

To a good approximation, the atmosphere is a perfect gas. The physical chemistry of such a phase is given by the equation of state:

$$pV = RT \sum_B n_B \qquad (3.15)$$

where the symbols have their conventional definition.[a] The equation implies that each gaseous species B(g) can be assigned a partial pressure $p_B = n_B \, RT/V$, which is the

[a] Equation 3.15 combines Boyle-Mariotte's law (1662; 1679), $pV = K'$; Dalton's law (1801), $p = \Sigma_B \, p_B$; Gay-Lussac's law (1802), $K' = K \, T$; and Avogadro's hypothesis (1811), $K = n \, R$. The "ideal gas law," that is, the formulation of Equation 3.15, is due to Guldberg (1868). A gas mixture that obeys this equation is called a *perfect gas*.

$\theta\ /\ ^\circ C$	0	5	10	20	25	30
Table 3.24 Hydrosphere-atmosphere, part 2						
$10^3\varphi$	29.18	25.68	22.84	18.68	17.08	15.64
$y(O_2)$	0.3491	0.3469	0.3447	0.3403	0.3382	0.3360
$10^6 w(O_2)$	14.36	12.55	11.1	8.97	8.13	7.42

Data for water-saturated atmospheric air in equilibrium with air-saturated water as a function of the temperature, θ.
φ is the volume of air that can be dissolved in a certain volume of water.
y is the mole fraction of oxygen in the air that is dissolved in the water phase.
w is the mass fraction of oxygen in water.

pressure that it would exert if it were alone in the volume. The partial pressure is proportional to the mole fraction y_B of B:

$$p_B = y_B p \tag{3.16}$$

Equation 3.15 also implies that the atmospheric gas mixture has a characteristic molar mass M_{air} as given in Table 3.18, and that pressure p and density ρ are proportional:

$$p = \frac{n}{V}RT = \frac{m}{V}\frac{RT}{M_{air}} = \rho\frac{RT}{M_{air}} \tag{3.17}$$

Equation 3.17 has been used extensively for determinations of the molar mass of gases and gas mixtures, because it is experimentally easy to measure ρ, p, and T simultaneously.

Real gases are imperfect; they do not obey the equation of state (Equation 3.15), but rather empirical ones of which van der Waals' equation is an example (see Equation 2.138 and the example). However, the assumption that the atmosphere is a perfect gas is sufficient for the present discussions.

$w_{salinity}$	0 °C	10 °C	20 °C	30 °C
Table 3.25 Hydrosphere-atmosphere, part 3				
0.000	450	347	280	232
0.010	420	326	264	220
0.020	392	306	248	208
0.030	366	287	234	197
0.035	354	278	228	192

Solubility, $c\ /\ \mu mol\,l^{-1}$, of oxygen in air-saturated water in equilibrium with water-saturated air as a function of temperature θ and salinity $w_{salinity}$ (see Section 3.2b). The pressure of the air is $p_{total} = 1.000$ bar, and the mole fraction of oxygen in dry atmospheric air is $x(O_2) = 0.2095$.

Altitude dependence of temperature and pressure

The lowest layer of the atmosphere, extending to 10–15 km, is called the troposphere (see Figure 3.4). The troposphere mixes rapidly because of convection: sunlight heats the surface at the bottom, and as we will discuss in Appendix 4, water vapor leads to vertical instability. In addition to rapid vertical mixing, air is mixed horizontally, transporting a great deal of heat from the equator to the poles. The troposphere moves because of wind and weather, governed by surface warming, pressure gradients, Coriolis forces, the surface morphology (elevation, friction), and gravitation. The average surface temperature of the Earth is 287 K,[249a] and the temperature of the troposphere decreases monotonically to the tropopause, reaching a minimum of about 220 K.[a]

A simple expression of the pressure p as a function of altitude z was given in Equation 2.81, and data from ref. 129 give a good fit to the expression

$$\ln(p/\text{bar}) = -\frac{z}{7.27\,\text{km}} + 1.04 \qquad (3.18)$$

The stratosphere is the layer of the atmosphere above the tropopause, extending to about 50 km.[b] The stratosphere is the location of the ozone layer and would not exist if it was not for ozone. Ozone absorbs solar ultraviolet radiation and photodissociates, and the products re-form ozone. The net result is that sunlight is converted to heat, which gives an increase of temperature with height. This temperature profile leads to a stable structure in which air does not mix by convection. Air is pumped into the stratosphere across the tropopause by convective storms and reenters the troposphere at the poles after a few years.

During the great discovery voyages the eastern trades in the Tropics and the westerlies in the temperate regions were discovered and used for transport. At that time Hadley (1735) proposed the model for the wind systems as shown in Figure 3.5.

Hadley circulation begins when equatorial air is heated by the Sun.[c] This causes the air to expand, and its decreased density leads to buoyant lifting and cooling approximated by isentropic expansion (see Appendix 4). This in turn liberates humidity as tropical rain, releasing the latent heat of condensation and generating more lift. The low pressure at the equator gives rise to winds toward the equator, and the Coriolis force creates NE and SE trades instead of N and S winds. After ascending to the tropopause, the dry air moves toward the poles and descends at latitudes of ±30°. The resulting increase of pressure and temperature lead to a temperature inversion and a global region with high pressure. This explains the occurrence and location of the great deserts (Sahara, Gobi, Kalahari, Gibson, Great Victoria) and the name "Horse Latitudes" at sea, because the lack of winds forced the crews on ships stuck in those regions to sacrifice their horses in order to save potable water for themselves. The result is two vortices, the so-called Hadley cells, one north and one south of the equator. Such active vortices are marked with "a" in Figure 3.5. Active vortices are created around the poles as well, giving rise to the passive vortices (marked "p" in

[a] According to the World Meteorological Organization (WMO), the tropopause is the lowest level at which the drop in temperature with altitude decreases to 2 °C/km or less.
[b] Lat. *stratum* ≙ blanket, "layer."
[c] See Figure 10.3A.

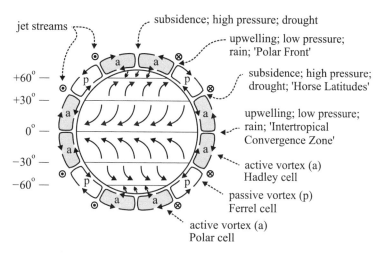

jet streams

subsidence; high pressure; drought

upwelling; low pressure; rain; 'Polar Front'

subsidence; high pressure; drought; 'Horse Latitudes'

upwelling; low pressure; rain; 'Intertropical Convergence Zone'

active vortex (a) Hadley cell

passive vortex (p) Ferrel cell

active vortex (a) Polar cell

$+60°$

$+30°$

$0°$

$-30°$

$-60°$

Figure 3.5 *Hadley's idealized model for tropospheric air movements*
The six main vortices of the troposphere. The active vortices are shaded. The directions of winds are shown as they are experienced on the surface of the Earth. The winds of each vortex are coiled: air parcels on the Northern Hemisphere move in right-handed helices, whereas the paths of air parcels on the Southern Hemisphere are left handed. The direction of the jet streams are marked with $\odot$ and $\otimes$ (moving out of and into the page, respectively).

Figure 3.5) with ensuing westerly winds in the two temperate regions. Jet streams are generated between the vortices.

The Hadley model is an idealized picture of atmospheric circulation and indicates average large-scale motion. The detailed motion is chaotic, especially the motions of high- and low-pressure systems at midlatitude.

3.4 Biota

Biota are defined as living and dead biological organisms. Biological organisms are composed of one or more cells, which are the basic entities of biology. Based on the structure of the cell, living organisms are divided into two classes: prokaryotes, which lack a cell nucleus, and eukaryotes, which have a nucleus. With few exceptions, all multicellular life forms are eukaryotes. Cellular life is divided into the bacteria and archaea (unicellular, prokaryotic), and the eukaryotic forms: protista, fungi, plantae, and animalia.

a. Chemical composition of biota

Elemental composition
As a first approximation, the composition of dry biological tissue can be described by the empirical formula CH_2O, with carbon occurring in the same formal oxidation

Table 3.26 Biota, part1					
	Element	Mole fraction		Element	Mole fraction
1	H	0.498	6	K	0.46×10^{-3}
2	O	0.249	7	Si	0.33×10^{-3}
3	C	0.249	8	Mg	0.31×10^{-3}
4	N	2.70×10^{-3}	9	P	0.30×10^{-3}
5	Ca	0.73×10^{-3}	10	S	0.17×10^{-3}

The 10 most common elements of dry biological tissue; average mole fraction. Note that the first three elements, corresponding to the empirical formula for a carbohydrate, CH_2O, comprise 98.7 % of the mass.

state – zero – as in carbohydrates such as glucose, $C_6H_{12}O_6$. Biological systems are much more complicated than the simple carbohydrate suggested by the formula, but nonetheless this is the main component of Table 3.26. Free glucose is rare, but plants are rich in related polymers such as starch and cellulose, which have the same empirical formula (see Figure 8.5).

The elemental composition of mammals is shown in Table 3.27, which gives the mole fractions of the elements found in humans. Note the significant difference between the composition of living material and that of the inorganic environment, such as the lithosphere or the hydrosphere.

Microorganisms break down most of the dead plant material that falls onto soil. When the origin of such decayed vegetable matter can no longer be traced, it is

Table 3.27 Biota, part 2					
	Element	Mole fraction		Element	Mole fraction
1	H	0.628	16	Cu	180×10^{-9}
2	O	0.257	17	B	90×10^{-9}
3	C	0.095	18	V	36×10^{-9}
4	N	0.014	19	Mn	36×10^{-9}
5	Ca	0.0024	20	Sn	27×10^{-9}
6	P	0.0020	21	Se	27×10^{-9}
7	S	0.49×10^{-3}	22	Li	27×10^{-9}
8	Na	0.41×10^{-3}	23	Ni	18×10^{-9}
9	K	0.32×10^{-3}	24	Cr	9.0×10^{-9}
10	Cl	0.27×10^{-3}	25	Co	4.5×10^{-9}
11	Mg	0.13×10^{-3}	26	Mo	4.5×10^{-9}
12	Fe	6.8×10^{-6}	27	I	1.8×10^{-9}
13	Si	4.5×10^{-6}			
14	F	3.6×10^{-6}			
15	Zn	3.2×10^{-6}			

The distribution of the 27 most abundant elements in an adult human, presented as decreasing mole fraction. Note that the first six elements account for 0.998 of the total, corresponding to the approximate empirical formula for mammals, given in Table 3.28.

Table 3.28 Biota, part 3	
Empirical formula	Species
CH_2O (biota)	Dry biomass (see Table 3.26)
$C_{10}H_{18}O_9$	Saccharides
$C_8H_6O_2$	Lipids
$C_{14}H_{12}O_7N_2$	Proteins
$C_{43}H_{285}O_{117}N_7CaP$	Mammals
$C_{106}H_{263}O_{110}N_{16}P$	Redfield's formula for aquatic biomass[68]
$C_{187}H_{186}O_{89}N_9S$	Humic acids[89]
$C_{135}H_{182}O_{95}N_5S$	Fulvic acids[89]
$C_{18}H_{19}O_9N$	Organic substance in sewage[105]

Average chemical composition of some organic materials.

called humus. The classical method for characterizing humus is through fractionation (Berzelius, 1830): it is first treated with a base (NaOH) resulting in an insoluble part, humin, and a solution. Acid (HCl) is added to the solution, causing humic acids to precipitate while fulvic acids remain in solution. Although all three fractions, humin, humic acids, and fulvic acids, are complex mixtures whose composition and chemical functional groups depend on where the soil was found, it is nevertheless possible to present average formulas (Table 3.28 see also Table 6.10).

Empirical formulas

It appears from Table 3.26 that the empirical formula CH_2O approximates the composition of dry plant material quite well, because it accounts for 99.6 % of the total on a mole fraction basis. From Table 3.27, we see that in mammals the mole fraction of C, H, and O is 98.0 %. Nevertheless, the formula of a carbohydrate does not exactly match the composition. In order to get a more accurate description, one may use the empirical formula $C_{43}H_{285}O_{117}N_7CaP$, which accounts for 99.8 % of the total on a mole fraction basis.

Table 3.28 collects a number of useful empirical formulas, which serve practical purposes connected with calculations on the aquatic environment. Examples include algal blooms in freshwater and organic material in sewage (gray waste water).

Diagenesis of biota

The natural diagenic series for the processing of organic material in the crust concentrates carbon. The series begins with peat and brown lignitic coal containing about 30 % C, then sub-bituminous and bituminous coal with between 40 % and 60 % C,[a] and finally anthracite with approximately 80 % to 85 % C (mole fraction). Diagenesis of organic material may also result in petroleum deposits and kerogen (see Table 8.9). Three conditions must be present for oil reservoirs to form: first, a source rock rich in

---

[a] The fraction of coal that is soluble in carbon disulfide is called bituminous coal; see Table 8.9.

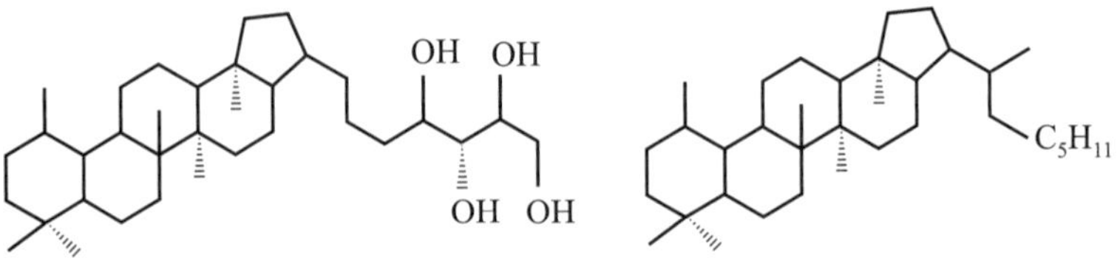

Figure 3.6   *Hopane derivatives*
Left: hopanetetrol from a prokaryotic cell. Right: hopane from bituminous coal.

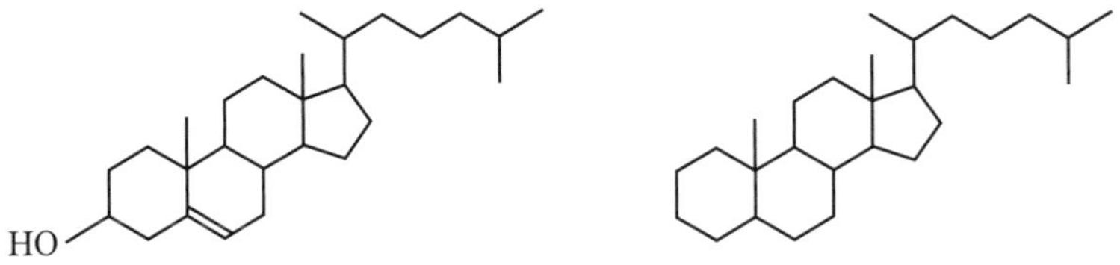

Figure 3.7   *Gonane derivatives*
Left: cholesterol (a sterol) from a eukaryote. Right: sterane from bituminous coal.

organic material buried deep enough for subterranean heat and pressure to change it into oil; second, a porous and permeable reservoir rock in which the oil can accumulate; and finally, a cap rock that prevents it from escaping to the surface. The chemical sequence is that proteins are broken down first, followed by saccharides and finally lignin. Beyond bituminous coal, the carbon content increases along with aromaticity.

There is a great deal of interest in the origin of coal and oil, for obvious environmental, economic, and industrial reasons. One of the most useful instruments in this respect is a gas chromatograph connected to a mass spectrometer.[a] The distinguishing characteristics include the distribution of stable isotopes, the occurrence of aromatic versus aliphatic hydrocarbons, and the observation of biomarkers. Two examples of these are shown in Figures 3.6 and 3.7. There are characteristic chemical structures called biomarkers in the molecules that make cell walls stiff, and these molecules differ between prokaryotes and eukaryotes. Whereas sterols (Figure 3.7), occur in the cell membranes of all eukaryotes, they are not found in prokaryotes, which use hopane derivatives (Figure 3.6). Both hopane and sterane are found in coal. The presence of enormous quantities of statistically improbable plant-based molecules in petroleum and coal is clear evidence of their biological (as opposed to abiological) origin. In addition to the presence of biomarkers, petroleum is known to be optically active; chiral centers produced in photosynthesis are preserved during diagenesis. Hopane from aerobic microorganisms contains tetrol, $-C_8H_{11}(OH)_4$. In hopane from anaerobic microorganisms, one finds the tetrol part replaced by isopropenyl, $CH_2=CH(CH_3)-$, which this cell type can synthesize without oxidative reaction steps. Overall, GCMS analysis shows that coal is primarily derived from land plants, whereas petroleum is derived from algae and zooplankton that accumulated on the sea bottom.

---

[a] The technique of GCMS is described in depth in refs. 29 and 88.

b. The cell

Structure

A cell is an open chemical system with an outer membrane that separates it from the surroundings. Living cells have metabolism, they can reproduce and grow, they can send and receive chemical signals, and, through time, they evolve. All cells contain proteins, lipids, polysaccharides, nucleic acids, and water, and these molecules are found in two locations with distinct roles. The first, the cytoplasm, contains the collections of molecules concerned with growth and function, whereas the second, the nuclear material, including strands of DNA, contains information for syntheses and for making new cells.

Prokaryotes[a] have no cell nucleus, that is, they lack a membrane to enclose the nuclear material. In eukaryotes,[b] DNA is found in a region enclosed by a membrane. Eukaryotic cells also contain specialized subunits (organelles) that serve specific purposes: mitochondria, which generate energy in the form of ATP; the endoplasmic reticulum, which is important in synthesizing lipids and proteins; the Golgi apparatus, which processes proteins; and vacuoles, which, in plant cells, hold sap. Both prokaryotes and eukaryotes contain ribosomes, and one of biology's recent discoveries is that the nucleotides in ribosomal RNA can be used to measure evolutionary relation. All organisms that are visible to the naked human eye are multicellular eukaryotes. In addition, eukaryotes include microorganisms (algae, fungi, and protozoa).

From the point of view of evolution, living organisms may be divided into three domains: bacteria, archaea, and eukarya (Woese, 1976). Bacteria and archaea are unicellular prokaryotic microorganisms. They are abundant and ubiquitous, and they have had an immense influence on the development of the surface of the Earth. The eukaria are divided into protists that are unicellular or have only simple multicellular groupings without specialized tissues, and multicellular fungi, plantae, and animalia. We will not discuss the "internal" chemistry of bioorganisms in depth,[54,18,61] but we will consider many aspects of the way they interact with their environments.

Photosynthesis and respiration

The essence of photosynthesis and respiration is the stoichiometric relation

$$CO_2(aq) + H_2O(l) = CH_2O(biota) + O_2(g) \qquad (3.19)$$

The forward reaction, the reduction of C^{IV} to C^0, is photosynthesis driven by energy from light. It requires a considerable amount of energy: If CH_2O is taken to be $(1/6)$ mol α-D-glucose, $\frac{1}{6} C_6H_{12}O_6$, then $\Delta_r G'^\circ = +478.7 \text{ kJ mol}^{-1}$. The process occurs in phototrophic cells of green plants, algae, and green and purple bacteria and converts carbon dioxide into cell material.[18] The reverse reaction is respiration, the process whereby cells oxidize organic matter to carbon dioxide.

---

[a] Gk. $\pi\rho o \;\hat{=}\;$ before, $\kappa\alpha\rho\upsilon o\nu \;\hat{=}\;$ nut.
[b] Gk. $\varepsilon\upsilon- \;\hat{=}\;$ good, true.

Photosynthesis is the series of enzymatic reactions by which light energy is stored in the phosphate bond energy of adenosine triphosphate (ATP) and, as reducing power, in the form of nicotinamide dinucleotide phosphate (NADPH). In the same processes, water is oxidized, releasing dioxygen.[54,61] In a parallel series of reactions, CO_2 is captured from the atmosphere and reduced to form carbohydrates, the major end products being starch and cellulose. Free glucose is not present in most higher plant tissues, but it is formed in the primary processes. There are two major reaction paths by which CO_2 enters a cell, the C_3 mechanism where primarily trioses are formed, and the C_4 mechanism where tetroses are the first species to be formed. Correspondingly, there are two large classes of photosynthetic plants: C_3 plants, mostly found in temperate climates, and C_4 plants, found in tropical and subtropical regions. The first series of reactions in C_3 plants is

$$6\,CO_2 \ + \ 18\,ATP \ + \ 12\,NADPH \ + \ 12\,H^+ \ + \ 12\,H_2O$$
$$\rightarrow \ C_6H_{12}O_6 \ + \ 18\,ADP \ + \ 18\,P_i \ + \ 12\,NADP^+ \qquad (3.20)$$

where ADP is adenosine diphosphate and P_i dihydrogen phosphate. The equation describes the so-called Calvin cycle, which comprises 14 reactions. The corresponding process in C_4 plants is less effective, the total cycle being

$$6\,CO_2 \ + \ 30\,ATP \ + \ 12\,NADPH \ + \ 12\,H^+ \ + \ 24\,H_2O$$
$$\rightarrow \ C_6H_{12}O_6 \ + \ 30\,ADP \ + \ 30\,P_i \ + \ 12\,NADP^+ \qquad (3.21)$$

but this is counterbalanced by greater efficiency at higher light intensities. The C_4 mechanism is found in both monocots and dicots,[a] and C_4 plants include more than 100 species, among them sugar cane, maize, and several desert plants.

The more detailed molecular explanation is as follows. The first step in the C_3 mechanism is the capture of CO_2 to form a derivative of glyceric acid, $CH_2OH-CHOH-COOH$. This reaction is catalyzed by the protein RuBisCO,[b] which is very abundant in green plants. However, RuBisCO has a low affinity for CO_2, and it may work as an oxygenase instead, that is, oxidizing the substrate using atmospheric O_2. This problem is aggravated by heat and drought. Nature has solved this by developing particular cells that efficiently capture CO_2 to form a derivative of oxaloacetic acid, $HOOC-CO-CH_2-COOH$. This C_4 species is then decarboxylated to enhance CO_2 concentrations in the vicinity of RuBisCO. Besides this additional mechanism and morphological changes, the biochemistry of C_4 plants is largely the same as for C_3 plants.

Hexoses and related carbohydrates store energy, which is then used by the plants themselves and by other life forms. To illustrate, consider an adult human with a mass of 70 kg. Over the course of an average day, this person processes 2.5 kg of water. About 300 g of the total is provided by respiration, approximated by oxidation of α-D-glucose, the reverse of reaction 3.19, and resulting in the release of 7.9 MJ of energy. Taken all at once, this would increase the temperature of the body by about

---

[a] Flowering plants, *angiosperms*, are divided into monocotyledons, "monocots," with one cotyledon or seed-eaf, and dicotlydons, "dicots," with two. Monocots diverged from the dicots about 0.1 Ga BP and include grasses, orchids, and onions. Most agricultural biomass is made by monocots, including the grasses rice, wheat, maize, and sugarcane.

[b] The abbreviation is derived from the name ribulose-1,5-biphosphate carboxylase/oxygenase.

Table 3.29 Summary of aerobic and anaerobic catabolic respiration mechanisms

a. Aerobic respiration; important in all oxygenated environments
$$CH_2O(aq) + O_2(g) \rightarrow CO_2(g) + H_2O(l)$$
$A^{\circ\prime} = 478.3\,kJ/mol$

b. Denitrification; important in soils and the ocean
$$CH_2O(aq) + \tfrac{4}{5}NO_3^-(aq) \rightarrow \tfrac{1}{5}CO_2(g) + \tfrac{4}{5}HCO_3^- + \tfrac{2}{5}N_2(g) + \tfrac{3}{5}H_2O(l)$$
$A^{\circ\prime} = 447.9\,kJ/mol$

c. Manganese reduction; occurs in some marine sediments
$$CH_2O(aq) + 2\,MnO_2(s) + 3\,H^+ \rightarrow HCO_3^- + 2\,Mn^{2+} + 2\,H_2O(l)$$
$A^{\circ\prime} = 319.9\,kJ/mol$

d. Iron reduction; soils and marine sediments
$$CH_2O(aq) + 4\,Fe(OH)_3(s) + 7\,H^+ \rightarrow HCO_3^- + 4\,Fe^{2+} + 4\,H_2O(l)$$
$A^{\circ\prime} = 114.0\,kJ/mol$

e. Sulfate reduction; anaerobic marine sediments
$$CH_2O(aq) + \tfrac{1}{2}SO_4^{2-}(aq) \rightarrow HCO_3^- + \tfrac{1}{2}H_2S(aq)$$
$A^{\circ\prime} = 79.7\,kJ/mol$

f. Methanogenesis; waterlogged soils, freshwater wetlands, low-sulfate marine sediment
$$CH_2O(aq) + \tfrac{1}{2}H_2O(aq) \rightarrow \tfrac{1}{2}HCO_3^- + \tfrac{1}{2}CH_4(g) + \tfrac{1}{2}H^+$$
$A^{\circ\prime} = 65.6\,kJ/mol$

Standard affinity $A^{\circ\prime}$ at pH 7. CH_2O is taken to be α-D-glucose.

27 K, but – as we know – body temperature is regulated and the gradual release of energy yields slightly more than 90 W. As shown in a detailed view, the process is mediated by adenosine triphosphate, ATP, with the net reaction

$$C_6H_{12}O_6 + 6\,O_2 + 38\,ADP + 38\,P_i \rightarrow 6\,CO_2 + 6\,H_2O + 38\,ATP \quad (3.22)$$

ATP is used incrementally, as needed. The oxidation and release of ATP occurs in 14 steps, each catalyzed by a specific enzyme. The main point here is to illustrate the flux of substance and energy through the open system.

Metabolism

The chemical transformations occurring in a cell are known collectively as metabolism. Metabolism is divided into the *anabolic* processes that synthesize molecules and eventually new cells, and the *catabolic* processes that break them apart. In catabolic pathways, nutrients such as saccharides, lipids, and proteins are converted into simpler molecules such as lactic acid, acetic acid, carbon dioxide, ammonia, and urea. Anabolic pathways form ATP via the phosphorylation of ADP. ATP is responsible for transporting energy within cells; ADP is its low-energy reaction partner. In anabolic processes, larger molecules, such as those mentioned earlier and nucleotides, are built from smaller ones.[54]

The majority of organisms obtain energy from respiration. In respiration, organic compounds are oxidized using an inorganic oxidant. A common oxidant is molecular oxygen ($O^0 \rightarrow O^{-II}$), in which case the process is called aerobic respiration (see Equation 3.22). Other oxidants are used in so-called anaerobic oxidation, for example, $N^V \rightarrow N^0$, $Mn^{IV} \rightarrow Mn^{II}$, $Fe^{III} \rightarrow Fe^{II}$, and $S^{VI} \rightarrow S^{-II}$, as shown in Table 3.29.

Table 3.30 Chemical classification of cells		
	Distinctive property	Cell type
Carbon source	Carbon dioxide	Autotrophic
	Other	Heterotrophic
Energy source	Light	Phototrophic
	Chemical bonds	Chemotrophic
Reduction substrate	Electron donor: inorganic	Lithotrophic
	Electron donor: organic	Organotrophic
Respiration	Electron acceptor: oxygen	Aerobic
	Other inorganic species	Anaerobic
Fermentation	Disproportionation	Anaerobic

The chemical state determines what kind of microbiological life can survive in a given environment. Groundwater pollution, for example, can drastically alter the redox potential, which may be high on the edge of a pollution fan and low in the center.[a] The distribution of bacteria changes in step with the changes in the chemical environment. Important regions of the biosphere are anoxic: life is present, but oxygen is not.

Some microorganisms obtain their energy through anaerobic fermentation (i.e., they disproportionate organic material). These organisms form ATP through a unique mechanism, for example:

$$C_6H_{12}O_6 + 2\,ADP + 2\,P_i \;\rightarrow\; 2\,C_2H_5OH + 2\,CO_2 + 2\,ATP \quad (3.23)$$
(alcohol fermentation)

$$C_6H_{12}O_6 + 2\,ADP + 2\,P_i \;\rightarrow\; 2\,CH_3 \cdot CHOH \cdot CO_2^- + 2\,H^+ + 2\,ATP$$
(glycolysis) $\qquad\qquad\qquad\qquad\qquad\qquad\qquad\qquad\qquad\qquad (3.24)$

As a third example, the last disproportionation reaction in Table 3.29 occurs only in prokaryotic microorganisms.

A distinction can also be made between assimilating and dissimilating processes. An assimilating process is a reaction in which a microorganism reduces an inorganic species, such as sulfate, nitrate, or carbon dioxide, with the condition that the products are used in the organism's own biosynthesis, for example, nitrate $\rightarrow$ amine. Many organisms display assimilative metabolism; examples include archaea, bacteria, algae, fungi, and plants. Dissimilative processes, such as those found in prokaryotic cells, may produce products in excess, which are given off to the environment, such as nitrate $\rightarrow$ nitrogen(0).

Table 3.30 summarizes the classical microbiological classification of cells. The first section of the table describes the resources necessary for the cell to live, and the second its method of catabolism: respiration or fermentation. The carbon source for

[a] See discussion of E-pH diagrams in Section 5.3a.

an autotrophic cell is atmospheric CO_2 (see Equations 7.32 and 7.33). A heterotrophic cell obtains carbon from organic molecules. Heterotrophic cells are further characterized by their method of respiration. Some cells are phototrophic, meaning that they can convert solar into chemical energy. Others are chemotrophs, meaning that their metabolism depends on obtaining energy by converting chemical substrates generated by phototrophic cells. The third characteristic parameter is the reductant used in a given cell's metabolism. Chemotrophic cells need organic molecules as reduction substrates, whereas lithotrophic cells use inorganic electron donors, such as hydrogen, ammonia, hydrogen sulfide, or sulfur.

Examples

1. Photolithotrophic organisms are also autotrophic. They receive carbon from CO_2 and energy from light, and they use H_2O, H_2S, or $S(0)$ as electron donors. Examples include photosynthetic plant cells, most algaes, and photosynthetic bacteria.

2. Photoorganotrophic organisms obtain their carbon from organic compounds and their energy from light, and they use organic compounds as electron donors. Examples include purple nonsulfur bacteria.

3. Chemolithotrophic organisms are also autotrophic. They obtain their carbon from CO_2 and energy from redox reactions (often with oxygen as the electron acceptor), and they use H_2, H_2S, $S(0)$, $Fe(II)$, or NH_3 as electron donors. Examples include hydrogen-, sulfur-, iron-, and nitrogen-fixing bacteria (see Chapter 7).

4. Chemo-organotrophic organisms obtain carbon from organic compounds and energy from organic redox reactions, using glucose as an electron donor. Examples include cells from advanced animals, the nonphotosynthetic cells of plants, the dark reactions occurring in photosynthetic cells, and the majority of microorganisms.

5. The process (Equation 3.19) is driven by solar energy and carried out by phototrophic cells. The reciprocal process is the respiration occurring in organotrophic, heterotrophic cells.

4

Chemistry of the atmosphere

The atmosphere interacts directly with the lithosphere, hydrosphere, biota, and society. Noble gases given off by radioactive decay in the core and crust have accumulated since the planet formed, with the exception of He, which has escaped to space. The exchange of water between the atmosphere and the oceans and the land is a key component of weather and the planetary heat engine (see Figure 3.3). The biosphere has dramatically changed the atmosphere's composition, including oxygen and the majority of trace gases. Human activities, such as agriculture, industry, and transportation, have left a clear mark on the composition and chemistry of the atmosphere. A summary of the gases found in the atmosphere is given in Section 3.3.

The state of the atmosphere affects us all. Crop yields depend on climate, rain, and nutrients present in rain, and on trace gases such as ozone and carbon dioxide. There are also direct effects: WHO[a] estimates that between 2.5 and 11% of human deaths are due to exposure to air pollution. The thickness of the ozone layer at midlatitudes has decreased dramatically over the past generation, not to mention the Antarctic ozone hole. Climate change has important implications for the global ecosystem and public.

The temperature and pressure profiles of the atmosphere have been described in Chapters 2 and 3. In this chapter we discuss the chemistry of the atmosphere, focusing on the troposphere and the stratosphere because these are the most important regions in terms of environmental impact (see Figure 3.4). The formalisms of chemical kinetics were described in Section 2.4 and are used throughout the following discussion.

Atmospheric chemistry largely consists of reactions between molecules and radicals,[b] and the main types of reactions are oxidation and photolysis. Because the oxidation chemistry is driven by radicals generated by photolysis, we begin with a discussion of a general photolytic reaction:

$$A + h\nu \quad \rightarrow \quad X^{\bullet} + Y^{\bullet} \tag{4.1}$$

where $h\nu$ denotes solar energy. A closed-shell molecule A is broken into the radical fragments $X^{\bullet}$ and $Y^{\bullet}$. The concentrations of radicals are typically many orders of magnitude less than those of closed-shell molecules, and so the most common reaction involving a radical is with a closed-shell molecule,

$$A^{\bullet} + B \quad \rightarrow \quad X^{\bullet} + Y \tag{4.2}$$

---

[a] The World Health Organization is a specialized agency of the United Nations. See http://www.who.int/en.
[b] A radical is a molecular entity with an odd number of electrons (not including paramagnetic coordination compounds). In this chapter, radicals (including chlorine atoms) will be marked with the superscript $^{\bullet}$; the only exceptions are the two stable radicals NO and NO_2, collectively referred to as NO_x.

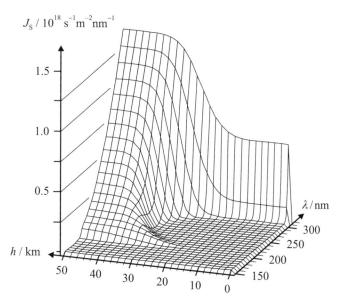

Figure 4.1 *Actinic flux density as a function of altitude and wavelength*
Solar photon flux density J_S as a function of wavelength λ and altitude h for an overhead Sun. The calculations are based on the absorption spectra of oxygen and ozone and the distribution as a function of altitude.

Here the radical A• reacts with closed-shell molecule B to generate a new radical X• and closed-shell molecule Y. Because radicals have an odd number of electrons and closed-shell molecules have an even number, a product with an odd number of electrons must be produced. Further, in a bimolecular reaction such as this, bonds are exchanged.

Radicals persist until they are removed, for example,

$$A^{\bullet} + B^{\bullet} + M \;\rightarrow\; X + M \tag{4.3}$$

Here two radicals join together to form the stable molecule X, with the excess energy being removed by collision with molecule M, typically N_2 or O_2. A radical generated in Equation 4.1 may lead to millions of reactions before being removed by processes such as Equation 4.3. Atmospheric chemistry is typified by different kinds of chain reactions involving initiation (Equation 4.1), propagation (Equation 4.2), and termination (Equation 4.3).

Sunlight starts the process in Equation 4.1 and is thus central to atmospheric chemistry. The quantity of light available to molecules at a particular point in the atmosphere that, on absorption, drives photochemical processes is called the actinic flux density.[a,b] The solar actinic flux density as a function of altitude and wavelength is shown in Figure 4.1. At the top of the atmosphere the spectral distribution of the incoming sunlight resembles that of a blackbody at 5780 K, the temperature of the Sun (see Figure 10.1). Approaching the surface, light is attenuated by molecular

[a] Gk. αχτινα ≙ ray.
[b] SI-unit: $W\,m^{-2}\,nm^{-1}$.
   Actinic flux density is often called "actinic flux."

absorptions and scattering by molecules and clouds. The most important absorbing species in the ultraviolet region are molecular oxygen ($\lambda < 190$ nm) and ozone (215 nm $< \lambda < 295$ nm; see Figure 4.6). Very little light at wavelengths shorter than 300 nm reaches the surface.

4.1 Tropospheric chemistry

a. The hydroxyl radical

Hydroxyl, $HO^\bullet$, is the most important radical in the troposphere in terms of its impact on the concentrations of trace gases.[a] Its reactivity is driven by its ability to abstract a hydrogen atom from virtually any reaction partner, resulting in the formation of a water molecule and a new radical. In addition, hydroxyl adds to unsaturated compounds and reacts with CO, giving CO_2. Formation of hydroxyl begins with the photolysis of ozone:

$$O_3 + h\nu \,(\lambda < 310\,\text{nm}) \quad \rightarrow \quad O_2(^1\Delta_g) + O(^1D) \tag{4.4}$$

Ozone is present in the troposphere due to transport from the stratosphere and in situ production, discussed later. Figure 4.1 shows that there is some actinic flux density at wavelengths shorter than 310 nm, even at the surface of the Earth.[b] The photolysis produces a dioxygen molecule and an oxygen atom in excited electronic states, characterized by the spectroscopic term symbols $^1\Delta_g$ and ^1D, respectively.[c]

The metastable oxygen atom is typically relaxed by collision with molecules in the atmosphere:

$$O(^1D) + M \quad \rightarrow \quad O(^3P) + M \tag{4.5}$$

Here, ^3P specifies the spectroscopic state of an oxygen atom in its electronic ground state. In a few percent of the collisions, $O(^1D)$ reacts with water to produce hydroxyl, $HO^\bullet$:

$$O(^1D) + H_2O \quad \rightarrow \quad 2HO^\bullet \tag{4.6}$$

Note that this reaction produces the radical hydroxyl, $HO^\bullet$, and *not* the anion hydroxide, HO^-. The lifetime of $O(^1D)$ is very short throughout the atmosphere. Therefore, its concentration is in a local steady state, relative to slowly changing factors such as pressure, transport, and actinic flux density. We will use the steady-state approximation on $O(^1D)$ to calculate the rate of production of hydroxyl in the atmosphere:[d]

$$
\begin{aligned}
d[O(^1D)]/dt &= r_{4.4} - r_{4.5} - r_{4.6} \\
&= j_{4.4}[O_3] - k_{4.5}[O(^1D)]\,[M] - k_{4.6}[O(^1D)][H_2O] \approx 0 \quad (4.7)
\end{aligned}
$$

---

[a] Alternative names:[124] oxidanyl or hydridooxygen($\cdot$).

[b] The threshold energy ($\lambda \approx 310$ nm) corresponds to 386 kJ/mol.

[c] Only the term symbols of the oxygen atom will be used in the present text, and only as a label to mark that approximately 190 kJ/mol is released by the process (see Equation 4.5).

[d] The variable j is used to indicate a photolysis rate constant and k an ordinary rate constant; subindices point to the appropriate reaction equation; see footnote b in Chapter 2, p. 42. Choice of units: Unit($j_{4.4}$) = s^{-1}; Unit($k_{4.6}$) = Unit($k_{4.5}$) = cm^{-3} s^{-1}, using number concentration; see Table A1.7.

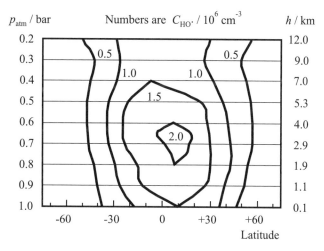

Figure 4.2 *Tropospheric distribution of* HO•
Concentration isobars indicated using number concentrations in units of 10^6 cm^{-3}.
Latitude given from $-75°$ to $+75°$. Redrawn from ref. 213.

Equation 4.7 is an application of the mass balance equation: Change in concentration
depends on rate of production minus rate of loss. The steady-state approximation
allows us to set d[O(^{1}D)]/d$t = 0$, which in turn allows us to solve for [O(^{1}D)]:

$$[O(^1D)] \;=\; j_{4.4}[O_3]/(k_{4.5}[M] \,+\, k_{4.6}[H_2O]) \tag{4.8}$$

The rate of production of HO• is given by (use Equation 2.82)

$$r_{HO•} \;=\; 2k_{4.6}[O(^1D)][H_2O] \tag{4.9}$$

Substituting Equation 4.8 into Equation 4.9,

$$r_{HO•} \;=\; 2k_{4.6}j_{4.4}[O_3][H_2O]/(k_{4.5}[M] \,+\, k_{4.6}[H_2O])$$
$$\approx\; 2k_{4.6}j_{4.4}[O_3][H_2O]/k_{4.5}[M] \tag{4.10}$$

The expression was simplified because $k_{4.5}$ [M] $\gg k_{4.6}$ [H$_2$O].[a] Equation 4.10 says
that the production of hydroxyl is promoted by sunlight (through $j_{4.4}$), ozone, water,
and low pressure. One might expect the concentration of hydroxyl to increase with
altitude because the actinic flux density will increase (see Figure 4.1) and the pressure
decrease. However, water vapor concentrations fall sharply with altitude because of
decreasing temperature. Model results for atmospheric hydroxyl concentrations are
shown in Figure 4.2. The maximum is seen near the equator in the middle of the
troposphere, reflecting a balance between the factors that have been mentioned.
The maximum is shifted toward the Northern Hemisphere because of anthropogenic
pollution (NO$_x$ and hydrocarbons) described later. The global mean tropospheric
hydroxyl concentration is estimated to be 1.2×10^6 cm^{-3}.

---

[a] Although $k_{4.5} \sim k_{4.6}$, [M] $>>$ [H$_2$O] throughout the atmosphere.

Table 4.1 Global budget for atmospheric carbon monoxide[a]		
Sources/Tmol a^{-1}		
Methane	29	
Isoprene	10	
Industrial NMHC	3.9	
Biomass NMHC	1.1	
Acetone	0.7	
Subtotal, in situ oxidation		44
Vegetation	5.4	
Oceans	1.8	
Biomass burning	25	
Fossil and domestic fuel	23	
Subtotal, direct emissions		55
Total sources		99
Sinks/Tmol a^{-1}		
Surface deposition	6.8	
Hydroxyl reaction	68	
Total sinks		75

NMHC is an acronym for nonmethane hydrocarbons (see Section 4.1), which in this case excludes isoprene. There are significant uncertainties, especially in the source terms.

Oxidation of carbon monoxide

The most common reaction partner for hydroxyl, HO$^{\bullet}$, is carbon monoxide, CO. CO is produced by the atmospheric oxidation of virtually all hydrocarbons, including methane and isoprene, as shown in Table 4.1. The following reaction removes most CO from the atmosphere:

$$CO + HO^{\bullet} \rightarrow CO_2 + H^{\bullet} \tag{4.11}$$

This reaction consumes about two-thirds of hydroxyl on a global basis and is the source of one-sixth of the CO_2 in the atmosphere; it could therefore be considered to be the single most important reaction in the atmosphere. Hydrogen very quickly reacts with dioxygen:

$$H^{\bullet} + O_2 + M \rightarrow HO_2^{\bullet} + M \tag{4.12}$$

In a polluted environment, dioxidanyl, HO$_2^{\bullet}$,[b] will react with nitrogen oxide:

$$HO_2^{\bullet} + NO \rightarrow HO^{\bullet} + NO_2 \tag{4.13}$$

This reaction regenerates hydroxyl (oxidanyl). In addition, it produces the red-brown gas nitrogen dioxide, NO_2. This is easily photolyzed in sunlight:

$$NO_2 + h\nu\,(\lambda < 415\,\mathrm{nm}) \rightarrow NO + O \tag{4.14}$$

---

[a] Based on IPCC 2001.[230]
[b] Alternative name:[124] hydridodioxygen($\cdot$); the name perhydroxyl is obsolete.[185]

followed by

$$O + O_2 + M \rightarrow O_3 + M \tag{4.15}$$

The sequence Equations 4.11–4.15 is a good example of an atmospheric chain reaction, and the net reaction (sum of Equations 4.11–4.15) is

$$CO + 2O_2 + h\nu \rightarrow CO_2 + O_3 \tag{4.16}$$

The presence of nitrogen oxides catalyzes the production of ozone during the oxidation of carbon monoxide. Nitrogen oxides are produced from atmospheric N_2 and O_2 at high temperatures, for example, in lightning or when fossil fuels are burned (see Equation 8.13 and Figure 8.5). In remote environments where sufficient concentrations of NO are not available, dioxidanyl reacts with itself:

$$2 HO_2^{\bullet} \rightarrow H_2O_2 + O_2 \tag{4.17}$$

Hydrogen peroxide is removed from the atmosphere via photolysis, producing hydroxyl,

$$H_2O_2 + h\nu \rightarrow 2 HO^{\bullet} \tag{4.18}$$

or via uptake into aerosols where it may be involved in other reactions, for example, the oxidation of $SO_2(aq)$ to H^+ (aq) and HSO_4^-(aq) (see Equation 4.51).

Oxidation of methane

Methane is the most abundant hydrocarbon in the atmosphere. It is an important greenhouse gas and is a key precursor for CO. Its global budget is shown in Table 4.2. Methane reacts with hydroxyl, first leading to methyl:

$$CH_4 + HO^{\bullet} \rightarrow CH_3^{\bullet} + H_2O \tag{4.19}$$

which reacts with oxygen to give methperoxyl:

$$CH_3^{\bullet} + O_2 + M \rightarrow CH_3O_2^{\bullet} + M \tag{4.20}$$

In high-NO_x conditions, methoxyl is first produced, which then reacts with oxygen to give formaldehyde and dioxidanyl:

$$CH_3O_2^{\bullet} + NO \rightarrow CH_3O^{\bullet} + NO_2 \tag{4.21}$$

$$CH_3O^{\bullet} + O_2 \rightarrow HCHO + HO_2^{\bullet} \tag{4.22}$$

The formaldehyde produced forms CO via a number of processes, the most important being[a]

$$HCHO + h\nu \rightarrow H^{\bullet} + HCO^{\bullet} \tag{4.23}$$

$$HCHO + h\nu \rightarrow H_2 + CO \tag{4.24}$$

$$HCHO + HO^{\bullet} \rightarrow H_2O + HCO^{\bullet} \tag{4.25}$$

$$HCO^{\bullet} + O_2 \rightarrow CO + HO_2^{\bullet} \tag{4.26}$$

---

[a] The name of the radical $HCO^{\bullet}$ is formyl.

Table 4.2 Global budget of atmospheric methane[a]

Sources/Tmol a^{-1}		
Wetlands	9.1	
Termites	1.4	
Subtotal, natural sources		10.5
Coal mining	3.0	
Gas, oil, industry	2.3	
Ruminants (cows, sheep . . .)	11.8	
Rice agriculture	7.0	
Biomass burning	2.7	
Subtotal, anthropogenic sources		26.8
Total sources		37.3
Sinks/Tmol a^{-1}		
Soil	1.9	
Troposphere, OH$^{\bullet}$	31.9	
Stratosphere, O(^{1}D), Cl$^{\bullet}$, HO$^{\bullet}$	2.5	
Total sinks		36.3

Reaction 4.24 is the source of over half of atmospheric hydrogen. In low-NO$_x$ conditions there is not enough NO available for reaction 4.21 to be dominant. In this case, peroxy radicals are removed by reactions such as

$$CH_3O_2^{\bullet} + HO_2^{\bullet} \rightarrow CH_3OOH + O_2 \tag{4.27}$$

The hydroperoxymethane[b] formed initially reacts further:

$$CH_3OOH + h\nu \rightarrow CH_3O^{\bullet} + HO^{\bullet} \tag{4.28}$$

$$CH_3OOH + HO^{\bullet} \rightarrow HCHO + HO^{\bullet} + H_2O \tag{4.29}$$

We can now write the overall process (see Equation 4.30, later) for methane oxidation in high-NO$_x$ conditions:

$$CH_4 + HO^{\bullet} \rightarrow CH_3^{\bullet} + H_2O \tag{4.19}$$

$$CH_3^{\bullet} + O_2 + M \rightarrow CH_3O_2^{\bullet} + M \tag{4.20}$$

$$CH_3O_2^{\bullet} + NO \rightarrow CH_3O^{\bullet} + NO_2 \tag{4.21}$$

$$CH_3O^{\bullet} + O_2 \rightarrow HCHO + HO_2^{\bullet} \tag{4.22}$$

(For illustration we assume that formaldehyde only reacts with hydroxyl.)

$$HCHO + HO^{\bullet} \rightarrow H_2O + HCO^{\bullet} \tag{4.25}$$

$$HCO^{\bullet} + O_2 \rightarrow CO + HO_2^{\bullet} \tag{4.26}$$

$$CO + HO^{\bullet} \rightarrow CO_2 + H^{\bullet} \tag{4.11}$$

$$H^{\bullet} + O_2 + M \rightarrow HO_2^{\bullet} + M \tag{4.12}$$

[a] Sources based on ref. 151, sinks on ref. 231.
[b] An obsolete name is methyl hydroperoxide.

$$HO_2^{\bullet} + NO \rightarrow HO^{\bullet} + NO_2 \quad \text{(three times)} \tag{4.13}$$

$$NO_2 + h\nu \,(\lambda < 415\,\text{nm}) \rightarrow NO + O \quad \text{(four times)} \tag{4.14}$$

$$O + O_2 + M \rightarrow O_3 + M \quad \text{(four times)} \tag{4.15}$$

Sum:

$$CH_4 + 8O_2 + h\nu = CO_2 + 2H_2O + 4O_3 \tag{4.30}$$

Next, consider the oxidation in low-NO_x conditions resulting in Equation 4.31:

$$CH_4 + HO^{\bullet} \rightarrow CH_3^{\bullet} + H_2O \tag{4.19}$$

$$CH_3^{\bullet} + O_2 + M \rightarrow CH_3O_2^{\bullet} + M \tag{4.20}$$

$$CH_3O_2^{\bullet} + HO_2^{\bullet} \rightarrow CH_3OOH + O_2 \tag{4.27}$$

$$CH_3OOH + HO^{\bullet} \rightarrow HCHO + HO^{\bullet} + H_2O \tag{4.29}$$

$$HCHO + OH^{\bullet} \rightarrow H_2O + HCO^{\bullet} \tag{4.25}$$

$$HCO^{\bullet} + O_2 \rightarrow CO + HO_2^{\bullet} \tag{4.26}$$

$$CO + HO^{\bullet} \rightarrow CO_2 + H^{\bullet} \tag{4.11}$$

$$H^{\bullet} + O_2 + M \rightarrow HO_2^{\bullet} + M \tag{4.12}$$

$$HO_2^{\bullet} \rightarrow \tfrac{1}{2}H_2O_2 + \tfrac{1}{2}O_2 \tag{4.17}$$

Sum:

$$CH_4 + 3\,OH^{\bullet} + (3/2)O_2 = 3\,H_2O + CO_2 + \tfrac{1}{2}H_2O_2 \tag{4.31}$$

Although an accurate atmospheric model must include hundreds of reactions, this scheme captures some of the basic elements of atmospheric reactivity. There is a clear difference between the two mechanisms, Equations 4.30 and 4.31. The first uses light and molecular oxygen to produce CO_2, water, and ozone. The ozone could subsequently be photolyzed to produce hydroxyl as shown in the reactions 4.4–4.6. NO_x acts as a catalyst and serves to increase atmospheric reactivity, removing hydrocarbon air pollution, at the same time as pollution is generated in the form of ozone. This is in contrast to the low-NO_x mechanism that consumes hydroxyl, reducing atmospheric reactivity.

Tropospheric chain reactions: HO_x and NO_x

As stated in the introduction, tropospheric chemistry can be understood in terms of chain reactions of radicals. The reaction cycles consume reduced trace gases such as hydrocarbons and produce a variety of oxidation products, including ozone and particles, together called smog. The key to understanding atmospheric reactivity lies in the families of radicals, the two most important being HO_x and NO_x. The name HO_x is used for $HO^{\bullet}$, $HO_2^{\bullet}$, and organic oxy and peroxy radicals such as $CH_3O^{\bullet}$ and $CH_3O_2^{\bullet}$. The two stable radicals NO and NO_2 are denoted NO_x. The radicals in each family are closely linked by chemical interconversion, which is why it makes sense to use a special name for each group.

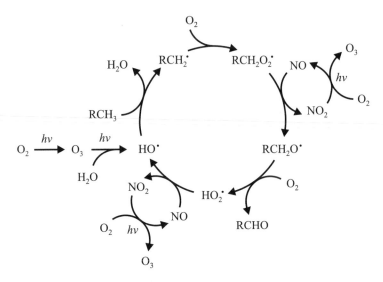

Figure 4.3 *Reaction cycles in the troposphere involving* HO_x *and* NO_x
The middle cycle includes all fast radical reactions. The overall reaction is the
oxidation of hydrocarbons (Equation 4.38). The two termination reactions are
Equations 4.17 and 4.39.

An overview of central features of tropospheric photochemistry is shown in
Figure 4.3. The process is initiated by the formation of HO_x in the form of hydroxyl.
The overall process that produces $HO^\bullet$ is the photolysis of oxygen (via ozone) to
react with water (see reactions 4.4–4.6):

$$\tfrac{1}{2}O_2 + H_2O + h\nu \;\rightarrow\; 2\,HO^\bullet \tag{4.32}$$

Hydroxyl can react by abstracting a hydrogen atom from a hydrocarbon:

$$HO^\bullet + RCH_3 \;\rightarrow\; H_2O + RCH_2^\bullet \tag{4.33}$$

$$RCH_2^\bullet + O_2 + M \;\rightarrow\; RCH_2O_2^\bullet + M \tag{4.34}$$

$$RCH_2O_2^\bullet + NO \;\rightarrow\; RCH_2O^\bullet + NO_2 \tag{4.35}$$

At this point we have generated NO_2, which will be photolyzed by sunlight (reac-
tion 4.14), reforming NO and leading to the formation of an ozone molecule by
Equation 4.15.
RCH$_2$O$^\bullet$ is oxidized to an aldehyde,

$$RCH_2O^\bullet + O_2 \;\rightarrow\; RCHO + HO_2^\bullet \tag{4.36}$$

and this reaction is followed by the reaction (Equation 4.13)

$$HO_2^\bullet + NO \;\rightarrow\; HO^\bullet + NO_2 \tag{4.37}$$

forming a second NO_2 that will also be photolyzed, producing ozone. Finally, the
cycle closes by reforming hydroxyl, which is now available to react with another
hydrocarbon. The HO_x radicals act as catalysts, promoting the oxidation of hydrocarbons; in this example the hydrocarbon is converted into an aldehyde. (The aldehyde
could also participate in the reaction cycle by Equation 4.33.) The oxidation scheme

illustrated in Figure 4.3 converts volatile hydrocarbons into semivolatile oxygenated hydrocarbons that have a greater potential for condensing to build smog particles:

$$RCH_3 + \tfrac{1}{2}O_2 \quad \rightarrow \quad RCHO + H_2O \tag{4.38}$$

Every time the HO_x cycle goes around once, two ozone molecules are formed. Overall the scheme is catalyzed by both HO_x and NO_x.

The efficiency of the cycle is limited by the availability of NO and by reactions removing HO_x radicals. If there is not sufficient NO, $HO_2{}^{\bullet}$ radicals will build up to such a concentration that they react with themselves:

$$2\,HO_2{}^{\bullet} \quad \rightarrow \quad H_2O_2 + O_2 \tag{4.17}$$

In the troposphere, the hydrogen peroxide will be photolyzed, be removed by reaction with $HO^{\bullet}$ or react in the aqueous phase with oxidizable species as SO_2, NO_x, or aldehydes, or be washed out. When NO_x is available, HO_x is removed via

$$HO^{\bullet} + NO_2 + M \quad \rightarrow \quad HNO_3 + M \tag{4.39}$$

The nitric acid will be deposited. Reaction 4.39 is responsible for about half of anthropogenic acid rain, the other half coming from sulfuric acid (see Equation 4.50).

Reaction 4.36 shows how aldehydes are formed in the troposphere. These compounds are involved in the distribution of NO_x over long distances. To illustrate, consider acetaldehyde, CH_3CHO:

$$CH_3CHO + HO^{\bullet} \quad \rightarrow \quad CH_3CO^{\bullet} + H_2O \tag{4.40}$$

$$CH_3CO^{\bullet} + O_2 \quad \rightarrow \quad CH_3C(O)OO^{\bullet} \tag{4.41}$$

$$CH_3C(O)OO^{\bullet} + NO_2 + M \quad \rightarrow \quad CH_3C(O)OONO_2 + M \tag{4.42}$$

Reaction 4.41 produces acetoperoxyl, and in reaction 4.42 this radical reacts with NO_2 to form nitric peroxymethanoic anhydride.[a] Larger analogues are also formed from other aldehyde precursors. The anhydride is susceptible to pyrolysis:

$$CH_3C(O)OONO_2 + Q \quad \rightarrow \quad CH_3C(O)OO^{\bullet} + NO_2 \tag{4.43}$$

At low temperatures in the mid to upper troposphere, it has a lifetime of weeks to months, whereas near the surface it decomposes within hours. Nitric peroxyethanoic anhydride forms from its precursors when the temperature drops in rising air columns (see the atmospheric lapse rate, Appendix 4). The molecule remains stable until the air warms on descent, releasing the NO_2. This method of transport has resulted in high-NO_x conditions over most of the Northern Hemisphere.

Ozone production

Tropospheric ozone damages plants and causes respiratory problems in humans. In the United States health alerts are issued if the average ozone concentration is greater than 84 ppb for an 8-hour period; this limit is often exceeded on sunny summer days in the populated areas on the East and West Coasts. Ozone pollution is a problem

---

[a] This species is an anhydride. However, it is often named (incorrectly) in the same way as an ester: peroxyacetyl nitrate. The acronym PAN, which derives from this name, is also used for polyacrylonitrile.

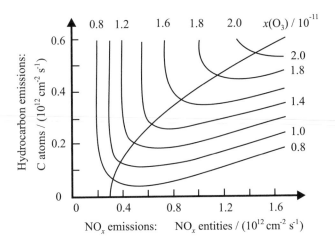

Figure 4.4 *Ozone production*
Ozone concentrations as a function of hydrocarbon and NO_x emissions. Upper left: NO_x is rate limiting; lower right: hydrocarbons are rate limiting. Based on modeling work of ref. 207.

in most populated regions of the planet. Poor air quality is linked to anthropogenic NO_x, biogenic and anthropogenic emissions of hydrocarbons, and meteorological conditions.

It is of considerable interest to determine the chemical factors limiting the production of tropospheric ozone. In particular, chain termination reactions will consume HO_x radicals and limit the efficiency of the process. As we have seen, there are two reactions that remove HO_x from the troposphere, reactions 4.17 and 4.39. If $r_{4.17} = 2k_{4.17}[HO_2]^2$ is larger than $r_{4.39} = k_{4.39}[OH^\bullet][NO_2][M]$, then the atmospheric conditions are said to be "low-NO_x." Under these conditions, ozone formation is limited by the availability of NO_x; there is not enough NO_x to convert $HO_2^\bullet$ into $HO^\bullet$. If the opposite is true, that is, $r_{4.39} > r_{4.17}$, the conditions are said to be "high-NO_x." Under these conditions, the formation of ozone is limited by the availability of hydrocarbons.

There are more reactions than those shown in Figure 4.4. An example is the oxidation of CO (reactions 4.11 and 4.12) that converts $HO^\bullet$ directly to $HO_2^\bullet$. Also, the NO/NO_2 equilibrium is affected by the presence of O_3 through

$$NO + O_3 \rightarrow NO_2 + O_2 \tag{4.44}$$

Ozone photolysis provides the seed for the HO_x cycle, but once it is running, the majority of hydroxyl is produced by reaction 4.13.

Figure 4.4 shows that the best strategy for controlling ozone air pollution in one region may be to focus on NO_x, whereas in another region decreasing NO_x may only make the problem worse. In the left portion of the figure, increasing hydrocarbon emissions have no effect on ozone levels, because ozone generation is limited by the rate at which hydroxyl is generated by reaction 4.13. Adding NO_x, however, would increase the rate of ozone production. In the lower right portion, adding NO_x would decrease ozone production. The reason for this seeming paradox is that NO_x removes

Table 4.3 Emissions of reduced species to the atmosphere	
Chemical species, B	$\dot{m}_B/\text{Tg a}^{-1}$
Hydrogen,[1] H_2	77
Ammonia,[2] NH_3	70
Methane,[3] CH_4	596
Isoprene,[4] C_5H_8	503
Terpenes,[4] $C_{10}H_{16}$	127
Other organic species[4,5]	260
Carbon monoxide[6]	2780

[1] Includes emission and in situ production.[244]

[2] Based on ref. 147.

[3] See Table 4.2

[4] Based on ref. 173.

[5] Assume average molecular formula of C_6H_6.

[6] Includes emission and in situ production. See Table 4.1.

HO_x radicals from the system via $HO^{\bullet} + NO_2$. In this regime, the best strategy is to focus on hydrocarbon emissions. In the 1970s, the importance of emissions of hydrocarbons by plants was not known. When these additional sources were considered, increased attention was given to limiting NO_x.

Concluding remarks

In this section we have focused mainly on the reactivity of hydroxyl, $HO^{\bullet}$, but many other radicals are active, often depending on specific circumstances. In the dark, NO_2 reacts with O_3 to generate $NO_3^{\bullet}$, the main nighttime oxidant, which reacts by addition and H-abstraction. In marine air, the combination of sea salt, nitrogen oxides, and sunlight generates Cl atoms, that is, $Cl^{\bullet}$; globally a few percent of all methane is removed by $Cl^{\bullet}$ rather than $HO^{\bullet}$. $Br^{\bullet}$ and $I^{\bullet}$ are also important oxidants in marine environments. High $Br^{\bullet}$ concentrations are generated when cracks in Arctic sea ice refreeze.[a] This $Br^{\bullet}$ causes surface level ozone depletion (Section 4.2.d) and oxidizes Hg^0 to Hg^{II}, thereby introducing mercury into the Arctic ecosystem.

b. Nonmethane hydrocarbons

NMHC is the acronym used for nonmethane hydrocarbons. The emission of reduced species into the atmosphere is summarized in Table 4.3. NMHC has an important effect on the atmospheric oxidation potential and on particle formation. The table is a global average, and local variation is significant. Although many NMHCs are produced by anthropogenic sources, more are produced by biological sources, mainly plants. NMHCs are important for the formation of ozone and particles. We will present the oxidation of isoprene as an example of NMHC chemistry. Structurally,

[a] Seawater salts are expelled when ice is formed.

Table 4.4 Reactions removing isoprene from the atmosphere

Reactant	$k/cm^3\ s^{-1}$	c/cm^{-3} [1]	k'/s^{-1} [2]	τ/s [3]	τ/h [3]
$HO^\bullet$	1.0×10^{-10}	1.5×10^6	1.5×10^{-4}	6.6×10^3	1.8
O_3	1.3×10^{-18}	7.0×10^{11}	9.1×10^{-7}	1.1×10^6	3.1×10^2
$NO_3^\bullet$	6.8×10^{-13}	4.8×10^8	3.3×10^{-4}	3.1×10^3	0.9

Rate constants, k, refer to one-to-one reactions between isoprene and the appropriate radical in the gas phase at 298 K.

[1] Reactant concentrations, c: $HO^\bullet$, typical daytime concentration; O_3, typical Northern Hemisphere concentration; $NO_3^\bullet$, typical nighttime concentration.

[2] First-order removal constant, $k' = kc$.

[3] The lifetime, τ, may be considered as the time in which the concentration falls to $1/e$ of its initial value, $\tau = 1/k'$ (see Equation 2.56).

isoprene is a half of a terpene, and it is a key intermediate in plant biochemical synthesis. However, isoprene is volatile and escapes from plants, especially when they are growing rapidly in warm weather.

Isoprene[a] can react with a number of atmospheric radicals as shown in Table 4.4. The mechanism is the addition of the radical to a double bond, rather than hydrogen abstraction. This is an alternative reaction mechanism available for unsaturated compounds. It is seen that the reaction with nitrooxidanyl,[b] $NO_3^\bullet$, is the most effective at removing isoprene. Its slower reaction rate is more than compensated by a higher radical concentration. Nitrooxidanyl is important for the nighttime chemistry of the troposphere:

$$NO_2 + O_3 \rightarrow NO_3^\bullet + O_2 \tag{4.45}$$

This reaction is prevented during the day by photolysis:

$$NO_3^\bullet + h\nu \rightarrow NO_2 + O \tag{4.46}$$

The reaction of isoprene with hydroxyl is certainly the fastest of Table 4.4, occurring near the collision-limited rate. The gross reaction is an oxidation of isoprene that gives butenone and methyl propenal as the major products:

$$\tag{4.47}$$

Figure 4.5a and its continuation, Figure 4.5b, show the results of mechanistic studies of the reactions in Equation 4.47, using quantum methods. In the first step, hydroxyl prefers to bind to either of the two double bonds, and the unpaired electron is localized close to the other carbon atom. The next step is the addition of an

[a] Isoprene (methyl butadiene) is $CH_2{=}C(CH_3){-}CH{=}CH_2$.
[b] An alternative name is trioxidonitrogen($\bullet$).

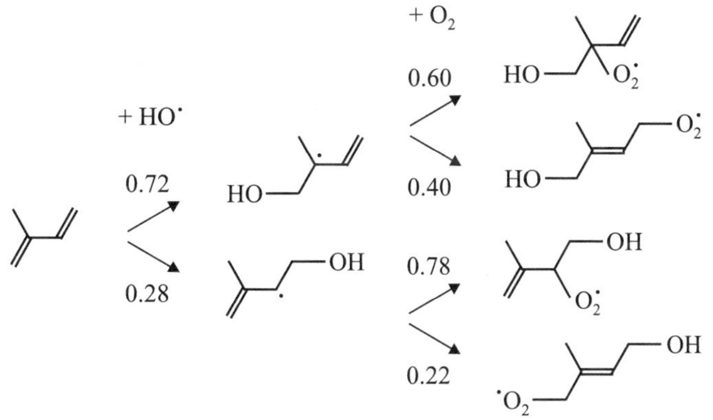

Figure 4.5a *Oxidation of isoprene in the atmosphere*
The major products for reactions initiated by hydroxyl. The numbers show the branching ratios.[218] The scheme is continued in Figure 4.5b.

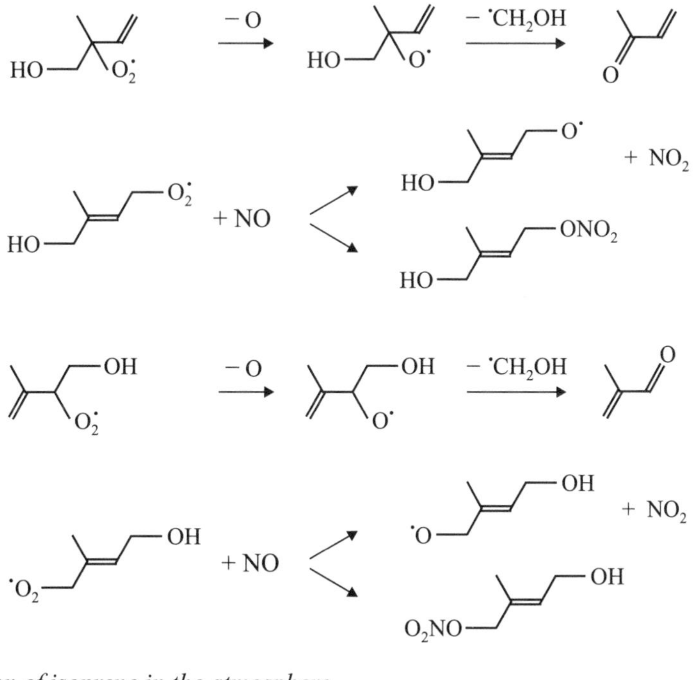

Figure 4.5b *Oxidation of isoprene in the atmosphere*
Continuation of Figure 4.5a. The reactions are discused in the text.[218]

oxygen molecule to the unpaired electron. This is a reaction pattern that we have seen previously:

$$RCH_2^{\bullet} + O_2 + M \rightarrow RCH_2O_2^{\bullet} + M \qquad (4.34)$$

It may seem unusual that a stable, long-lived atmospheric molecule such as O_2 has such an affinity for unpaired electrons. The key lies in the electronic structure of molecular oxygen, an exception that is not described correctly by the Lewis diagram

(::O=O::). The normal way of thinking of molecular oxygen is that there is a σ bond and a π bond and that the nonbonding electrons lie in sp^2 hybrid orbitals localized on each oxygen atom. However, the bond length of the molecule is short enough to cause a repulsive interaction between the nonbonding electrons in the p orbitals on opposite atoms. The electronic state in which the π bond is broken and the nonbonding electrons lie in orthogonal atomic p orbitals is lower in energy. This arrangement gives dioxygen two unpaired electrons (in atomic p orbitals that are orthogonal to one another), leading to a triplet ground state. The unpaired spins make O_2 paramagnetic, and its inherent diradical character leads to reactions of the type shown earlier. Addition of molecular oxygen to a radical is a recurring pattern in atmospheric chemistry, as we have seen in the reactions in Equations 4.12, 4.15, 4.20, 4.34, and 4.41.

Other characteristic reactions seen in Figure 4.5b are the reactions of a peroxy radical with NO to form an oxy radical and NO_2 or an organic nitrate. An additional type of reaction (not shown in the figure) is the spontaneous fragmentation of an intermediate to form smaller species. In many cases, these are carbonyl compounds. A simple beginning, the addition of $HO^\bullet$ to a double bond in isoprene, leads to more than a dozen reaction products. When terpenes are considered, even more reaction products are seen. However, one characteristic of atmospheric oxidation is that a volatile species acquires polar functional groups such as carbonyl or nitrate moieties. These lead to a reduction in vapor pressure relative to the precursor and an increased solubility and affinity for the condensed phase. The formation of particles is a characteristic of atmospheric photochemistry, especially of biogenic emissions, but also of anthropogenic emissions. Particles are hazardous to health and have an important effect on climate (Chapter 10). They are the subject of the next section.

c. Tropospheric aerosols

An aerosol is a suspension of liquid and/or solid particles in a gas.[a] In order for a particle to be held aloft by turbulent air currents, it must have a small terminal velocity, determined by the balance between the forces of gravity and aerodynamic friction. The gravitational force on the particle depends on its mass, proportional to volume or the third power of radius. The drag depends on the cross section, proportional to the second power of radius. Thus, larger particles will have relatively less aerodynamic drag and more gravitational force, allowing them to fall out of suspension. For example, raindrops fall whereas cigar smoke hangs in the air. Particles with diameters larger than 10 μm have settling speeds greater than 10 m h^{-1} and do not remain suspended.

Nature has evolved a strategy to prevent foreign material from reaching the lungs. When air is inhaled through the nose and sinus cavity, it is caused to spin. The resulting centrifugal force causes heavier particles to impact on the mucous membrane, preventing them from reaching the lungs. However, just as in the balance between the forces of gravity and drag, the centrifugal force is not able to remove small particles

---

[a] Gk. $\alpha\eta\rho \doteq$ air; Lat. *solutio* $\doteq$ solution.

from the airflow. Small airborne particles are a significant threat to human health because they can deposit in the lungs and enter the bloodstream. Particles may contain carcinogens and other hazardous substances. Elevated exposure to aerosols is associated with increased mortality.

Atmospheric particles play an important role in climate. It is very difficult for a mixture of water vapor and clean air to form water droplets, even at water vapor partial pressures several hundred percent above the saturation vapor pressure. However, the presence of so-called cloud condensation nuclei allows water droplets to form at relative humidities only slightly above 100%. Cloud formation would be very difficult indeed if it were not for atmospheric particles. In addition, the number of cloud condensation nuclei present in an air parcel can change the radiative properties of a cloud. Imagine two clouds, each with the same volume of liquid phase water. The cloud formed from clean air (few condensation nuclei) will have fewer, larger droplets, whereas the cloud formed from air from a polluted region with many aerosols will have more and smaller droplets. Even though it has the same total amount of liquid-phase water, the polluted cloud with many droplets presents a larger cross section and will scatter more light than the clean cloud, leading to a "whiter" cloud and a cooler climate. The radiative impacts of clouds are discussed further in Chapter 10.

Direct or primary sources of atmospheric particles include sea spray and dust blown from the surface, and emissions from construction and industry. Secondary aerosols are those formed within the atmosphere from precursors, for example, the oxidation products of isoprene or terpenes.

d. Henry's law and deposition

Henry's law (see Section 2.3b and Section 3.3b) describes the equilibrium between the gas and liquid phase concentrations of a given species. In an atmospheric context, it can be used to determine whether a given species will be found in a water droplet if a cloud is present, or in the gas phase. It can also be used to determine which compounds are susceptible to washing out of the atmosphere in rain. The Henry's law constants of some atmospheric gases are shown in Tables 3.22 and 4.5.

Table 4.5 Henry's law constants for some gases in water at 298.15 K

Species	$k_H/(\text{mol kg}^{-1}\,\text{bar}^{-1})$	Species	$k_H/(\text{mol kg}^{-1}\,\text{bar}^{-1})$
NO	1.9×10^{-3}	$CHCl_3$	0.25
C_2H_4	4.8×10^{-3}	$S(CH_3)_2$	0.6
NO_2	1.0×10^{-2}	HCl	1.1
O_3	1.1×10^{-2}	SO_2	1.23
N_2O	2.5×10^{-2}	Octanol	4.0×10^1
CO_2	3.4×10^{-2}	Propanol	1.3×10^2
H_2S	0.1	Acetic acid	5.5×10^3

See Ref. 243, which also gives values of d ln $k_H/\text{d}(T^{-1})$ (Equation 2.173).

It is clear that there are important inorganic and organic components of atmospheric aerosols. For example, emissions such as NO_x and SO_2 may be oxidized:

$$HO^\bullet + NO_2 + M \rightarrow HNO_3(g) + M \tag{4.39a}$$

$$HNO_3(g) + \text{aqueous aerosol} \rightarrow H^+(aq) + NO_3^-(aq) \tag{4.39b}$$

and

$$SO_2 + HO^\bullet + M \rightarrow HSO_3^\bullet + M \tag{4.48}$$

$$HSO_3^\bullet + O_2 \rightarrow SO_3 + HO_2^\bullet \tag{4.49}$$

$$SO_3 + H_2O + M \rightarrow H_2SO_4(g) + M \tag{4.50a}$$

$$H_2SO_4(g) + \text{aqueous aerosol} \rightarrow H^+(aq) + HSO_4^-(aq) \tag{4.50b}$$

Sulfuric and nitric acid produced by the reactions 4.39b and 4.50b are key components of acid rain. Other species may neutralize acidity, for example, $CaCO_3$ from mineral dust and NH_3 produced in an agricultural area. Ammonium sulfate is an important aerosol component.

Gas-phase oxidation of SO_2 competes with a variety of liquid-phase reactions. One of the most important oxidants in this regard is hydrogen peroxide in aqueous solution:

$$SO_2 + H_2O_2(aq) \rightarrow HSO_4^-(aq) + H^+(aq) \tag{4.51}$$

Because of this reaction, SO_2 and H_2O_2 are not found together when a cloud is present.

There have been efforts to control anthropogenic emissions of both NO_x and SO_2 because of the negative impacts of acid rain on the terrestrial and aquatic environments. For example, NO_x emissions per vehicle have decreased dramatically because of catalytic converters. However, the distance driven has increased at the same time, so total NO_x emissions have remained nearly the same for 20 to 30 years. At the same time, efforts have been made to use less brown coal, which is high in sulfur, and to eliminate the use of diesel oil containing sulfur. Again, these efforts have been counteracted by increased demand, resulting in little overall improvement.

4.2 Stratospheric chemistry

The preceding section began with the photolysis of ozone. As we will see, ozone is even more central to the stratosphere because it is responsible for the stratosphere itself. The concentration of ozone in the stratosphere is controlled by radicals. The ozone layer shields life on Earth from harmful UV radiation; Figure 4.1 showed that there is virtually no radiation at the surface at wavelengths shorter than 300 nm. The absorption spectrum of ozone is shown in Figure 4.6.[a]

---

[a] A typical amount of ozone in the atmospheric column is 134 mmol/m^2. Then the decadic absorbance at 250 nm is $A \approx 300 \times 0.134 \approx 40$, meaning that the fraction 10^{-40} of the radiant power from the Sun reaches the surface of the Earth; see Figure 4.1, Equation 10.15, and Equation 10.16.

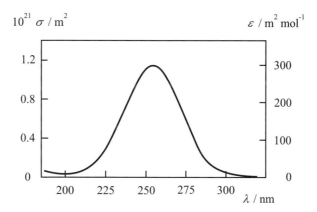

Figure 4.6 *Absorption spectrum of ozone*
Cross section, σ, and molar absorption coefficient, ε, for O_3 (g) at 295 K. The relation is $\sigma\, N_A = \varepsilon \ln 10$, where N_A is the Avogadro constant (see Equation 10.15).

a. The Chapman mechanism: O_x

In the early part of the 20th century, the solar spectrum was measured over the course of sunsets, and large changes in the UV light flux were observed. It was concluded that the cause was stratospheric ozone, which absorbs ultraviolet radiation (Figure 4.6). The existence of the ozone layer was first explained by Sidney Chapman (1930),[151a] using a four-step photochemical mechanism:[a]

$$O_2 + h\nu\,(\lambda < 240\,\text{nm}) \;\rightarrow\; 2\,O \qquad (4.52)$$

$$O + O_2 + M \;\rightarrow\; O_3 + M \qquad (4.53)$$

$$O_3 + h\nu\,(\lambda < 750\,\text{nm}) \;\rightarrow\; O + O_2 \qquad (4.54)$$

$$O + O_3 \;\rightarrow\; 2\,O_2 \qquad (4.55)$$

In the first step, molecular oxygen is photolyzed, producing two oxygen atoms. This process is relatively slow because there is little sunlight at the wavelengths at which O_2 absorbs. The atoms quickly combine with molecular oxygen to produce ozone (Equation 4.53). Ozone is more easily photolyzed than molecular oxygen, producing an oxygen atom that also quickly forms ozone via Equation 4.53. The oxygen species O and O_3 are collectively called "odd oxygen" and referred to as O_x. The cycle, Equations 4.53–4.54, converts sunlight into heat, giving the stratosphere the temperature inversion that is the reason for its stability, that is, its stratification. The odd oxygen radical chain is terminated as dioxygen in the reaction in Equation 4.55. We now analyze the mechanism to see what it can tell us about the ozone layer. The partitioning of odd oxygen into O and O_3, $[O]/[O_3]$, will be determined by the two fast reactions 4.53 and 4.54. The total amount of O_x, $[O_x] = [O] + [O_3]$, will be determined by the slow reactions 4.52 and 4.55.

---

[a] Rate constants and other conditions are given in Table 4.6.

Table 4.6 The Chapman cycle		
Altitude	15 km	45 km
T/K	215	267
$[M]/cm^{-3}$	4×10^{18}	6×10^{16}
$[O]/cm^{-3}$	1×10^7	1×10^9
$[O_3]/cm^{-3}$	2×10^{12}	1×10^{11}
$j_{4.52}/s^{-1}$	$<10^{-14}$	5×10^{-10}
$k_{4.53}[O_2][M]/s^{-1}$	4.5×10^3	0.6
$j_{4.54}/s^{-1}$	6×10^{-4}	5×10^{-3}
$k_{4.55}[O_3]/s^{-1}$	1.1×10^{-3}	3.6×10^{-4}

Typical conditions for 40° N, March 15, local noon.[223]

Using the steady-state approximation for [O], we can determine the ratio $[O]/[O_3]$:

$$\begin{aligned} d[O]/dt &= (2r_{4.52} + r_{4.54}) - (r_{4.53} + r_{4.55}) \\ &\approx r_{4.54} - r_{4.53} \approx 0 \end{aligned} \quad (4.56)$$

Here we have used the approximations $r_{4.54} \gg 2r_{4.52}$ and $r_{4.53} \gg r_{4.55}$. Writing out the rates, one finds that $j_{4.54}[O_3] = k_{4.53}[O][O_2][M]$. Solving:

$$[O]/[O_3] = j_{4.54}/(k_{4.53}[O_2][M]) = j_{4.54}(z)/(k_{4.53} x(O_2) C_{air}(z)^2) \quad (4.57)$$

The right-hand side of Equation 4.57 is written to emphasize that $[O]/[O_3]$ depends on the altitude due to $j_{4.54}(z)$ and $C_{air}(z)$; the concentration of molecular oxygen is equal to its mole fraction $x(O_2)$ times the concentration of air, $C_{air}(z)$. Setting in numbers, one finds that $[O]/[O_3]$ is less than a part in a thousand in the lower stratosphere. The pressure dependence means that oxygen atoms are more common in the upper stratosphere (Table 4.6) than in the lower. The approximation $[O_x] \approx [O_3]$ is often used because oxygen atoms are so rare relative to ozone.

Next we will use the steady-state approximation on O_x to determine the concentration of ozone in the stratosphere. Two units of O_x are made in Equation 4.52, and two units are destroyed in Equation 4.55:

$$d[O_x]/dt = 2j_{4.52}[O_2] - 2k_{4.55}[O][O_3] \approx 0 \quad (4.58)$$

Solving for $[O_3]$:

$$[O_3] = j_{4.52}[O_2]/(k_{4.55}[O]) \quad (4.59)$$

Substituting [O] from Equation 4.57 into Equation 4.59 and solving for $[O_3]$ yields

$$[O_3] = \{j_{4.52} k_{4.53}/(j_{4.54} k_{4.55})\}^{1/2} x(O_2) C_{air}(z)^{3/2} \quad (4.60)$$

This equation predicts that there will be a minimum in the ozone concentration at the top of the stratosphere, due to the pressure dependence of $C_{air}(z)$. There will be a second minimum at the bottom of the stratosphere because there is not enough light to photolyze molecular oxygen. In between there is a maximum, the ozone layer.

Chapman's mechanism was the first to explain the ozone layer, and it is the basis of the current understanding of the stratosphere. Measurements of the ozone column

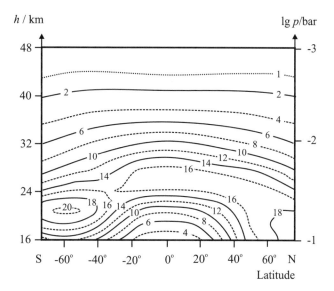

Figure 4.7 *The ozone layer*
The ozone layer as measured by the Nimbus-7 satellite, a SBUV instrument, in Dobson units per km; see footnote *a* on p. 159. Adapted from ref. 182.

show that there is only about half as much ozone present as predicted by Chapman's mechanism, and the observed maximum is lower in altitude than predicted. The subsequent sections will show that there are many mechanisms for destroying ozone besides Equation 4.55. Also, the Chapman mechanism assumes photochemical steady state. In the atmosphere, the lifetime of ozone ranges from minutes and hours in the upper stratosphere to years at the bottom. This means that ozone can be transported over long distances and stored in the lower stratosphere, leading to higher concentrations than predicted by Chapman. The structure of the ozone layer is shown in Figure 4.7.[a]

The stratosphere is not static, but its circulation follows roughly the pattern shown in Figure 3.5: Air is lifted into the stratosphere by convective storms in the tropics. After a journey taking a few years, it subsides back into the troposphere at the poles. In the absence of extraordinary chemical effects, ozone concentrations are expected to reach a maximum in the lower stratosphere near the poles.

b. The radicals HO$_x$

The O(^{1}D) atoms generated in the photolysis of ozone in reaction 4.4 have nearly 190 kJ/mol of electronic energy that can be used to overcome reaction barriers.[b] This

---

[a] The amount of ozone is measured as the content of a column with a cross section of 1 m^2 extending through the atmosphere. On average, this amount is 134 mmol. For historical reasons, the non-SI unit *Dobson unit*, denoted DU, may be used: 1 DU = 446.2 μmol m^{-2}, and the average concentration of ozone is 300 DU = 134 mmol m^{-2}.

[b] The spectroscopic transition O(^{1}D$_2$) → O(^{3}P$_2$) occurs at 1.58677 μm^{-1} (λ = 630.2 nm) corresponding to 189.820 kJ/mol.[131]

excited electronic state has a reactivity that the $O(^3P)$ ground state does not have. One reaction has been described:

$$O(^1D) + H_2O \rightarrow 2HO^{\bullet} \tag{4.6}$$

In addition, $O(^1D)$ can react with methane and hydrogen:

$$O(^1D) + CH_4 \rightarrow HO^{\bullet} + CH_3^{\bullet} \tag{4.61}$$

$$O(^1D) + H_2 \rightarrow HO^{\bullet} + H^{\bullet} \tag{4.62}$$

The molecules H_2O, CH_4, and H_2 account for virtually all stratospheric hydroxyl formation. In the late 1950s, a new mechanism for removing ozone was recognized,[141]

$$HO^{\bullet} + O_3 \rightarrow HO_2^{\bullet} + O_2 \tag{4.63}$$

$$HO_2^{\bullet} + O_3 \rightarrow OH^{\bullet} + 2O_2 \tag{4.64}$$

the net reaction being

$$2O_3 = 3O_2 \tag{4.65}$$

The HO_x radicals thus catalyze ozone depletion. They are removed by radical-radical reactions such as

$$HO^{\bullet} + HO_2^{\bullet} \rightarrow H_2O + O_2 \tag{4.66}$$

HO_x chemistry is an important sink for ozone, especially in the lower stratosphere. However, it cannot explain all of the difference between the Chapman mechanism and reality.

c. The radicals NO_x

In the late 1960s, plans were under way in the United States, Europe, and the Soviet Union to build supersonic jets that would fly in the lower stratosphere, reducing travel time between the major cities of the world. Hal Johnston and Paul Crutzen recognized that nitrogen oxides generated at the high temperatures found in engines[a] could catalyze ozone depletion:

$$O_3 + NO \rightarrow O_2 + NO_2 \tag{4.67}$$

$$NO_2 + O \rightarrow NO + O_2 \tag{4.68}$$

The net reaction is

$$O + O_3 + 2O_2 \tag{4.69}$$

We have mentioned a competing reaction that results in a null cycle:

$$O_3 + NO \rightarrow O_2 + NO_2 \tag{4.70}$$

$$NO_2 + h\nu \, (\lambda < 415 \, nm) \rightarrow NO + O \tag{4.14}$$

$$O + O_2 + M \rightarrow O_3 + M \tag{4.53}$$

[a] See Equation 7.13 and the ensuing discussion on the thermodynamics.

There is no net reaction.

Although the null cycle reduces the efficiency of the NO_x cycle, it turns out that the NO_x radical reactions are the most important reactions removing stratospheric ozone.

The NO_x reactions are terminated by a HO_x-NO_x cross reaction:

$$NO_2 + HO^{\bullet} + M \rightarrow HNO_3 + M \tag{4.39}$$

At night when $HO^{\bullet}$ radicals are not available, the following sequence occurs:

$$NO_2 + O_3 \rightarrow NO_3^{\bullet} + O_2 \tag{4.45}$$

$$NO_2 + NO_3^{\bullet} + M \rightarrow N_2O_5 + M \tag{4.71}$$

During the day, both $NO_3^{\bullet}$ and N_2O_5 are quickly photolyzed. Thus, reactions 4.39 and 4.71 are a way of storing the radicals at night. As an acid anhydride, dinitrogen pentoxide can be hydrolyzed on aerosol particles:

$$N_2O_5 + H_2O(l, \text{aerosol}) \rightarrow 2H^+(aq) + 2NO_3^-(aq) \tag{4.72}$$

The nitric acid produced in reaction 4.72 will typically stay with the particle. If the particle grows large enough, it will fall out of the stratosphere. Some NO_x may be reformed from nitric acid:

$$HNO_3 + hv \, (\lambda < 230 \, nm) \rightarrow NO_2 + HO^{\bullet} \tag{4.73}$$

$$HNO_3 + HO^{\bullet} \rightarrow NO_3^{\bullet} + H_2O \tag{4.74}$$

The main source of stratospheric NO_x is dinitrogen oxide, N_2O, which is produced as a by-product of nitrification and denitrification reactions by soil bacteria.

Whereas 90% of N_2O is photolyzed, producing N_2 and $O(^1D)$, 6% reacts via Equation 4.75:

$$N_2O + O(^1D) \rightarrow 2NO \tag{4.75}$$

The remaining 4% of N_2O reacts with $O(^1D)$ to produce N_2 and O_2.

Supersonic jets and ozone

Harold Johnston of the University of California at Berkeley testified before Congress (1971) that construction of a fleet of supersonic jets would have catastrophic consequences for the ozone layer and thereby for life on the surface of the planet. The United States decided not to initiate such a project. In Europe, a joint French-British group built the Concorde jet, and the Russians built the Tupolev Tu-144 supersonic transport. However, these projects never achieved the scale of subsonic mass air transit and therefore caused only insignificant ozone depletion. On the other hand, they brought the role of NO_x to the attention of the atmospheric community.

d. The radicals ClO_x and coupling of the cycles

In 1970 James Lovelock invented the electron capture detector and used it to determine the concentrations of chlorofluorocarbons (CFCs) in the atmosphere for the first time. He calculated that the atmospheric accumulation of CFCs was equal to the total amount that had been manufactured up to that time. It seemed as if there was no mechanism for removing them from the environment. The CFCs were invented around 1930 by T. Midgley, working at General Motors. They represented a tremendous advance because they could be used to make refrigerators more efficient and cheaper, improving food storage and thereby public health. They were nontoxic and inert. In addition, the CFCs found use as blowing agents in producing foam, as propellants in aerosol sprays, and as industrial solvents and cleaning agents. During the 1970s and 1980s, atmospheric concentrations of $CFCl_3$ and CF_2Cl_2 increased at a rate of a few percent per year.

In 1974 Molina and Rowland proposed a mechanism for removing the CFCs from the atmosphere. They asserted that although the compounds are inert in the troposphere (and nonsoluble in the oceans), they will eventually be transported to the stratosphere and photolyzed in the UV window around 200 nm (see Figure 4.1):

$$CF_2Cl_2 \; + \; hv \; (\lambda < 210 \, nm) \quad \rightarrow \quad CF_2Cl^{\bullet} \; + \; Cl^{\bullet} \qquad (4.76)$$

Molina and Rowland warned that Cl atoms, $Cl^{\bullet}$, would remove ozone in a catalytic process:

$$Cl^{\bullet} \; + \; O_3 \quad \rightarrow \quad ClO^{\bullet} \; + \; O_2 \qquad (4.77)$$

$$ClO^{\bullet} \; + \; O \quad \rightarrow \quad Cl^{\bullet} \; + \; O_2 \qquad (4.78)$$

The net reaction is

$$O_3 \; + \; O \; = \; 2\,O_2 \qquad (4.79)$$

The ClO_x reaction cycle is most important in the upper stratosphere, because reaction 4.78 relies on the oxygen atom. Reaction 4.77 occurs relatively quickly, whereas 4.78 is much slower, simply because the concentrations of chlorine monoxide and oxygen atoms are small. Thus, $[ClO^{\bullet}] >> [Cl^{\bullet}]$, and the rate-limiting step is Equation 4.78. This mechanism results in a few percent reduction in stratospheric ozone at midlatitudes. The cycle is quite efficient, and one chlorine atom can destroy approximately 10^5 ozone molecules before being removed:

$$Cl^{\bullet} \; + \; CH_4 \quad \rightarrow \quad HCl \; + \; CH_3^{\bullet} \qquad (4.80)$$

$$ClO^{\bullet} \; + \; NO_2 \; + \; M \quad \rightarrow \quad ClNO_3 \; + \; M \qquad (4.81)$$

In the nonpolar stratosphere, roughly 90% of inorganic chlorine is in the nonreactive forms HCl or $ClNO_3$, the balance being $ClO_x \equiv Cl^{\bullet} + ClO^{\bullet}$. These reservoir molecules can slowly regenerate radicals:

$$HCl \; + \; HO^{\bullet} \quad \rightarrow \quad Cl^{\bullet} \; + \; H_2O \qquad (4.82)$$

$$ClNO_3 \; + \; hv \quad \rightarrow \quad Cl^{\bullet} \; + \; NO_3^{\bullet} \quad and \quad ClO^{\bullet} \; + \; NO_2 \qquad (4.83)$$

The catalytic cycles interact with one another, as the following two examples show:

1. Coupled NO_x/ClO_x cycle:

$$Cl^{\bullet} + O_3 \rightarrow ClO^{\bullet} + O_2 \tag{4.84}$$

$$NO + O_3 \rightarrow NO_2 + O_2 \tag{4.85}$$

$$ClO^{\bullet} + NO_2 + M \rightarrow ClNO_3 + M \tag{4.86}$$

$$ClNO_3 + hv \rightarrow Cl^{\bullet} + NO_3^{\bullet} \tag{4.87}$$

$$NO_3^{\bullet} + hv \rightarrow NO + O_2 \tag{4.88}$$

Net reaction:

$$2\,O_3 = 3\,O_2 \tag{4.89}$$

2. Coupled HO_x/ClO_x cycle:

$$Cl^{\bullet} + O_3 \rightarrow ClO^{\bullet} + O_2 \tag{4.90}$$

$$HO^{\bullet} + O_3 \rightarrow HO_2^{\bullet} + O_2 \tag{4.91}$$

$$ClO^{\bullet} + HO_2^{\bullet} \rightarrow HOCl + O_2 \tag{4.92}$$

$$HOCl + hv \rightarrow HO^{\bullet} + Cl^{\bullet} \tag{4.93}$$

Net reaction:

$$2\,O_3 = 3\,O_2 \tag{4.94}$$

Computer models are necessary to understand the interlocking cycles resulting in ozone depletion. For example, in some conditions more NO_x can result in less ozone via the reactions 4.67 and 4.68, whereas in other conditions it can suppress ozone depletion by removing HO_x and ClO_x radicals through reactions 4.39 and 4.81, respectively.

e. The ozone hole

In 1985, Joe Farland of the British Antarctic Survey published a paper that shook the scientific world. The data and subsequent measurements showing a dramatic depletion in the ozone column over Antarctica are reproduced in Figure 4.8.

It was known that chlorine could remove ozone (see Section 4.2b) and that atmospheric chlorine loading had increased during the same period that the ozone was shown to be falling. The trouble was that the mechanism of Molina and Rowland depended on the reaction of chlorine monoxide with oxygen atoms, and there simply were not enough oxygen atoms available in the mid- to lower stratosphere, where the ozone was being destroyed, especially at that time of year: October is the Antarctic springtime, and light levels are low. Neither could changes in atmospheric circulation explain the observed ozone depletion.

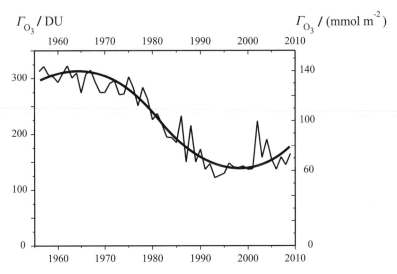

Γ_{O_3} / DU

Γ_{O_3} / (mmol m^{-2})

Figure 4.8   *Antarctic ozone measurements 1956–2009*
The October mean ozone column measured over the British Antarctic Survey's research station at Halley Bay, Antarctica, shows the rapid decline in ozone caused by increased stratospheric chlorine levels. In the past few years, the severity of the ozone hole has begun to decrease because of the limits set in place by the Montreal Protocol on Substances That Deplete the Ozone Layer.[245] 1 DU = 446.2 μmol m^{-2}; see the footnote *a* on p. 159.

Measurements by the NASA ER-2 aircraft (a U-2 spy plane modified for environmental research) showed a near-perfect anticorrelation between the chlorine monoxide radical and ozone concentrations. The data are shown in Figure 4.9. Ozone levels dropped when the plane entered the polar vortex. At the same time, concentrations of the chlorine monoxide radical increased to 500 times the normal value. The polar vortex is a feature of atmospheric circulation caused by high wind speeds around the Antarctic continent and reinforced by cold surface temperatures (see Chapter 2). The circulation blocks air within the vortex from mixing with the rest of the atmosphere, resulting in the coldest temperatures anywhere in the troposphere or stratosphere. Altitude profiles of the ozone hole are shown in Figure 4.10. The loss of ozone is almost total in the lower stratosphere, where it should normally be at its highest. The figure also shows how quickly the ozone hole forms over the course of a half-dozen weeks.

Mario Molina, coauthor of the 1974 paper describing the "classic" ozone depletion process, proposed a new reaction mechanism involving dioxygen dichloride and chlorine($^{\bullet}$), but not the oxygen atom:

$$2\,ClO^{\bullet} \;\rightarrow\; ClOOCl \tag{4.95}$$

$$ClOOCl + h\nu \;\rightarrow\; Cl^{\bullet} + ClOO^{\bullet} \;\rightarrow\; 2\,Cl^{\bullet} + O_2 \tag{4.96}$$

$$Cl^{\bullet} + O_3 \;\rightarrow\; ClO^{\bullet} + O_2 \quad \text{(two times)} \tag{4.77}$$

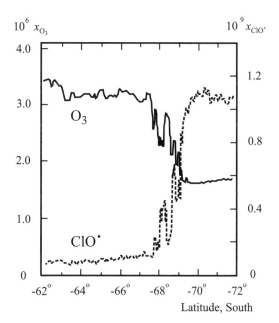

Figure 4.9 *Anticorrelation of* ClO$^{\bullet}$ *and* O$_3$ *concentrations*
In situ data from a flight over Antarctica starting from Punta Arenas Chile on
September 16, 1987. Mole fractions of O$_3$ (solid line, left scale) and of ClO$^{\bullet}$
(broken line, right scale). Redrawn from ref. 138.

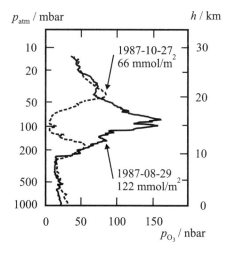

Figure 4.10 *Ozone pressure versus altitude*
Vertical profiles of the ozone layer at McMurdo Station, Antarctica (77° S), showing
the change in the ozone column between August 29, 1987 (solid line) and October
27, 1987 (broken line). Abscissa: ozone partial pressure; the amount of ozone in the
columns are shown. Redrawn from ref. 176.

Net reaction:

$$2\,O_3 + hv = 3\,O_2 \tag{4.97}$$

This process was able to explain ozone depletion in the lower stratosphere. Given high enough concentrations, chlorine monoxide radicals will react with themselves, forming the unstable dimer. The difficulty was in explaining why chlorine monoxide concentrations were so high in the first place; under normal conditions in the stratosphere, most chlorine should be carried by HCl and $ClNO_3$.

The anormalous chemical state in the Antarctic stratosphere was explained via a heterogeneous catalytic reaction[231] that occurs on the surface of stratospheric ice particles, called polar stratospheric clouds (PSCs). For example,

$$HCl + ClNO_3 \;\rightarrow\; Cl_2(g) + H^+ + NO_3^- \tag{4.98}$$

The stratosphere is very dry, and ice particles can form only at extremely low temperatures, in the presence of seed nuclei containing sulfuric and nitric acid. During the winter when temperatures are low, PSCs form and process the chlorine reservoirs HCl and $ClNO_3$, producing chlorine gas. Nitric acid stays behind in the condensed phase. In the springtime, when the Sun returns, the chlorine is photolyzed:

$$Cl_2 + hv \;\rightarrow\; 2Cl^\bullet \tag{4.99}$$

One feature of the reaction mechanism is that the stratosphere becomes depleted in nitrogen oxides, because they are converted to nitric acid in particles. This prevents the normal reaction that would limit $ClO^\bullet$ concentrations:

$$ClO^\bullet + NO_2 + M \;\rightarrow\; ClNO_3 + M \tag{4.81}$$

The temperature profile of the stratosphere depends on O_3. Without ozone, temperatures stay low, meaning that PSCs persist even after sunrise. This inhibits the recovery of the normal ozone distribution. Later in the season, the vortex breaks up and the ozone hole disappears via dilution.

f. Midlatitude ozone depletion

If there is going to be an ozone hole, Antarctica is likely the "best place to put it" in terms of minimizing impact on the biosphere. Studies have shown detrimental effects not only for plankton and lichen but also for people living in Australia and Chile who are affected by decreased ozone levels. Midlatitude ozone depletion shown in Figure 4.11 is a much larger public health problem and a larger concern for the biosphere.

There are two main causes of the decrease at midlatitudes shown in the figure. The first is that climate change has affected the stratosphere. Water vapor in the stratosphere has increased by about 50% over the past 50 a, because of a combination of increased transport via convective storms and increased in situ production from methane. In addition, although greenhouse theory predicts that surface temperatures will increase, it also shows that stratospheric temperatures will decrease, as part of the

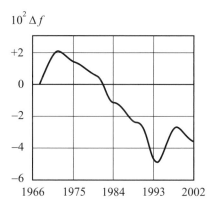

Figure 4.11 *Trend in the global ozone column at mid-latitudes*
Trend in the global ozone column from 60° S to 60° N, over the period 1967 to
2002. Δf is the percent deviation from July 1967.[250,256]

same effect. These two factors lead to increased aerosol loading in the stratosphere.
This allows dinitrogen pentoxide to hydrolyze:

$$N_2O_5 + H_2O(l, \text{aerosol}) \quad \rightarrow \quad 2\,H^+(aq) + 2\,NO_3^-(aq) \qquad (4.72)$$

This represents a sink for NO_x, as described in Section 4.1. This in turn makes the
stratosphere more sensitive to ozone depletion by the HO_x and ClO_x cycles, because
the NO_x reactions that remove these radicals are reduced:

$$NO_2 + HO^\bullet + M \quad \rightarrow \quad HNO_3 + M \qquad (4.39)$$

$$ClO^\bullet + NO_2 + M \quad \rightarrow \quad ClNO_3 + M \qquad (4.81)$$

The hydrolysis of N_2O_5 thus makes the midlatitude stratosphere more sensitive
to increases in chlorine loading, in addition to enhancing HO_x-catalyzed ozone
depletion.

Another concern is the potential formation of a Northern Hemisphere ozone hole,
driven by increased water vapor concentrations and decreased temperatures. Relative
to Antarctica, the major difference is that there is much more weather in the Northern
Hemisphere, and so a vortex like the one over the South Pole does not form. Temperatures are sometimes low enough to produce PSCs, but they tend not to persist, and
only smaller ozone holes with a shorter depth and duration have been observed.

The use of CFCs was regulated by the Montreal Protocol of 1987 and banned by the
Copenhagen Amendment in 1992. However, CFCs are not easily degraded and remain
in the atmosphere for around a hundred years. Several types of CFC replacements
have been developed that serve the same functions as CFCs, while at the same time
breaking down in the troposphere. One result is that they transport much less chlorine to the stratosphere, in addition to having smaller global warming potentials (see
Chapter 10). An example of such a compound is 1,1,1,2-tetrafluoroethane, CF_3CH_2F,[a]
which was developed specifically as a replacement for dichlorodifluoromethane (see
Equation 4.76). It contains no chlorine and so cannot result in chlorine-catalyzed

---

[a] Trade name: HFC-134a.

ozone depletion. However, CF bond stretching has a frequency that could potentially make it a powerful greenhouse gas. CF_3CH_2F reacts with hydroxyl in the troposphere:

$$F_3CCH_2F + OH^\bullet \rightarrow CF_3CHF^\bullet + H_2O \qquad (4.100)$$

The radical undergoes a number of reactions leading to its removal from the atmosphere. The relatively high reactivity of the CH bond gives this compound a shorter lifetime in the atmosphere and reduces its impact as a greenhouse gas.

As a result of an international effort to control emissions of chlorine-containing gases, the ozone hole phenomenon is predicted to stop around mid-century.

Compilations of reaction rates for gas phase chemical reactions available through the Internet: refs. 236, 241, 242, 243b.

5 Chemistry of the hydrosphere

A few properties of the hydrosphere, including its interaction with the other spheres, were described in Section 3.2. In this chapter we present fundamental aspects of the hydrosphere's chemistry that are critical for a thorough understanding of the environment. Pollution of water and its treatment are discussed in Section 9.5.

Chemistry in an aqueous solution always takes place with a great surplus of water. At room temperature, the amount concentration of water is $c_{aq} \approx 55.5$ M, which may be compared with $c_{NaCl} \approx 0.5$ M in seawater (see Table 3.13).[a] Seawater is a fairly concentrated solution, and even here each ion is surrounded by about 100 water molecules. Therefore, the physicochemical properties of water itself underlie all chemistry in the hydrosphere. We shall see that water exerts a leveling effect, not only on the apparent strengths of acids and bases, but also on the range of electrode potentials that are observed, allowing species to be stable in solution. In this chapter we discuss acid-base chemistry, coordination chemistry, redox chemistry, and the thermodynamics of species in aqueous solution.

Section 5.1 deals with the acid-base chemistry of aqueous solutions using Liebig's definition of an acid and the Brønsted-Lowry definition of a base: an acid is a hydron donor, and a base is a hydron acceptor. Important concepts and tools for describing environmental systems include alkalinity, formation functions, and speciation diagrams. We will see that titration of simple acid-base systems is a useful model for describing more complex systems with regard to the determination of the isoelectric point (ampholytes) or, as will be discussed in Section 6.3, the point of zero charge (colloids and surface adsorption).

On the other hand, as shown in Section 5.2, the discussion of the coordination chemistry of aqueous solutions profits from using the Lewis acid-base definition: an acid is an electron pair acceptor, and a base is an electron pair donor. In particular, the ensuing hard/soft acid-base classification is an important qualitative concept, for example in the description of how chemical species move in the environment.

Section 5.3 discusses the electrolytic properties of aqueous solutions. This includes the limits of water-splitting into (1) hydron + hydroxide and (2) hydrogen + oxygen as defining the scope of the acid-base and redox chemistry of natural waters. Calculations of activities of strong electrolytes are important for environmental applications, and a guide on this concludes the section.

We have not included a proper discussion of reaction rates in aqueous solutions in this chapter. It should be noted, however, that the rate of normal neutralization of a strong acid with a strong base is diffusion controlled, which makes it among the

---

[a] Symbols for concentrations are collected in Appendix 1, Table A1.7.

fastest reactions known in aqueous systems; k(first order) $\approx 10^{12}$ s^{-1}.[104] Reaction of a metal-aqua ion that exchanges a water ligand with another entity, and the substitution of one water molecule with another one, a water exchange reaction, are conceptually simple reactions. The rate constants for this class of reaction range over approximately 22 orders of magnitude: the most robust complexes, such as $Ir(H_2O)_6^{3+}$, have $k \approx 10^{-12}$ s^{-1}; the more common ions $Al(H_2O)_6^{3+}$, $k \approx 1$ s^{-1}; $Fe(H_2O)_6^{3+}$, $k \approx 100$ s^{-1}; $Fe(H_2O)_6^{2+}$ and $Zn(H_2O)_6^{2+}$, $k \approx 10^7$ s^{-1}; and aquaions of alkali metals, $k \approx 10^{10}$ s^{-1}. As an example, decisive parameters for the chemistry of magnesium and calcium in life processes include the different rate constants for water exchange in $Mg(aq)_6^{2+}$, $k \approx 10^5$ s^{-1}, and $Ca(aq)_6^{2+}$, $k \approx 10^8$ s^{-1}.[a]

Except for Section 5.3b, thermodynamic standard states will be chosen close to environmental conditions such that concentrations enter all equilibrium constants.

5.1 Acid-base chemistry

Acids seem to be fundamental to human existence because acidity is one of the five tastes sensed by the tongue.[b] Their opponents, bases, neutralize acids, resulting in a bland taste. Acids are common in Nature and include organic acids and the mineral acids. In comparison, bases are rare; only carbonates, calcium hydroxide, and sodium hydroxide have been known since ancient times. The historical development of chemistry reflects our growing understanding of the nature of acids and how and where they are formed.

In 1675 Lémery wrote a textbook of chemistry that was widely used for the following 100 years. He asserted that the properties of matter are derived from the shapes of atoms, and atoms of acids were believed to have spikes that broke off when they were neutralized.[c] Many chemists had observed that the combustion of carbon, sulfur, phosphorus, and certain other elements created products with acidic properties. However, it was Lavoisier (1776) who made the general observation that such reactions resulted in an uptake of oxygen. For some time it was believed that acids must contain oxygen, and in the following year he coined the word *oxygen*.[d] However, soon after, acids were found (HCl, HCN, H_2S) that contained no oxygen, and oxides (CaO, K_2O) with basic properties were identified. Such observations led Liebig (1838) to tie acidity to the hydron, H$^+$, and in Arrhenius' (1884) theory of water, hydroxide, HO$^-$, was considered to be the only base. Brønsted and Lowry (1923) extended the concept of a base to include all chemical species that can take up a hydron, and finally, Lewis (1938) extended the concept of an acid to include all chemical species that can take up a base.

---

[a] See ref. 17, Chapters 10 and 11.
[b] Acid, sweet, salt, bitter, umami.
[c] Until the 1960s, the phrase *at afstumpe en syre*, meaning "to make the acid blunt," was still used in Danish.
[d] Gk. οξυς $\hat{=}$ sharp; γενναω $\hat{=}$ create.

Table 5.1 The ion product of water as function of temperature			
θ / °C	pK_w	θ / °C	pK_w
0	14.941	30	13.833
4	14.773	37	13.619
10	14.533	50	13.26
25	13.995	100	12.26

A table of pK_w between 0 °C and 100 °C at 0.1013 MPa. The figures are calculated from the empirical expression $pK_w = -4470.99/(T/\mathrm{K}) + 6.0875 - 0.01706 \times (T/\mathrm{K})$; T is the thermodynamic temperature.[30]

a. Acid-base properties of water

Definition. An acid is a chemical species capable of donating a hydron; a base is a chemical species capable of accepting a hydron (Brønsted and Lowry, 1923).

Consider an acid HA^q (q is an integer that denotes the ionic charge) that donates a hydron to the acceptor, water:

$$HA^q(aq) + H_2O(l) \;\rightleftharpoons\; H_3O^+(aq) + A^{q-1}(aq) \tag{5.1}$$

The pairs of species (HA^q, A^{q-1}) and (H_3O^+, H_2O) are called conjugate acid-base pairs; the conjugate acid (base) always carries one unit of positive charge more (less) than the base (acid). The reaction of a base B^q with water gives the conjugate acid, HB^{q+1}:

$$B^q(aq) + H_2O(l) \;\rightleftharpoons\; HO^-(aq) + AB^{q+1}(aq) \tag{5.2}$$

Thus, the conjugate acid of water is oxonium H_3O^+, and its conjugate base is hydroxide HO^-. Hence, water plays a dual role: it is an acid as well as a base. In general, a chemical species that behaves both as an acid and as a base is called amphoteric,[a] but the term amphiprotic is reserved for such a solvent. The amphiprotic properties of the solvent water can be written in a succinct way:

$$2\,H_2O(l) \;\rightleftharpoons\; H_3O^+(aq) + HO^-(aq) \tag{5.3}$$

with the equilibrium constant

$$K_w = [H_3O^+][HO^-] \tag{5.4}$$

K_w is called the *ion product of water*. Table 5.1 shows selected values of $pK_w = -\lg(K_w/\mathrm{M}^2)$ for liquid water at 1 bar ($= 0.1$ MPa).

The formula H_3O^+ for the oxonium ion does not give an adequate picture of the structure of the hydrated hydron. Actually, the naked hydron has a large affinity for water forming oligoaquahydrogen(1+), $[H(H_2O)_n]^+$, but the affinity for further addition of water is small after n has reached 8. For this reason, the simple notation H^+ is preferred, which does not make any assumptions about the exact composition

---

[a] Gk. $\alpha\mu\varphi o\tau o\varsigma \;\hat{=}\;$ two-handled.

Table 5.2 Thermodynamic functions for the dissociation of water		
pK_{w}	13.997 ± 3	
$\Delta_{\mathrm{r}} G°$	79.886	kJ mol^{-1}
$\Delta_{\mathrm{r}} H°$	55.76	kJ mol^{-1}
$\Delta_{\mathrm{r}} S°$	80.71	J (mol K)$^{-1}$
$\Delta_{\mathrm{r}} C_p°$	223.6	J (mol K)$^{-1}$
d($\Delta_{\mathrm{r}} C_p°$)/dT	3.8	J (mol K^2)$^{-1}$
$\Delta_{\mathrm{r}} V°$	−22.12	cm^3 mol^{-1}

The reaction, indicated by the subindex r, is Equation 5.5. $T = 298.15$ K; $p = 1$ bar.

of oxonium in aqueous solution. Now, because the concentration of hydron [H$^+$] is an observable physical quantity, it is customary to write Equation 5.3 in the following way:

$$H_2O(l) \;\rightleftharpoons\; H^+(aq) + HO^-(aq) \tag{5.5}$$

This does not change the definition of the ion product, Equation 5.4, but usually, thermodynamic quantities refer to this equation rather than to Equation 5.3 (Table 5.2).

Returning to the ion product of water, Table 5.1 shows its temperature dependence in the liquid state at 1 bar. Using the van't Hoff equation, Equation 2.173, and the enthalpy for the process, Table 5.2, one can see that the dependence is quite large. As an example, at $(\theta, p) = (-35\,°C, 1\ bar)$ one finds p$K_{\mathrm{w}} \approx 17$. It is obvious that such a large change relative to room temperature is of great significance for those forms of life that have evolved in cold environments. However, for conditions that are common in the upper mantle of the Earth, $(T, p) = (1300\ K, 10\ GPa)$, the mole fraction of hydron in water is 1.2×10^{-3}, to be compared with the value of 6.4×10^{-10} at standard temperature and pressure, STP: $(T, p) = (273.15\ K, 1.013\ bar)$.[a] Extrapolation of available data indicates that dissociation increases at still higher temperatures, finally reaching almost complete dissociation. Under such conditions, water is an aggressive species with a chemistry comparable to that of the molten hydroxides.[22] The ion product is also dependent on the ionic strength.[b] Some representative values for seawater are shown in Figure 5.1.

b. An acid and its conjugate base

In this section we consider generic expressions for the chemical equilibrium of an acid HA and its conjugate base A$^-$ in aqueous solution.

---

[a] $x(H^+) = (10^{-14.9/2})/55.5 = 6.4 \times 10^{-10}$.
[b] See Equation 5.56 and the following discussion.

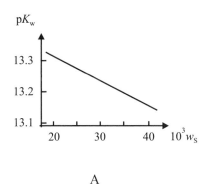
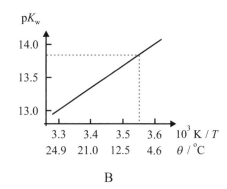

A   B

Figure 5.1 *The ion product of water in seawater*[175]
Panel A. pK_w as a function of salinity at temperature $\theta = 25\,^\circ$C.
Panel B. pK_w as a function of temperature at salinity $w_S = 35 \times 10^{-3}$.
The quality of the data allows readings from the plots, for example, from panel B: at
$\theta = 8.5\,^\circ$C, that is,
$T^{-1} = 0.00355\ \text{K}^{-1}$, $pK_w = 13.8$.

The strength of an acid

The strength of an acid is measured by the equilibrium constant for the dissociation
of the acid in water,

$$HA(aq) + H_2O(l) \rightleftharpoons H^+(aq) + A^-(aq) \tag{5.6}$$

It is understood that the entity HA may have an ionic charge and that the formula
for A^- must then be modified accordingly (see Equation 5.1). The acidity constant
is defined as the equilibrium constant

$$K_a = \frac{[H^+][A^-]}{[HA]} \tag{5.7}$$

The interpretation of the components of Equation 5.7 depends on the applications. For
thermodynamic uses, activities enter the equation, and the constant may be dimensionless, with the experimental conditions and standard states specified separately.
For the present discussion, which aims at environmental applications, the constant
has the dimension of concentration. The value of the constant may be measured at
a given ionic strength, or it may be taken from a table and corrected for use in the
system of interest.

It is common practice to use the acidity exponent

$$pK_a \overset{\text{def}}{=} -\lg(K_a/\text{M}) \tag{5.8}$$

in place of the acidity constant.[a] This definition parallels the classical pH definition
(S. P. L. Sørensen, 1909),

$$pH \overset{\text{def}}{=} -\lg([H^+]/\text{M}) \tag{5.9}$$

---

[a] We follow IUPAC[126] in using "lg" for the decadic logarithm (Briggs' logarithm, base 10) and "ln" for
the natural logarithm (Napierian logarithm, base e).

Note that the definition of the operator "p" (for *power* exponent) includes the definite unit of concentration $M = mol\ dm^{-3} = mol/l$.

In analogy to Equation 5.7, the basicity constant is defined using the reaction

$$A^-(aq) + H_2O(l) \rightleftharpoons HA(aq) + HO^-(aq) \qquad (5.10)$$

as the equilibrium constant

$$K_b = \frac{[HO^-][HA]}{[A^-]} \qquad (5.11)$$

This applies to all bases, but the conjugate base A^- of the acid HA is chosen here to illustrate that the product of the two constants

$$K_a K_b = K_w \qquad (5.12)$$

is the ion product of water. Note, however, that today, acidity constants are preferred for all practical applications of acid-base equilibria.

The more complete the reaction of an acid with water[a] the greater the acidity constant and the stronger the acid. For a certain strength of an acid, reaction 5.6 is considered as complete, and stronger acids cannot be distinguished in dilute aqueous solution. In practice, with $c_{HA} \approx 0.1$ M, this happens for $[H^+] > 10\ [HA]$, that is, for $pK_a \approx 0$. Reactions 5.3 and 5.6 show that the strongest acid available in water is oxonium. The ion product of water is also its acidity constant, and Equation 5.4 shows that the weakest acid that can be detected as an acid in water has $pK_a \approx pK_w$.

Similarly, the strongest base in aqueous solution is hydroxide, and the weakest base is water itself.

It is seen that the acidity and basicity of water exerts a profound effect on the strength of acids and bases available in aqueous solution; this is the so-called leveling effect of an amphiprotic solvent. Thus, the strengths of acids stronger than oxonium are reduced to the strength of that ion in water: the sequence $K_{HClO_4} > K_{HI} > K_{HBr} > K_{HNO_3} > K_{HCl}$ for the acidity constants of selected strong acids is known from other, independent experiments. Nevertheless, these species have equal acidities in aqueous solution. It follows from Equation 5.12 that their conjugate bases are weaker bases than water. Similarly, the sequence $K_{NaH} > K_{NaNH_2} > K_{NaOCH_2CH_3}$ for basicity constants is known, but one cannot observe any difference in water, and their conjugate acids, H_2, NH_3, and CH_3CH_2OH, are weaker acids than water. Acidity exponents of some common acids are given in Table 5.3 together with names of the acids and their conjugate bases.

Weak acids

Acidity constants of weak acids are determined via acid-base titrations in aqueous solution. With a proper experimental setup, a simultaneous determination of the ion product of water can be achieved.

Here, we consider a simple, robust method for determining acidity constants: pH is measured in a dilute aqueous solution of an acid HA with a stoichiometric concentration c_{HA}, and a soluble salt of the acid (i.e., the conjugate base), which is

---

[a] For unit activity of the acid. The dissociation of weak acids increases with increasing dilution.

Table 5.3 Acidity exponents of some common acids					
#	Acid	pK_a	#	Acid	pK_a
1.	HCl	−3	*16.*	$C_6H_5SO_3H$	−4
2.	HNO_3	−1	*17.*	CH_3COOH	4.76
3.	H_2SO_4	−3	*18.*	CH_3CH_2COOH	4.87
4.	HSO_4^-	1.9	*19.*	$C_6H_5CH_2COOH$	4.31
5.	H_3PO_4	2.1	*20.*	C_6H_5COOH	4.18
6.	$H_2PO_4^-$	7.2	*21.*	$(COOH)_2$	1.2
7.	HPO_4^{2-}	12.3	*22.*	$HOOC\cdot COO^-$	4.2
8.	H_2CO_3	6.4	*23.*	$H_3N^+\cdot CH_2\cdot COOH$	2.3
9.	HCO_3^-	10.3	*24.*	$H_3N^+\cdot CH_2\cdot COO^-$	9.6
10.	H_2S	7.2	*25.*	CH_3CH_2OH	18
11.	HS^-	12.9	*26.*	C_6H_5OH	9.9
12.	NH_4^+	9.25	*27.*	$CH_3CH_2NH_3^+$	10.6
13.	H_4SiO_4	10	*28.*	$C_6H_6NH_3^+$	4.6
14.	$[Fe(H_2O)_6]^{3+}$	2.2	*29.*	HCOOH	3.8
15.	$[Al(H_2O)_6]^{3+}$	4.9			

Strong acids: $pK_a < 0$; weak acids: $0 < pK_a < 14$; aprotic in aqueous solution: $pK_a > 14$. The names of the acids and their conjugate bases are[a] (1) hydrochloric acid – chloride; (2) nitric acid – nitrate; (3) sulfuric acid – hydrogensulfate; (4) hydrogensulfate – sulfate; (5) phosphoric acid – dihydrogenphosphate; (6) dihydrogenphosphate – hydrogen-phosphate; (7) hydrogenphosphate – phosphate; (8) carbonic acid – hydrogencarbon-ate; (9) hydrogencarbonate – carbonate; (10) dihydrogensulfide – hydrogensulfide; (11) hydrogensulfide – sulfide; (12) ammonium – ammonia; (13) silicic acid – trihydroxidooxidosilicate(1−); (14) hexaaquairon(3+) – pentaaquahydroxidoiron(2+); (15) hexaaquaaluminium(3+) – pentaaquahydroxidoaluminium(2+); (16) benzenesulfonic acid – benzenesulfonate; (17) acetic acid – acetate; (18) propionic acid – propionate; (19) phenylacetic acid – phenylacetate; (20) benzoic acid – benzoate; (21) oxalic acid – hydrogenoxalate; (22) hydrogenoxalate – oxalate; (23) glycinium – glycine; (24) glycine – glycinate; (25) ethanol – ethanolate; (26) phenol – phenolate; (27) ethaneammonium – ethaneamine; (28) anilinium – aniline; (29) formic acid – formate.

taken here to be the potassium salt KA, having the stoichiometric concentration c_{KA}. pK_w is taken from tables such as Table 5.1.

The stoichiometric conditions are[b]

$$c_{HA} + c_{KA} = [HA] + [A^-] \tag{5.13}$$

and

$$c_{KA} + [H^+] = [A^-] + [HO^-] \tag{5.14}$$

When these relations are substituted into the acidity constant, Equation 5.7, one obtains

$$K_a = [H^+]\frac{c_{KA} + [H^+] - [HO^-]}{c_{HA} - [H^+] + [HO^-]} \tag{5.15}$$

---

[a] Note that the generic name *ion* is not used; that is, hydrogensulfate and dihydrogenphosphate are the names of the chemical species HSO_4^- and $H_2PO_4^-$, respectively.

[b] Symbols for concentrations are collected in Appendix 1, Table A1.7.

In the present case, where known amounts of HA and KA have been added to the water phase, the relations

$$c_{HA} > 10 \left| [HO^-] - [H^+] \right| \quad \text{and} \quad c_{KA} > 10 \left| [H^+] - [HO^-] \right| \quad (5.16)$$

and the condition

$$\frac{10}{1} > \frac{c_{HA}}{c_{KA}} > \frac{1}{10} \tag{5.17}$$

are easily fulfilled simultaneously. Then, to a good approximation, the acidity constant is obtained from a single pH measurement on the solution of known composition, limited by the conditions (Equations 5.16 and 5.17):

$$K_a \approx [H^+] \frac{c_{KA}}{c_{HA}} \tag{5.18}$$

Notes: (1) Equation 5.18 contains the same approximations as the common buffer equation 5.26. (2) Equation 5.15 is the master equation for all "pH formulas" in standard textbooks of general chemistry.

Degree of dissociation

A physical quantity representing the extent of the dissociation of an acid was introduced by Ostwald (1888). We shall see that such a concept is very important for understanding the complex systems that arise in seemingly simple environmental problems.

Definition. The degree of dissociation of an acid is the mole fraction of the conjugate base:

$$\alpha_0 = \frac{[A^-]}{c} \tag{5.19}$$

where $c = c_{HA} + c_{KA}$ is the total concentration of the conjugate acid-base system (see Equation 5.13). The index 0 indicates that no hydrons have been taken up by the base. Similarly, the mole fraction of the conjugate acid (one hydron taken up by the base) is given by

$$\alpha_1 = \frac{[HA]}{c} \tag{5.20}$$

with the obvious condition that

$$\alpha_0 + \alpha_1 = 1 \tag{5.21}$$

Using Equations 5.7–5.9, one can obtain expressions for the two fractions in terms of $[H^+]$ and K_a or, equivalently, in terms of pH and pK_a,

$$\alpha_0 = \frac{K_a}{[H^+] + K_a} = \frac{1}{1 + 10^{pK_a - pH}} \tag{5.22}$$

$$\alpha_1 = \frac{[H^+]}{[H^+] + K_a} = \frac{1}{1 + 10^{pH - pK_a}} \tag{5.23}$$

For historical reasons, acid-base chemistry is described in terms of dissociation constants. Therefore, the mole fraction α_0 is still called the "degree of dissociation." With the development of modern theories of electrolytes and an increased understanding of the chemistry of coordination compounds, focus shifted to complex formation and association constants.[9] In many respects a description based on the uptake of hydrons leads to a lucid understanding of acid-base chemistry. Accordingly, the α_i quantities will be called degrees of formation: α_0 is the degree of formation of species A^-, and α_1 is the degree of formation of species HA.

The α_i quantities are used in Bjerrum diagrams and speciation diagrams where they are plotted against the concentration of free (i.e., not complex bound) hydron or ligand.

Alkalinity

Definition. The alkalinity A of an aqueous solution is the concentration of titratable base.

Consider an aqueous solution of an acid-base pair, subject to the stoichiometric conditions given by Equations 5.13 and 5.14. Here, the titratable base is the added amount of potassium salt KA whose concentration is given by Equation 5.14. Owing to Equation 5.19, one has

$$A = [HO^-] - [H^+] + c\alpha_0 \tag{5.24}$$

which defines the alkalinity of a solution of a conjugate acid-base pair. Addition of weak acid does not change the alkalinity because it is not titratable with strong acid.

Alkalinity is an equilibrium property that can be determined by an acid-base titration. The end point of the titration, a definite value of pH, is reached when an equivalent amount of the titrant has been added according to the reaction equation. The titration of a strong base has an end point at $[H^+] = \sqrt{K_w}$, whereas titrations of weaker bases have end points at greater hydron concentrations. A negative value of the alkalinity means that the solution contains a surplus of acid, which has to be titrated with base in order to obtain $A = 0$. Analogously, titrations of strong acids also have an end point at $[H^+] = \sqrt{K_w}$, and weaker acids titrate to lower hydron concentrations. Examples are shown in Figures 5.2 and 5.3.

Some remarks on nomenclature: owing to the symmetry of the expressions, one could equally well have chosen the concentration of titratable acid, $c_{HA} = [H^+] - [HO^-] + c\alpha_1$, as the basic physical quantity. Such a choice was widespread around World War I: c_{HA} was called the acidity, pH was called the reaction, and the word *basicity* was used for the alkalinity, Equation 5.24. However, since then the usage has changed. Nowadays, the pH of a solution is called its *acidity* or *degree of acidity*, and, with the appearance of environmental chemistry, the word *alkalinity* has been given its present meaning, namely, the definition in Equation 5.24.

Buffer value

Definition. The buffer value of an aqueous solution is the derivative of the alkalinity with respect to pH.

From Equation 5.24, one obtains

$$\beta \overset{\text{def}}{=} \frac{dA}{d\,\text{pH}} \tag{5.25a}$$

$$= \{c\alpha_0\alpha_1 + [H^+] + [HO^-]\} \times \ln 10 \tag{5.25b}$$

Further differentiation shows that $d\beta/d\text{pH} = 0$ for $\alpha_1 = \alpha_0$, implying that the buffer value has an extremum (more precisely, it is a maximum) for $\alpha_1 = \alpha_0 = 1/2$ with $\beta = (c/4)\ln 10 \approx 0.58c$. This happens at $\text{pH} = pK_a$. In general, at pH values where $|\text{pH} - pK_a| < 1$, the buffer equation

$$\text{pH} \approx pK_a + \lg \frac{c_{KA}}{c_{HA}} \tag{5.26}$$

gives a good approximation of pH (see Equation 5.18). In medicinal and biochemical literature, this equation is sometimes called the Henderson-Hasselbalch equation (Henderson, 1908, and Hasselbalch, 1927).

Note that the buffer value is also large at low and high pH values, because it is proportional to the hydron concentration as well as the hydroxide concentration.

Example 1. A conjugate acid-base pair

The basic properties of a conjugate acid-base pair in aqueous solution are shown in Figure 5.2 with acetic acid–acetate as the example. The experimental data c, pK_a, and pK_w are given in the legend. A titration of 0.10 M acetate with strong acid starts at $\text{pH} = 8.88$, $A(8.88) = 0.10$ M, and the end point occurs at $\text{pH} = 2.88$, $A(2.88) = 0.00$ M. Negative values of A mean that the end point has been exceeded.

Example 2. Several conjugate acid-base pairs

Figure 5.3 shows an acid-base titration. The titrand is a basic solution of ammonia and acetate and the titrant (titrator) is a strong acid. Consider some characteristic values of the alkalinity as a function of pH, $A(\text{pH})$. The titration starts at $A(13.00) = 0.4$ M, then $A(11.0) = 0.3$ M and $A(7.0) = 0.1$ M, and the end point is $A(2.88) = 0.0$ M. In terms of hydron uptake, one may say that the titrand has taken up first one, then two, and then one equivalent of hydrons in a stepwise manner, where one "equivalent" is equal to 0.1 M.

c. Oligovalent acids

In this section, the principles and methods that were used in the discussion of a monovalent acid-base system will be used to describe di- and trivalent acids. Indeed, it is possible to generalize the results to include polyvalent acids as shown in Appendix A2.1, but most tables of equilibrium constants use the formalism of complex formation first introduced for coordination compounds[9] (see Section 5.2 and Appendix A2.2).

Analysis of titration curves reveals that weak polyvalent acids give off hydrons as a function of one independent variable, the concentration of free hydron. The term *free hydron* refers to hydron that is not the bound to the acid, but rather to solvent molecules. It is easily seen from the following expressions that once the equilibrium

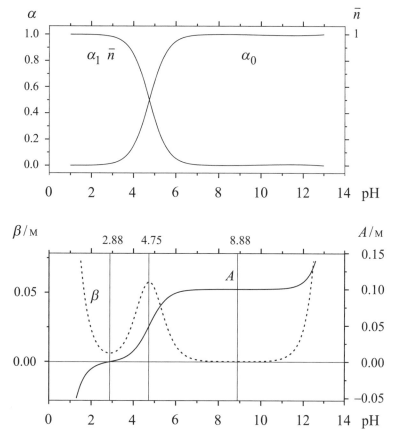

Figure 5.2 *The acetic acid–acetate system*
Titration of 0.10 M potassium acetate at 25 °C with strong acid. $pK_a = 4.75$; $pK_w = 14.00$.
Upper panel: Speciation diagram $\alpha_i(\mathrm{pH})$ and formation function $\bar{n}$ (pH) (note: $\bar{n}\,\alpha_1$).
Lower panel: Alkalinity $A(\mathrm{pH})$ (solid line, right scale) and buffer value $\beta(\mathrm{pH})$ (broken line, left scale). A solution of 0.10 M CH_3COOK has $A(8.88) = 0.10$ M, and 0.10 M CH_3COOH has $A(2.88) = 0.00$ M.

constants and $\mathrm{pH} = -\lg\{[H^+]\,/\,\text{M}\}$ are known, the composition of the system is also known. Because the constants are normally taken from tables, this explains why the determination or an assessment of pH is central to all work with aqueous solutions.

A divalent acid H_2A has two dissociation constants.[a] They are stepwise constants, and each equilibrium step deals with one conjugate acid-base pair:

$$H_2A \;\rightleftharpoons\; H^+ + HA^- \qquad\qquad K_{a1} = \frac{[H^+][HA^-]}{[H_2A]} \qquad (5.27a)$$

$$HA^- \;\rightleftharpoons\; H^+ + A^{2-} \qquad\qquad K_{a2} = \frac{[H^+][A^{2-}]}{[HA^-]} \qquad (5.27b)$$

---

[a] A possible ionic charge q on the species H_2A is suppressed in the equations to follow, because the equations are independent of such charge.

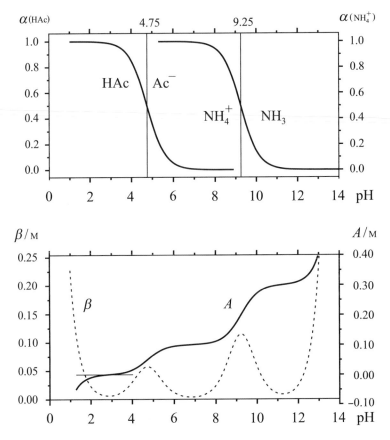

Figure 5.3 *Two conjugate acid-base pairs*
Titration of a mixture of 0.10 M acetate, 0.20 M ammonia, and 0.10 M hydroxide with strong acid.
$pK_a(HAc) = 4.25$, $pK_a(NH_4^+) = 9.25$, $pK_w = 14.00$.
Upper panel: Speciation diagram $\alpha_1(pH)$ for the two conjugate acid-base pairs.
Lower panel: Alkalinity $A(pH)$ (solid line, right scale) and buffer value $\beta(pH)$ (broken line, left scale).
The titration starts at $A(13.0) = 0.40$ M and ends at $pH = 2.88$.

The stoichiometric concentration of the acid-base system is given by

$$c \;=\; [H_2A] + [HA^-] + [A^{2-}] \tag{5.28a}$$

$$=\; [H_2A]\left(1 + K_{a1}[H^+]^{-1} + K_{a1}K_{a2}[H^+]^{-2}\right) \tag{5.28b}$$

$$\overset{\text{def}}{=}\; [H_2A]\,\sigma \tag{5.28c}$$

where the sum $\sigma = \sigma([H^+])$ is a function of the concentration of free hydron. In analogy to the definition for the monovalent acid-base system, one defines the mole fraction of each species through the relations

$$\alpha_0 \;=\; \frac{[A^{2-}]}{c} \;=\; \frac{K_{a1}K_{a2}}{\sigma[H^+]^2} \tag{5.29a}$$

$$\alpha_1 = \frac{[HA^-]}{c} = \frac{K_{a1}}{\sigma[H^+]} \tag{5.29b}$$

$$\alpha_2 = \frac{[H_2A]}{c} = \frac{1}{\sigma} \tag{5.29c}$$

which are subject to the condition

$$\sum_{i=0}^{2} \alpha_i = 1 \tag{5.30}$$

and the functions, expressed through the independent variables, are $\alpha_i = \alpha_i\,([H^+], c)$. The alkalinity may then be given by

$$A = [HO^-] - [H^+] + [HA^-] + 2[A^{2-}] \tag{5.31a}$$

$$= [HO^-] - [H^+] + c\,(\alpha_1 + 2\alpha_0) \tag{5.31b}$$

$$\overset{\text{def}}{=} [HO^-] - [H^+] + c\,d \tag{5.31c}$$

where the sum d, which is a function of the concentration of free hydron, is called the dissociation function of the acid-base system.

Finally, the buffer value is given by

$$\beta = \frac{dA}{d\,\mathrm{pH}}$$
$$= \left[c\left(4\alpha_0\alpha_2 + \alpha_1(1 - \alpha_1)\right) + [H^+] + [HO^-]\right] \times \ln 10 \tag{5.32}$$

Applications of Equations 5.29, 5.31, and 5.32 are shown for the system carbon dioxide–water in Figure 5.5. Figure 5.4 for phosphoric acid can be obtained by a straightforward extension of these equations.

Determination of acidity constants[a]
Given a solution of a weak divalent acid of known concentration c to which strong acid and strong base have been added such that the concentrations are c_a and c_b, respectively,

$$c_b - c_a = [HO^-] - [H^+] + c\,d \tag{5.33}$$

Introducing the ion product of water K_w leads to a nonlinear equation in $[H^+]$:

$$[H^+] - \frac{K_w}{[H^+]} - c\,d + c_a - c_b = 0 \tag{5.34}$$

The numerical method used to determine an acidity constant based on experimental data is to adjust the constants such that Equation 5.34 is fulfilled for measured values of $[H^+]$ in a series of solutions.[197]

For practical work, suppose a volume V_a containing the stoichiometric concentrations c' and c_a' of the two acids is mixed with a volume V_b containing the stoichiometric concentration c_b' of the base. Then the volume fractions are $\varphi_a = V_a/(V_a +$

---

[a] Expressions of s and d for a general weak polyvalent acid are given in Appendix A2 together with appropriate Mathematica routines.

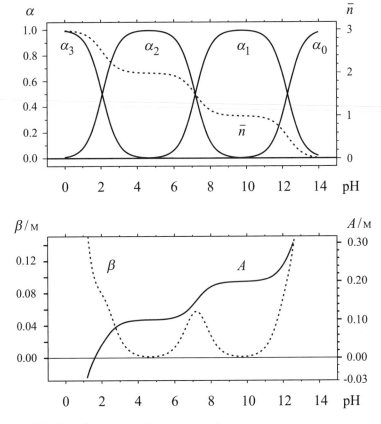

Figure 5.4 *Titration of 0.10 M phosphate with strong acid*
$pK_{a1} = 2.1$, $pK_{a2} = 7.2$, $pK_{a3} = 12.3$, $pK_w = 14.0$.
Upper panel: Speciation diagram $\alpha_i(pH)$ (solid lines, left scale) and formation function $\bar{n}(pH)$, that is, average number of bound hydrons per PO_4^{3-} species (broken line, right scale, see Equation A2.25).[a]
Lower panel: Alkalinity $A(pH)$ (solid line, right scale) and buffer value $\beta(pH)$ (broken line, left scale). As seen, phosphate can only be titrated to dihydrogen phosphate by an endpoint titration.

V_b) and $\varphi_b = V_b/(V_a + V_b)$, and the concentrations of Equation 5.34 are $c = c'\,\varphi_a$, $c_a = c_a'\,\varphi_a$, and $c_b = c_b'\,\varphi_b$.

Isoelectric point

The isoelectric point, iep, is the pH value at which the net charge of all species in an acid-base system is zero.[123] The word was originally coined for research in electrophoresis[b] but is now used generally for amphoions in acid-base chemistry, as will be discussed for glycine p. 184. The solubility reaches a minimum at the isoelectric point and increases on both sides because the electrically neutral species gains a charge (see footnote c on p. 184).

---

[a] Speciation diagrams are easily drawn by hand: Use the pK's, Equations 5.17 and 5.18, together with the general forms of Equations 5.22 and 5.23.

[b] Electrophoresis is a way of separating ions according to charge using a static electric field.

Table 5.4 pK_a values for normal dicarboxylic acids								
i	0	1	2	3	4	5	6	7	8
pK_{a1}	1.42	2.83	4.16	4.31	4.43	4.49	4.52	4.60	4.62
pK_{a2}	4.35	5.69	5.61	5.41	5.41	5.58	5.55	5.56	5.60

Formula: $HOOC \cdot [CH_2]_i \cdot COOH$, $0 \leq i \leq 8$.[129, 146]

Consecutive acidity constants

We shall discuss the variation of the size of stepwise acid dissociation constants in the series K_{a1}, K_{a2} (and K_{a3}) for some acids:

1. The oxoacids of the main group elements H_pXO_{p+q} may be written using the general formula $XO_q(OH)_p$, which shows the molecular structure more correctly. The larger q is, the greater the electron-withdrawing power of the electronegative oxygens (inductive effect), and the more easily the first hydron is released. For neutral oxoacids, a rough estimate of the value of pK_{a1} is given by

$$pK_{a1} = 7 - 5q \tag{5.35}$$

For example, for the acids $HClO$, $HClO_2$, $HClO_3$, and $HClO_4$, pK_{a1} values of 7, 2, -3, and -8 are calculated, respectively.[a] In most cases the value of pK_a increases by 5 units when going from one pK_a value to the next one in a series.

Phosphoric acid H_3PO_4 illustrates these rules. The structural formula is $PO(OH)_3$ (trihydroxidooxidophosphorus) and the (experimentally obtained) acidity exponents are $pK_{a1} = 2.1$, $pK_{a2} = 7.2$ and $pK_{a3} = 12.3$. Figure 5.4 shows the variation of several parameters during a titration of 0.10 M phosphate solution with a strong acid.

Phosphonic acid, H_3PO_3, with the (tetrahedral) geometry $PH(O)(OH)_2$ and $q = 1$ (hydridodihydroxidooxidophosphorus), obeys Equation 5.35, having $pK_{a1} = 1.3$ and $pK_{a2} = 6.7$. Other acids with $q = 1$ include HNO_2 and H_2SO_3. Carbonic acid, $CO(OH)_2$, is discussed separately later.

2. The normal dicarboxylic acids are divalent acids with the general formula $HOOC \cdot [CH_2]_i \cdot COOH$ ($= H_2A$) with $i \geq 0$. The two consecutive equilibria are given by Equations 5.27. The acidity exponents for $0 \leq i \leq 8$ are given in Table 5.4. With increasing spatial distance between the two carboxylic groups, the difference $pK_{a2} - pK_{a1}$ decreases. This is due to a simple electrostatic effect: the negative charge makes HA^- a weaker acid than the similar acyclic monocarboxylic acid. For statistical reasons, the difference is expected to approach $\lg 4 = 0.6$ for large values of i.[b] Freehand drawings of the degrees of formation α_i as a function of pH (a speciation diagram) show that the region of existence of HA^- decreases with decreasing difference between the two pK values, that is, with an increasing value of i.

---

[a] The experimental values are 7.5, 2.0, -1 and -7.
[b] The divalent acid H_2A may lose a hydron from two different positions, and the anion A^{2-} may gain a hydron at two positions.[146]

3. Unsymmetrical conjugate acid-base pairs.[146] A large number of acids (bases) do not have a symmetrical conjugate base (acid). By this is meant that the species HA^- of Equations 5.27 is actually a set of isomeric species, for example, HA'^- and HA''^-. Examples include carbonic acid (see later discussion), unsymmetrically substituted succinic acids such as L-malic acid,[a] and the common amino acids.[b] The general equilibrium pattern is shown in Equation 5.36:

$$H_2A \begin{array}{c} K_1' \quad HA'^- \quad K_2' \\ \nearrow \qquad \searrow \\ \Updownarrow K \qquad A^{2-} \\ \searrow \qquad \nearrow \\ K_1'' \quad HA''^- \quad K_2'' \end{array} \qquad (5.36)$$

In these cases the two acid constants of Equations 5.27 are related to the more fundamental constants of Equation 5.36 by the following equations:

$$K_{a1} = K_1' + K_1'' \qquad (5.37)$$

$$\frac{1}{K_{a2}} = \frac{1}{K_2'} + \frac{1}{K_2''} \qquad (5.38)$$

$$K = \frac{K_1'}{K_2''} = \frac{K_2''}{K_2'} \qquad (5.39)$$

Example

Glycine (aminoethanoic acid, Gly). The species are identified as follows:

$$H_2A = {^+H_3NCH_2COOH}$$
$$HA'^- = {^+H_3NCH_2COO^-}, \quad HA''^- = H_2NCH_2COOH$$
$$HA^{2-} = H_2NCH_2COO^-$$

The constants at 25 °C are found to be

$$K_{a1} = 4.47 \times 10^{-3} M^{-1}, \quad K_{a2} = 2.45 \times 10^{-10} M^{-1}, \quad K = 400 \qquad (5.40)$$

and the system of equilibrium constants, Equation 5.36, is completely described. The isoelectric point pH_{iep} is the pH at which the net electric charge of all species containing A is zero. Hence, $pH_{iep} = 5.98$, and the amphoion (zwitterion), $H_3N^+CH_2COO^-$, is the dominant form at this pH.[c] To a good approximation, the isoelectric point is independent of dilution, but detailed calculations give the values

$$c_{Gly} = 10^{-1} M, \quad pH_{iep} = 5.99$$
$$c_{Gly} = 10^{-2} M, \quad pH_{iep} = 6.06 \qquad (5.41)$$
$$c_{Gly} = 10^{-3} M, \quad pH_{iep} = 6.34$$

showing that pH_{iep} approaches $\sqrt{K_w}$ with increasing dilution.

---

[a] (S)-Hydroxybutanedioic acid.
[b] See Appendix A1.9.
[c] For $c_{Gly} = 0.10000$ M, $[^+H_3NCH_2 COO^-] + [H_2NCH_2 COOH] = 0.09996$ M, $[^+H_3NCH_2 COOH] = [H_2NCH_2 COO^-] = 0.00002$ M.

4. Examples of acids that do not follow the rules just discussed:
   a. Boric acid, H_3BO_3, which is $B(OH)_3$ (trihydroxidoboron), is an antibase whose acidity arises by complex formation with hydroxide:

$$B(OH)_3 + H_2O \quad \rightleftharpoons \quad H^+ + B(OH)_4^- \qquad\qquad pK_a = 9.2 \qquad (5.42)$$

   b. Silicic acid, H_4SiO_4, or rather $Si(OH)_4$ (tetrahydroxidosilicon), is a monovalent acid, $pK_a = 9.9$, which only exists in very dilute aqueous solutions. Nevertheless, salts of tetraoxidosilicate(4−) are found in Nature; see Section 3.1b.
   c. The consecutive acid constants of tetraaquamercury(II) reflect an interesting aspect of the chemistry of mercury(II):

$$Hg(H_2O)_4{}^{2+} \quad \rightleftharpoons \quad H^+ + Hg(H_2O)_3(OH)^+ \qquad pK_{a1} = 3.59 \quad (5.43a)$$
$$Hg(H_2O)_3(OH)^+ \quad \rightleftharpoons \quad H^+ + Hg(OH)_2 + 2\,H_2O \quad pK_{a2} = 2.59 \quad (5.43b)$$

Here, the second acidity exponent is less than the first! This is due to the change from tetrahedral to linear coordination, the energetically more favorable geometry.

The carbon dioxide–water system

The behavior of carbonic acid in natural waters is an important example of a divalent acid-base system. The rate constants presented in this section show that surface water and liquid water in the atmosphere are in equilibrium with atmospheric carbon dioxide.

The chemical equilibria are shown in Equations 5.44 and 5.45. Because only CO_2 can exist in both phases, the state of aggregation is only marked for this species.

$$(5.44)$$

$$HCO_3{}^- \quad \rightleftharpoons \quad H^+ + CO_3{}^{2-} \qquad\qquad K_{a2} \qquad (5.45)$$

As regards the algebraic description, it is noted that Equation 5.44 has the same structure as Equation 5.36. The left-hand portion represents Henry's law (Equation 2.205 and Appendix A1.7):

$$k_H^{-1} = \frac{[CO_2(aq)] + [H_2CO_3]}{p(CO_2(g))} = K_0' + K_0'' \qquad (5.46)$$

The Henry's law constant may be obtained from Table 5.5. The right-hand side of Equation 5.44 defines the first acidity constant K_{a1} of carbonic acid,

$$\frac{1}{K_{a1}} = \frac{1}{K_1'} + \frac{1}{K_1''} \qquad (5.47)$$

whereas the acidity constant for hydrogencarbonate K_{a2} refers to Equation 5.45 in the usual way. Finally, the constant K is given by

$$K = \frac{[CO_2(aq)]}{[H_2CO_3]} = \frac{K_0'}{K_0''} = \frac{K_1''}{K_1'} \qquad (5.48)$$

Hence, the equilibria are completely described by the constants k_H, K_{a1}, K, and K_{a2}. In order to measure values for these quantities, the experimental demands include the determination of the CO_2 partial pressure, an acid-base titration, and a spectroscopic method for measuring the ratio K.

We now explore some of the constants by considering reaction rates. The exchange rate of carbon dioxide between the surface of the sea and the atmosphere is an important process that is very difficult to assess, much less to measure. The problem is discussed in Chapter 7, but as an estimate, the amount flux density may be set to $\approx 3 \times 10^{-7}$ mol m^{-2} s^{-1} (in both directions) leading to a lifetime[a] of carbon dioxide in the atmosphere of around 17 a (see Equation 7.1).

Gaseous carbon dioxide, $CO_2(g)$, reacts with water and forms an adduct, aqueous carbon dioxide, $CO_2(aq)$, which in turn reacts with the coordinated water to form carbonic acid, $H_2CO_3(aq)$:

$$CO_2(aq) \underset{k_{-1}}{\overset{k_{+1}}{\rightleftharpoons}} H_2CO_3(aq) \tag{5.49}$$

$$k_{+1} = 3.0 \times 10^{-2}\,\text{s}^{-1}, \qquad k_{-1} = 11.9\,\text{s}^{-1} \tag{5.50}$$

Here k_{+1} and k_{-1} denote the rate constants for the reaction from left to right and from right to left, respectively. They are pseudo-first-order constants because one of the reactants, water, is present in great excess. At equilibrium, the reaction rates are the same in both directions, and the equilibrium constant is the ratio of the rate constants:

$$K = \frac{[CO_2(aq)]}{[H_2CO_3(aq)]} = \frac{k_{-1}}{k_{+1}} \approx 400 \tag{5.51}$$

Similarly, the acidity constant of carbonic acid may be expressed as the ratio between two rate constants:

$$H_2CO_3(aq) \underset{k_{-1}}{\overset{k_{+1}}{\rightleftharpoons}} H^+ + HCO_3^- \tag{5.52}$$

$$k_{+2} = 8 \times 10^6\,\text{s}^{-1} \qquad k_{-2} = 4.7 \times 10^{10}\,\text{M}^{-1}\,\text{s}^{-1} \tag{5.53}$$

$$K_1'' = \frac{[H^+][HCO_3^-]}{[H_2CO_3]} = \frac{k_{+2}}{k_{-2}} = 10^{-3.78}\,\text{M} \tag{5.54}$$

The reaction from right to left is very fast. It can be shown that this rate is diffusion limited: each encounter leads to reaction. As previously described, the recombination of H^+ and HO^-, with a rate constant of 1.4×10^{11} M^{-1} s^{-1} at 25 °C, is one of the fastest reactions observed in aqueous solutions.

The acidity constant of aquated carbon dioxide is $K_1' = K_1''/K \approx 10^{-6.38}$ M, and this is approximately equal to the value of K_{a1}. The fact that the dominant species in aqueous solutions of carbon dioxide is not carbonic acid explains the deviation from the empirical rule, Equation 5.35.

The constants used to describe carbon dioxide in natural waters must take into account the ionic strength I (see Table 5.5).

[a] See Equation 2.53.

Table 5.5 Equilibrium constants for the system[135] $CO_2 - HCO_3^- - CO_3^{2-} - H_2O$

	$\lg(k_H / \text{bar M}^{-1})$	pK_{a1}	pK_{a1}
$I = 0$ M; $\theta = 5\,°C$	1.19	6.52	10.49
$I = 0$ M; $\theta = 25\,°C$	1.46	6.35	10.33
$I = 1$ M; $\theta = 25\,°C$	1.51	6.02	9.57
$w_{\text{salinity}} = 35\,‰; \theta = 5\,°C$	1.27	6.14	9.39
$w_{\text{salinity}} = 35\,‰; \theta = 25\,°C$	1.53	6.00	9.11

k_H is the Henry's law constant, Equation 5.46, K_{a1} and K_{a2} the acidity constants, and I the ionic strength (Equation 5.56).

Finally, we note that the solubility of carbon dioxide is given by

$$c_{CO_2} = (1 + 10^{pH-pK_{a1}} + 10^{2pH-pK_{a1}-pK_{a2}})\, k_H^{-1}\, p_{CO_2} \qquad (5.55)$$

as a function of pH and p_{CO_2}.

Seawater

Figure 5.5 shows the alkalinity as a function of pH for the titration of a 0.785 mM carbonate solution with strong acid. This concentration mimics the behavior of the pH-determining processes of seawater. As seen, hydrogencarbonate is the main component of carbonic acid species in seawater at pK 8.1. The underlying reasons for this and for the remarkable buffering capacity of seawater are debated.[108] See Section 5.2b, example 3.

d. Polyvalent acids

Polyvalent acids play an important role in the chemistry of particles with large specific surface areas. Examples include sols of metal hydroxides, minerals, and organic matter, and gels of proteins (see Section 6.3d).

Quantitative treatments of the acid-base properties of polyvalent acids can be made using the expressions presented in Appendix A2.1. However, a good understanding can often be obtained from speciation diagrams and curves of alkalinity and buffer value made as freehand drawings from known acidity constants.

Example: A polyvalent acid

In order to illustrate the main characteristics of the acid-base properties of a polyvalent acid, we will consider the titration of a pentavalent base $C_{30}H_{38}O_7N^{4-}$ (L^{4-}) with strong acid. The neutral species is $C_{30}H_{42}O_7N$, $M_w = 528$ g/mol, and it can accommodate a hydron, giving the cation $C_{30}H_{43}O_7N^+$, which has the following five acidic groups:

1. An anilinium ion $C_6H_5 \cdot NH_2^+ \cdot CH_2 \cdot CH_2 \cdot R$ $pK_1 = 4.40$
2. Three carboxylic acid groups $R \cdot (CH_2)_3 \cdot COOH$ $pK_{2,3,4} = 4.86$
3. A phenol $1,4\text{-}(R \cdot (CH_2)_3) \cdot C_6H_4 \cdot OH$ $pK_5 = 10.19$

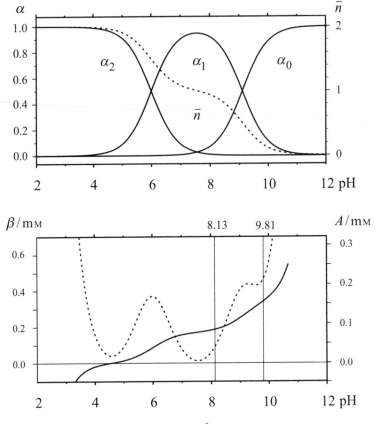

Figure 5.5 *The chemistry of* CO_2–H_2CO_3–HCO_3^-–CO_3^{2-}–H_2O
Titration of 0.785 mM Na_2CO_3 in 0.5 M NaCl-solution at 25 °C with strong acid.
$pK_{a1} = 6.00$; $pK_{a2} = 9.11$; $pK_w = 13.70$.
Upper panel: Speciation diagram $\alpha_i(pH)$ (solid lines, left scale) and formation function $\bar{n}(pH)$
(broken line, right scale).
Lower panel: Alkalinity $A(pH)$ (solid line, right scale; $A(9.81) = 1.57$ mM) and buffer value $\beta(pH)$ (broken line, left scale; $\beta(8.13) = 0.17$ mM).

For later use (see Table 6.5) it is noted that the cation-exchange capacity of the species $C_{30}H_{42}O_7N$ is $b_{cec} = 9.47$ mol/kg at pH 3.

Figure 5.6 shows relevant acid-base data for titration of $c = 0.05$ M Na_4L where c is the stoichiometric concentration. The starting point is given by $A(pH) = 0.25$ M, $\bar{n} = 0$, at pH = 11.33, and the size of α_1 shows that some of the salt has reacted with water. The end point at $A(pH) = 0.00$ M, $\bar{n} = 5$, pH = 2.85 is well defined, because the value $\beta = 0.006$ M is quite low. The isoelectric point, which has $A(pH) = 0.05$ M, $\bar{n} = 4$, occurs at $pH_{iep} = 4.454$. At this point the composition of the solution is $[H_4L] = \alpha_4 c = 0.0201$ M, $[H_5L^+] = \alpha_5 c = 0.0177$ M $= [H_3L^-] + 2[H_2L^{2-}] + 3[HL^{3-}] + 4[L^{4-}] = c(\alpha_3 + 2\alpha_2 + 3\alpha_1 + 4\alpha_0)$ as illustrated in Figure 5.6.[a]

---

[a] All calculations of this example were performed using the Mathematica routines shown in Appendix A2.1.

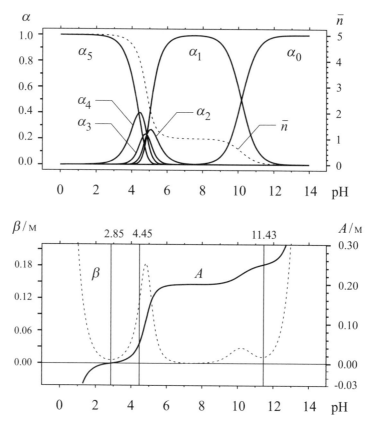

Figure 5.6 *Titration of a pentavalent base*
Titration of 0.05 M L^{4-} with strong acid. Formulas and constants are given in the text.
Upper panel: Speciation diagram $\alpha_i(pH)$ (solid lines, left scale) and the formation function $\bar{n}(pH)$ (broken line, right scale).
Lower panel: Alkalinity $A(pH)$ (solid line, right scale) and buffer value $\beta(pH)$ (broken line, left scale).

5.2 Coordination chemistry

This section discusses three items: (a) the thermodynamics of coordination compounds, (b) a set of empirical rules for their formation, and (c) the coordination chemistry of natural waters.

The choice of thermodynamic standard state is a problem of great practical importance. By tradition, many equilibrium constants are determined by extrapolation of the observed data to a standard state of infinite dilution (see Equation 2.149). When properties are to be estimated on this basis, there are two choices: (1) convert the constants to the actual state, giving so-called stoichiometric constants, and then use actual concentrations (an example is given in Table 5.6), or (2) calculate the activities of the chemical species in the actual state and use the original constants (see Section 5.3b).

Table 5.6 Corrections for constants of formation due to ionic strength				
Type	1:1	1:2	1:3	2:2	2:3
$q = 1$	-0.3	-0.6	-0.9	-1.2	-1.8
$q = 2$	-0.4	-1.0	-1.5	-2.0	-2.8

The correction concerns the logarithm of a constant of formation that has been determined at zero ionic strength and is to be applied in seawater with an ionic strength of 0.7 M. The correction is a function of the number of ligands q and the type of electrolyte. As an example, $AlCl_2^+$ is a 1:3 type with $q = 2$; thus, a table value of $\lg(\beta_2 / M^{-2}) = 2.2$ at $I = 0$ M is to be corrected by the number -1.5 to give $\lg(\beta_2 / M^{-2}) = 2.2 - 1.5 = 0.7$ at $I = 0.7$ M.

Many constants have been determined using a constant salt medium as the standard state (see the examples in Section 2.3a, Choices of standard state), and the user may choose those that fit the actual problem in the best way. Ideally, measurements should be carried out and equilibrium constants determined using the same standard state as that of the applications.

At several places in this section, ionic strength I is used rather than species concentrations. The reason is that this quantity enters the Debye-Hückel limiting law directly (see Equation 5.105). The definition is (Lewis, 1921)

$$I = \frac{1}{2} \sum_i c_i z_i^2 \tag{5.56}$$

where c_i is the concentration of the ion i and z_i its charge. For a 1:1 strong electrolyte B, $I = c(B)$, but for other types of electrolytes (1:2, 2:2, etc.), the ionic strength is greater than the mean value of the concentrations.

a. Complex formation

The formation in solution of a species ML_n from a solvated metal ion M and a ligand L takes place in successive steps such that one can identify stepwise equilibrium constants, so-called consecutive constants.[a] The reaction equations and the stepwise formation constants K_i are

$$M + L \rightleftharpoons ML \qquad K_1 = \frac{[ML]}{[M][L]} \tag{5.57a}$$

$$ML + L \rightleftharpoons ML_2 \qquad K_2 = \frac{[ML_2]}{[ML][L]} \tag{5.57b}$$

$$\cdots \qquad\qquad \cdots$$

$$ML_{n-1} + L \rightleftharpoons ML_n \qquad K_n = \frac{[ML_n]}{[ML_{n-1}][L]} \tag{5.57c}$$

To this is added the stoichiometric concentrations:

$$c_M = [M] + [ML] + [ML_2] + \cdots + [ML_n] \tag{5.58a}$$

and

$$c_L = [L] + [ML] + 2[ML_2] + \cdots + n[ML_n] \tag{5.58b}$$

[a] Note that M and L may carry a charge, but this is ignored here.

giving a total of $n + 2$ equations for n equilibrium constants and $n + 2$ chemical species.

The basic problem is the determination of the constants. In some cases, direct spectroscopic determination of each species is possible. For example, ^{51}V-, ^{13}C-, and ^{1}H nuclear magnetic resonance (NMR) was used to unravel the vanadocitrate system (see Table 5.8). However, most work relies on classical methods because of the lack of a convenient spectroscopic method. In general, the experimental determination of the concentration of free ligand [L] for a series of different solutions (with known values of c_M and c_L) makes possible a numerical treatment of the data, eventually leading to a consistent set of constants.[a]

In order to calculate the speciation of complexes, most studies use tables of equilibrium constants (Equations 5.57, including corrections for ionic strength), the known stoichiometry (Equations 5.58), and the concentration [L]. In this way the problem is reduced to an assessment or an experimental measurement of the magnitude of [L].

For technical reasons, literature values of complex constants are often given as gross constants β, so-called cumulative constants. Thermodynamic consistency of Equations 5.57 demands that the ith equilibrium step be characterized by a stepwise formation constant K_i as well as a gross constant $\beta_i = K_1 K_2 \ldots K_i$. Note that the relation $\Delta G_i^\circ = -RT \ln K_i^\circ$ implies that a $\Delta G_{i,\mathrm{tot}}^\circ = \Delta G_1^\circ + \Delta G_2^\circ + \cdots + \Delta G_i^\circ$ exists for which $\Delta G_{i,\mathrm{tot}}^\circ = -RT \ln \beta_i^\circ$.

A given series of stepwise constants usually decreases in magnitude, $K_1 > K_2 > \ldots > K_n$, which might be expected from consideration of statistics, steric factors, and, for charged ligands, electrostatic interactions.[9] The following two examples illustrate these observations and show how such data are presented in the literature.

Table 5.7 Formation constants for chloro complexes[135]

	K_1 / M^{-1}	K_2 / K_1	K_3 / K_1	K_4 / K_1
Zn^{2+}	2.7	1.4	1.2	0.2
Cd^{2+}	95.0	4.4×10^{-2}	6.6×10^{-3}	2.1×10^{-3}
Hg^{2+}	5.5×10^6	0.55	1.4×10^{-6}	1.8×10^{-6}
	$\lg(K_1/\mathrm{M}^{-1})$	$\lg(K_2/\mathrm{M}^{-1})$	$\lg(K_3/\mathrm{M}^{-1})$	$\lg(K_4/\mathrm{M}^{-1})$
Zn^{2+}	0.43	0.58	0.51	−0.23
Cd^{2+}	1.98	0.62	−0.20	−0.70
Hg^{2+}	6.74	6.48	0.89	1.00

Temperature 25 °C. Ionic strength 0 for Zn and Cd complexes, and 0.5 M for Hg complexes. The data of the lower panel are redundant but convenient for the understanding of Figure 5.7.

Example 1: Chloro complexes of group 12 cations

Table 5.7 gives the consecutive formation constants for chloro complexes of the aqua ions of the divalent cations of group 12 of the periodic table. Figure 5.7, which shows the speciation curves of the three systems, is drawn on the basis of these data.

[a] See Appendix 3.

The general trends mentioned earlier are illustrated, and the range of magnitudes of the constants of the zinc and cadmium systems are as observed for most coordination compounds. The mercury system differs from the norm because it has a large value of K_1 and a large range of existence of $HgCl_2(aq)$. The rationale for the first is that mercury(II) is a soft acid (electron acceptor); see Table 5.10. The second is because of a change of coordination geometry from linear to tetrahedral.

It is interesting to compare this with the terminology of acid-base chemistry: a weak acid is associated with a conjugate base having a great affinity for the hydron ligand.

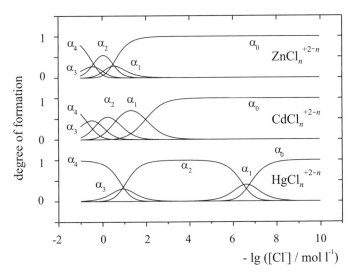

Figure 5.7 *Mole fractions of chloro complexes*
Data of Table 5.7: Degree of formation α as function of pL; see Appendix 2.

Example 2: Hydron and citrato complexes of vanadate(V)

Crude oil from the North Sea contains up to 0.5 % of vanadium. This originates from certain marine organisms such as ascidians (V^{III}) and red algae (V^{V}).[87] When untreated oil is burned, vanadium(V) is released and as the 1990s analyses showed, vanadium concentrations in Scandinavian trees increased. This prompted series of experiments aimed at the understanding of the environmental chemistry of vanadium(V).

Dihydrogenvanadate(V) forms esters with alcohols and may simultaneously form complexes with carboxylate, and complex formation with various hydroxyacids has been investigated. Table 5.8 shows the results of a study where citric acid was taken as a model for fulvic acids. The ionic strength was fixed at $I = 0.6$ M with NaCl, so the results are relevant to seawater as well as for certain saline soil solutions.

$H_p M_q L_r$	p	q	r	$\lg(\beta_{pqr} / M^{1-p-q-r})$
Table 5.8 The vanadocitrate system[160]				
HVO_4^{2-}	-1	1	0	-7.92
$V_2O_7^{4-}$	-2	2	0	-15.17
$HV_2O_7^{3-}$	-1	2	0	-5.25
$H_2V_2O^{2-}$	0	2	0	277.
$V_4O_{13}^{6-}$	-2	4	0	-8.88
$HV_4O_{13}^{5-}$	-1	4	0	0.22
$V_4O_{12}^{4-}$	0	4	0	10.
$V_5O_{15}^{5-}$	0	5	0	12.4
$V_{10}O_{28}^{6-}$	4	10	0	52.1
$HV_{10}O_{28}^{5-}$	5	10	0	58.1
$H_2V_{10}O_{28}^{4-}$	6	10	0	61.9
$H_3V_{10}O_{28}^{3-}$	7	10	0	63.5
VO^{2+}	2	1	0	6.96
$C_6H_6O_7^{2-}$	1	0	1	5.217
$C_6H_7O_7^{-}$	2	0	1	9.298
$C_6H_8O_7$	3	0	1	12.067
	2	1	1	14.1
	3	1	1	18.1
	1	2	1	12.84
	2	2	1	19.68
	3	2	1	24.12
	4	2	2	31.3
	5	2	2	35.3
	6	2	2	39.2

$\beta_{pqr} = [H_p M_q L_r] / [H]^p [M]^q [L]^r$; $H = [H^+]$; $M = H_2VO_4^-$ (aq); $L = C_6H_5O_7^{3-}$
Temperature 25.0 °C; $I = 0.6$ M. See Appendix A2.2.

The acid-base chemistry of vanadate, hydrogenvanadate, and dihydrogenvanadate is very similar to that of the corresponding phosphates, but in acid solution these species oligomerize. Decavanadates dominate, and no analogues of orthophosphoric acid are known. Also, oxidovanadium(IV) is stable in aqueous solution.[a]

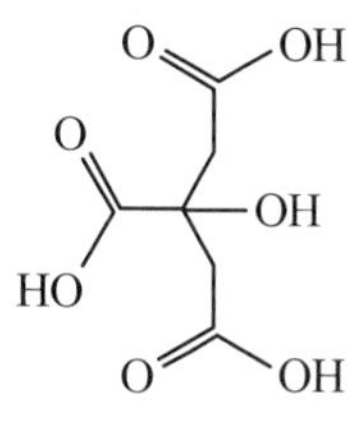

(5.59)

citric acid

---

[a] As VO^{2+} has an unpaired electron, the ^{51}V-NMR spectra cannot be measured, but this species is EPR (Electron Paramagnetic Resonance) active.

The acidity constants of citric acid are given as gross association constants for the stepwise formation of the acids from the dehydronated base. The reaction equation is

$$C_6H_5O_7^{3-} \; \underset{-H^+}{\overset{+H^+}{\rightleftharpoons}} \; C_6H_6O_7^{2-} \; \underset{-H^+}{\overset{+H^+}{\rightleftharpoons}} \; C_6H_7O_7^- \; \underset{-H^+}{\overset{+H^+}{\rightleftharpoons}} \; C_6H_8O_7$$
$$\quad\quad\quad\quad K_1 \quad\quad\quad\quad\quad K_2 \quad\quad\quad\quad\quad K_3$$

$$(5.60)$$

with the stability constants:

$$\lg(\beta_1/\text{M}^{-1}) = \;\; 5.217 \qquad \lg(K_1/\text{M}^{-1}) = 5.217 = pK_{a3} \qquad (5.61a)$$
$$\lg(\beta_2/\text{M}^{-1}) = \;\; 9.298 \qquad \lg(K_2/\text{M}^{-1}) = 4.081 = pK_{a2} \qquad (5.61b)$$
$$\lg(\beta_3/\text{M}^{-1}) = 12.067 \qquad \lg(K_3/\text{M}^{-1}) = 2.769 = pK_{a1} \qquad (5.61c)$$

Note that the definition of the degree of formation α_i of an acid is the same as for a complex. This ensures that the formation function $\bar{n}$ and the dissociation function d are independent of whether the equilibrium constants are associative or dissociative.

b. Lewis acids and bases

Empirical rules for which type of metal ions (electron acceptor, Lewis acid) will have favorable interactions with which ligands (electron donor, base) are of great use when assessing transport of chemical species in Nature. A useful starting point is the qualitative description ionic versus covalent bonding. One approach is to consider the bond to be covalent and then estimate the consequence of increasing charge displacement from one atom toward another. In effect, this is describing chemical bonding in terms of electronegativity.

Another approach is to consider the bond as being ionic and then consider the addition of covalent bonding. This point of view was advanced by Fajans (1923), who formulated Fajans' rules based on polarizability data.[a] Polarization or deformation of an electron shell is increased by (1) high charge and small size of the cation and by (2) high charge and large size of the anion. Some additional rules concerning the electron configuration account for the fact that cations of the same charge and size may differ in polarizability: for example, transition metals versus alkaline earth metals.

A great majority of all bases coordinate with metal ions, with the ligating atom being from group 15, 16, or 17 of the periodic table, and with properties that depend on the number of the period. For example, hydron forms compounds with elements from period 2 that are the weakest acids, whereas the elements of the later periods bind hydrons less effectively. Generally, metal ions may be divided into two classes: the *hard* ones, with chemical behavior similar to the hydron, and, in contrast, the second class, the *soft* ones. Table 5.9 shows affinity sequences for electron acceptors. Table 5.10 collects and classifies the most common Lewis acids and bases.[199]

---

[a] Polarizability of a chemical species is measured as the permittivity; see Chapter 5.3.

Table 5.9 Affinity sequences for electron acceptors			
Group	Oxidation state	Affinity sequence for a hard acid	Affinity sequence for a soft acid
17	$-I$	$F \gg Cl > Br < I$	$F \ll Cl < Br < I$
16	$-II$	$O \gg S > Se > Te$	$O \ll S < Se \sim Te$
15	$-III$	$N \gg P > As > Sb$	$N \ll P > As > Sb$

The listed ligators are the bases; a detailed classification of the acids (hard and soft) is given in Table 5.10.

Table 5.10 Qualitative classification of Lewis acids and bases		
	Lewis acids (M)	Lewis bases (L)
Hard	H^+, Li^+, Na^+, K^+, Be^{2+}, Mg^{2+} Ca^{2+}, Cr^{2+}, Al^{3+}, Ti^{3+}, Cr^{3+} Fe^{3+}, Co^{3+}, Ga^{3+}, Tl^{3+}, Zr^{4+} La^{3+}, SO_3	F^-, O^{2-}, $\underline{H}O^-$, $H_2\underline{O}$, $\underline{N}H_3$ Cl^-, $C\underline{O}_3^{2-}$, $N\underline{O}_3^-$ $P\underline{O}_4^{3-}$, $S\underline{O}_4^{2-}$
Borderline	Fe^{2+}, Co^{2+}, Ni^{2+}, Zn^{2+} Pb^{2+}, Sn^{2+}, Cu^{2+}, SO_2	$\underline{N}O_2^-$, $S\underline{O}_3^{2-}$, Br^-, $SC\underline{N}^-$ $\underline{N}_3^-$
Soft	Cu^+, Ag^+, Au^+, Tl^+, Hg^{2+} Pt^{2+}, Cd^{2+}, Pd^{2+}, Pb^{2+} Tl^{3+}, Au^{3+}, Cs^+	$\underline{C}N^-$, $R_2\underline{S}$, $R\underline{S}^-$, $\underline{C}O$, I^- $\underline{S}CN^-$, $R_3\underline{P}$, S^{2-}

The ligator atom is underlined.

The hardest acid is the hydron, H^+, and the softest is methylmercury(II), CH_3Hg^+. Therefore, the reaction

$$HB^+(aq) + CH_3Hg^+(aq) \rightarrow CH_3HgB^+(aq) + H^+(aq) \quad (5.62)$$

has been used for quantitative determinations of the relative softness of a series of bases B.[37] Further, this type of reaction is of general interest because half of the mercury in the oceans is found as methylmercury(II), which biomagnifies in fish[142,180,214] (see later discussion).

The empirical rule for the affinity of the reaction of a Lewis acid with a base is that hard acids prefer hard bases and soft acids prefer soft bases.

The affinity of acid-base reactions may be analyzed using Equation 2.170,

$$A° = -\Delta H° + T\Delta S° = RT \ln K° \quad (5.63)$$

where the subindex "r" has been omitted. The following examples illustrate the general behavior:

1. Hard-hard interaction, entropy dominated:
   Ionic character of the bonding: $\Delta H°$ small; $\Delta S° \gg 0$, for example,

   $H^+ + F^- \rightarrow HF(aq) \quad \Delta H° = +51$ kJ/mol; $T\Delta S° = 121$ kJ/mol

Table 5.11 Key complex formation constants for seawater[135]

	Cl^-	SO_4^{2-}	HCO_3^-	CO_3^{2-}	F^-	HO^-
Na^+	.	.	.	.	.	-0.3
K^+	-0.7	0.4	.	.	.	-0.6
Mg^{2+}	-1.0	1.01	0.21	2.20	1.3	2
Ca^{2+}	.	1.03	0.29	1.89	0.6	0.7
Sr^{2+}	.	1.14	.	.	0.1	0.2
Co^{2+}	-0.14	0.23	.	.	0.4	3.9
Ni^{2+}	-0.21	0.57	.	6.9	0.5	3.8
Cu^{2+}	0.09	0.95	.	6.7	0.9	6.3
Zn^{2+}	0.43	0.93	.	4.0	0.7	4.4
Pb^{2+}	4.47	2.0	.	.	1.4	6.0

Values given as: $\lg(K_1 / \text{M}^{-1})$ at temperature 25.0 °C and ionic strength 0.7 M.

2. Soft-soft interaction, enthalpy dominated:
   Covalent character of the bonding: $\Delta H° \ll 0$, ΔS small, for example,

$$Hg^{2+} + I^- \rightarrow HgI^+ \qquad \Delta H° = -315 \text{ kJ/mol}; \qquad T\Delta S° \approx 0 \text{ kJ/mol}$$

c. Coordination chemistry of natural waters

In Section 5.1c the carbon dioxide–hydrogencarbonate–carbonate equilibria in aqueous solution were discussed. Here we consider the formation of complexes and precipitates in natural waters, concluding with the behavior of trace metals.

Seawater: Common complexes

Seawater is a rather concentrated aqueous solution of salts containing chiefly Cl^-, Na^+, Mg^{2+}, and SO_4^{2-}. The average composition is given in Table 3.13; seawater is largely a 0.5 M solution with respect to NaCl. Formation constants for the dominant complex-forming species are given in Table 5.11. The data in Tables 3.13 and 5.11 lead to an estimate of the most abundant chemical species in seawater, Table 5.12. Note that the ionic strength calculated using the analytical composition $I_{analysis} = 0.697$ M is the quantity to be used in all applications.

Table 5.13 gives solubility products for sparingly soluble salts whose precipitation puts an upper limit on the abundances of the species involved. The data of Tables 5.12 and 5.13 are not quite consistent. This is in part due to differences in temperature and ionic strength. One example is that the solubility of calcium carbonate increases with decreasing temperature and increasing pressure. The effects of ionic strength on solubility products are discussed in Section 5.3. We refer to Table 5.6 that shows corrections for equilibrium constants determined at low ionic strength but which are to be used at the high ionic strengths found in seawater.

	aq	Cl^-	SO_4^{2-}	HCO_3^-	CO_3^{2-}	sum
Table 5.12 Speciation of complexes in seawater						
aq	.	555.9	18.5	0.66	0.10	.
Na^+	479.9	.	.	.	.	479.9
Mg^{2+}	43.5	2.4	8.2	0.04	0.67	54.9
Ca^{2+}	8.7	.	1.7	0.01	0.06	10.5
K^+	9.1	1.0	0.4	.	.	10.5
sum	.	559.3	28.9	0.72	0.84	.

Amount concentration, c/mM, of the most common species in seawater.
The values were calculated from the composition given in Table 3.13 and the formation constants of the complexes, given in Table 5.11.
Mass density 1.02476 g ml^{-1}; pH 8.1; pK_a (HCO_3^-) = 9.1.

Examples: $c_{K^+} = [K^+] + [KCl] + [KSO_4^-] = 10.5$ mmol/l

$$c_{Cl^-} = [Cl^-] + [MgCl^+] + [KCl] = 559.3 \text{ mmol/l}$$

	SO_4^{2-}	CO_3^{2-}	F^-	HO^-
Table 5.13 Solubility products of some Mg^{II} and Ca^{II} salts[135]				
Mg^{2+}	.	$10^{-7.46}$ M^2	$10^{-8.18}$ M^3	$10^{-11.15}$ M^3
Ca^{2+}	$10^{-4.62}$ M^2	$10^{-8.35}$ M^2	$10^{-10.41}$ M^3	$10^{-5.19}$ M^3

Temperature 25.0 °C; ionic strength 0 M.

Seawater: Solubility of carbonates

According to Table 5.13, the solubility of calcium carbonate in water is $[Ca^{2+}] = [CO_3^{2-}] = 6.68 \times 10^{-5}$ M. The observed solubility in seawater is somewhat higher because of the higher ionic strength and the effects of atmospheric carbon dioxide. The reaction equation is

$$CaCO_3(s) + H_2O(l) + CO_2(g) \rightleftharpoons Ca^{2+}(aq) + 2\,HCO_3^-(aq) \quad (5.64)$$

(see Equation 3.9). The value of the equilibrium constant K may be found as a product of appropriate, known constants:

$$K = \frac{[Ca^{2+}][HCO_3^-]^2}{p_{CO_2}} = \frac{L_{CaCO_3} K_{a1}}{K_{a2} k_H} \quad (5.65)$$

where K_{a1} and K_{a2} are the acidity constants, L the solubility product, and k_H the Henry's law constant (see Equation 2.205 and footnote a on p. 89).

Example 1. Estimation of the solubility of calcite in seawater

The solubility product of $CaCO_3$ (calcite)[135d] at 20 °C at a salinity of 19‰ is $L = 10^{-8.01}$ M^2. The Henry constant and the acidity constants of carbonic acid at

$\theta = 25\,^{\circ}\text{C}$ and $I = 1$ M may be taken from Table 5.5 ($k_H = 32.4$ bar M^{-1}). The acidity exponents are taken from ref. 175, $pK_{a1} = 5.95$ and $pK_{a2} = 9.10$, leading to $K = 4.27 \times 10^{-7}$ M^3 bar^{-1}. The solubility is measured using the concentration of Ca^{2+}, and from Equation 5.64, $[\text{HCO}_3{}^-] = 2[\text{Ca}^{2+}]$. Assuming the partial pressure $p_{\text{CO}_2} = 33$ Pa gives the result

$$[\text{Ca}^{2+}] = \sqrt[3]{\frac{K\,p_{\text{CO}_2}}{4}} \approx 3.28 \times 10^{-4}\ \text{M}; \quad [\text{HCO}_3^-] \approx 6.55 \times 10^{-4}\ \text{M} \qquad (5.66)$$

Thus, relative to Table 5.13, the solubility of calcite is increased by a factor of 5 in the presence of atmospheric carbon dioxide (see Equation 5.64).

Example 2. Estimation of the solubility of dolomite in seawater

Under the same conditions, the mineral dolomite, $\text{CaMg}(\text{CO}_3)_2$, has the solubility product $L = 1.0 \times 10^{-17}$ M^4. The solubility is

$$[\text{Ca}^{2+}] = [\text{Mg}^{2+}] = \sqrt[6]{L_{\text{CaMg}(\text{CO}_3)_2} \frac{K_{a1}^2}{K_{a2}^2} \frac{p_{\text{CO}_2}^2}{4^4\,k_H^2}} \approx 1.4 \times 10^{-4}\ \text{M}$$

According to Table 5.12, seawater contains the stoichiometric concentrations $c_{\text{Ca}} = 10.5$ mM and $c_{\text{Mg}} = 54.9$ mM, so the contribution of Ca^{2+} and Mg^{2+} from calcite and dolomite is small.

Example 3. Alkalinity and buffer value of seawater

We now discuss Table 5.12.

1. Measurements of pH and $pK_a(\text{HCO}_3{}^-)$ lead to the value of the ratio $[\text{CO}_3{}^{2-}]/[\text{HCO}_3{}^-] = 0.15$, which deviates from the value 0.10 found in a more concentrated solution. Conversely, the sum $[\text{CO}_3{}^{2-}] + [\text{HCO}_3{}^-]$ and the ratio $[\text{CO}_3{}^{2-}]/[\text{HCO}_3{}^-]$ determine pH.
2. The alkalinity, given in Table 3.13, is $A = 2.4$ mM $= [\text{HCO}_3{}^-] + [\text{MgHCO}_3{}^+] + [\text{CaHCO}_3{}^+] + 2[\text{CO}_3{}^{2-}] + 2[\text{MgCO}_3] + 2[\text{CaCO}_3] + pK_w/[\text{H}^+] - [\text{H}^+]$.
3. The buffer value $\beta = \text{d}A/\text{d(pH)} = 0.98$ mM is obtained using Equation 5.32 (also see Equation A2.10).

Example 4. Hardness of seawater

The hardness of the seawater (Table 5.12) is $w_{\text{CaO}} = 588.0 \times 10^{-6} = 59\ ^{\circ}\text{dH}$; see Section 3.2b.

Seawater: Trace elements

Table 5.14 shows the most abundant trace elements in seawater.

1. Group 12 elements: Zn, Cd, Hg

These elements are all found as M(II).

Element	c / nM	Element	c / nM
U (48)	14.0	Se (68)	1.0
Fe (4)	8.0	Co (30)	0.1
Ni (22)	5.0	Cd (65)	0.1
Zn (24)	5.0	Pb (36)	0.05
Mn (12)	4.0	Hg (67)	0.02
Cu (25)	4.0		

Table 5.14 Some trace elements of seawater[108]

The numbers in parentheses refer to the position of the element in Table 3.2, which shows the abundance of the elements in the crust.

Hexaaquazinc(II) is a weak acid, $pK_{a1} = 9.6$; accordingly, the formation constant of $Zn(OH)^+$ is rather high, $K_1 = 10^{4.4}$ M^{-1} (see Table 5.11). The speciation is given by $\alpha(Zn^{2+}) = 0.41$; $\alpha(ZnCl^+) = 0.24$; $\alpha(ZnCO_3) = 0.28$; $\alpha(Zn(OH)^+) = 0.02$; $\alpha(ZnSO_4) = 0.05$. Solid phases such as $Zn_5(OH)_6(CO_3)_2$ with $L = 10^{-86.0}$ M^{11} ($I = 0.7$ M) exists, but because its solubility is $s \approx 2.4 \times 10^{-9}$ M, this has no influence on the speciation in solution.[a]

In seawater, cadmium(II) aquaions only form chloro complexes because CO_3^{2-}, SO_4^{2-}, and Br^- cannot compete; either their formation constants or their concentrations are too small. Table 5.7 (see Figure 5.7) gives $\alpha(Cd^{2+}) = 0.03$; $\alpha(CdCl^+) = 0.34$; $\alpha(CdCl_2) = 0.51$; $\alpha(CdCl_3^-) = 0.12$.

Equilibrium constants for mercury(II) complexes may be found elsewhere[135c,d] and give the following speciation values:[108] $\alpha(CH_3HgCl) = 0.60$; $\alpha(HgCl_2) = 0.01$; $\alpha(HgCl_3^-) = 0.05$; $\alpha(HgCl_4^{2-}) = 0.26$; $\alpha(HgBrCl) = 0.01$; $\alpha(HgBrCl_2^-) = 0.02$; $\alpha(HgBrCl_3^{2-}) = 0.05$.

2. The transition metals Mn, Fe, Co, Ni, Cu

Manganese is found only as the dispersed dioxide, $MnO_2(s)$. Species with manganese in higher oxidation states are reduced (by dissolved organic matter) and species with lower oxidation states are oxidized (by oxygen).

Iron is found as dispersed, hydrous oxides, $FeO(OH)(s)$. Dissolved oxygen will oxidize iron(II), and iron(III) is too hard of an acid to form coordination compounds with chloride. Table 5.16 shows that at equilibrium at pH 8.1, $[Fe^{3+}]_{max} = 10^{-25.0}$ M and the concentration of all other species is much smaller. A minor part of the 8 nM of iron(III) species may be found as iron(III) complexes with dissolved organic matter.

The iron-manganese nodules found on the ocean floor are composed of hydrous oxides of Mn(IV) and Fe(III).

The speciation of cobalt(II) and nickel(II) in seawater is very similar to that of zinc(II). However, the formation constants of $Co(OH)^+$ and $Ni(OH)^+$ are small, and very little of these species are present.

---

[a] If the solubility is set equal to s, then $L = (5s)^5 (6s)^6 (2s)^2$.

Copper(II) behaves differently: the carbonato complex dominates, $\alpha(CuCO_3) = 0.77$, followed by the monohydroxo complex, $\alpha(Cu(OH)^+) = 0.09$; the chloro and sulfato complexes are of minor importance. Solid phases such as malachite, $L(Cu_2(OH)_2CO_3) = 10^{-31.4}$ M^5, and azurite, $L(Cu_3(OH)_2(CO_3)_2) = 10^{-42.1}$ M^7 (both measured in 0.2 M NaClO$_4$) do not limit its solubility.

3. The elements Se, Pb, U

Selenium is essential to life. It is surprising that its chemical state in seawater is very far from chemical equilibrium. For example, the standard electrode potential $E^\circ = +50$ mV for the process

$$SeO_4^{2-} + H_2O + 2e^- = SeO_3^{2-} + 2\,HO^- \tag{5.67}$$

and $pK_{a2} = 8.0$ for the reaction

$$HSeO_3^- \;\rightleftharpoons\; H^+ + SeO_3^{2-} \tag{5.68}$$

indicate that at the pH of seawater, the equilibrium value of the ratio $r = [Se(VI)]/[Se(IV)]$ should be $r \approx 10^{12}$, so selenate(VI), a very weak base, should be extremely stable. Nevertheless, in the Pacific Ocean the observed ratio r was 0.75 in surface water and 0.50 in the deep ocean.[209] The reason, as has been observed in independent experiments, is that the rate of oxidation of Se(IV) is very slow. Such kinetically stable species are called *robust* species.

In seawater, lead occurs as Pb(II). Table 5.11 shows that $[PbCl^+]/[PbSO_4] \approx 300$ and $[PbCl^+]/[Pb^{2+}] \approx 1.7 \times 10^4$, so lead(II) is found only as a chloro complex.

Uranium(VI) is always present in complexes of (the linear) dioxidouranium(VI), UO_2^{2+}. The high abundance of U may be explained as being due to very stable carbonato complexes,[153,163] $UO_2(CO_3)_3^{4-}$ and $(UO_2)_3(CO_3)_6^{6-}$. They do not react with water or hydroxide, and no mixed hydroxo-carbonato species have been observed.

Soil solution: Sparingly soluble hydroxides

Tables 5.15 and 5.16 show the all-important equilibrium data for species containing Al(III), Fe(III), and HO$^-$. As the general chemistry of these species has been discussed previously, we will only make a few remarks on the data.

The aluminium data are for an ionic strength of $I = 0.1$ M and $I = 0$ M. In general, *with increasing ionic strength, solubility products increase and complex formation constants decrease*; see Table 5.6.

Generally the iron(III) hydroxides are less soluble than the aluminium hydroxides and all calculations show that the concentration of soluble iron(III) species are very low in natural waters.

5.3 Electrolytic properties

The electrolytic properties of water set the boundary conditions for most of the chemistry of aqueous solutions and aerosols and are thus central to a description of the

Table 5.15 Aluminium-hydroxide complexes[135]

$M_p L_q$	p	q	$\lg(\beta_{pq} / M^{1-p-q})$	I / M
$Al(OH)^{2+}$	1	1	8.5	0.1
$Al(OH)_2^+$	1	2	17.6	0.1
$Al(OH)_3$	1	3	25.7	0.1
$Al(OH)_4^-$	1	4	31.	0
$Al_2(OH)_2^{4+}$	2	2	20.3	0
$Al_3(OH)_4^{5+}$	3	4	42.1	0

$\beta_{pq} = [M_p L_q] / [M]^p [L]^q$; $M = Al^{III}(aq)$; $L = HO^-$. Temperature 25.0 °C.
Solubility product for $I = 0$ M, with the solid phase in brackets:

$$[Al^{3+}][HO^-]^3 = 10^{-33.5} \, M^4 \qquad (\alpha\text{-}Al(OH)_3, \text{ hydrargillite or gibbsite}).$$

Further, the following constant is known ($I = 0$ M):

$$\frac{[Al_{13}O_4(OH)_{24}^{7+}][H^+]^8}{[Al^{3+}]^{13}[OH^-]^{32}} = 10^{349.2} \, M^{-36}.$$

Table 5.16 Iron(III)-hydroxide complexes[135]

$M_p L_q$	p	q	$\lg(\beta_{pq} / M^{1-p-q})$	I / M
$Fe(OH)^{2+}$	1	1	11.	0.5
$Fe(OH)_2^+$	1	2	22.3	0
$Fe(OH)_4^-$	1	4	34.4	0
$Fe_2(OH)_2^{4+}$	2	2	24.7	0.5
$Fe_3(OH)_4^{5+}$	3	4	49.7	0

$\beta_{pq} = [M_p L_q] / [M]^p [L]^q$; $M = Fe^{III}(aq)$; $L = HO^-$. Temperature 25.0 °C.
Solubility products for $I = 0$ M, with the solid phase in brackets:

$[Fe^{3+}][HO^-]^3 = 10^{-38.8} \, M^{-4} \qquad (Fe(OH)_3)$

$[Fe^{3+}][HO^-]^3 = 10^{-41.5} \, M^{-4} \qquad (\alpha\text{-}FeO(OH), \text{ goethite})$

$[Fe^{3+}][HO^-]^3 = 10^{-42.7} \, M^{-4} \qquad (\alpha\text{-}Fe_2O_3, \text{ hematite})$

environment. Electrolytic properties are important for (a) redox chemistry at electrodes, often metallic, and (b) transport of ions of solutes and solvent (conductivity).

a. Redox chemistry of natural waters

The quantitative description of redox chemistry is based on describing equilibrium at an electrode using Nernst's equation. Consider a general electrochemical reaction

$$z\,e^- + \sum_R \nu_R R = \sum_P \nu_P P \qquad (5.69)$$

where the reactant species R represents the oxidized system and the product species P the reduced system. The electrode potential at equilibrium is given by

$$E_{eq} = E° - \frac{RT}{zF} \sum_i v_i \ln a_i \tag{5.70a}$$

$$= E^{°'} - \frac{RT}{zF} \sum_i v_i \ln \frac{c_i}{M} \tag{5.70b}$$

where v_i are positive for products and negative for reactants, and p_i / bar may replace c_i/ M when relevant. The standard electrode potential $E°$ is the value of E_{eq} in Equation 5.70a when all activities a_i are unity. The formal potential $E^{°'}$ is measured at known values of the concentrations c_i or partial pressures p_i of the various species; its value depends on the composition of the electrolyte solution and the gases present, if any. The other symbols have their usual meaning, and the value of the constant $(R\ 288.2\ K\ /\ F)$ $\ln 10 \approx 57.18$ mV at 15 °C will be used in the following general presentation.

The redox chemistry of water is based on the electrode equilibria

$$2\,H^+(aq) + 2\,e^-(Pt) = H_2(g) \qquad\qquad E° \overset{def}{=} 0\,V \tag{5.71}$$

$$\tfrac{1}{2}\,O_2(g) + 2\,H^+(aq) + 2e^-(Pt) = H_2O(1) \qquad E° = 1.2288\,V \tag{5.72}$$

to which the Nernst equation may be applied:

$$E/V = 0.0296\,\lg\{p(H_2)/\,bar\} = -0.0592\,pH \tag{5.73}$$

and

$$E/V = 1.2288 + 0.0148\,\lg\{p(O_2)/bar\}$$
$$= 1.2288 - 0.0572\,pH \tag{5.74}$$

For the pH dependence, it is assumed that the partial pressures of the two gases are 1 bar. Note that no reversible oxygen electrode exists, meaning that all statements about such potentials are based on indirect data.[130]

Fundamental properties of E-pH diagrams
Water may be decomposed in two quite different ways:

$$H_2O(1) = H^+(aq) + HO^-(aq) \tag{5.75}$$

$$H_2O(1) = H_2(g) + \tfrac{1}{2}O_2(g) \tag{5.76}$$

These equations are connected through the two electrode reactions, Equations 5.71 and 5.72. This connection is used to determine the potential of a reversible electrode as a function of acidity (pH) using Nernst's equation. Because such diagrams were first developed by Pourbaix (1938), they are often called Pourbaix diagrams or simply E-pH diagrams.[76] When they were developed, their purpose was partly to demonstrate the importance of connecting pH determinations and electrode potentials, and partly to develop tools for the study of corrosion of metals.

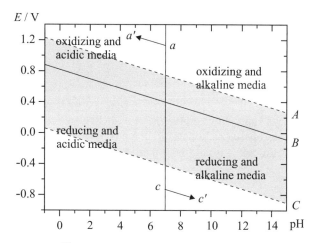

E-pH diagram of water[76]
Drawn for 25 °C, 1 bar where the ion product of water $K_w = 10^{-14.00}$ M^2 and the Nernst factor
$RT \ln 10/F = 0.0572$ V at $T = (273.16 + 15)$ K.

A Pourbaix diagram for water at 25 °C and 1 bar is shown in Figure 5.8. The straight line C indicates the potential of the reversible hydrogen electrode (Equation 5.73), and line A shows the potential of a (hypothetical) reversible oxygen electrode (Equation 5.74). The vertical line at pH $= 7.00$ corresponds to the equilibrium of Equation 5.75, and line B,

$$E/V = 0.8162 - 0.0572 \, \text{pH} \qquad (5.77)$$

corresponds to the equilibrium of Equation 5.76, $p(H_2) = 2p(O_2)$. The two lines intersect at the "neutral point" ($E = 0.4021$ V, pH $= 7.00$) and divide the E-pH plane into four domains with properties as shown in Figure 5.8.

A redox pair whose (reversible) electrode potential lies between lines A and C is stable in aqueous solution. One example is the equilibrium Fe^{III}–Fe^{II} with $E_{eq} \approx E^\circ = 0.77$ V in acidic solution; this is represented by a line parallel to the pH-axis as shown later in Figure 5.10. If a redox pair is reversibly subject to conditions near line C, it will be reduced by hydrogen. In contrast, if the pair is subject to conditions near line A, it will be oxidized by oxygen. Redox pairs with electrode potentials below line C will be oxidized by water, generating hydrogen, for example, $Cr^{II} \rightarrow Cr^{III}$ with $E_{eq} \approx E^\circ = -0.4$ V (acidic solution), whereas pairs with electrode potentials above line A will be reduced by water, generating oxygen, for example, $Co^{III} \rightarrow Co^{II}$ with $E_{eq} = E^\circ \approx 1.8$ V (acidic solution).[130] Nernst's equation describes systems at equilibrium, but reaction inhibitors (irreversible processes) may expand the range of stability approximately 0.4 V above line A and approximately 0.4 V below line C. This phenomenon is known as *overvoltage*.

Redox reactions that are independent of pH are depicted in an E-pH diagram as a straight line parallel to the pH-axis, for example, Fe^{III}–Fe^{II} in strong acid solution; see Figure 5.10. If pH is increased, such a line approaches line A, and the pair becomes more oxidizing relative to water. However, traces of oxygen react with Fe^{II} according to

$$2\,Fe^{2+}(aq)\;+\;{}^{1}\!/\!_{2}O_2(g)\;+\;2\,H^+\;\rightleftharpoons\;2\,Fe^{3+}(aq)+H_2O(l) \qquad (5.78)$$

with the standard affinity $A° = +\,90.7$ kJ/mol. For example, at pH 4 and $p(O_2) = 10^{-4}$ bar, the ratio is $[Fe^{3+}]/[Fe^{2+}] = 896$.

Acid-base reactions without redox activity are shown as lines parallel to the E-axis, for example, the acid-base equilibrium HCO_3^- - CO_3^{2-} at a pH value given by[a]

$$pH = pK_{a2} + \lg\frac{[CO_3^{2-}]}{[HCO_3^-]} \qquad (5.79)$$

Therefore, in order to draw an E-pH diagram, one must assume fixed concentrations for the chemical species involved. In this context the qualitative nature of the diagrams must be emphasized.

Electrolytic decomposition of water requires a pH-independent potential difference of 1.2288 V between the cathode and the anode, given full reversibility. In practice one observes an overvoltage: the actual decomposition voltage is higher than the reversible one. The most important contribution to irreversibility is the activation energy required to unite nascent atoms to form gas molecules. Points a and c on Figure 5.8 represent the conditions around the anode and the cathode respectively, in a well-stirred solution of a strong electrolyte that gives rise to conductivity but which does not react with the electrodes. Suppose that the electrolyte was not stirred; then the conditions in the vicinity of the electrodes would change toward points a' and c' as indicated in the figure. A practical example from the chlor-alkali industry is discussed in Section 8.3a.

E-pH diagrams for natural waters

E-pH values for a majority of natural waters are found in specific regions of the Pourbaix diagram (Figure 5.9). Seawaters (the area marked by C) possess high pH values, normally between 7.7 and 8.3. Normal surface waters with oxygen from the atmosphere have high electrode potentials (calculated) and pH values around 8.1 (measured). When low potentials and low pH values are observed, it may be due to natural causes. For example, values as low as ($E = -183$ mV, pH = 7.5) have been measured in bays in the southern part of the Baltic Sea. This is due to sulfur-hydrogensulfide chemistry resulting from the activity of sulfate-reducing bacteria.

Soil waters (the area marked by A) are normally acidic to some extent and with greatly varying values of the electrode potentials. Table 9.4 shows the range of potentials of an inert electrode (platinum blank) at pH 7 as a function of the dominant bacterial process in a given sample of soil. Variations of potential reflect aerobic and anaerobic conditions in soil water depending on whether the soil is above or below the water table.

---

[a] The $Fe^{3+}(aq)$ - $Fe(OH)_3(s)$ equilibrium gives a similar line representing an equilibrium independent of E; shown in Figure 5.10.

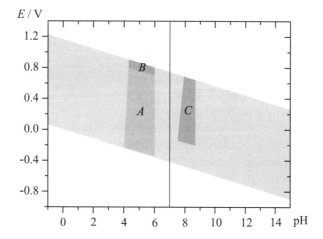

Figure 5.9   *E-pH diagram for natural waters*
Approximate *E*-pH limits for soil water (*A*), rain water (*B*), and seawater (*C*).

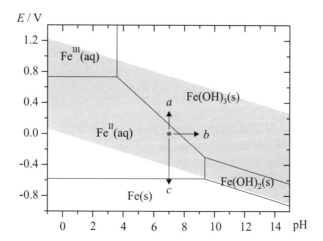

Figure 5.10   *E-pH diagram of the iron-water system*
The asterisk and the significance of the letters *a*, *b*, and *c* are discussed in the text.

Rain water is distilled water with a high content of dissolved oxygen and carbon dioxide; accordingly its *E*-pH properties correspond to the area marked by *B* in Figure 5.9.

We now discuss the Pourbaix diagrams of the simple iron-water system, and then for the same system in natural environments.

E-*pH* diagram of the iron-water system

The Pourbaix diagram for the iron-water system, Figure 5.10, can be constructed from the following constants:

$$E^{\circ}_{Fe^{2+}|Fe} = -0.440\,V \tag{5.80a}$$

$$E^{\circ}_{Fe^{3+}|Fe^{2+}} = +0.771V \tag{5.80b}$$

$$L_{Fe(OH)_2} = 7.9 \times 10^{-15} \text{ M}^3 \tag{5.80c}$$

$$L_{Fe(OH)_3} = 1.1 \times 10^{-36} \text{ M}^4 \tag{5.80d}$$

and by taking the sum of the concentrations of all iron species to be 10^{-5} M.

In such a simple system, only two iron species are found: $Fe^{II}(aq)$ and $Fe(OH)_3(s)$. This is in agreement with observations: groundwater transports Fe^{II} in reducing environments, whereas Fe^{III} is immobile under oxidizing conditions (see Table 5.16). However, small amounts of other species may disturb this simple picture. For example, CO_2 causes precipitation of iron(II)carbonate even at pH 5, because the solubility product is $L(FeCO_3) = 3.13 \times 10^{-11}$ M^2. (Acid constants for carbonic acid are given in Table 5.5.[a])

The Pourbaix diagram also illustrates that iron at $(E, \text{pH}) = (0, 7)$, marked by an asterisk, may be protected against corrosion by (a) raising the potential of the metal (anodic protection), (b) increasing the pH (pH-protection), or (c) lowering the potential (cathodic protection; the metal is given a negative potential using DC voltage).

Iron may also be protected by electroplating. Here the surface is covered by a layer of $Zn(OH)_2 \cdot ZnCO_3$ that prevents water and oxygen from reaching the iron. A so-called protection anode can be made using a piece of zinc or magnesium in contact with the iron, acting as a short-circuited galvanic element: the anode is dissolved instead of Fe.

E-pH diagram of stagnant soil water

We conclude the discussion of Pourbaix diagrams by considering the mineral pyrite FeS_2 in equilibrium with various natural waters. Consider an example applying Figure 5.11 to environmental systems. A sample of pyrite is placed in bog water at conditions marked by position A in the diagram, and the following changes are implemented assuming equilibrium conditions. First the acidity is increased to point B, and pyrite dissolves according to

$$FeS_2(s) + 2 H^+(aq) \rightarrow Fe^{2+}(aq) + H_2S(aq) + \tfrac{1}{8} S_8(s) \tag{5.81}$$

because the disulfide disproportionates. Next the amount of oxygen is increased to point C, and hydrogen sulfide is oxidized to sulfate. Finally, pH and the content of oxygen are increased to point D, and iron(II) is oxidized to iron(III). Part of this chemistry may be catalyzed by bacteria, as will be discussed in Section 7.5b.

b. Aqueous solutions of electrolytes

We now summarize some of the main concepts of electrolytic conductivity.[80] This will clarify the physical basis for the fact that the activity of ions in liquid solutions is different from their concentration.

---

[a] Note: The electrode to be used for the reaction behind Equation 5.80a is solid iron, and the electrode to be used for the reaction behind Equation 5.80b is smooth platinum.

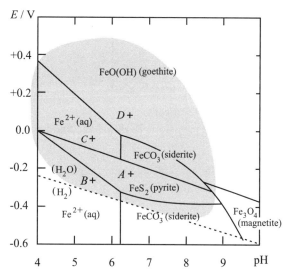

Figure 5.11 *E-pH diagram for the Fe-H₂O-CO₂-H₂S system*
The phase limits are subject to the following conditions (where appropriate):
$[Fe^{2+}] = 10^{-6}$ M
$[H_2S] + [HS^-] + [S^{2-}] = 10^{-6}$ M
$[CO_2(aq)] + [H_2CO_3] + [HCO_3^-] + [CO_3^{2-}] = 0.4$ M
Solubility product for siderite: $L(FeCO_3) = 10^{-10.7}$ M²
Acidity constant for $H_2CO_3(aq) + CO_2(aq)$: $K_{a1} = 10^{-6.3}$ M
Acidity constant for $HCO_3^-(aq)$: $K_{a2} = 10^{-10.3}$ M
Acidity constant for H_2S: $K_{a1} = 10^{-6.9}$ M
Solubility product for pyrite: $L(FeS_2) = 10^{-10.7}$ M²
Solubility product for goethite, Table 5.16.

$$H_2S_2(aq) \rightarrow H_2S(aq) + (1/8)\, S_8 \qquad\qquad A_r > 0$$

$$SO_4^{2-} + 10\,H^+ + 8e^- \;\rightleftharpoons\; H_2S + 4H_2O \qquad E^\circ(pH = 5) = -0.15V$$

$$FeO(OH) + 3H^+ + e^- \;\rightleftharpoons\; Fe^{2+}(aq) + 2H_2O \qquad E^\circ(pH = 5) = +0.20V$$

The shaded area covers values of *E*-pH, which are found in soil solutions. The
points *A*, *B*, *C*, and *D* are discussed in the text. Calculation is based on ref. 25, cited
in ref. 73.

Electrical conductors

Conductors of electrical current are physicochemical systems that contain movable
charge carriers. The carriers are electrons (in metals and semimetals) or ions (in
electrolytes), and this section discusses the general physical laws for this area of
science.

When a conductor is subject to an electric potential difference $\Delta\varphi$, the ensuing
current *I* is inversely proportional to the electrical resistance *R* according to Ohm's
law (1827),

$$\Delta\varphi = RI \qquad\qquad (5.82)$$

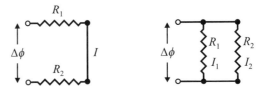

Figure 5.12 *Configurations of resistors*
Left panel: series resistors. Right panel: parallel resistors.

This relationship is applicable to currents in uniform and homogeneous conductors.[a] The current is the flux of charge,

$$I = \left(\frac{\partial Q}{\partial t} \right)_{T,p,n_i} \tag{5.83}$$

where Q denotes the charge and t the time.[b] Resistors may be connected in two different configurations as shown in Figure 5.12. Application of Equation 5.82 gives, for the total resistance R_s of a series circuit of the resistors R_1 and R_2,

$$\Delta \varphi = (R_1 + R_2)I = R_s I \tag{5.84}$$

and for the total resistance R_p of a parallel connection of the two resistors,

$$I = \frac{\Delta \varphi}{R_1} + \frac{\Delta \varphi}{R_2} = \left(\frac{1}{R_1} + \frac{1}{R_2} \right) \Delta \varphi = \frac{1}{R_p} \Delta \varphi \tag{5.85}$$

Resistance is a characteristic property of a conducting system. Experiments show that the resistance in a uniform, homogeneous rod of length l and cross-section A perpendicular to the direction of the electrical current is given by

$$R = \frac{\rho l}{A} = \frac{l}{\kappa A} \tag{5.86}$$

where ρ is the resistivity and $\kappa = \rho^{-1}$ the conductance of the material. ρ and κ are material constants that depend only on the kind of substance and its thermodynamic state, that is, T and p. Introducing the spatial quantities l and A in Equation 5.82 yields

$$\frac{\Delta \varphi}{l} = \rho \frac{I}{A} \quad \text{or} \quad E = \rho j \tag{5.87}$$

where $E = \Delta \varphi / l$ is the electric field strength and $j = I/A$ the current density.[c] The quantities E and j are vector fields, but in the present context these properties need not be expressed explicitly. Recall, however, that the direction of the current is that of the electric field; that is, positive charge carriers flow in the same direction as the field, and negative charge carriers flow in the opposite direction.

---

[a] Static electric fields and the proportionality between $\Delta \varphi$ and I are assumed throughout. Note, however, that electrolytic conductivity is measured using alternating fields (in the range 10^2–10^5 Hz) in order to avoid polarization at the electrodes.

[b] Electric current is one of the seven SI base quantities; see Table 5.6.

[c] Note that the current density is the flux density of electric charge; see Section 2.2a and Appendix A1.5.

Table 5.17 Conductivities at 20 °C

	κ / S m^{-1}		κ / S m^{-1}
Ag	6.29×10^7	NaCl (aq, sat) [1]	22.7
Cu	5.99×10^7	seawater [2]	4.8
Al	3.77×10^7	0.01 M KCl [3]	0.125
Fe	1.03×10^7	H_2O	5.5×10^{-6}
Hg	1.04×10^6	SiO_2	1×10^{-13}
Ge	2.17	S_8	5×10^{-16}

[1] $w_{NaCl} = 0.293$ (≈ 5.85 м).
[2] $w_{salinity} = 0.035$ (≈ 0.5 м NaCl; see Section 3.2b.
[3] At 25.00 °C, $\kappa = 0.140877$ S m$^{-1} \approx 1409$ µS cm^{-1}.[183]

Quite often the electric field strength is the independent variable and the current density the dependent one; hence, the relationship

$$j = \kappa E \tag{5.88}$$

is useful. Rewriting Equation 5.85 to deal with conductivities shows that the total conductivity of a system is the sum of the individual contributions:

$$\kappa = \sum_i \kappa_i \tag{5.89}$$

where i runs over the number of conductors. It follows that for a fixed electric field strength E, an additional conductor will increase the current density j. Therefore, it is feasible to define the transport number t_n of the nth conductor as the fraction of current carried by this conductor,

$$t_n = \frac{\kappa_n}{\kappa} = \frac{\kappa_n}{\sum_i \kappa_i} = \frac{j_n}{\sum_i j_i} \tag{5.90}$$

Materials may be classified according to their conductivities (Table 5.17). Conductivities of so-called conductors fall in the range of 10^7 S m^{-1}, those of semiconductors and electrolytes are around 1 S m^{-1}, and those of insulators are very small.

We have placed the SI units of frequently used electrical quantities in Appendix A1, Table A1.6. It may be helpful to note that the unit "siemens per meter" of conductivity is the same as "ampere per volt per metre."

Electrolytic conductors

In order to illustrate the direction of the electric field strength, Figure 5.13 shows the direction of migration of cations in solutions and formal carriers of positive charge in metallic conductors. An electrolytic cell (right-hand side of Figure 5.13) consists of two inert metallic electrodes placed in a liquid electrolyte. The electrode processes for aqueous solutions of electrolytes that do not react with the electrodes have been discussed (see Equations 5.71 and 5.72). In the present context, the focus is moved

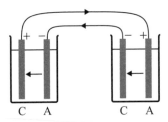

Figure 5.13   Galvanic cell (left) and electrolytic cell (right)
The arrows show the direction of the flow of positive charge (the conventional current). C and A denote cathodes and anodes, respectively.

from the electrodes to the solute itself, the measuring device being a conductivity cell that is basically an electrolytic cell; see footnote a, p. 208.

The *molar conductivity* Λ of a solute is defined as the conductivity κ divided by the amount concentration c,

$$\Lambda \stackrel{\text{def}}{=} \frac{\kappa}{c} \tag{5.91}$$

that is, molar conductivity and concentration are inversely proportional. The solute embraces an ensemble of ions, and Equation 5.91 applies to the electrically neutral ensemble.

Consider an ion B subject to an electric field strength E. Experiments show that such a species migrates with speed v_B

$$v_B = E u_B \tag{5.92}$$

proportional to E with a direction that depends on the sign of the charge of B as discussed earlier. The constant of proportionality u_B is called the electric mobility of B. The contribution i_B to the current density is given by

$$i_B = c_B \, |z_B| \, F v_B \tag{5.93}$$

where $|z_B|$ is the charge number of B and c_B its concentration. This expression follows directly from the meaning of Equation 5.88. The conductivity κ_B of B is

$$\kappa_B = c_B \, |z_B| \, F u_B \tag{5.94}$$

Modern nomenclature defines the *ionic conductivity* λ_B of B by[126]

$$\lambda_B \stackrel{\text{def}}{=} |z_B| F u_B \tag{5.95}$$

such that

$$\kappa_B = c_B \, \lambda_B \tag{5.96}$$

With these definitions, the conductivity of a solution is the sum of the ionic conductivities (see Equation 5.89), and Equation 5.91 applies with

$$\kappa = \sum_B \kappa_B = \sum_B \lambda_B c_B \tag{5.97}$$

It follows from Equation 5.90 that the fraction of electric current carried by a particular ionic species is given by the transport number of that species:

$$t_B = \frac{\kappa_B}{\kappa} = \frac{\lambda_B \, c_B}{\sum\limits_B \lambda_B \, c_B} \qquad (5.98)$$

Example 1

In the following four examples, the concentration c has the same value.

a. KCl (aq) of concentration c has the conductivity $\kappa_a = (\lambda_{K^+} + \lambda_{Cl^-}) \, c$.
b. $BaCl_2$ (aq) of concentration c has the conductivity $\kappa_b = (\lambda_{Ba^{2+}} + 2\lambda_{Cl^-}) \, c$.
c. $LaCl_3$ (aq) of concentration c has the conductivity $\kappa_c = (\lambda_{La^{3+}} + 3\lambda_{Cl^-}) \, c$.
d. $\frac{1}{3}LaCl_3$ (aq) of concentration c has the conductivity
$$\kappa_d = c(\lambda_{\frac{1}{3}La^{3+}} + \lambda_{Cl^-}) = c\left(\tfrac{1}{3}\lambda_{La^{3+}} + \lambda_{Cl^-}\right) = \frac{\kappa_c}{3}.$$

Example 2

The general content of the previous example: Consider the two chemical species B and $(1/b)$B. The following statements are true:[a]

a. If $c(B)$ is the amount concentration of species B, then

$$c\left(\frac{1}{b}B\right) = b \, c(B) \qquad (5.99)$$

is the amount concentration of species $(1/b)$B.
b. If $\Lambda(B)$ is the molar conductivity of species B, then

$$\Lambda\left(\frac{1}{b}B\right) = \frac{1}{b}\Lambda(B) \qquad (5.100)$$

is the molar conductivity of species $(1/b)$B.

Obviously, Equation 5.91 demands that Equations 5.99 and 5.100 be fulfilled because c and Λ are inversely proportional.

Example 3

In order to compare conductivities, an ion's charge is normalized so that the ionic conductivities of K^+, Ba^{2+}, and La^{3+} would be quoted as $\lambda(K^+)$, $\lambda[(1/2)Ba^{2+}]$, and $\lambda[(1/3)La^{3+}]$. Formerly the term *equivalent conductivity* was used for such quantities; instead of Equation 5.95, equivalent conductivity, here denoted λ'_B, was defined by $\lambda'_B = \lambda_B/|z_B| = F \, u_B$.

Thermodynamic properties

The fundamental behavior of an electrolyte is expressed in Equation 5.91: conductivity and concentration are proportional, $\kappa = \Lambda \, c$. Figure 5.14 shows that this is not

---

[a] See Equation 2.13 and footnote a, p. 120.

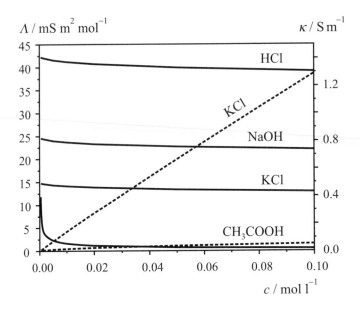

$\Lambda\,/\,mS\,m^{2}\,mol^{-1}$ $\kappa\,/\,S\,m^{-1}$

Figure 5.14   *Conductivities of electrolytes*
Examples of Equation 5.91, $\theta = 25\,^{\circ}\text{C}$.
Conductivity $\kappa(c)$ of KCl and CH_3COOH, broken lines, right-hand scale.
Molar conductivity $\Lambda(c)$ of several electrolytes, solid lines, left-hand scale; some
limiting values are given in the legend of Table 5.18.

quite true: Λ decreases slightly with increasing concentration, and the two examples
of $\kappa(c)$ show a weak curvature. Further, the qualitatively different behavior of molar
conductivities at low concentrations allows the classification of electrolytes into two
groups: strong electrolytes, such as salts, and weak electrolytes, such as coordination
compounds and carboxylic acids.

   A large set of experimental data shows that ionic conductivities can be defined and
that Equation 5.97 is fulfilled to a good approximation. As an example, the differences

$$\Lambda^{\infty}(\text{KCl}) - \Lambda^{\infty}(\text{NaCl}) = \Lambda^{\infty}(\text{KNO}_3) - \Lambda^{\infty}(\text{NaNO}_3)$$
$$= \Lambda^{\infty}(\tfrac{1}{2}\text{K}_2\text{SO}_4) - \Lambda^{\infty}(\tfrac{1}{2}\text{Na}_2\text{SO}_4) = 2.34\ \text{mS m}^{-1}\ \text{mol}^{-1}\ (25\,^{\circ}\text{C}) \qquad (5.101)$$

are independent of the nature of the anion. Here Λ^{∞} denotes the limiting value of
Λ at infinite dilution. Relationships like Equation 5.101 imply that the migration of
a particular ion at infinite dilution is independent of that of all others (Kohlrausch,
1876). Further, the presentation of Equation 5.97 and the discussion of partial molar
quantities (see Section 2.3a) indicate that λ_B is the partial molar conductivity of the
ion B, that is,

$$\lambda_B = \left(\frac{\partial \kappa}{\partial c_B}\right)_{p,T,C \neq B} \qquad (5.102)$$

Table 5.18 Ionic conductivities at infinite dilution at 25 °C [129]

λ^∞ / mS m^2 mol^{-1}		λ^∞ / mS m^2 mol^{-1}	
H$^+$	34.965	HO$^-$	19.85
K$^+$	7.348	I$^-$	7.68
Na$^+$	5.008	Br$^-$	7.81
Li$^+$	3.866	Cl$^-$	7.631
½Ba^{2+}	6.364	½SO$_4{}^{2-}$	7.99
N(C$_4$H$_9$)$_4{}^+$	1.95	CH$_3$COO$^-$	4.09

Examples:
Λ^∞ (CH$_3$COOH) = 39.06 mS m^2 mol^{-1}, Λ^∞ (KCl) = 14.979 mS m^2 mol^{-1}.

and analogous to Equation 2.126, the differential of the conductivity is

$$d\kappa = \sum_B \lambda_B \, dc_B \qquad (5.103)$$

In actual applications of this expression and of Equation 5.97, it must be remembered that because the solution is electrically neutral, some of the c_B's are not independent.

Selected values of the limiting value of λ_B at infinite dilution,

$$\lambda_B^\infty = \lim_{c_B \to 0} \lambda_B$$

are presented in Table 5.18.

Figure 5.15 shows molar conductivities of several strong electrolytes at low concentrations as functions of the square root of concentration. They vary according to the expression

$$\Lambda = \Lambda^\infty - a\sqrt{c} \qquad (5.104)$$

as discovered empirically by Kohlrausch (1900). The figure also shows that the slope a is to a certain extent determined by the charges of the constituent ions; thus, uni-univalent, uni-divalent, and uni-trivalent salts fall into different groups. The data discussed up to now show that electric mobilities diminish with increasing concentration of ions in a solution, and as the ion's charge increases. This implies that the interionic interactions are of a purely electrostatic nature (Coulomb interaction): each ion is surrounded by an ionic atmosphere of opposite sign. Generally, this will impede the migration of the ions. It also gives rise to an increase of the viscosity of dilute solutions of salts relative to that of pure water. According to the arguments of Debye and Hückel (1923),[a] this is the basis for the quantitative calculation of activity coefficients of ions in solution. Such calculations, later improved by Onsager (1927), also account for the functional form and the constants of Equation 5.104.

---

[a] Details are given in Appendix 3.

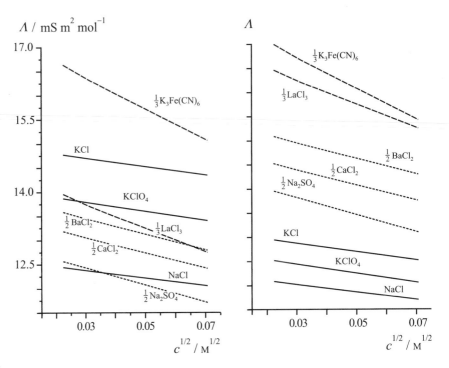

Figure 5.15 *Molar conductivity*
Left panel: Molar conductivities at $\theta = 25\,°C$ of some equivalent species as a function of the square root of the amount concentration; see Equation 5.104.
Right panel: The conductivities collected according to the size of the slopes.
Type of electrolyte: 1:1 ———; 1:2 -------; 1:3 ———.

The different behavior of weak electrolytes is due to their pronounced tendency to form molecules (i.e., formation of covalent bonding, usually between carboxylate and hydron) at higher concentrations.

Activity of electrolytes

Activities and choices of standard states were discussed in Section 2.3a. In particular, some choices for ionic equilibria in aqueous solution were described.

As seen earlier, ions behave ideally only at infinite dilution. At increased concentration, the electrostatic interactions increase and the activity of an ion decreases. The screening power of the solvent is expressed by the permittivity as discussed in Appendix A3. In water, a 1:1 electrolyte is effectively screened at concentrations less than 0.01 M, but at higher concentrations the solvent sphere is to some extent replaced by an ionic atmosphere, partly of opposite charge. The ideality of very dilute solutions motivated the choice of the standard state of a solute to be "a 1 M solution with the properties of an infinitely dilute solution" (see Equation 2.149). However, thermodynamic properties including equilibrium constants were measured in real solutions and then *extrapolated* to infinite dilution. Accordingly, estimates of solution properties must include a transformation of data from the standard state to the actual conditions. This is reviewed next and discussed in some detail in Appendix 3, which also gives references to recent computer programs. Note, however, that in order to get reliable results, data ought to be measured using the standard state of the application.

The activity of ions depends on concentration and charge, and both effects are taken in consideration by defining the ionic strength I (see Equation 5.56). The Debye-Hückel limiting law (1923) yields the activity coefficient γ_B of the ion B as

$$\lg \gamma_B = -0.509 \, z_B^2 \sqrt{I/M} \tag{5.105}$$

at $\theta = 25\,°C$ and $I < 0.01$ M. At higher ionic strength, $0.01\,M < I < 0.1\,M$ the expression

$$\lg \gamma_B = -0.509 \, z_B^2 \sqrt{I/M}/(1+\sqrt{I/M}) \tag{5.106}$$

is a useful approximation (see Equation A3.21). The activity is then given by the middle part of Equation 2.145,

$$a_B = (c_B/M) \, \gamma_B \tag{5.107}$$

Example 4

The solubility product of AgCl is $L° = a_{Ag^+} a_{Cl^-} = 1.56 \times 10^{-10}$, the standard state being that of infinite dilution at 25 °C. The solubility s_{AgCl} of AgCl in 0.050 M KCl at the same temperature is calculated as follows: the ionic strength is $I = 0.050$ M, and according to Equation 5.106, $\lg \gamma = -0.509 \times 1^2 \times 0.224/1.244 = -0.0930$. Therefore, $L° M^2 = [Ag^+][Cl^-]\gamma_{Ag^+}\gamma_{Cl^-} = [Ag^+] \times 0.050\,M \times 0.807^2$, and $s_{AgCl} = [Ag^+] = 4.79 \times 10^{-9}$ M.

The Debye-Hückel model considers the energy difference between an ion solely surrounded by solvent and the same ion surrounded by an ionic atmosphere. However, experimentally it is not possible to transfer a single ion between the two kinds of shells; an ion can only be moved together with a counterion. The formal solution to this problem is to let the activity coefficient be shared equally between the two ions. For a 1:1 electrolyte, the mean value of the chemical potential is

$$\begin{aligned}
\mu &= \tfrac{1}{2}(\mu_+ + \mu_-) \\
&= \tfrac{1}{2}\mu_+{}° + \tfrac{1}{2}RT\ln(c_+/c°) + \tfrac{1}{2}\mu_-{}° + \tfrac{1}{2}RT\ln(c_-/c°) + RT\ln(\gamma_+\gamma_-)^{\frac{1}{2}}
\end{aligned} \tag{5.108}$$

where the relations $a_+ = \gamma_+ \, c_+/c°$ and $a_- = \gamma_- \, c_-/c°$ have been used (see Equation 5.107). In order to equalize the two coefficients, one defines the mean activity coefficient $\gamma_\pm$ as

$$\gamma_\pm = \sqrt{\gamma_+\gamma_-} \tag{5.109}$$

to obtain $a_+ = \gamma_\pm \, c_+/c^\circ$ and $a_- = \gamma_\pm \, c_-/c^\circ$ for the two ions. Following the same reasoning, for the general case of a salt C_pA_q that dissociates into the ions $pC + qA$ in aqueous solution, one obtains the mean activity coefficient $\gamma_\pm$ by the geometrical mean

$$\gamma_\pm = \sqrt[p+q]{\gamma_+^p \gamma_-^q} \tag{5.110}$$

However, actual calculations using Equations 5.105 and 5.106 are unchanged.

6 Chemistry of the pedosphere

The pedosphere consists of soil and constitutes the interface between the lithosphere, the hydrosphere, and the atmosphere. Instead of attempting a short definition of *soil*, we shall look at the main disciplines of soil science: edaphology and pedology,[a] both of which were founded in the 19th century. Edaphology deals with the conditions necessary for the growth of plants (Liebig, around 1840), that is, their nutrition, in particular the types of chemical species they consume and produce. Pedology is the study of soil genesis, morphology, and classification (Dokuchaev, 1883).

The upper layer of the Earth's crust consists of fragments of rocks, mainly silicates, combined with water, air, and organic material. The largest pieces of rock, boulders and cobbles, have sizes ranging from meters to decimeters, whereas small fragments are coarse gravel in the centimeter range, followed in order of decreasing size by fine gravel, sand, silt, and clay (maximum size 4 μm), as shown later in Table 6.3. This heterogeneous, solid material extending from the outer surface, in contact with the atmosphere or ocean, downward to the solid bedrock, is called the regolith.[b] In some cases the regolith results from weathering of bedrock at a given location, but more often it has been transported to a given site; in a few cases there is no regolith. Material at a given site may have been deposited by ice, water, or wind, individually or in combination. The regolith is often stratified; for example, Figure 6.1 shows layers formed by water and air in material deposited by glaciers.

Soil is a heterogeneous mixture with highly variable composition. The solid substance includes mineral particles of very different sizes and air, water, salts, colloids, and dead organic material, as well as living organisms. Soil formation begins when the solid crust comes in contact with weather and wind and continues for hundreds, thousands (e.g., postglacial soils), or millions of years (e.g., tropical soils). Besides time, many factors are important in soil genesis, including the original mineral material, topography, climate, and, not least, living organisms.

Soil chemistry[a,89,94] is basically the chemical interaction of the species of the soil solution with soil air and the solid surfaces of soil particles, which includes colloids. The large surface of soil particles forms the basis for the occurrence and survival of microorganisms.[110] The present chapter encompasses a survey of the structure and chemical composition of soils, soil adsorption phenomena, and exchange reactions. We build this discussion on the chemistry of coordination compounds and acids and bases known from general chemistry in aqueous solution. We conclude the chapter with sections on standard soil analyses.

---

[a] Gk. $\pi\acute{\epsilon}\delta o\nu \,\hat{=}\,$ soil; $\epsilon\delta\alpha\varphi o\varsigma \,\hat{=}\,$ ground; $\lambda o\gamma\iota\alpha \,\hat{=}\,$ study. The two disciplines include studies of plants and microorganisms in ocean sediments.

[b] Gk. $\rho\eta\gamma o\varsigma \,\hat{=}\,$ blanket; $\lambda\iota\theta o\varsigma \,\hat{=}\,$ stone, rock.

Table 6.1 Horizons of soil profile		
Horizon	Depth (approx.)	Description
O	0.5-1.5 dm	Organic horizon Top: Plant litter Middle: Fermented plant matter Bottom: Humus
A	0.5-1.0 dm	Mineral horizon Thin layer of clay minerals mixed with dark-colored humus, pH 3-4
E	1.0-2.5 dm	Eluvial horizon Clay particles, humus, iron oxides, and aluminium oxides have been leached out. Color is typically ash gray
B		Illuvial horizon This accumulation zone contains at least one of the following: clay particles, humus, and iron and aluminium oxides
B_h	0.5-1.0 dm	Stained black by a high humus content
B_{fe}	0.5 dm	Stained black by humus; high content of hydrated iron(III) oxides
B_s	1.5-3.5 dm	High concentration of the iron(III) and aluminium(III) oxides Fe_2O_3 and Al_2O_3; the layer is reddish-brown; pH 5
C		Weathered bedrock that has not been affected by soil-forming processes Spodosols contain a layer of sand that can be colored red by hydrated iron(III) oxides
R		Unweathered bedrock

A spodosol. The height of the layers varies; pH increases with depth. The terminology is standard in soil science[248]: O = organic; h = humus; fe = content of Fe(III); s = sesquioxide = $(3/2)O^{-II}$, referring to the formulas $FeO_{1.5}$ and $AlO_{1.5}$.

6.1 Structure of soil

A main result of pedology is that – despite significant local variation – it is possible to give a general description of the structure of soil, which normally is clearly stratified.[11, 13, 84]

a. Soil profile

Table 6.1 shows an idealized soil profile (vertical cut), with soil horizons (layers), being characteristic of the soil type called spodosol, which can be found in moors. The top layer, the O horizon, contains 20 to 40 % organic material by mass, mixed with clay minerals and sand. Underneath the organic horizon is the mineral soil, the

A horizon, which is often dark because of humus. Humus is organic material that has reached the end state of weathering. At this point it is weathered to the extent that the material's origin from a certain plant cannot be determined. In cultivated soils this horizon shows the traces of tillage: liming and fertilization. Next follows the eluvial horizon,[a] called the *E* horizon, which is depleted of soluble species such as iron- and aluminium-containing minerals, colloidal clay, and soluble organic material. It is grayish to the extent that only colorless quartz is left. The soluble material accumulates in the illuvial horizon,[b] called the *B* horizon, in which the upper layer is often colored dark by humus. Iron species are often concentrated in the lower part of the *B* horizon, around the water table, called the gley or *G* horizon, not marked in Table 6.1. The color is orange to brown (Fe^{III} hydroxides) above and blue to green (Fe^{II} hydroxides) below the water table as determined by the redox conditions: Water above the water table is in contact with atmospheric oxygen, resulting in an oxidizing environment; below, the environment is reducing. A layer, called hardpan, can accumulate under dry conditions; this consists of grains of sand cemented with precipitated iron(III) hydroxides. The lowest layers consist of the original weathered material, the *C* horizon, and native bedrock, the *R* horizon.

Soil is porous, and more than half of its volume can be filled by water and/or air. Near the surface, the pores in the soil are only partly filled by water; this is the region of abundant life. Farther down, the pores are completely filled with water. The dividing surface is called the water table, and the water below it is called *groundwater*. Groundwater is supplied by the fraction of precipitation that does not transpire via plants, evaporate, or flow away. In climates with an excess of precipitation, water seeps downward, transporting colloidal particles and ions. In the same way, during dry periods or in arid climates, colloidal particles and ions flow to the surface, where they are deposited as salty mixtures when water evaporates. Groundwater levels change in response to precipitation, evaporation, patterns of flow, and withdrawal from wells.

In well-drained soil, there are three especially important soil-forming processes.

1. The formation of spodosol in temperate regions is dependent on the leaching of the upper layer by acidic water. The acidity comes from oxidized organic material, and the hydron concentration can be up to 10^{-4} M. At this concentration Al and Fe(III) oxides are soluble and clay particles are peptized, that is, they are caused to go into a colloidal suspension. This material is transported downward, where it reprecipitates, leading to the development of both *A* and *B* horizons.

2. In humid tropical climates, a different process occurs that is due to extensive leaching by precipitation with a higher pH. At higher pH quartz becomes more soluble and is removed from the surface layer. Minerals with a low Si content such as kaolinite[c] and Al and Fe(III) oxides (bauxite) therefore evolve, leading to a hard cement-like layer called laterite that in some areas is used as a building material.

---

[a] Lat. *eluvio* $\hat{=}$ flood.
[b] Lat. *illuvies* $\hat{=}$ dirt, mud.
[c] See Table 6.4.

Order	Short description	100 A	Weathering
Table 6.2 Soil orders			
Entisol	new soil, no profile development	13	Small
Vertisol	clay-like soil that cracks when dry	2	Small
Inceptisol	young soil, no B-horizon	16	Small
Aridisol	desert soil, little organic material	19	Medium
Mollisol	semi aridic grassland, high organic content	9	Medium
Spodosol	B-horizon with amorphous Al_2O_3 and humus	5	Medium
Alfisol	B-horizon: clay minerals; >35 % base saturation[1]	15	Medium
Ultisol	B-horizon: clay minerals; < 35% base saturation[1]	9	Large
Oxisol	tropical soil with goethite and bauxite	9	Large
Histosol	organic soil; peat	1	
Gelisol	soil with permafrost		
Andisol	volcanic ash		

[1] See footnote a on page 239.
The U.S. Department of Agriculture: soil taxonomy.[248c, 247] The upper 9 of the 12 orders are listed according to increasing degree of weathering. In this system, orders are divided into suborders and subgroups, not shown here. Approximate proportion of the land surface, A, covered by each soil order is given in percent.

3. In drier climates, calcium and magnesium salts accumulate at the surface as, for example, dolomite or gypsum.[a] It is the high rate of evaporation at the surface that causes these minerals to precipitate.

Soil can evolve in many different ways and, as shown in Table 6.2, there is a highly developed taxonomy. Modern soil classification is based on the degree of weathering and the morphology.

b. Regolith and groundwater

Figure 6.1 shows a vertical cut through the regolith, showing some of the typical conditions in which water is found on the Earth.[3, 77] Note that the vertical scale is enlarged relative to the horizontal one. The top layer is the moraine, M1 and M2, composed of sand and clay soils. Typically, the pores in the upper layer, M2, are only partially filled with water. The water table lies farther down, and below this the pores are completely filled. Three dense layers of clay soil are marked C1, C2, and C3. C1 is impermeable to water, whereas layers C2 and C3 merely slow its flow.[b] The three regions S1 to S3 are sand layers in which water moves easily. The watershed lies far to the left, and the water in layers S1 and S2 flows to the right; S3 is a sand lens. A reservoir within which water flows relatively easily is called an *aquifer*. The figure shows three aquifers, S1, S2, and M1. The water in aquifer M1 flows from both sides toward the river.

[a] Calcite, $CaCO_3$; dolomite, $(Ca,Mg)CO_3$; magnesite, $MgCO_3$; gypsum, $CaSO_4 \cdot 2H_2O$.
[b] C1 is called an aquiclude, C2 and C3 aquitards.

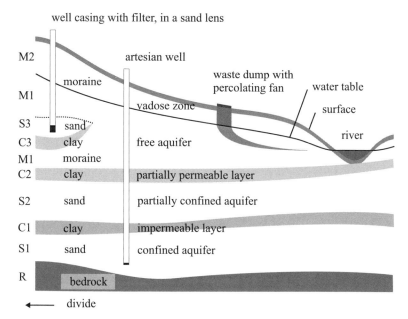

well casing with filter, in a sand lens

Profile of a regolith

The region between the surface and the water table is often called the unsaturated or vadose zone.[a] Water in this zone comes from precipitation. Water is retained in the vadose zone through adhesion to soil particles. The region can become saturated, for example, after rainfall or if a sand lens S3 exists as indicated in Figure 6.1. The vadose zone serves as an interface between the atmosphere and the layer in which water flows. It contains most of the living organisms of the soil: bacteria, fungi, small animals (e.g., nematodes, earthworms, insects), and plant roots.[106]

The flow of water through the surface is called *infiltration*. When precipitation exceeds surface flow, evaporation, and transpiration from plants, infiltration results in advection of ions and colloids into the groundwater. Similarly, in dry periods and in arid climates these species are transported toward the surface, where they deposit as salts and flocculates.

The area under the water table is called the saturated zone, although it is possible to find air pockets, especially near the water table. The height of the water table is not constant but varies because of variations in precipitation. It may be recognized as follows: a hole dug into the vadose zone will remain dry, while one dug into the saturated zone will fill with water. An aquifer such as M1 that has the water table as a boundary is called a free aquifer. In Figure 6.1, there are also aquifers in the sand layers S1 and S2. S1 is confined by the bedrock and an aquiclude; this can result in an excess pressure in the aquifer such that water will rise into a well dug here. Such a system is called a confined aquifer, and the well, an artesian well. The pressure in an artesian well comes about because the water table that recharges the well lies above the elevation of the well, or because of geological pressure on an impermeable layer.

[a] Lat. *vadum* $\hat{=}$ shallow water.

In the United States, groundwater supplies 25 % of all water used. Of this, 70 % is pumped for irrigation, 15 % for industry, and 15 % for households. One source is the Ogallala Aquifer, which lies underneath eight states in central North America, an area of semiarid grassland. It is estimated that precipitation replaces only one tenth of the water that is pumped from this reservoir. In this region the water table is falling, and many wells have gone dry.

The rate at which water flows through aquifers and aquitards varies over many orders of magnitude. In a gravel aquifer in sloping terrain, water can move at a rate of 50 m d^{-1}, whereas the rate of flow around a typical well may be only 1 m d^{-1}. The rate of flow through an aquitard can be less than 10^{-10} m d^{-1}, corresponding to roughly 1 m in 30 Ma.

There is a great practical interest in determining the flux density at which an aquifer is recharged. In addition, one would like to know what happens to water once it has passed the water table. Figure 6.1 shows an example of a waste dump from which pollutants spread into the groundwater. The fate of this flow and similar subjects will be discussed in greater detail in the following sections. As shown in the figure, the so-called percolation fan decreases in size and becomes dilute. Assuming that the pollutants do not decompose (which of course they could do), slices parallel to the surface would show that the pollution spreads; see the discussion around Figure 6.3. The chemical reactions involved will be discussed in connection with Table 9.4; see also Table 3.29 and Figure 5.9.

6.2 Physics of soil water

The quantitative description of how fluids (e.g., water, oil, natural gas) flow near the surface of the Earth is of considerable technical and economic interest, and many of the expressions that are used have been known for centuries; see Section 2.2.

A detailed physical description of the transport of water in the pedosphere depends on whether the soil is saturated or unsaturated. In the saturated zone, transport involves only the condensed phase, whereas in the unsaturated zone, transport can occur in the liquid or the gas phase. In the condensed phase, the forces that drive the motion are determined as differences in hydraulic potential, whereas in the gas phase, they are determined by differences in temperature.

a. The saturated zone

Bernoulli's equation was originally derived and used to describe the flow of water in tubes, but it deals with physical quantities that are fundamental to the description of the flow of water in soil. The equation assumes that the flow is laminar, a condition that groundwater fulfills. For details, see Chapter 2.

The equation that is used is a part of Equation 2.88:[a]

$$h_h = h_z + h_p \tag{6.1}$$

---

[a] The quantities of this equation have dimension of length.

Table 6.3 Particles in soil and some properties			
	Diameter / mm	φ_p	μ_D / (m d^{-1})
Gravel	>2	≈0.2	2000 - 200
Coarse sand and			
Medium sand	2 - 0.25	≈0.3	300 - 20
Fine sand	0.25 - 0.125	≈0.4	30 - 1
Very fine sand	0.125 - 0.0625		
Silt	0.0625 - 0.0039	≈0.5	10 - 0.01
Clay	0.0039 - 0.0002	0.6 - 0.7	1 - 10^{-5}
Colloids	< 0.0005		
Limestone		0.1 - 0.2	1 - 10^{-2}
Sandstone		0.05 - 0.3	1 - 10^{-3}
Shale			10^{-7}

The porosity φ_p and the hydraulic conductivity μ_D (water at 10 °C) for different classes of soil particle size and for three types of sedimentary rock. $\mu_D = 1$ m d^{-1} corresponds to the Darcy permeability $A_D = 1.5$ μm^2; see Equations 2.96 and 2.97. The soil textural classes are noted according to the USDA system;[248c] for convenience, some sediments are also shown. Diameters of colloid particles range between 500 nm and 5 nm; see Table 6.8.

where the hydraulic head h_h is the sum of two terms, the elevation head h_z and the pressure head h_p; the velocity head h_u plays no role when considering water in soils.

Differences in the concentrations of a dissolved species B can occur in groundwater, giving rise to a difference in the chemical potential and therefore to a flow of B within the fluid phase. A distinction can be made between two types of processes, diffusion and osmosis, which are both manifestations of the same phenomenon, namely, that the dilution of a solution occurs spontaneously. Diffusion results in a flow of dissolved species into areas of lower concentration. In osmosis, which can only occur across a membrane that allows the solvent to pass but blocks dissolved species, a different process occurs: the solvent moves to the region with the higher concentration of dissolved species, which are thereby dilluted.

Diffusion does not play any significant role in the spreading of dissolved material in flowing groundwater, but eddy diffusion (turbulent mixing processes) uses the same formalism (see Equation 2.103). Osmotic pressure does not play a role in the flow of groundwater, but it is a part of the process whereby plants obtain water from the soil. Therefore, it may be appropriate to add a term representing the osmotic pressure head h_o to the right side of Equation 6.1. Similarly, it is normal to add the osmotic potential of the form π dV, where π is the osmotic pressure, to Equation 2.90.

Flows in groundwater may be described by expressing Bernoulli's equation, Equation 6.1, in terms of Darcy's equation:

$$u_D = -\mu_D \, \text{\textbf{grad}} \, h_h \tag{6.2}$$

See Equation 2.96. Table 6.3 shows examples of hydraulic conductivities μ_D for water. The conversion factor between Equation 2.96 and Equation 2.97 is given in the legend. Table 2.1 gives the relation for other temperatures.

b. The vadose zone

Porosity

Liquid water is found above the water table in the unsaturated zone. It is held in place by attractive forces between water and the surface of the soil particles. These intermolecular forces, treated here using a classical macroscopic point of view, describe the physical environment of water in porous soil particles.

Consider a volume of soil V_{total} as it occurs in a natural bed. The porosity of the soil φ_p is defined as the ratio of the pore volume (the volume that is not occupied by solid phase) to the total volume:

$$\varphi_p = \frac{V_{pore}}{V_{total}} \tag{6.3}$$

Because the solid soil particles make up the remaining volume, one has

$$\varphi_s = \frac{V_{particles}}{V_{total}} \quad \text{and} \quad \varphi_s + \varphi_p = 1 \tag{6.4}$$

The amount of water found in the pores is usually specified as a fraction relative to the total volume of the system:

$$\varphi_w = \frac{V_{H_2O}}{V_{total}} \tag{6.5}$$

The amount of water may also be given as a mass fraction,

$$w_w = \frac{m_{H_2O}}{m_{total}} \tag{6.6}$$

where the denominator includes the soil water. The parameter w_w is the easiest to determine experimentally. The next steps would be to determine the density of the liquid phase (ca. 1.0 g/ml) and of the dry soil particles (ca. 2.6 g/ml), and then calculate the volume fraction. The volume fraction is preferred in practice because it gives a clear indication of the amount of water in a given soil sample.

Capillary forces

Laplace's equation, Equation 2.99, for water in a capillary tube is

$$p_m \approx -\frac{4\gamma}{d} \quad \text{or, equivalently} \quad h_m \approx -\frac{4\gamma}{\rho g d} \tag{6.7}$$

where γ is the surface tension and d the diameter of the tube. The factor $(\rho\, g)$ connects the matrix head h_m (a height) to the formal pressure p_m. The index m represents the soil matrix. The word *matrix* is used in the sense of a solid phase (the regolith) in which different processes occur, for example, adsorption of water. The quantity p_m is called the matrix pressure.[a] Accordingly, the corresponding quantity h_m is called

---

[a] This is *not* a genuine pressure. The matrix head has no connection to the vapor pressure of water, and its value may be considerably larger than the elevation head $h = p\,/\,(g\,\rho) = 10^5/(9.81\times10^3)$ m = 10.2 m, which is equivalent to the pressure of the atmosphere. Equation 6.7 deals with surface tension.

the matrix head, which is the equivalent head that should be entered in the right-hand side of Equation 6.1 instead of a pressure head:

$$h_h = h_z + h_m \tag{6.8}$$

The shapes of soil pores are irregular, and they do not have a well-defined diameter. Therefore, one purpose of Equation 6.7 is to show the physical nature of capillary forces and to give order-of-magnitude estimates of pore diameters, so-called equivalent diameters.

The capillary action of soil pores retains water within the vadose zone, and the forces holding the water increase as the pore diameter decreases. When water is removed from soil, large pores are emptied first, followed by progressively smaller ones. Because of capillarity, the roots of plants must work harder to obtain the remaining water from the pores of dry soil. At a certain point only the tiniest pores will contain liquid water. When the matrix pressure decreases to about -15 bar, plants wilt; this pressure is defined as the wilting point.

When Equation 6.8 is applied to water in the soils, the zero point of the hydraulic head is chosen to be at the water table. Everything else being equal, the pressure will increase from zero as the point of observation is moved downward and decrease as it is moved upward, giving rise to a formally negative pressure. The scalar fields $h_p(r)$ and $h_m(r)$ may be considered as a single field that is a continuous function of r, but not differentiable in the surface $z = 0$.

Matrix pressure is measured using a tensiometer. The primitive version of this instrument consists of a porous porcelain cup placed in the soil. The cup is filled with water and connected to an above-ground pressure gauge via a water-filled tube. Tensiometer measurements show a negative matrix pressure when the cup is placed in the vadose zone. Because the water column in the tube must be continuous, it is not possible to measure pressures lower than about -0.8 bar using this technique. Figure 6.2 shows experimentally determined relationships between matrix pressure and water content for three soil types.

The amount of soil water that is available to plants is of great practical importance. An early method for measuring it was to weigh the amount of water that was left in the soil after watering, that is, after the excess water had seeped from the bottom of the sample. However, it was not easy to define exactly when this had occurred. Today, *soil water* is defined as the amount of water at a matrix pressure of $p_m = -0.1$ bar $= -10$ kPa. The *field capacity* is the mass of water that a soil sample can release when the matrix pressure is changed from $p_m = -10$ kPa to the wilting limit, where $p_m = -1.5$ MPa. Some pores are so large that they do not contribute to the field capacity of soil. In this case, the capillary force is less than the effect of gravity. Water in pores larger than about 30 μm is not held by the soil and easily drains away and is therefore not available for plants.

Release and uptake of soil water is a phenomenon with hysteresis. This is shown in Figure 6.2 for one of the soils: if a certain amount of water m_1 is released at the matrix pressure $p_1 = -4$ bar, say, then a greater pressure $p_2 = -3$ bar is required to refill the pores. From Equation 6.7, the relation between the equivalent diameters is $d_1 < d_2$, which reflects the irregular shape of the pores: the cross section of their interior is greater than that of their holes.

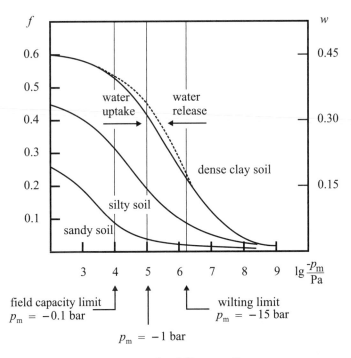

Figure 6.2 *Matrix pressure-water content curves for different soils*
f and *w* are the volume and mass fraction of water in the soil, respectively. The hysteresis has been drawn for a dense clay soil.

c. Flowing groundwater

The preceding discussion has made it clear that the hydraulic conductivity μ_D of soil depends on the size of the pores, whereas the porosity φ_p does not by itself govern water's ability to flow. Some saturated soils (aquitards) are more or less impermeable to water, whereas other saturated soils (aquifers) have high hydraulic conductivities. A confined reservoir is defined as an aquifer that is surrounded by an impermeable layer of soil, as opposed to a free reservoir, whose upper limit is the water table. The pressure head in a confined reservoir can be larger than that of a water column extending to the surface, because the confining surface presses the water using the full weight of the overhead column of regolith. If a well is drilled into such a confined reservoir, the water will flow out of the tube: we have an artesian well. Such behavior would not occur in a well drilled into a free reservoir.

In order to discuss the main principles involved with the flow of groundwater,[3] we will consider a free aquifer with a rectangular cross section as shown in Figure 6.3. It is assumed that the matrix is isotropic and that water is supplied through infiltration at a constant flux density. This is due to the amount of precipitation that does not evaporate or run off the surface. The mass flux density $J_m = 1$ kg m^{-2} d^{-1} corresponds to an excess precipitation (a rate) of $u_n = J_m/\rho = 1$ mm d^{-1}, if we fix the density of water to $\rho = 1$ g/ml.

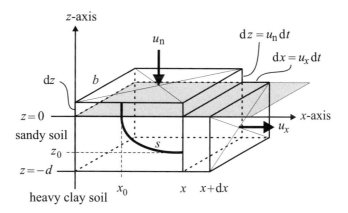

Figure 6.3 *A free aquifer*
Description: The confining boundary is far to the left and the drain far to the right. The free aquifer is placed above an aquitard of clay, $z < -d$. The upper limit is the water table located at $z = 0$. The path s shows the projection onto the zx plane of a fluid's motion through the aquifer.

The physical system is a tube with length x, width b, and depth $z = d$. Mass balance dictates that the amounts of water flowing in and out the system must be equal. Provided that water is incompressible, one then has

$$b x u_n dt = b d u_x dt \tag{6.9}$$

For the velocity of water in the x direction at the point x,

$$\boldsymbol{u}_x = \boldsymbol{u}_n \frac{x}{d} \tag{6.10}$$

Equation 6.10 implies the relationship

$$\frac{x}{x_0} = \frac{d}{d - z_0} \tag{6.11}$$

Two characteristic differential equations should be considered. Darcy's equation, Equation 6.2, is used in the first:

$$u_x = u_n \frac{x}{d} = \mu_D \frac{dh_h}{dx} \tag{6.12}$$

on integration,

$$h_h(x) = \frac{u_n}{2 \mu_D d} x^2 \tag{6.13}$$

This equation shows that the hydraulic head in a free aquifer grows as the square of the length of the aquifer. The sign is correct because it is assumed that water (see Figure 6.3) flows to the right. Note that the result, Equation 6.13, is in agreement with Bernoulli's equation, because in this example the farther one moves with the flow, the more water enters the aquifer.

The second equation is obtained by introducing time dependence:

$$u_x = \frac{dx}{dt} \leq \frac{u_n}{d} x \tag{6.14}$$

Rearranging gives

$$\frac{\mathrm{d}x}{x} = \mathrm{d}\ln x = \frac{u_n}{d}\,\mathrm{d}t \tag{6.15}$$

This equation can be integrated. Water that enters the aquifer at x_0 at time $t = t_0$ reaches point x at time $\Delta t = t - t_0$:

$$x = x_0 e^{\frac{u_n}{d}\Delta t} \tag{6.16}$$

that is, the path grows exponentially with time. The depth z_0 that water reaches after a certain time Δt can be found by substituting Equation 6.11 into Equation 6.16:

$$z_0 = d\left(1 - \exp\left(-\frac{u_n}{d}\,\Delta t\right)\right) \tag{6.17}$$

This equation shows that separate layers of groundwater in the free aquifer do not mix; each layer has a characteristic age that increases with depth. This remarkable result has been verified by experiments. One example was provided by measurements of the concentration of tritium in groundwater. Tritium was released by above-ground nuclear testing in the 1960s.[a] Other examples can be found among industrial compounds that produce unique, soluble products when they break down in the atmosphere. These compounds can be found in newer water samples but not in old, so-called geological water.

Equation 6.13 shows how quickly the hydrostatic pressure grows. With the exception of Equation 6.13, Equations 6.9 through 6.17 are independent of the material-specific hydraulic conductivity μ_D. This only means that the present discussion concerning the physics of groundwater assumes that the hydraulic gradient is small enough that Equation 6.12 is valid. At larger Darcy's velocities, other phenomena would start to occur, for example, water could flow upward through the unsaturated zone as a spring. (In certain regions, such as Denmark, the distance between the source and the drain is seldom larger than a few kilometers.)

6.3 Chemistry of soils

Soil is a heterogeneous mixture of solid and fluid phases, and the chemistry is that of the phases *and* the interfaces.[b]

The gaseous phase is called the *soil air*. It includes the same chemical species as the atmosphere, but in different ratios. The mole fraction of oxygen $y(O_2)$ is lower. Solely because of the presence of water it may drop (from 0.21 to 0.18; see the principles laid down in Table 3.21. After floods, $y(O_2)$ may be half this value, and near plant roots values as low as 0.02 are often observed. The mole fraction of carbon dioxide $y(CO_2)$ is markedly higher than the atmospheric value (which is 3.8×10^{-4}), and it can approach 0.1 near plant roots, with obvious implications for the acid-base chemistry of the soil solution and the solubility of carbonates. Under anaerobic conditions, microorganisms produce CH_4, NH_3, NO, N_2O, and H_2S, which therefore

---

[a] Its half-life of 12.3 a means that very little is left today.
[b] Soil chemistry: refs. 10, 89, 90, 91, 94. Chemistry of surfaces: refs. 1, 7, 65.

are constituents of the soil air. The mole fraction of water is $y(H_2O) \approx 0.01$–0.02 in moist soil; see Table 3.15.

The water phase is called the *soil solution*. While the saturated zone may contain water at up to two-thirds of the volume (see Table 6.3), the vadose zone may also contain a considerable amount of water. As an example, the topsoil (25 cm) of typical Danish farmland holds 525 m^3 ha^{-1} of water (volume fraction $\varphi(H_2O) = 0.210$).[a] This solution contains molecular species such as electrolytes, coordination compounds, organic molecules, dissolved gases, and colloid particles. Although some of this chemistry is discussed later, it was found more appropriate to discuss major parts elsewhere in this book; see Chapter 5.

The regolith is the solid phase consisting of soil minerals. It gives mechanical support to the fluid phases and soil organic matter.[b] Soil organic matter is a generic term that includes biomolecules and humic substances in the liquid phase, microorganisms growing on the solid surfaces, and more advanced life forms such as worms and plants. The basic chemistry of the regolith consists of weathering reactions, as shown by the examples in Table 3.11, and this gives rise to the inorganic content of the soil solution. As the solid phase decomposes through weathering and approaches the size of colloids, the surface area increases, and as a consequence, the number of chemically active sites also increases. Generally, soil minerals carry a negative surface charge and thus attract aprotic cations and, with greater affinity, Brønsted and Lewis acids. In fact, soil surfaces have cation-exchange properties that are even more pronounced for fixed organic molecules (humic substances) with basic functional groups such as carboxylates, phenolates, and amines.

For millennia, agriculture has been essential to human existence, and soon after the birth of scientific chemistry, systematic investigations of the prerequisites for plant growth took place (edaphology). In the first part of the 19th century, the essential components of the natural cycles were described. These included photosynthesis, respiration and fermentation, the role of microorganisms, and the necessity of supplying fertilizer to farmland.[57] In the 1830s, Liebig showed that phosphorus (as dihydrogenphosphate) is an essential plant nutrient, and subsequently, bones from soldiers' graves (from the Napoleonic Wars) were ground and spread onto fields. This source was rapidly depleted and was replaced by "superphosphate" and other products made from phosphate rocks.[c] Fixed nitrogen was also found to be a key nutrient. Early in the 20th century, industrial nitrogen fixation was invented (the Birkeland-Eyde synthesis of Norway saltpeter and the Haber-Bosch synthesis of ammonia). Today, the amount and production rate of artificialy fixed nitrogen is of the same order as naturally fixed nitrogen; see Table 7.6. The present world population would not be possible without this enormous production of fertilizer.

Note: Fertilizers supply the soil with substances that are also found under natural circumstances. However, proper use of fertilizer requires chemical soil analyses and guidelines regarding the amount and timing of application.

---

[a] 1 ha = 10^4 m^2; $\rho_{tot} = 1.4$ g cm^{-3}; $w_w = 0.15$, see Equation 6.6.
[b] Soil organic matter: in specialist literature, the acronym is SOM.
[c] See Sections 7.3 (N) and 7.4 (P).

Table 6.4 Soil minerals: Sheet silicates		
Mica minerals		
a. Kaolinites-serpentines:		
	Kaolinite	$Al_4[Si_4O_{10}](OH)_8$
	Serpentine	$Mg_6[Si_4O_{10}](OH)_8$
b. Talcs:	Pyrophyllite	$Al_2[Si_4O_{10}](OH)_2$
	Talc	$Mg_3[Si_4O_{10}](OH)_2$
c. Micas:	Muscovite	$KAl_2[Si_3AlO_{10}](OH)_2$
	Phlogopite	$KMg_3[Si_3AlO_{10}](OH)_2$
d. Chlorites:	Chlorite	$Mg_3(OH)_6 \cdot Mg_3[Si_4O_{10}](OH)_2$
Clay minerals		
a. Illites:	Illite	$K_{1-x}Al_2[Si_{3+x}Al_{1-x}O_{10}](OH)_2 \cdot (H_2O)_n$
b. Smectites:	Montmorillonite	$Na_{<1}Al_{<2}[(Si,Al)_4O_{10}](OH)_2 \cdot (H_2O)_n$
c. Vermiculites:	Vermiculite	$Mg_{<1}Mg_{\leq3}[(Si,Al)_4O_{10}](OH)_2 \cdot (H_2O)_n$
Amorphous clay minerals		
Allophane, imogolite		
This group consists of amorphous silicates with a significant ion-exchange capacity		

Only a few representative examples are given for each group of minerals.

Substances that control pests and plant diseases are called *pesticides*. In contrast to fertilizers, the pesticides humans make are normally not found in Nature.[a] Modern pesticides are organic compounds that are applied in small amounts (a few g/ha) because their poisonous effect is due to a specific interaction between the chemical species and the target organism. Environmental issues include the rate of destruction and the fate of the products, and such investigations require technically advanced analytical chemistry.

a. Structure of soil minerals

Table 6.4 gives an overview of the main minerals found in soil. The structures of these minerals, the phyllosilicates, are shown in Figures 6.4 to 6.13.

Phyllosilicates[b] are sheet silicates, as discussed in Chapter 3. The structure of the anion, $(Si_2O_5{}^{2-})_n$, shown in Figure 6.4, is shared by all members of the class. Each Si atom is tetrahedrally coordinated with four oxygen atoms, of which three are shared with their neighbor Si, leading to the stoichiometric formula $Si_2O_5{}^{2-}$. It is a unifying feature of the phyllosilicate structure that other anions, namely, hydroxide and/or fluoride, always occur in their structure; their positions are also marked in Figure 6.4. In many instances, the consequence of the layered structure is a platelike morphology and perfect {0 0 1} cleavage. Flakes, easily pulled free with a knife, are observed perpendicular to the *c*-axis shown in the figures. In some cases, however, the anion may be rolled up (e.g., chrysotile), giving rise to a fibrous macroscopic structure.

Soil minerals are grouped into three classes: the mica minerals, the clay minerals, and the amorphous clay minerals. The structures within each subgroup of the mica

---

[a] The role of antibiotics, produced by actinomycetes in soil, is not fully understood; see Section 6.3c.
[b] Gk. φύλλον $\hat{=}$ leaf.

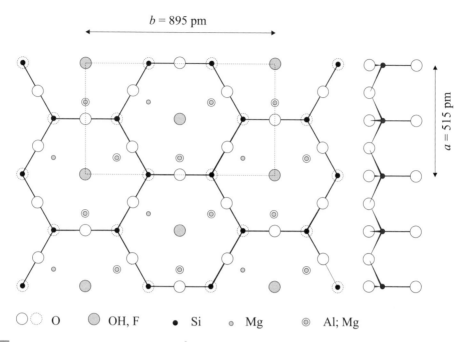

The phyllosilicate anion, $(Si_2O_5^{2-})_n$

The formula of the silicate anion in the unit cell is $[\ Si_4O_4\ (\mu\text{-}O)_6\ ]^{4-}$, and in addition the cell contains $2\ (HO, F)^-$. The figure shows the geometry of the anion common to all phyllosilicates and the characteristic positions of the cations. To the left is the projection onto the $\{001\}$ plane, which is also the plane of the silicon atoms. To the right, the projection is onto the $\{010\}$ plane. The left side shows the positions of atoms in planes parallel to the Si atoms. A layer of oxygen is nearest the eye, and each oxygen atom is coordinated with two silicon atoms in the layer just below. Next comes a layer of oxygen, each atom coordinated to one silicon atom and located under it. These three layers are seen sideways on the figure to the right. Hydroxide or fluoride are also situated in the bottom plane; these anions constitute an integral part of the phyllosilicate structure. Below this layer is a layer of octahedrally coordinated metal ions; only three of their six ligands, namely, the ones just mentioned (2 O bound to Si + 1 OH/F), are shown.

minerals are rather well defined, apart from isomorphic substitution: $Na^I \rightarrow Ca^{II}$, $K^I \rightarrow Mg^{II}$, $Mg^{II} \rightarrow (Al, Fe)^{III}$ concurrent with $Si^{IV} \rightarrow Al^{III}$, and so forth, which is not shown in Table 6.4. The term *clay mineral* is a generic name for hydrated mica minerals; they are the most widespread of the soil minerals, and frequently the particle size is only a few micrometers; see Table 6.3. This small particle size, when combined with a high water content, gives rise to the greasy appearance and the plastic mechanical behavior of clay.

Kaolinites-serpentines

The structure of the mineral kaolinite is shown in Figures 6.5 and 6.6. The negative charge of the silicate anion is compensated by a corresponding positively charged

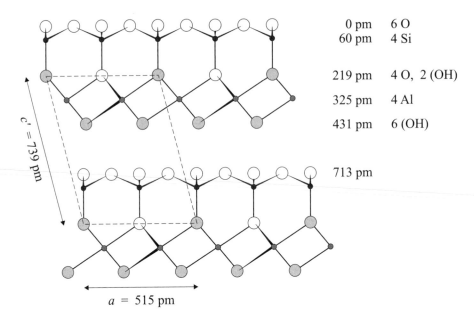

0 pm	6 O
60 pm	4 Si
219 pm	4 O, 2 (OH)
325 pm	4 Al
431 pm	6 (OH)
713 pm	

$c' = 739$ pm

$a = 515$ pm

Figure 6.5 *Kaolinite*, $Al_4[Si_4O_{10}](OH)_8$
Triclinic unit-cell, $a = 515$ pm, $b = 895$ pm, $c = 739$ pm, $\alpha = 91.8°$, $\beta = 104.8°$, $\gamma = 90°$, $Z = 1$. In the figure, $c' = c \sin \alpha$; the elements are marked to the right and shown in detail in Figure 6.6.
For comparison, the dimensions of the (pseudohexagonal) unit cell for the serpentine named clino-chrysotile, $Mg_6[Si_4O_{10}](OH)_8$, are $a = 0.53$ nm, $b = 0.92$ nm, $c = 1.47$ nm, $\beta = 93°$, $Z = 4$.

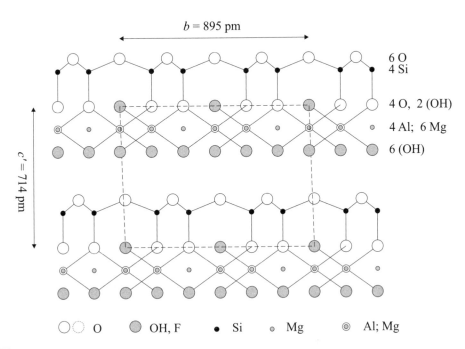

$b = 895$ pm

$c' = 714$ pm

	6 O
	4 Si
	4 O, 2 (OH)
	4 Al; 6 Mg
	6 (OH)

○◐ O   ● OH, F   • Si   ∘ Mg   ◎ Al; Mg

Figure 6.6 *Kaolinite*, $Al_4[Si_4O_{10}](OH)_8$
Triclinic unit cell; the dimensions of the cell are given in Figure 6.5. In the present figure, $c' = c \sin \beta$.
The positions of Mg in serpentine are marked.

	a	b		
	$\begin{bmatrix} O_6 \\ Si_4 \\ O_4(OH)_2 \\ Al_4 \\ (OH)_6 \end{bmatrix}$	$\begin{bmatrix} O_6 \\ Si_4 \\ O_4(OH)_2 \\ Mg_6 \\ (OH)_6 \end{bmatrix}$	Si,	tetrahedral coordination
			Al, Mg	octahedral coordination

Figure 6.7 Representations of 1:1 minerals
a: Kaolinite, $Al_4[Si_4O_{10}](OH)_8$.
b: Serpentine (chrysotile), $Mg_6[Si_4O_{10}](OH)_8$.

layer made up of Al^{3+} and HO^- species. One may imagine that the silicate sheet is fused to a gibbsite sheet that has achieved positive charge through the loss of an appropriate amount of HO^-. The structure of the mineral gibbsite, $Al(OH)_3$, is shown in Figure 6.8, and it appears that Al^{III} is in octahedral coordination. Thus, the fused structure consists of a tetrahedrally coordinated Si sheet and an octahedrally coordinated Al sheet. Such a structure is called a 1:1 structure.

The structure of the serpentines may similarly be thought of as a tetrahedrally coordinated Si sheet fused together with an octahedrally coordinated Mg sheet derived from the mineral brucite, $Mg(OH)_2$ (Figure 6.9). The unit cell dimensions for the serpentine mineral chrysotile are given in Figure 6.5. This is also a 1:1 structure. The principle of the two structure types is shown in Figure 6.7.

The gibbsite structure is called *dioctahedral* because each oxygen atom is coordinated with two metal ions; for a similar reason, the brucite structure is called *trioctahedral*. This classification is transferred to the soil minerals just discussed:

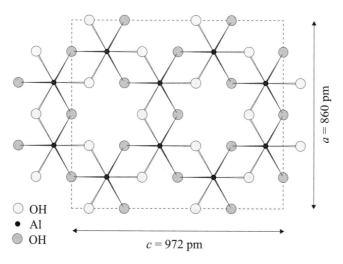

○ OH
● Al
◐ OH

$a = 860$ pm

$c = 972$ pm

Figure 6.8 *Gibbsite (hydrargillite)*, $Al(OH)_3$
Monoclinic unit cell, $a = 864$ pm, $b = 507$ pm, $c = 972$ pm, $\beta = 94.57°$, $Z = 8$.
The coordination number of oxygen is 2: dioctahedral structure.

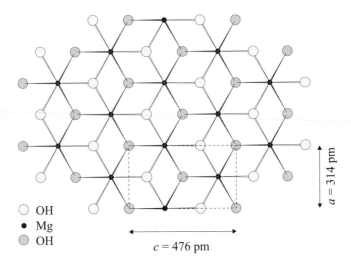

○ OH
● Mg
◐ OH

$c = 476$ pm

$a = 314$ pm

Figure 6.9 *Brucite*, $Mg(OH)_2$
Trigonal (D_{3d}). Unit cell: $a = 314$ pm, $c = 476$ pm, $Z = 2$.
The coordination number of oxygen is 3: trioctahedral structure.

kaolinite has a dioctahedral 1:1 structure, whereas chrysotile has a trioctahedral 1:1 structure.

Talcs

The structure of the mineral pyrophyllite is shown in Figure 6.10. This structure may be envisaged as being analogous to the structure of kaolinite. A sheet of octahedrally coordinated Al^{III} is sandwiched between two sheets of linked SiO_4 tetrahedra. Two-thirds of the available octahedral sites are occupied by Al^{III} and the remainder are empty. Thus, pyrophyllite has a dioctahedral 2:1 structure.

The structure of the mineral talc is not shown, but the unit cell dimensions and an indication of the position of the trioctahedral brucite sheet are given in Figure 6.10. Talc is said to possess a trioctahedral 2:1 structure. An isomorphic substitution may take place in the tetrahedral sheet, $Si^{IV} \rightarrow Al^{III}$, and concurrently in the octahedral sheet, $Mg^{II} \rightarrow Al^{III}$, showing the great versatility of aluminium.

The principles of the structures of the talc minerals are shown in Figures 6.13a and 6.13b.

Micas

The structure of the mineral phlogopite is shown in Figure 6.11. It may be described as a potassium-expanded talc structure where the substitution $Si^{IV} \rightarrow Al^{III}$ is matched by the insertion of a potassium ion. It is noted that the octahedral sites are filled and that the coordination number of the relatively large K^+ is 12. The dark mica mineral biotite has the mica structure of phlogopite, but some of the Mg^{II} has been replaced by Fe^{II}, which explains the color.

In a similar way, the mineral muscovite may be described as potassium expanded pyrophyllite. A rich mineral chemistry is derived from muscovite via isomorphic

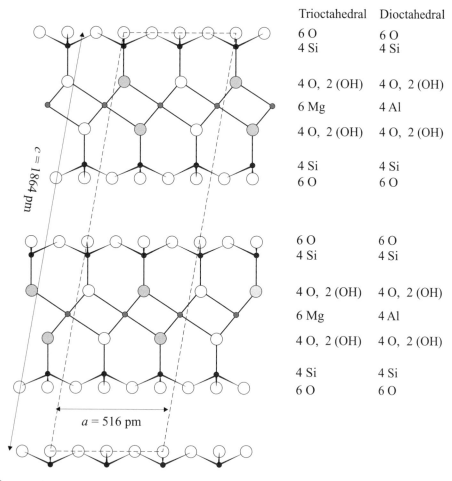

Trioctahedral	Dioctahedral
6 O	6 O
4 Si	4 Si
4 O, 2 (OH)	4 O, 2 (OH)
6 Mg	4 Al
4 O, 2 (OH)	4 O, 2 (OH)
4 Si	4 Si
6 O	6 O
6 O	6 O
4 Si	4 Si
4 O, 2 (OH)	4 O, 2 (OH)
6 Mg	4 Al
4 O, 2 (OH)	4 O, 2 (OH)
4 Si	4 Si
6 O	6 O

Figure 6.10 *Pyrophyllite*, $Al_4[Si_8O_{20}](OH)_4$
Monoclinic unit cell, $a = 516$ pm, $b = 890$ pm, $c = 1864$ pm, $\beta = 99.92°$, $Z = 2$.
For comparison, the dimensions of the monoclinic unit cell for talc;
$Mg_6[Si_8O_{20}](OH)_4$, are: $a = 528$ pm, $b = 915$ pm, $c = 1.89$ nm, $\beta = 100°$, $Z = 2$.

substitutions: K $\rightarrow$ Na, Rb, Cs, Ca, Ba, and so forth; octahedral Al $\rightarrow$ Mg, Ti, Cr, Fe, Li, and so on; and Si or tetrahedral Al $\rightarrow$ Fe^{III} and Ti.

Chlorites

The structure of the mineral chlorite is shown in Figure 6.12. It can be described as a brucite-expanded talc structure. The chemical composition of the chlorites is extremely versatile, as is evident from the formula:

$$(Mg, Fe, Al, Mn)_3(OH)_6 \cdot (Mg, Fe, Al, Mn)_3[(Si, Al, Fe)_4O_{10}](OH, F)_2$$

Chlorites are very common soil minerals.

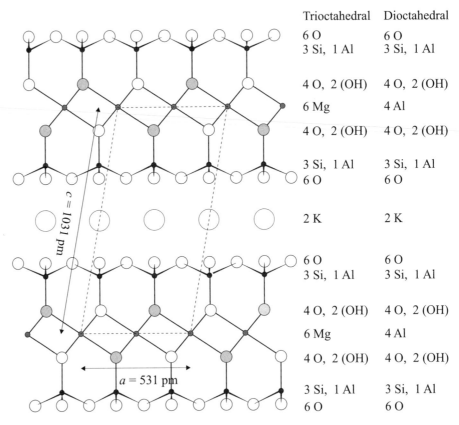

Trioctahedral | Dioctahedral

6 O | 6 O
3 Si, 1 Al | 3 Si, 1 Al

4 O, 2 (OH) | 4 O, 2 (OH)
6 Mg | 4 Al
4 O, 2 (OH) | 4 O, 2 (OH)

3 Si, 1 Al | 3 Si, 1 Al
6 O | 6 O

2 K | 2 K

6 O | 6 O
3 Si, 1 Al | 3 Si, 1 Al

4 O, 2 (OH) | 4 O, 2 (OH)
6 Mg | 4 Al
4 O, 2 (OH) | 4 O, 2 (OH)

3 Si, 1 Al | 3 Si, 1 Al
6 O | 6 O

Figure 6.11 *Phlogopite*, $K_2Mg_6[Si_6Al_2O_{20}](OH)_4$
Monoclinic unit cell, $a = 531$ pm, $b = 920$ pm, $c = 1031$ pm, $\beta = 99.90°$, $Z = 1$.
For comparison, the dimensions of the monoclinic unit cell for muscovite:
$K_2Al_4[Si_6Al_2O_{20}](OH)_4$,
are: $a = 519$ pm, $b = 904$ pm, $c = 2008$ pm, $\beta = 95.50°$, $Z = 2$.
Note that the coordination number of potassium is 12.

Clay minerals

The structure of clay minerals corresponds to that of the micas, but here the large cations between the talc layers have been replaced by a sheet of water molecules. For example, the structure of montmorillonite can be described as water-expanded pyrophyllite. The lattice constant c varies with the water content, and by replacing water with substances such as glycerol, it may swell to many times the original size. Hard-dried vermiculite, a phyllosilicate, is commonly used to absorb spilled chemicals. The illites are the most widespread soil minerals; the basic structure of this group is shown in Figure 6.13c.

b. The soil solution

The major inorganic constituents of soil water are the same as those found in most lakes and rivers; see Table 3.14. Cations and anions can form ion pairs, and under

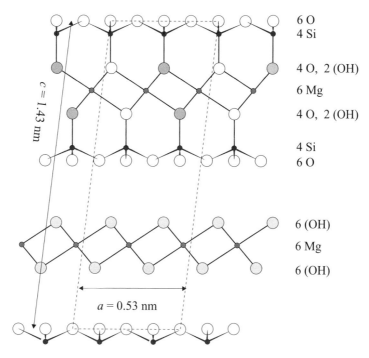

6 O
4 Si

4 O, 2 (OH)

6 Mg

4 O, 2 (OH)

4 Si
6 O

6 (OH)

6 Mg

6 (OH)

$c = 1.43$ nm

$a = 0.53$ nm

Figure 6.12 *Chlorite*
Chlorite, $Mg_{12}[Si_8O_{20}](OH)_{16} = Mg_6(OH)_{12} \cdot Mg_6[Si_8O_{20}](OH)_4$.
Monoclinic unit-cell, $a = 0.53$ nm, $b = 0.92$ nm, $c = 1.43$ nm, $\beta = 97°$, $Z = 1$.

saline conditions the constants of formation resemble those of coordination compounds in seawater as shown in Table 5.11.

Organic entities of small molecular mass include carboxylates, commonly with concentrations up to 5 mM;[a] at pH 7 one finds formate, acetate, oxalate, tartrate, and citrate (in order of decreasing concentrations). As seen from the acidity exponents (see Table 5.3 and Table 5.8 (citric acid)), the conjugate bases of the acids are found in Nature. The monovalent acids are mainly made by microorganisms and the polyvalent acids by the roots of fungi and plants.

The most common amino acids[b] are found at concentrations less than 3 mM as amphoions; see Section 5.1c. These species are the final product in the decomposition of proteins. Similarly, the decomposition of cellulose and lignin gives oligosaccharides and derivatives of phenols such as coniferyl alcohol,[c] respectively. The industrial chemistry of these polymeric compounds is discussed in Section 8.2.

Acid-base chemistry, coordination chemistry, and redox chemistry applicable to the soil solution are discussed in detail in Chapter 5, and Figure 6.17 shows the acid-base properties of a spodosol fulvic acid. The pH of a soil solution is governed by various buffer systems because water itself has no significant buffer value in natural

[a] Note: 1 mM = 1 mmol / l = 1 mol m^{-3}.
[b] Gly, Ala, Asp, Glu, Arg, Lys. See Appendix A1.9.
[c] 3-(4-hydroxy-3-methoxyphenyl) prop-2-en-1-ol.

a	b	c

$$
a\quad
\begin{bmatrix}
O_6 \\
Si_4 \\
O_4(OH)_2 \\
Al_4 \\
O_4(OH)_2 \\
Si_4 \\
O_6
\end{bmatrix}
\qquad
b\quad
\begin{bmatrix}
O_6 \\
Si_4 \\
O_4(OH)_2 \\
Mg_6 \\
O_4(OH)_2 \\
Si_4 \\
O_6
\end{bmatrix}
\qquad
c\quad
\begin{bmatrix}
O_6 \\
Si_{3+y}Al_{1-y} \\
O_4(OH)_2 \\
Al_4 \\
O_4(OH)_2 \\
Si_{3+y}Al_{1-y} \\
O_6
\end{bmatrix}^{2y-2}
$$

Figure 6.13 *Representations of 2:1 minerals*
a. Pyrophyllite, $Al_4[Si_8O_{20}](OH)_4$.
b. Talc, $Mg_6[Si_8O_{20}](OH)_4$.
c. Illite, $K_{2-x}Al_4[Si_{6+x}Al_{2-x}O_{20}](OH)_4$; x in this latter formula is equal to $2y$ in the above drawing.

waters. For agricultural purposes, the hydron concentration of a soil is determined as the amount per mass that is liberated by ion exchange with ammonium chloride.

The salinity $c_{\text{soil-salinity}}$ is another important parameter in the edaphological description of a soil solution, because too high a concentration of ions is harmful to plant growth. The salinity of a soil solution is measured by its ionic strength I,

$$c_{\text{soil-salinity}} = I \tag{6.18}$$

(see Equation 5.56). A good farm soil (e.g., in Denmark) has $c_{\text{soil-salinity}}$ between 1 mM and 10 mM. Weakly salt sensitive crops are affected at $c_{\text{soil-salinity}} > 15$ mM, and moderately salt sensitive at $c_{\text{soil-salinity}} > 30$ mM. Most crops cannot survive $c_{\text{soil-salinity}} \approx 60$ mM, which is 1/10 of that of seawater. In the field the salinity may be assessed by measuring the conductivity χ because of the empirical proportionality

$$I/\text{mM} = 0.1467\chi/(\text{mS m}^{-1}), \tag{6.19}$$

accurate within this range of salinities.[a]

c. Soil adsorption phenomena

Two hundred years ago Reuss (1809) discovered electroosmosis, the fact that clay minerals carry electric charge. The subsequent understanding that interfaces normally carry an electrical charge was an important advance in colloid and surface science, and it is a key factor in the interpretation of the chemistry of soils. *Electrified surface chemistry* is an established field, and, for example, the journal *Physical Chemistry Chemical Physics* recently devoted a full issue to this subject alone.[b]

In this section we present three examples that show how the surface charge is detected and explained, and how it is utilized in agriculture.

[a] The empirical expression $\lg(I/\text{M}) = -0.832 + 1.009\,\lg\{\chi/(\text{mS m}^{-1})\}$ is useful up to ionic strength $I \approx 0.3$ M.[190]
[b] *PCCP 12* (2010) 15149-15320.

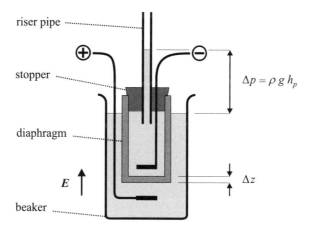

Figure 6.14   *Electroosmosis I*
Experimental demonstration of electroosmosis. A porous diaphragm, here a cylinder of (unglazed) clay, is immersed in water with two electrodes (zinc) placed on either side of the wall. Application of an electric field strength $E \approx 2000$ V m^{-1} with polarity as shown causes water to rise in the riser pipe, placed in a stopper. The water will continue to rise until it reaches the steady-state pressure head.

Electroosmosis

The surface of clay minerals is negatively charged, which is readily demonstrated in an electrokinetic experiment such as that of electroosmosis, shown in Figure 6.14. At the outset the water level in the beaker is equal to that in the riser pipe. Application of an external electric field of strength E with the direction shown causes an electroosmotic flow with a net transport of water through the diaphragm. As a hydraulic head develops, a steady state is established and *net* transport ceases.

This phenomenon was investigated quantitatively in the 1850s and explained by postulating the existence of an electrical double layer at the interface. The solid cannot move, but the liquid is displaced in the direction of the electric field, and therefore it must have a positive charge. Von Helmholtz (1879) made calculations based on the assumption that the double layer is an electrical capacitor with parallel plates no more than a molecular distance apart.[a] Accordingly, the potential gradient at the interface should be large, but Gouy (1909), Chapman (1913), and Stern (1924) modified this theory, proposing the diffuse double layer, shown in Figure 6.15.

The left side of Figure 6.15 shows an enlarged picture of a mineral pore, and the right side shows the variation of the electric potential of the double layer. At some distance b from the interface, the electric perturbation from the wall is effectively shielded by the water, and only the bulk properties of the liquid remain. The position of water molecules very close to the wall may be fixed, but there must exist a shear plane near the wall. The difference between the potential at this plane φ_{sp} and the

---

[a] The plates of a parallel-plate capacitor have opposite electric charge, Q. Let the electric potential between the plates be φ, then the capacitance is $C = Q/\varphi$ with the SI-unit F = C/V. For such a capacitor, $C = \varepsilon A/d$ where ε is the permittivity, A the area of a plate, and d the distance between them.

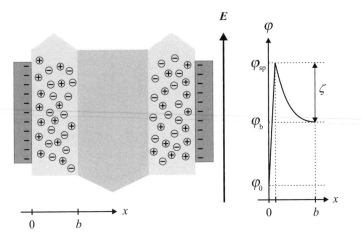

Figure 6.15 *Electroosmosis II*
Left panel: The movement of water in a capillary channel in clay subject to an electric field. The surface charge of clay is negative. Therefore, an aqueous layer between a shear plane close to the wall of the pore and the bulk moves in the direction of the electric field strength, E. A counterflow arises in the center of the tube, eventually stopping the net rise of water.
Right panel: The electric potential φ as a function of the distance from the solid surface. The potential is negative at the interface (φ_0), positive at the shear plane (φ_{sp}), and zero in the bulk (φ_b). The ζ-potential is marked; here it is positive.

potential of the bulk φ_b is called the ζ potential. This is not a genuine phase boundary potential, because it is developed wholly within the liquid. However, it is a kind of material constant, as shown in the following macroscopic treatment.

The mass flux density $\boldsymbol{J}_{m,E}$ due to the electric field is given by (see Section 2.1d)

$$\boldsymbol{J}_{m,E} = \rho\boldsymbol{u}_E, \qquad \boldsymbol{u}_E = \zeta\frac{\varepsilon\boldsymbol{E}}{\eta}, \qquad \varepsilon = \varepsilon_r\varepsilon_0 \tag{6.20}$$

where ρ is the mass density of the liquid and $\boldsymbol{u}_E$ the velocity of the layer close to the wall of the pore. From electromagnetic theory (see Section 5.3b), the velocity is proportional to the permittivity ε of the liquid and the electric field strength E and is inversely proportional to the viscosity η.[a] The mass flux density $\boldsymbol{J}_{m,D}$ of the counterflow may be expressed using Darcy's equation, Equation 2.97,

$$\boldsymbol{J}_{m,D} = \rho\boldsymbol{u}_D, \qquad \boldsymbol{u}_D = \frac{A_D}{\eta}\boldsymbol{grad}\,p \tag{6.21}$$

and at steady state with no net flow, the flux densities are numerically equal, leading to an empirical value of the ζ-potential:

$$\zeta = \frac{A_D}{\varepsilon E}\frac{\Delta p}{\Delta z} \tag{6.22}$$

The following example gives a set of values for the variables in this relation.

[a] Note: The dielectric displacement $\boldsymbol{D} = \varepsilon\boldsymbol{E}$ and Unit(η) = Pa s. See Appendix A1.5.

Table 6.5 Cation-exchange capacity of minerals and humus			
Mineral	b_{cec} / mol kg^{-1}	Mineral	b_{cec} / mol kg^{-1}
Smectite	1.00	Goethite	0.10
Illite	0.30	Gibbsite	0.05
Kaolinite	0.07	Hematite	0.05
Chlorite	0.30	Allophane	0.70
Vermiculite	1.50	Imogolite	0.70
		Humus	2.00

Note: The unit cmol hg^{-1} is sometimes used:
1 mol/kg = 1 mmol/g = (100 cmol)/(10 hg) = 10 cmol/hg.

Example

Some typical values for the physical quantities in the experiment shown in Figure 6.14 are $h_p = 25$cm ($\Delta p = 2450$ Pa), $\Delta z = 3$ mm, $A_D = 0.1$ μm^2 (see Equation 2.98 and Table 6.3), $E = 2000$ Vm^{-1}, $\varepsilon = 7.1 \times 10^{-10}$ Fm^{-1} (see Appendices A1.6 and A3) $\Rightarrow \zeta = 58$mV.

Ion exchange

The question of whether the replacement of an ion in the electrical double layer is a genuine ion exchange is a matter of degree. Ion exchange occurs when the exchange capacity is "large enough," that is, more than 25 μmol g^{-1}; see Tables 6.5 and 6.6. Here the concentration is given as the exchangeable amount of substance per mass of ion exchange matrix. Multiplication with the Faraday constant F converts concentrations to electrical charge per mass: assume a cation-exchange capacity (cec)[a] of $b_{cec} = 1$ mmol g^{-1}. Then the electrical charge is $\rho_{cec} = F b_{cec} = 96.5$ C g^{-1}. Cation-exchange capacities for some soil minerals and humus are shown in Table 6.5 and for some soil orders in Table 6.6.

In the following discussion, the symbol $\equiv$R$^-$ is used to represent a generic anion site on the surface of the soil matrix.[b] Ion exchange between adsorbed ions A$^+$ on the cation-exchange matrix and dissolved ions B$^+$ is described by

$$A^+(\equiv R^-) + B^+(aq) \quad \rightleftharpoons \quad B^+(\equiv R^-) + A^+(aq) \tag{6.23}$$

The equilibrium constant may be written as the selectivity constant

$$K_{exchange} = \frac{[A^+]}{[B^+]} \tag{6.24}$$

because the activities of the solid phases, which are not known, can be included as part of $K_{exchange}$ and, for comparison under similar conditions, concentrations are adequate measures of solute activities. Even for values of $K_{exchange}$ near 1, effective

[a] For nomenclature of symbols, see Table A1.7.
[b] Analogously, $\equiv$R$^+$ will be used for a generic cation site, that is, an anion exchange site. R (for resin) is taken from standard laboratory terminology for ion exchange. IUPAC has not made any recommendations regarding this nomenclature (2007). Stumm[91] uses $\equiv$ML for a metal-ligand soil complex. Sposito[89] uses $\equiv$ as "bound to a surface," for example, $\equiv$NH$_3{}^+$ for a protonated amine.

Table 6.6 Cation-exchange capacity of soil orders			
Soil order	b_{cec} / mol kg^{-1}	Soil order	b_{cec} / mol kg^{-1}
Alfisol	0.12	Mollisols	0.22
Aridisol	0.16	Oxisols	0.05
Entisol	0.13	Spodosols	0.11
Histosol	1.4	Ultisols	0.06
Inceptisol	0.19	Vertisols	0.37

Average values of b_{cec} / mol kg^{-1} for a variety of soils, cf. Table 6.2.[89]

exchange can be achieved, because percolation of the soil solution through the profile effects many equilibrium steps. This corresponds to ion-exchange chromatography in the laboratory, and when carried out in the field, such experiments are called lysimeter experiments. Theories of chromatography are well suited for describing ion-exchange processes in the soil; see Section 2.2e.

The affinity of cations for a cation-exchange matrix $(\equiv R^-)_n$ decreases according to charge and size of the aqua ion: $Th^{4+} > Al^{3+} > Ca^{2+} > Na^+$, $NH_4^+ \sim K^+ > Na^+ > Li^+$, and $Ba^{2+} > Ca^{2+} > Mg^{2+}$. The series of alkali metal ions illustrates an important point: the smallest naked ion, lithium, attracts the largest amount of water, which prevents close contact with the matrix.[a] Analogous series are found for the affinities of anions to an anion-exchange matrix, $(\equiv R^+)_n$: $SO_4^{2-} > Cl^-$ and $I^- > Cl^- > F^-$. The ion-exchange capacity of a soil sample can be determined experimentally through ion exchange with an NH_4Cl solution, followed by quantitative analysis for each metal.

For agricultural applications, a distinction is made between the acidic cations H^+, Al^{3+}, and Fe^{3+} and cations that are aprotic in water solution, Na^+, K^+, Mg^{2+}, Ca^{2+}, and Fe^{2+}. Members of the former group are acids, the pK_{a1}'s for aqua ions of the trivalent metals being in the range 3–5. Tables 5.15 and 5.16 show this and also that the hydroxides of these ions are sparingly soluble. Note that members of the latter, aprotic group have no acid-base properties and they are found as aqua ions at the pH values of soil water.[b]

A soil sample may be characterized according to the nature of the cations that can be exchanged:

$$b_{acid} = b(H^+) + 3\,b(Al^{3+}) + 3\,b(Fe^{3+}) \tag{6.25}$$

$$b_{aprot} = b(Na^+) + b(K^+) + 2\,b(Mg^{2+}) + 2\,b(Ca^{2+}) + 2\,b(Fe^{2+}) \tag{6.26}$$

The total cation-exchange capacity of soil is the sum of the two terms, and the mole fraction of aprotic cations is[c]

$$x_{aprot} = b_{aprot}/(b_{acid} + b_{aprot}) \tag{6.27}$$

---

[a] Use of Coulomb's law: the force between two charges is proportional to the product of the charges and inversely proportional to the square of the distance between the charges; see Equation A3.1.

[b] The acids $Mg(H_2O)_6^{2+}$ ($pK_{a1} = 11.4$) and $Fe(H_2O)_6^{2+}$ ($pK_{a1} = 9.5$) are not dissociated in soil solutions.

[c] This fraction is called *base saturation*.

Acidification of a soil gives rise to a decrease in the value of x_{aprot}: the aprotic cations Na^+, K^+, Mg^{2+}, and Ca^{2+} are replaced by the acidic ions H^+ and Al^{3+}. To understand this qualitative statement, consider Equation 6.23 with A^+ representing an aprotic cation and B^+ a hydron. If $[H^+]$ in the soil solution is increased, then the aprotic, economically valuable cations are washed out. Acids are produced in *all* of the naturally occurring processes in which atmospheric oxygen is reduced by a nonmetal, producing, for example, H_2CO_3, HNO_3, and H_2SO_4. Therefore, certain natural environments can become quite acidic as a consequence of weathering. The immediate effect of acid rain is that hydrons are retained and aprotic cations are lost from the topsoil. Over the long term, acid rain can wash away the ions necessary for plant growth. Also, acid can dissolve gibbsite, $Al(OH)_3$, and bauxite, $AlO(OH)$, which are virtually insoluble at pH 7, releasing aluminium ions that are poisonous for almost all life forms; see Appendix A2.2.

Acidification can be fought by liming. Limestone, calcium carbonate, is mixed into the topsoil, resulting in the following reaction:

$$2H^+(\equiv R^-) + 2CaCO_3 \;\rightleftharpoons\; Ca^{2+}(\equiv R^-)_2 + Ca(HCO_3)_2 \quad (6.28)$$

Similar processes occur for other acidic cations:

$$2Al^{3+}(\equiv R^-)_3 + 6CaCO_3 + 3H_2O$$
$$\rightleftharpoons 3Ca^{2+}(\equiv R^-)_2 + 3Ca(HCO_3)_2 + 2Al(OH)_3 \quad (6.29)$$

As seen, only half of the calcium binds to the soil in the latter reaction. For this reason, calcium hydroxide is sometimes used:

$$2H^+(\equiv R^-) + Ca(OH)_2 \;\rightleftharpoons\; Ca^{2+}(\equiv R^-)_2 + 2H_2O \quad (6.30)$$

and

$$2Al^{3+}(\equiv R^-)_3 + 3Ca(OH)_2 \;\rightleftharpoons\; 3Ca^{2+}(\equiv R^-)_2 + 2Al(OH)_3 \quad (6.31)$$

Bound Ca^{2+} may react with carbon dioxide if soils become acidic:

$$Ca^{2+}(\equiv R^-)_2 + CO_2 + H_2O \;\rightleftharpoons\; 2H^+(\equiv R^-) + CaCO_3 \quad (6.32)$$

Liquid ammonia, $NH_3(l)$, is frequently applied directly to the soil as a fertilizer. Here it reacts with soil water giving $NH_3(aq)$, and in turn ammonium:

$$NH_3(aq) + H_2O(l) \;\rightarrow\; NH_4^+(aq) + HO^-(aq) \quad (6.33)$$

Ammonium reacts rather quickly with oxygen to give nitrate in bacterial processes, as discussed in Section 7.3a, the total process being

$$NH_4^+(aq) + 2O_2(aq) \;\rightarrow\; NO_3^{\,-}(aq) + 2H^+(aq) + H_2O(l) \quad (6.34)$$

This is called *nitrification*. The transformation of nitrogen from a cation to an anion increases its mobility, because the negative charge of the surface of the soil particles cannot retain anions. During the growing season, evapotranspiration[a] exceeds rainfall, water moves upward, and nitrate is available for plants. In the wintertime, rainfall

---

[a] Evapotranspiration = evaporation from the surface of the field, including transpiration through plants.

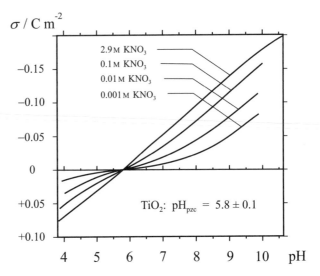

Figure 6.16 *Titration of rutile powder with strong acid and base*
Rutile powder, $TiO_2(s)$: specific surface area 19.8 m^2 g^{-1}; concentration of
suspension in water 0.190 M. Titration (with strong acid, 0.10 M HNO_3, and strong
base, 0.10 M KOH) at four different concentrations of KNO_3 is shown. The data
have been treated as explained in the text. Data from ref. 217.

normally exceeds evapotranspiration, and nitrate is leached. This is unfortunate,
because fertilizer is expensive and nitrate is unwanted in groundwater.

Point of zero charge

The point of zero charge, pzc, is a key concept in surface chemistry.[189] It is the
value of the negative logarithm of the concentration (units of mol/l) of the potential
determining ion that gives a surface charge of zero.[a, 123] In the special case of an
acid-base titration of a powder with a large surface area, the point of zero charge
occurs at a definite value of pH, denoted pH_{pzc}. For example, Figure 6.16 shows the
acid-base titration of a suspension of rutile, TiO_2, in water.[217]

The primary data[20] are the simultaneous values of alkalinity, the amount of added
strong base or acid, and pH. Their transformation into electric charge per area requires
an independent determination of the surface area.[b]

The point of zero charge is easily recognizable on Figure 6.16, and the following
general conclusions emerge. Let $\equiv TiO_2$ represent a neutral surface site which acts
as an amphoteric species. The conjugate acid is $\equiv TiO_2H^+$, and the conjugate base
is $\equiv TiO_2(OH)^-$. The Ti-O bond is polarized: $[\delta(+)$ Ti-O $\delta(-)]$, such that the oxygen

---

[a] Potential determining ions are defined as those species that, by virtue of their equilibrium distribution
between the solid and the liquid phase, determine the potential difference between these phases. In the
present case, H^+ and HO^- are the potential determining ions.

[b] The Langmuir adsorption isotherm (1916) was derived in Section 2.2c, example 3. The surface of the
suspension of Figure 6.16 was determined from an analogous BET adsorption isotherm (Brunauer,
Emmett, Teller; 1938).[1, 216]

Table 6.7 Point of zero charge of some minerals in water			
Mineral	pH_{pzc}	Mineral	pH_{pzc}
Gibbsite, α-Al(OH)$_3$	5.0	Feldspars	2.0 - 2.4
Boehmite, γ-AlO(OH)	8.2	Albite	2.0
Hematite, α-Fe$_2$O$_3$	8.5	Chrysotile	>10
Goethite, α-FeO(OH)	7.8	Montmorillonite	2.5
Quartz, SiO$_2$	2.0	Kaolinite	4.6
Data from ref. 91.			

surface of rutile is negatively charged. Accordingly, in an aqueous suspension, water coordinates to the surface of rutile, forming complexes of the type $\equiv TiO_2 \cdot (HOH)_n$; normally, this type of water is not included in the reaction equations. Figure 6.16 therefore suggests the formal acid-base equilibria:

$$\equiv TiO_2H^+ \;\; \underset{+H^+}{\overset{-H^+}{\rightleftharpoons}} \;\; \equiv TiO_2 \;\; \underset{+H^+}{\overset{-H^+}{\rightleftharpoons}} \;\; \equiv TiO_2OH^- \qquad (6.35)$$
$$\hspace{3.5cm} K_{a1} \hspace{3.5cm} K_{a2}$$

with two acid constants. Note that no titration experiment can determine the number n of coordinate water, nor can it discriminate between adsorption of HO^- and desorption of H^+. Nevertheless, Equation 6.35 gives a picture of the chemistry on the surface of a nonstoichiometric colloid particle. In many cases, the data do not allow the determination of more than one constant, namely, the product $K_{a1}K_{a2}$. Modern surface techniques[a] may provide deeper insight regarding these reactions.

Table 6.7 gives examples of the point of zero charge for a selection of minerals in water, determined by association or dissociation of hydrons on mineral surfaces. These are not absolute numbers but may be used as a guide when considering surface properties of soil minerals. The surfaces of common rocks are negatively charged in natural environments. Generally, the more polarized the metal-oxygen bond, the lower the pH_{pzc}, with associated lower acidity exponents.

The electrified interface

A significant feature of the experiments depicted in Figure 6.16 is the dependence of the charge density on the ionic strength, I. In order to understand this, we first consider ion exchange and then some details of the experiment.

One defines the charge density of a surface possessing ion-exchange capacity as follows. Let b_{cec} and b_{aec} be the cation-exchange capacity and anion-exchange

[a] Present tools include various spectroscopic techniques: extended x-ray absorption fine structure spectroscopy (EXAFS); x-ray photoelectron spectroscopy (XPS); various reflectance methods, such as Fourier transform infrared spectroscopy (FTIR) and photoacoustic spectroscopy in IR *and* visible and nuclear magnetic resonance (NMR) and electron paramagnetic resonance (EPR) spectroscopy. Future tools (for surfaces under water) will include atomic force microscopy (AFM), and scanning tunneling microscopy (STM).

capacity, respectively, of a solid. Assuming a specific area $a = A/m$ (area A per mass m), the surface charge density σ is given by

$$\sigma = -F \frac{b_{cec} - b_{aec}}{a} \tag{6.36}$$

where F is the Faraday constant.[a] Introducing the surface concentration[b] $\Gamma = b/a$ gives

$$\sigma = -F(\Gamma_{cec} - \Gamma_{aec}) \tag{6.37}$$

which will be used subsequently.

According to Equation 6.35, the potential determining ions H^+ and HO^- are largely responsible for charging the surface. Analogous to the expression, Equation 6.37, one defines the surface charge density σ through

$$\sigma = F(\Gamma_{H^+} - \Gamma_{HO^-}) \tag{6.38}$$

where Γ_B denotes the surface excess of the species B. Operationally, each adsorbing H^+ is accompanied by a NO_3^- and each adsorbing HO^- by a K^+, so the operational definition is[189]

$$\sigma = F(\Gamma_{HNO_3} - \Gamma_{KOH}) \tag{6.39}$$

where it is understood that the ions NO_3^- and K^+ are not surface ions. This is the quantity depicted as the ordinate in Figure 6.16.

From Figure 6.16: For $pH > pH_{pzc}$, $\sigma < 0$; if $c(KNO_3)$ is increased, σ becomes more negative at a given pH because there are more cations to screen the surface charge. Assuming the simple Helmholtz model (see Footnote b on page 236), the double-layer capacitance increases with increasing c because potassium ions are not surface ions. Similarly, for $pH < pH_{pzc}$, $\sigma > 0$, and σ becomes more positive with increasing electrolyte concentration. The electrolyte has no influence only when $\sigma = 0$.

d. Soil colloid phenomena

Colloids

Heterogeneous systems consisting of two phases in which one phase, the disperse phase, is distributed within the other one, the dispersion medium, are called disperse systems. There are only eight types of disperse systems, as seen in Table 6.8, because gas mixtures are homogeneous. Disperse systems in which the diameters of the entities of the disperse phase are between 500 nm and 5 nm are called colloids (T. Graham, 1861).[c] The upper limit coincides with the wavelength of yellow light, and

---

[a]　σ has the SI-unit C m^{-2}.
[b]　See Appendix A1.7.
[c]　Gk. $\kappa o \lambda \lambda \alpha \triangleq$ glue; $\varepsilon \iota \delta \omega \triangleq$ be like.

Disperse phase	Dispersion medium	Name	Examples
Table 6.8 Disperse systems[26]			
s	s	Solid suspension	Some alloys,[1] ruby glass
s	l	Suspension	Solid particles in water[2]
s	g	Aerosol	Smoke, dust
l	s	Solid emulsion	Pearls
l	l	Emulsion	Fat in water (milk)
l	g	Aerosol	Fog, mist, cloud
g	s	Solid foam	Bread, insulation, pumice
g	l	Foam	Soap foam, whipped cream

[1] Examples include several iron-carbon phases; see Figure 8.3.
[2] Examples include silt in river water; see Section 9.5c.

particles below this size cannot be observed in a normal microscope. Solutions of particles less than 5 nm are genuine solutions, that is, they exhibit all the colligative properties of solutions;[a] colloids exhibit only osmotic pressure.

A colloidal suspension is called a sol, and this word is incorporated as the root of the term aerosol, used when the dispersing medium is the gas phase. Solvents attracting colloids are called lyophilic, and those repelling colloids lyophobic; see the discussion of thermodynamics of mixtures, Section 2.3b. If the dispersion medium is water, the terms hydrophilic and hydrophobic, respectively, are used instead. One characteristic of a lyophilic colloid is a significantly higher viscosity than that of the pure solvent, whereas the viscosity of a lyophobic colloid is virtually unchanged.

When a sol forms compact aggregates, the process is called a coagulation (if the aggregation is irreversible) or flocculation (if the aggregation is reversible). In technical chemistry, the aggregate itself is called a floc. The reverse of flocculation, that is, the dispersion into a sol, is called peptization. The empirical Schulze-Hardy rule states that hydrophobic colloids (e.g., sulfides, clay) are flocculated most efficiently by ions of opposite charge and high charge number; see Section 9.5c[b] In contrast, flocculation of hydrophilic colloids (e.g., silicic acid, proteins, soaps) requires large amounts of electrolytes. It turns out that small ions (with a large electric field) are more effective than larger ions. For example, in the series $Li^+ > Na^+ > K^+$, Li^+ is the most effective; see the discussion of the cation-exchange matrix above.[c]

[a] The presence of a solute causes the following changes of solvent properties: lowering of the vapor pressure, elevation of the boiling point, depression of the freezing point, and creation of an osmotic pressure.
[b] The general theory for the stability of lyophobic colloids is called the DLVO theory after its developers Derjaguin, Landau, Verwey, and Overbeek.
[c] Such series are called lyotropic series or the Hofmeister series.

Table 6.9 Number and mass of soil microflora and microfauna[75]		
	Number / kg^{-1}	Mass / $Mg\ ha^{-1}$
Bacteria	10^{11} - 10^{12}	0.4 - 4.5
Actinomycetes	10^{10} - 10^{11}	0.4 - 4.5
Fungi	10^8 - 10^9	1.0 - 11
Algae	10^7 - 10^8	0.06 - 0.6
Protozoa	10^7 - 10^8	0.02 - 0.2
Average values for Danish soils.[75]		

Soils

The chemistry of weathering processes is outlined in Table 3.11, and Table 6.3 summarizes the sizes of soil particles. However, despite the formal classifications, fine sand, silt, and clay (together with humus) are considered to be part of the soil colloids. The reason is that the layered structure of these particles results in colloidal behavior even for relatively large particles.

e. Soil organic matter

Soil organic matter[a] is normally synonymous with humic substances; see Section 3.4a. Living matter includes plants and soil organisms[99] like the microorganisms shown in Table 6.9. Their importance is manifested, for example, by the bacterially regulated redox potentials; see Table 9.4 and Table 3.29.

Worms play an important role in maintaining the structure of soil. Consider a well-developed soil horizon (see Table 6.1). There are around 10^7 worms (e.g., *Lumbricus terrestris*) per hectare with a mass of about 3 Mg ha^{-1}. Each worm can eat a third of its body weight per day, and over a season the field will contain 50 Mg ha^{-1} of worm excrement.[75] If we estimate the density of soil as 1.2 g/ml and the depth of the topsoil as 15 cm, this corresponds to 1.8 Gg of soil, and at full worm activity, soil will pass through a worm on average once every 5 years.

Humus

Humus is organic material found in the soil and weathered to such an extent that its origin cannot be recognized. In Section 3.4a, it was described how humus can be separated into fulvic acids (soluble in acid), humic acids (insoluble in acid, soluble in base), and humin (insoluble in base). Average values for the composition of humic and fulvic acids are given in Table 6.10. The average cation-exchange capacity of unfractionated humus is given in Table 6.5.

Table 6.10 shows the average content of chemical functional groups. Note: Humic acid has about the same amount of carboxylic groups as phenolic groups, whereas in fulvic acids this ratio approaches 3:1.

---

[a] In specialist literature, the acronym is SOM.

Fraction	Empirical formula	b_{RCOO^-}	b_{ArOH}	b_{ROH}	$b_{R/RCO}$	b_{R/OCH_3}
Table 6.10 Functional groups of acid fractions of humus [89]						
Humic acid	$C_{187}H_{186}O_{89}N_9S$	3.5	3.1	2.6	3.1	0.4
Fulvic acid	$C_{135}H_{182}O_{95}N_5S_2$	8.2	3.0	6.1	2.8	0.4

Unit of b is $\mathrm{mol\,kg^{-1}}$. R = aliphatic carbon, Ar = aromatic carbon, R′ and R″ mean both types of hydrocarbon.

Example: Fulvic acids

Fulvic acids are soluble polyvalent acids. Accordingly, their acid-base chemistry can be described through the formalisms of Chapter 5 without any constraint. In the following example, we discuss this chemistry using experimental data.

The (hydron) formation function for a spodosol fulvic acid as obtained from titration data is shown in Figure 6.17. The formation function $\bar{n}$ is simply the mole fraction of bound hydron with respect to the acid; see Equation A2.11. As the experiment shows, this fulvic acid has $\bar{n}_{max} = 8$ mol/kg of acidic functional groups, of which 6 mol/kg are mostly of carboxylic nature and 2 mol/kg of phenolic nature. For pK_a values, see Table 5.3. The curve shows with certainty that each member of the two types of groups is situated in a somewhat different chemical environment. If they had had only two pK_a values, then the plateau around $\bar{n} = 2$ mol/kg would have been very pronounced, and the remaining parts of the curve more steep. An independent experiment determined the cation-exchange capacity $b_{cec} = 8$ mol/kg.[89]

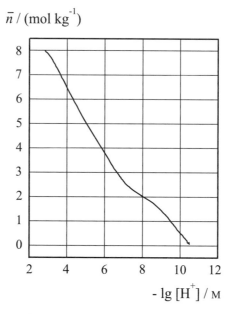

Figure 6.17 *The formation function of a soil fulvic acid*
The formation function of a spodosol fulvic acid; concentration 0.5 kg m^{-3} in 0.1 M KCl solution.
$b_{cec} = 8$ mol/kg. Data from ref. 89.

Global cycles of the elements

7.1 Biogeochemical cycles

The Wilson cycles circulate elements are circulating through the crust over long time periods, as discussed in Section 1.4. Cycling of material within and between the atmosphere, hydrosphere, and biosphere is typically much faster.

Biogeochemical cycles are flows of matter within and between the spheres as defined in the preceding Chapters 3 to 6. These flows are driven by solar energy, and they begin and end in the crust, which acts as the main reservoir of nearly all elements. Motions within the crust are driven by energy from radioactive decay, and the circulations of the atmosphere and ocean are driven by the sun. In fact, biogeochemical cycles are processes that take place within an open physicochemical system as depicted in Figure 7.1. The spheres are marked by ellipses, and their mutual interactions are indicated by arrows. We refer to the discussion of open systems in Section 2.2b, in particular with respect to the significance of the surroundings. The system can be treated in its entirety, but usually the focus is on a few chemical elements and their flow within and through the subsystems. Since 1971, the International Council for Science (ICSU) committee SCOPE[a] has published a series of reports on the global turnover of elements central to the environment. Because of their importance for life, the nonmetals C, N, P, and S have been taken up several times, but Cl, Pb, Hg, Cd, and As have also been investigated. In Sections 7.2 through 7.6 of this chapter, we will follow the approach of SCOPE, that is, little emphasis is given to the details of the biochemical processes involved.

Anthropogenic perturbations to the biogeochemical cycles are of growing importance. They include the use of fossil fuels, the emission of substances such as carbon dioxide and fertilizers, and the discharge of domestic and industrial wastes. The nature of waste has changed as the industrial revolution has proceeded, from natural refuse to ever more esoteric chemical substances. For this reason we will consider a few major industrial products from Table 8.7: coal, calcium carbonate, ammonia, and NP fertilizers, as well as sulfuric acid. However, a more detailed discussion of chemical industry and the use of pesticides is deferred to Chapters 8 and 9.

Biogeochemical cycles (Figure 7.1) are models of Nature that give a picture of reservoirs and fluxes of the species under consideration. The basic steps in establishing such models are to collect appropriate data, assess their reliability, and then place

---

[a] SCOPE = Scientific Committee on Problems of the Environment; see http://www.icsu-scope.org/ (2012).

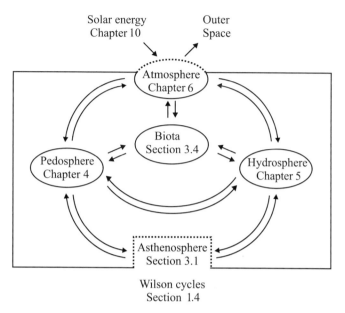

Solar energy
Chapter 10

Outer
Space

Atmosphere
Chapter 6

Biota
Section 3.4

Pedosphere
Chapter 4

Hydrosphere
Chapter 5

Asthenosphere
Section 3.1

Wilson cycles
Section 1.4

Figure 7.1   *Biogeochemical cycles*
Biogeochemical cycles as processes in an open physicochemical system; its
boundaries are marked with a broken line. The overall structure is discussed in the
text.

them into the framework of scientific knowledge. For example, we have constructed
(Section 7.6) the chlorine cycle mainly using data found elsewhere in the book.

7.2 Carbon

The carbon cycle describes the movements of carbon between biota (living or dead),
the atmosphere, the hydrosphere, and the lithosphere.

a. Reservoirs of carbon

The partitioning of carbon between the different reservoirs is shown in Table 7.1.
Note that although the mass of C in the crust shown here is consistent with Table 3.2,
values proposed in the literature span nearly an order of magnitude. Carbonates are
the most widespread form of carbon. Other forms include reduced carbon (e.g., peat,
brown coal and anthracite, mineral oil, and natural gas); deposits of the allotropes
graphite and diamond are rare.

The environmental chemistry of carbonates was discussed in Sections 1.4b and
5.2c, and the chemistry of the carbon dioxide-water system in Section 5.1c. Isotopic
analyses were discussed in Section 1.3c.

The origin of fossil carbon was discussed in Section 1.4c. A crude analysis of
coal can be made using carbon disulfide to separate the soluble "bitumen" from the

Table 7.1 Global distribution of carbon		
Reservoir	m_C	Species
Atmosphere	0.3 Pg	CO
	4 Pg	CH_4
	808 Pg	CO_2
Land surface	700 Pg	Dead biota
	833 Pg	Living biota (mostly plants)
	330 Pg	CO_2 in freshwater
Oceans, surface layer	2 Pg	Living biota
	500 Pg	HCO_3^-/CO_3^{2-}
Oceans, deep	1 Eg	Dissolved or suspended dead biota
	35 Eg	HCO_3^-/CO_3^{2-}
Crust	10 Eg	Fossil material
	20 Zg	Inorganic material, especially Ca and Mg carbonates

All figures refer to the mass of elementary carbon, for example, 808 Pg C is contained in 67.4 Pmol CO_2; see Table 3.18.

insoluble "kerogen." The composition of typical bituminous coal is shown in Table 8.9. When heated, it will separate into coal gas (see Table 8.10) and mechanically stable coke, which together with kerogen has important domestic and industrial uses; see Chapter 8. Kerogen is the most widespread geological form of coal. It is a polymeric carbonaceous material with a high content of H, O, and S as aromatics, carboxylic acids, and sulfides.

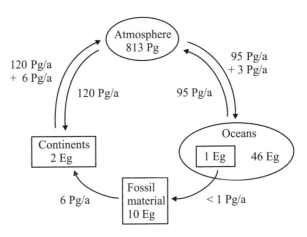

Figure 7.2 *Main features of the biogeochemical carbon cycle*
Reservoirs: mass of C; fluxes: mass of C per year. □ denotes carbon as biota. Approximately 6 Pg/a is due to fossil fuel combustion and production of ammonia and cement. The content of the atmosphere is consistent with Table 3.18; the fluxes are estimates, which may vary up to 15 %[109]; the figures for continents and oceans are estimates, literature values vary by up to a factor of 4.

b. Fluxes of carbon dioxide

The global biochemical carbon cycle is shown in Figure 7.2. Most of the carbon exchange between the Earth's surface and the atmosphere occurs via CO_2, which is removed from the atmosphere by photosynthesis and returned via respiration. In addition, there are large fluxes of CO_2 between the oceans and the atmosphere.

The fluxes shown in Figure 7.2 can be described as follows: looking at land-based processes, photosynthesis fixes 120 Pg/a C. However, this mass is reemitted to the atmosphere: 59 Pg/a C through cellular respiration, 59 Pg/a C from the breakdown of soil organic matter, and 2 Pg/a C from biomass burning. There is a net flux of approximately 6 Pg/a C into the atmosphere that is due to human activity. Over the ocean, there is a mass flow of 97 Pg/a C from the atmosphere (solvation of CO_2 in the water), accompanied by a mass flow of CO_2 out of the oceans of 95 Pg/a C; the net is an atmospheric loss of about 2 Pg/a C to the oceans. The remainder of the carbon, about 3 Pg/a C, accumulates in the atmosphere. Although a great deal of effort has gone into determining these numbers, there is some uncertainty, but not with regard to the annual increase of carbon dioxide in the atmosphere.

Table 7.2 The constants of the decay function, Equation 7.1			
i	0	1	2	3
a_i	0.217	0.259	0.338	0.186
τ_i / a	10^8	172.9	18.51	1.186

The function denotes a fraction, and at $t = 0$, $a_{CO_2}(0) = 1$.

According to the information in Figure 7.2, the average lifetime of CO_2 in the atmosphere is 3.7 a (see Equation 2.54). As discussed, the fluxes are sums of many contributions, but for the estimation of the global warming potential of gases (see Table 10.5) we have used a simple model of the decay of atmospheric CO_2 consisting of four terms:

$$a_{CO_2}(t) = \sum_{i=0}^{3} a_i \exp\left(-\frac{t}{\tau_i}\right) \tag{7.1}$$

The values of the constants are given in Table 7.2. The function is a sum of terms with the form of Equation 2.51, and Equation 2.56 gives the lifetime $\tau_{CO_2} = 1.11$ a. The lifetime $\tau_0 = 10^8$ a represents the very slow carbon cycle in the Earth's crust, and the corresponding exponential function is equal to $1.^{[109]}$ Roughly speaking, the lifetime $\tau_1 \approx 173$ a represents the slow mixing between the surface and deep oceans; $\tau_2 \approx 19$ a covers the uptake of CO_2 into terrestrial biomass, soils, and wetlands; and $\tau_3 \approx 1.2$ a the annual photosynthesis and exchange with the surface ocean.[231,a]

Atmospheric carbon dioxide has been recorded since 1958, as shown in Figure 7.3. The data show that the mass of CO_2 in the atmosphere is increasing. Further, the

---

[a] This reference is a revised version of the "Bern Carbon Cycle Model" used by the Intergovernmental Panel on Climate Change (IPCC).

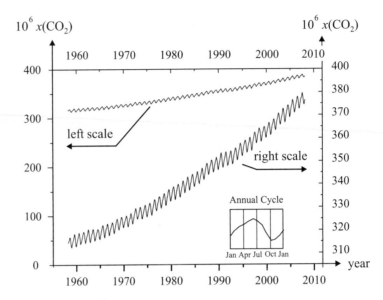

Figure 7.3 *The Keeling curve (2008)*

Time series of atmospheric carbon dioxide concentrations measured continuously at Mauna Loa, Hawaii, since 1958.[240] The net flux of CO_2 into the atmosphere is $\Delta x(CO_2)/\Delta t \approx 1.3 \times 10^{-6}$ a^{-1}, corresponding to the mass flux[a] $\dot{m}_C \approx 2.77$ Pg/a, an increase of 3.42 ‰ per year. At this rate of increase, the preindustrial mole fraction of CO_2, 280×10^{-6}, will have doubled early in the next century.[b]

seasonal cycle of growth and decay is reflected in the atmospheric concentration of CO_2, as shown in the inset of the figure. Analysis of ancient air trapped in the glaciers of Greenland and Antarctica shows that the amount of CO_2 in the atmosphere was low and constant in the period preceding the agricultural and industrial revolutions, but has increased since then; see Figure 10.6 (ice core record). Since 1860, anthropogenic CO_2 production (from fossil fuel and cement) has doubled every 15 to 20 years.[c]

The exchange of carbon between the atmosphere and the ocean is limited by the diffusion of CO_2. On average, the amount flux density is estimated to be 10 mol m^{-2} a^{-1} each way. As discussed in the chapter on aqueous chemistry, there is an extended equilibrium among CO_2 (aq), H_2CO_3, HCO_3^-, and CO_3^{2-},[d] resulting in the solubility of CO_2 being dependent on the pH of the seawater. The amount of dissolved inorganic carbon in the surface layer of the ocean (mixed by wind and waves) is essentially the equilibrium value, corresponding to the present partial pressure of CO_2 in the atmosphere. As most of the ocean was equilibrated with previously lower CO_2 concentrations, there is a net uptake of CO_2 by the oceans, limited by ocean circulation. The thermohaline circulation is pumped by deep water formation when cold salty water drops to the bottom near the poles. At the other end of the circulation

---

[a] Total amount of substance of the atmosphere: $n_{atm} = 177.3$ Emol, see Table 3.18; then $\Delta m(C)/\Delta t = 177.3$ Emol $\times 1.3 \times 10^{-6}$ a$^{-1} \times 12$ Eg/Emol $= 2.77$ Pg/a.

[b] Starting at 388×10^{-6} in 2010, 560×10^{-6} will be reached in 2117 at 0.342 % annual increase: $1.00342^n = (560/388) \Rightarrow n = 107$.

[c] Ref. 113 gives details. Essential data are: 1860: 0.1 Pg/a; 1976: 5 Pg/a; accumulated amount in 1976: 175.6 Pg.

[d] These four species comprise so-called dissolved inorganic carbon (DIC).

system, deep water (undersaturated in CO_2) comes to the surface. The result is a long characteristic time of nearly a thousand years for the oceans to come to equilibrium with the modern high-CO_2 atmosphere.

There is also a biological pump transferring CO_2 into the deep ocean. Photosynthetic plankton transform dissolved inorganic carbon into biomass. These plankton die and fall, where the reduced material is oxidized lower in the water column, consuming dissolved oxygen and increasing dissolved inorganic carbon, or the plankton may precipitate as sediment on the ocean floor. Over geological time periods, diagenesis will turn this carbon into sedimentary rock. The Wilson cycle eventually reexposes this material to the atmosphere, closing the cycle.

c. Fluxes of methane

Around half of atmospheric methane is generated by anaerobic respiration (methanogenesis, Table 3.29 f), from bacteria in freshwater, and from bacteria in the digestive systems of termites and grazing animals. Anthropogenic sources, including rice paddies, livestock, and leaks in the natural gas distribution system, are responsible for the increasing concentration of CH_4 in the atmosphere. Although the atmospheric concentration of methane has doubled relative to 1850, in the past 15 a the increase has slowed and almost stopped. The reason for this is not known with certainty but may be related to deforestation and improved natural gas pipelines in Russia.

d. Anthropogenic sources of atmospheric carbon dioxide

The industrial use of coal, oil, and carbonates is discussed in Chapter 8. Here we present some numbers to put these processes into perspective. Coal is the world's fastest-growing energy source. Its annual production is around 6.94 Pg (2009) and is increasing at 2 to 3 % per year.[222]

As an example, consider fluxes from Europe's largest coal-fired power plant, Drax, in England. It reached full capacity in 1996, generates (2005) 3.96 GW of power, and uses 11 Tg/a of coal. In addition to electricity, the plant produces 2 Tg/a of ash, 1 Gg/a of waste water, and 70 Gg/a of sludge. The SO_2 exhaust stream is 40 Gg/a, relative to 400 Gg/a if SO_2 were not removed. At the same time, 1 Tg/a of gypsum is produced.

In 2009, the world's annual production of natural gas was around 3.0×10^{12} m^3 (1 bar, 0 °C), and in addition, a substantial portion of synthesis gas came from the production of coke. The production of crude oil was 4.06 Pg, and both figures are a few percent lower than those for 2008. Around 85 % of oil was used for heating and transport. There are indications that traditional oil production has reached its maximum.[222]

Significant amounts of carbon dioxide are generated by cement manufacture. Although part of the release is due to the fossil fuel used to heat the kilns, more is produced by the overall process converting the raw materials, $CaCO_3$ (limestone), SiO_2 (sand), and water, into concrete:

$$6\,CaCO_3(s) \; + \; 2\,SiO_2(s) \; + \; 3\,H_2O(l) \quad \rightarrow \quad Ca_6Si_2O_7(OH)_6(s) \; + \; 6\,CO_2(g)$$

$$(7.2)$$

(see Table 8.8). The scheme is complete in the sense that carbonate occurs in the raw material but not in the cement itself.[a] Equation 7.2 embraces many reactions, the most important being the process called *limeburning*, which involves heating of limestone (chalk):[b]

$$CaCO_3(s) \xrightarrow{\Delta} CaO(s) + CO_2(g)$$
$$K_p(1167\,K) = 1\,bar \tag{7.3}$$

This single reaction contributes approximately 440 Tg/a to the CO_2-flux into the atmosphere.

The use of limeburning to produce mortar has been known since ancient times. First, CaO (quick lime) is converted into $Ca(OH)_2$ (slaked or hydrated lime) with water. Next, this is blended with sand to give mortar. Hardening involves the reformation of carbonate using atmospheric carbon dioxide:

$$Ca(OH)_2(aq) + CO_2(g) \rightarrow CaCO_3(s) + H_2O(l) \tag{7.4}$$

The removal of water from the interior of brick walls is slow; drying times of more than 800 a have been observed in Danish churches from the Middle Ages.

7.3 Nitrogen

The biogeochemical nitrogen cycle is mainly biochemical, with the atmospheric nitrogen pool acting as source as well as sink. We follow the approach of the SCOPE reports[109,118,119] and first discuss estimates of the global reservoirs, then the fluxes, and finally characteristic chemical processes.

Outline of history and use

Nitrogen is essential to life. In fact, the growth in human population to the present number (see p. 2) would not have been possible without the industrial production of fertilizer. As an introduction to this section, a few historical events related to nitrogen chemistry, including the production of nitrogen fertilizers, are listed in Table 7.3; they are discussed in greater detail in the text.

Tables 7.4 and 7.5 present an overview of masses of nitrogen in various reservoirs; key mass fluxes are given in Table 7.6.

a. Natural nitrogen fixation

The atmosphere contains the Earth's reservoir of nitrogen in the form of N_2 (see Table 7.4). The bond strength of dinitrogen[c] is large relative to that of other molecular bonds.

---

[a] Al^{III} is omitted from Equation 7.2. It is essential to the function of cement, but not for the present argument.

[b] Technical names: $CaCO_3$: chalk or limestone; mineral name: calcite. CaO: quick lime. $Ca(OH)_2$: slaked lime.

[c] $\Delta_r H° = 945$ kJ mol^{-1} for the dissociation $N_2(g) \rightarrow 2N(g)$ at 298.15 K; corresponding values for other diatomic molecules are H_2: 432 kJ mol^{-1}; O_2: 494 kJ mol^{-1}; F_2: 159 kJ mol^{-1}.

Table 7.3 Important events in the history of nitrogen chemistry

1772	Dinitrogen is identified and isolated; D. Rutherford.
1828	Urea is synthesized from ammonium cyanate; F. Wöhler.
1862	Recognition that nitrogen compounds are ubiquitous in agricultural soil; J. v. Liebig.
1895	First industrial fixation of nitrogen (cyanamide, see Equation 7.20); the Frank-Caro process.
1902	Nitrogen fixation as Norway saltpeter (see Equation 7.13); the Birkeland-Eyde process.
1902 - 08	Technical development of the burning of ammonia using the Ostwald process (see Equation 7.25).
1905 - 09	Nitrogen fixation, development of ammonia synthesis (see Equation 7.23); F. Haber.
1913	Industrial production of ammonia via the Haber process; C. Bosch.
1956 - 58	Herbicides simazine and atrazine developed (see Equation 9.9).
1992	The structure of the nitrogenase enzyme (see Equation 7.7); D. Rees (Caltech).

Table 7.4 Reservoirs of nitrogen[118, 119]

Reservoir	m_N	comments
Biota[1]	15 Pg	Plant biomass
	300 Pg	Organic matter in soil
Hydrosphere	22 Eg	Dissolved nitrogen, not biota
Atmosphere	3.9 Zg	138 Emol N_2, cf. Table 3.18
Lithosphere[2]	57.4 Zg	Table 3.2 gives 448 Eg in the crust

[1] Tables 3.26 and 7.1 give $\approx$70 Pg.
[2] Datum from 1981.[118]

Table 7.5 Nitrogen in the biosphere, excluding the atmosphere[109]

Reservoir	Land surface m_N / Pg	Ocean m_N / Pg	Total m_N / Pg
Salts	160	570	730
Dead biomass	3.3	800	803
Plants	12.0	0.3 ⎫	
Animals	0.7	0.2 ⎭	13

Partitioning of nitrogen in the biosphere (definition: Section 3.1).

Table 7.6 Nitrogen fixation

Method	$\dot{m}_N$ / Tg a^{-1}	Principal chemical process
Prokaryotic cells	158	$N_2 \rightarrow NH_4^+$
Fertilizer production (2005)	150	$N_2 \rightarrow NH_3$ and $N_2 \rightarrow NO_3^-$
Combustion	100	$N_2 \rightarrow NO_x \rightarrow NO_3^-$
Lightning	20	$N_2 \rightarrow NO_x \rightarrow NO_3^-$

Global mass fluxes of nitrogen, N.[165].

Because of this, in contrast to most other elements, a large fraction of N_2 is found in the atmosphere.

The processes whereby atmospheric nitrogen is changed into other chemical forms are known collectively as nitrogen fixation. Examples are given in Table 7.6, where two different reaction pathways are recognized. One form is oxidative, transforming N^0 to N^V; the other is reductive, changing N^0 to N^{-III}. In principle,

$$2 N_2 + 5 O_2 + 2 H_2O + aq \rightarrow 4 H^+(aq) + 4 NO_3^-(aq) \qquad (7.5)$$

$$2 N_2 + 3 CH_2O(biota) + 6 H_2O + aq \rightarrow$$

$$3 NH_4^+(aq) + NH_3(aq) + 3 HCO_3^-(aq) \qquad (7.6)$$

Here CH_2O(biota) represents "organic matter." This is an example of a more general rule: the oxidative branch yields an acid and the reductive branch yields a base.[a] Table 7.6 also gives an overview of the mass fluxes involved. The sum of the values given in the table is 430 Tg/a, and it is remarkable that the quantity of nitrogen fixed by human activity approaches that fixed by the natural system.

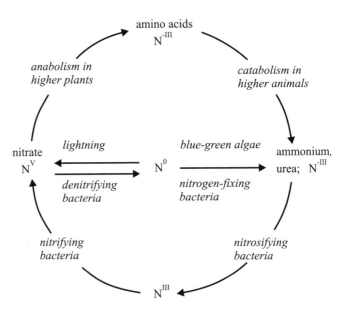

Figure 7.4   *Nitrogen cycles of the biosphere*
A summary based on ref. 109. The mass flux for natural nitrogen fixation is $\dot{m}_N \approx 0.16$ Pg/a.

Life processes

The processes that make up natural nitrogen fixation are collected and described qualitatively in Figure 7.4. The atmosphere is the major source and the sink for

[a] Assume $c_N = 0.10$ M; for H_2CO_3 $pK_{a1} = 6.5$ and $pK_{a2} = 10.5$; for NH_4^+ $pK_a = 9.3$; for water $pK_w = 14.0$. Then the solution of Equation 7.5 has pH 1.0, and the solution of Equation 7.6 has pH 8.8 with the alkalinity (see Equation 5.31) $A = 0.10$ M.

nitrogen that occurs in organic material; only a minor part is exchanged directly as nitrate with the crust.

Whereas natural oxidative nitrogen fixation is inorganic (heating of the atmosphere) as well as microbial, natural reductive nitrogen fixation is exclusively carried out by microorganisms that assimilate the fixed nitrogen. Summarizing:

$$N_2 + 8\,H_2O + 8\,e^- \xrightarrow{\text{many ATP}} 2\,NH_3 + H_2 + 8\,HO^- \quad (7.7)$$

This reaction is catalyzed by the enzyme nitrogenase, whose structure is known; however, the detailed chemistry of the reactions is not. The overall process is one of several reactions in Nature that generate hydrogen, which acts as the reducing agent in some of the processes of Figure 7.4.

Soil microorganisms use ammonium for nitrification, which takes place in three major steps:[61] *Nitrosomonas* and *Nitrosococcus* oxidize ammonium via hydroxyl-amine (azanol) to nitrite:[a]

$$NH_4^+ + \tfrac{1}{2}O_2 \;\rightarrow\; NH_2OH + H^+ \qquad A^{\circ\prime} = -15\,\text{kJ mol}^{-1} \quad (7.8)^b$$

$$NH_2OH + O_2 \;\rightarrow\; NO_2^- + H^+ + H_2O \qquad A^{\circ\prime} = +289\,\text{kJ mol}^{-1} \quad (7.9)$$

and *Nitrobacter* oxidizes nitrite to nitrate:

$$NO_2^- + \tfrac{1}{2}O_2 \;\rightarrow\; NO_3^- \qquad\qquad A^{\circ\prime} = +77\,\text{kJ mol}^{-1} \quad (7.10)$$

These processes are fast, and the microorganisms effectively scavenge their surroundings for ammonium.

Although plants are able to use nitrate as well as ammonium for anabolism, measurements of the level of the enzyme nitrate reductase, a measure of the relative importance of nitrate, indicate that nitrate is the most important source.

Denitrification, for example, as carried out by the ubiquitous bacteria *Pseudomonas aeruginosa*, has a very high affinity:

$$NO_3^- + 2\,H^+ + 5\,H_2 \;\rightarrow\; N_2(g) + 6\,H_2O \quad A^{\circ\prime} = +1121\,\text{kJ mol}^{-1} \quad (7.11)$$

A type of denitrification not shown in Figure 7.4 is carried out by several bacteria, including *Escherichia coli*:

$$NO_3^- + H_2 \;\rightarrow\; NO_2^- + H_2O \qquad\qquad A^{\circ\prime} = +161\,\text{kJ mol}^{-1} \quad (7.12)$$

Generally, however, microorganisms prefer ammonium rather than nitrate as the nitrogen source. Thus, of the 2,500 genera of fungi described (1978), only 20 have been reported to assimilate nitrate.[118]

Lightning

When air is heated to high temperature, it produces nitrogen oxides, species that have a profound impact on the chemistry of the troposphere and the stratosphere. The atmospheric chemistry of NO_x was discussed in Sections 4.1 and 4.2. Here we consider the thermodynamics of the series of reactions that take place when a mixture of nitrogen and oxygen is heated and reacts with water.

---

[a] See Equation 2.156: The affinity is given by $A^{\circ} = -\Delta G^{\circ}$; a prime indicates a standard state at pH $= 7$.
[b] For NH_3OH^+, $pK_a = 5.8$.

At high temperature, nitrogen burns in oxygen (is oxidized) to form nitrogen monoxide,

$$\tfrac{1}{2}\,N_2(g) + \tfrac{1}{2}\,O_2(g) \;\leftrightharpoons\; NO(g) \tag{7.13}$$

which reacts with more oxygen to form nitrogen dioxide:

$$NO(g) + \tfrac{1}{2}\,O_2(g) \;\to\; NO_2(g) \tag{7.14}$$

In contrast to the monoxide, the dioxide reacts with water and disproportionates into nitric acid and nitrogen monoxide:

$$(\tfrac{3}{2})\,NO_2(g) + (\tfrac{1}{2})\,H_2O(g) + aq \;\to\; H^+(aq) + NO_3^-(aq) + \tfrac{1}{2}\,NO(g) \tag{7.15}$$

The monoxide reacts further with oxygen and water until it is completely converted into nitric acid.

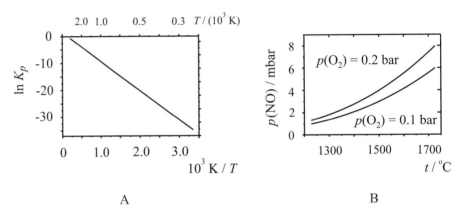

Figure 7.5 *Thermodynamics of reaction 7.13*
Panel A: The logarithm of the equilibrium constant K_p as a function of the reciprocal temperature; corresponding values of T on top of the frame.
Panel B: The partial pressure of NO as a function of temperature; upper curve: equilibrium in air; lower curve: equilibrium in an oven with a reduced oxygen content. In both cases the total pressure is 1 bar.

The thermodynamics of this chemistry is as follows: for reaction 7.13, the standard quantities of formation at 298.15 K and 1 bar are[167]

$$A_f^\circ = -86.45 \text{ kJ/mol}; \quad \Delta_f H^\circ = 90.21 \text{ kJ/mol}; \quad \Delta_f S^\circ = 12.6 \text{ J/(K mol)} \tag{7.16}$$

which by use of Equation 2.170 gives

$$\ln K_p = \frac{-1.085 \times 10^4 \text{ K}}{T} + 1.52; \quad K_p = \frac{p_{NO}}{\sqrt{p_{N_2} \cdot p_{O_2}}} \tag{7.17}$$

for the equilibrium constant, K_p. The temperature dependence has been determined[167] for the range 298 K $\leq T \leq$ 5000 K and is depicted in panel A of Figure 7.5. Panel B shows $p(NO)$ in equilibrium with an atmosphere containing $p(N_2)$ and $p(O_2)$ in the ratios 4:1 and 9:1 at the temperature of a rotary kiln used in cement manufacture and a total pressure of 1 bar. In the former case, $p(NO, 1450\,^\circ C) = 3.4$ mbar; in the latter case, $p(NO, 1450\,^\circ C) = 2.5$ mbar.

As can be seen, the formation affinity of nitrogen monoxide is negative, and very high temperatures are required in order to change its sign. On the other hand, formation of nitrogen dioxide according to Equation 7.14 has a positive standard reaction affinity, $A_r^° = 36.3$ kJ/mol.[a] In addition, the ensuing reaction, the formation of nitric acid according to Equation 7.15, is thermodynamically favorable with a standard reaction affinity of $A_r^° = 29.7$ kJ/mol.

Conclusion. From a thermodynamic point of view, one can see why the atmosphere has such an extreme stability: the affinity of formation of nitrogen monoxide from the elements is *negative*. The Ellingham diagram (see Figure 8.4) shows that the formations of a large number of oxides are associated with positive affinities. In contrast, nitrogen monoxide requires $T > 7000$ K in order to have a positive affinity of formation. But as shown earlier once nitrogen monoxide is formed, its conversion via the dioxide into nitric acid is very favorable. In order to discourage the formation of nitrogen oxides and the resulting acid rain from industrial activities, processes exposed to the atmosphere should be run at low temperatures and at a reduced partial pressure of oxygen.

b. Industrial nitrogen fixation

Because of the need to increase the yield of agricultural land and an enhanced understanding of the importance of nitrogen-containing compounds to soil fertility, fertilizer was added to soil, starting toward the end of the 1800s. At first sodium nitrate[b] was used, but in the 20th century, industrial products took over.

The cyanamide method was the first practical process for synthesizing nitrogen fertilizer. This is a reductive nitrogen fixation whose final step is shown in Equation 7.19. The first step is the classical limeburning reaction, Equation 7.3, where calcium carbonate is heated to produce calcium oxide. Next, the oxide is heated with coke to give calcium carbide, which is heated with nitrogen to produce calcium cyanamide:

$$CaO(s) + 3\,C(s) \quad \rightarrow \quad CaC_2(s) + CO(g) \tag{7.18}$$

$$CaC_2(s) + N_2(g) \quad \rightarrow \quad CaNCN(s) + C(s) \tag{7.19}$$

The cyanamide was spread onto the fields where it slowly hydrolyzed, yielding chalk and ammonium/ammonia:

$$CaNCN(s) + 4\,H_2O(l) + aq \quad \rightarrow$$
$$CaCO_3(s) + NH_4^+(aq) + NH_3(aq) + HO^-(aq) \tag{7.20}$$

The intermediate, calcium carbide, could also be reacted with water to give ethyne (acetylene), used as welding gas (flame temperature $\approx 3300\ °C$):

$$CaC_2(s) + 2\,H_2O(l) \quad \rightarrow \quad Ca(OH)_2 + HC{\equiv}CH(g) \tag{7.21}$$

$$C_2H_2(g) + (^3/_2)\,O_2(g) \quad \rightarrow \quad 2\,CO(g) + H_2O(g) \quad \Delta_r H^° = -1.3\,\text{MJ/mol} \tag{7.22}$$

---

[a] The standard state for this and the next affinity quoted is 298.15 K and 1 bar; the two numbers have been calculated using data from ref. 129.

[b] Common name of $NaNO_3$: Chile saltpeter.

However, the development of the Haber-Bosch method of synthesizing ammonia was a dramatic step in artificial reductive nitrogen fixation. As noted previously, the present human population would not be possible without the large-scale production of fertilizer (see Table 7.6). The formation of ammonia from the elements is exothermic:

$$(^1\!/_2)\,N_2(g) \; + \; (^3\!/_2)\,H_2(g) \quad \rightarrow \quad NH_3(g) \tag{7.23}$$

with $\Delta_f H^\circ = -45.94$ kJ/mol at 298.15 K and 1 bar.[a] Also, when the reaction has reached $\Delta\xi = 1$ mol, then the change in amount of gaseous substance is $\Delta n = -1$ mol. Accordingly, the synthesis must be carried out at relatively low temperature and high pressure;[b] actual conditions are 450 °C and 200 bar. The catalyst is magnetite (Fe_3O_4) doped with a few percent of potassium aluminium oxide. At the beginning of the 20th century, hydrogen was obtained from the electrolysis of water. Later, a better method was developed using methane and water:

$$CH_4 \; + \; 2\,H_2O \quad \rightarrow \quad CO_2 \; + \; 4\,H_2 \tag{7.24}$$

This process involves several steps, including a carbon monoxide intermediate. The starting material is natural gas, from which sulfur compounds must be removed because they poison the ammonia catalyst. The resulting gas (see Equation 7.23), contains 15 % ammonia, which is removed by cooling,[c] and the unreacted gas mixture reenters the plant. Ammonia is transported in pipelines over very long distances (megameters)[d] and is also shipped in the liquid phase. A road tanker carries between 30 and 130 m^3 of liquid. Ammonia manufacture is one of the cleanest large-scale industrial processes.

The first successful industrial method for oxidative nitrogen fixation was the formation of nitrogen monoxide according to Equation 7.13, which was performed by passing air through an electric discharge. Cheap electricity was necessary and was available in Norway in 1902 because of newly built hydroelectric plants. As shown earlier (see Equations 7.14 and 7.15), the synthesis yields nitric acid, which was eventually neutralized using calcium carbonate to give the fertilizer calcium nitrate (Norway saltpeter).

Today, nitric acid is produced by burning ammonia. Ammonia burns in oxygen to give nitrogen and water in the absence of a catalyst, but with a platinum catalyst it reacts to give nitrogen monoxide (the Ostwald process):

$$4\,NH_3(g) \; + \; 5\,O_2(g) \quad \rightarrow \quad 4\,NO(g) \; + \; 6\,H_2O(g) \tag{7.25}$$

Huge amounts of ammonia are oxidized in this way and converted into nitric acid through reactions 7.14 and 7.15. In contrast to the production of ammonia, this

---

[a] $\Delta_f G^\circ = -16.41$ kJ/mol, $\Delta_f S^\circ = -99.03$ J/(K mol) at (298.15 K, 1 bar),[243] then K_p (298 K) = 751 bar^{-1} and K_p (723 K) = 0.014 bar^{-1}; this latter "perfect gas value" differs less than 10 % from the actual value.

[b] Hint: For a qualitative argument, use Le Chatelier's principle.

[c] The boiling point of ammonia at 1 bar is -33.5 °C.

[d] For example, pipelines from southern Louisiana to northern Nebraska, USA.

process is not "clean," and a great technical effort has been made to remove nitrogen oxides, NO_x, from the flue gas.

7.4 Phosphorus

Table 7.7 shows a summary of the history of phosphorus chemistry within environmental chemistry. Phosphorus is essential to life as shown by the examples. In addition, calcium phosphates are necessary for the formation of bones and teeth. Both use hydroxylapatite, and in addition tooth enamel contains some fluorapatite, which enhances resistance to dental cavities.[a] The table illustrates humankind's progress in understanding phosphate fertilizers. Growing demand led to the conversion of phosphates into dihydrogenphosphates on an industrial scale.

Table 7.7 Summary of the history of phosphorus	
1669	Phosphorus isolated from urine; H. Brandt.
1769	Phosphate found in the joints of mammals; J. G. Gahn and C. W. Scheele.
1779	Phosphate found in minerals.
1820	Synthesis of phosphonic acids, $R \cdot PO(OH)2$; J. L. Lassaigne.
1830–40	It was shown that crushed animal joints are effective as fertilizer; J. v. Liebig. Use of the unit BPL (see the example in Section 8.1c).
1843	Superphosphate patented, active ingredient $Ca(H_2PO_4)_2 \cdot H_2O$.
1844	Annual production of phosphorus for matches is 750 kg; by 1851 this increased to 26.5 Mg.
1937	Production of the xenobiotic organophosphate insecticide parathion (see Equation 9.6); IG Farben.
1951	^{31}P-NMR chemical shift observed; W. C. Dickinson.
1952	Production of polyphosphate detergents exceeds the production of soap in the United States.
1971	The herbicide glyphosate (Roundup; see Equation 9.13) patented; Monsanto.

Understanding of the role of phosphorous in Nature was advanced by the discoveries of phospholipids (1811), nucleic acids (1869) and their structure (1962), ATP (1929), glycolysis (1932), and the citric acid cycle (1937).

A survey of the biogeochemical cycle of phosphorus is shown in Figure 7.6.

The outer cycle follows the Wilson cycle: the crust weathers to give P^V as phosphate to the soil; phosphates are washed out to the sea and reenter the crust as sparingly soluble phosphates via diagenesis. No volatile P^V compounds are known, and transport (as dust) through the atmosphere is not significant.

The small cycles, due to the biochemical turnover of phosphate, are fast. Figure 7.6 shows a terrestrial cycle (left-hand side) and an aquatic one (right-hand side). The source of phosphorus for marine organisms is inorganic phosphate P_i, which occurs

---

[a] The mineral apatite is $Ca_5(Cl,F,OH)(PO_4)_3$; see Table 3.6. Hydroxylapatite = pentacalcium hydroxide tris(phosphate), $Ca_5(OH)(PO_4)_3$. Fluorapatite = pentacalcium fluoride tris(phosphate), $Ca_5F(PO_4)_3$.

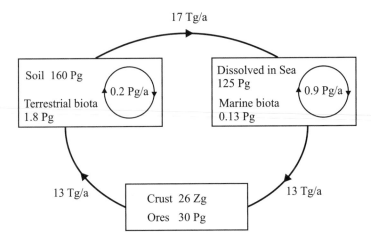

Figure 7.6 *The biogeochemical cycles of phosphorus*
There is significant uncertainty regarding the size of masses and mass fluxes
presented here.[27, 109, 117]

mainly as hydrogenphosphate, HPO_4^{2-}.[a] Only the isotope ^{31}P occurs naturally, and
studies made using the short-lived isotope ^{32}P have shown that phosphates are assim-
ilated in phytoplankton (algae) within hours of their release from land. Such fast
reactions are the reason why the turnover of P is high compared to the P content of
marine biota. Algae are eaten by herbivores that are eaten by carnivores (fish). All life
forms eventually die, and P^V reenters the cycle. Phosphate precipitates the cations
Ca^{2+}, Al^{3+}, and Fe^{3+}, which are either redissolved as hydrogen- or dihydrogenphosphates (using carbonic acid; $FePO_4$ also redissolves because Fe^{III} is reduced to
Fe^{II}) or buried through diagenesis.

The main reservoir of the land-based phosphorus cycle is apatites with the general
composition $Ca_5(F,OH)(PO_4)_3$; F^- and HO^- have nearly identical ionic radii such
that the crystals constitute a continuous isomorphous series, analogous to that of the
olivines (see Section 3.1b). Apatites are sparingly soluble in water, fluorapatite being
the least soluble, $[Ca^{2+}] \approx 3.1 \times 10^{-7}$ M for a saturated solution.[b] As noted earlier,
whereas phosphates of trivalent metal ions and calcium(II) are sparingly soluble,
hydrogenphosphates are more soluble, and dihydrogenphosphates are considered to
be "available soil phosphate."

The annual production of phosphate fertilizer in 1994 is shown in Table 8.6. The
basic process is the conversion of apatite into "superphosphate" using sulfuric acid:

$$2\,Ca_5F(PO_4)_3 + 7\,H_2SO_4 + H_2O \rightarrow$$
$$7\,CaSO_4 + 3\,Ca(H_2PO_4)_2 \cdot H_2O + 2\,HF \quad (7.26)$$

---

[a] pK (see Table 5.3, acids 6 and 7) corrected for salinity: $pK_{no\,6} = 6.5$ and $pK_{no\,7} = 11.0$.[135d] Using
pH 8 for seawater gives the following composition of P_i: ≈ 3 % $H_2PO_4^-$, ≈ 96 % HPO_4^{2-}, and
≈ 1 % PO_4^{3-}.

[b] The solubility product of fluorapatite is $L = 10^{-60}$ M^9 $= 5^5\,3^3\,[F^-]^9$.
The analogous lead apatites with phosphate or arsenate are less soluble: $L\{Pb_5Cl(PO_4)_3\} = 10^{-84}$ M^9
and $L\{Pb_5Cl(AsO_4)_3\} = 10^{-84}$ M^9.[55]

This fertilizer contains a large amount of gypsum. A more effective (but also more expensive) fertilizer is "triple superphosphate," obtained by treating apatite with phosphoric acid:

$$Ca_5F(PO_4)_3 + 7 H_3PO_4 + H_2O \rightarrow 5 Ca(H_2PO_4)_2 \cdot H_2O + HF \qquad (7.27)$$

Anthropogenic release of P mainly results from sewage and the excessive use of detergents and fertilizers. These phosphates may cause excess growth by phytoplankton, *algal bloom*. When the algae die, they decay, depleting dissolved oxygen and eventually causing the death of fish. The phenomenon[a] is called eutrophication,[b] and its prevention is a subject of legislation in several countries.

World production (2007) of phosphate-containing rocks was 147.0 Tg; in the same year the U.S. production was 30 Tg.[252]

7.5 Sulfur

Table 7.8 presents a summary of the history of sulfur chemistry with relevance to environmental chemistry.

Table 7.8 Key events in the history of sulfur chemistry	
1245	Black powder is produced in Europe (invented in China in the 1100s).
1661	Occurrence of sulfate aerosol pollution in London.
1746	Lead chamber process for the production of sulfuric acid; J. Roebuck.
1813	Sulfur is shown to be a component of animal blood; H. A. Vogel.
1822	Xanthogenates (R · OCS$_2$H); W. C. Zeise.
1831	Contact process for the production of sulfuric acid; P. Philips. The Pt catalyst was later replaced with V$_2$O$_5$ on asbestos.
1834	Mercaptans (thiols) described; W. C. Zeise.
1839	Vulcanized rubber; C. Goodyear.
1865	Large sulfur deposits found in Louisiana, USA.
1923	Concept of chalcophile metals; V. M. Goldschmidt.

Sulfur has been known since ancient times. It is the brimstone that together with fire was rained on the towns of Sodom and Gomorrah;[c] also, in the Odyssey, sulfur was used for fumigation.[d] Elemental sulfur occurs in volcanic areas, and sulfur occurs as sulfates of alkaline earth metals (Ca, Sr, Ba) and as sulfides of chalcophilic metals, for example, Pb, Bi, Cu, Cd, As, Sb, Sn, Mo, Hg, Co, Ni, Zn. In order to obtain the metals from the sulfide ores, the ore is crushed and the sulfide is concentrated by flotation or sedimentation. In classical mining, sulfides are converted to oxides by

---

[a] See Section 9.5a.
[b] Gk. $\varepsilon\upsilon \,\hat{=}\,$ well; $\tau\rho o\varphi\varepsilon\acute{\iota}\nu \,\hat{=}\,$ nourish.
[c] Book of Genesis, 19: 24.
[d] Homer's Odyssey, 22: 481.

roasting, and finally the oxide is reduced. Using zinc sulfide as the example, the final steps are

$$ZnS(s) + \tfrac{3}{2}O_2(g) \rightarrow ZnO(s) + SO_2(g) \tag{7.28}$$

$$2\,ZnO(s) + C(s) \rightarrow 2\,Zn(s) + CO_2(g) \tag{7.29}$$

For this reason sulfur dioxide has been a persistent problem in areas with such production. Today, virtually all SO_2 is reclaimed and used to produce sulfuric acid. At present the most important sources of sulfur dioxide pollution are from combustion of fossil fuels containing sulfur, for example, marine diesel and brown coal. Sulfur dioxide is oxidized in the atmosphere,

$$SO_2(g) + \tfrac{1}{2}O_2(g) + H_2O(l) \rightarrow H_2SO_4(\text{aerosol}) \tag{7.30}$$

giving rise to acid rain. The mass of sulfur that is exchanged between the atmosphere and other reservoirs is estimated to be between 250 and 550 Tg/a (1980 values), of which the anthropogenic contribution is around 30 %.

a. Natural sulfur cycles

The natural sulfur cycles consist of the biological cycles summarized in Figure 7.7. There is a resemblance to the nitrogen cycles, with the exception that the active reservoir does not lie in the atmosphere but in the lithosphere, as pyrite, FeS_2, and sulfur, S_8. When pyrite is exposed to moist air, it is oxidized directly to sulfate:

$$2\,FeS_2(s) + 7\,O_2(g) + 2\,H_2O(l) \rightarrow 2\,Fe^{2+}(aq) + 4\,SO_4{}^{2-}(aq) + 4\,H^+(aq) \tag{7.31}$$

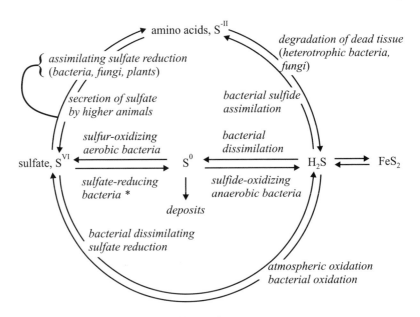

Figure 7.7 *Main processes in the biological sulfur cycle*
* The sulfur deposits found in the United States, Mexico, and Poland are believed to have been produced by anaerobic bacterial reduction of sedimentary calcium sulfate deposits.

(see Figure 5.11). Apart from this reaction and the syntheses of the sulfur containing amino acids (Cys, Met; cf. Appendix A1.9), which can be carried out by plants but not animals, most pathways of Figure 7.7 are catalyzed by microorganisms.

A set of oxidations of sulfur species is performed by photoautotrophic bacteria (see Table 3.30 and the examples given there):

$$2\,H_2S \,+\, CO_2 \quad \rightarrow \quad \tfrac{1}{4}S_8 \,+\, CH_2O(biota) \,+\, H_2O \tag{7.32}$$

$$\tfrac{1}{8}S_8 \,+\, \tfrac{3}{2}CO_2 \,+\, \tfrac{5}{2}H_2O \quad \rightarrow \quad SO_4{}^{2-} \,+\, \tfrac{3}{2}CH_2O(biota) \,+\, 2H^+ \tag{7.33}$$

and a different set of oxidations is carried out by chemolithotrophic bacteria:

$$H_2S \,+\, \tfrac{1}{2}O_2 \quad \rightarrow \quad \tfrac{1}{8}S_8 \,+\, H_2O \tag{7.34}$$

$$\tfrac{1}{8}S_8 \,+\, \tfrac{3}{2}O_2 \,+\, H_2O \quad \rightarrow \quad SO_4{}^{2-} \,+\, 2H^+ \tag{7.35}$$

Sulfate-reducing bacteria work under anaerobic conditions according to the reaction

$$SO_4{}^{2-} \,+\, 2\,CH_2O(biota) \quad \rightarrow \quad H_2S \,+\, 2\,HCO_3{}^- \tag{7.36}$$

Part of this hydrogen sulfide re-forms iron(II) disulfide, and part is reoxidized in the atmosphere. However, about half of the mass of sulfur that reaches the atmosphere comes from the oceans. Of this total, 30 % is in the form of dimethyl sulfide and the balance is sulfate aerosol. Dimethyl sulfide, DMS, is generated as dimethylsulfonium propionate, $[(CH_3)_2SH^+][C_2H_5 \cdot COO^-]$, which is used by marine algae to regulate their osmotic pressure. Dimethyl sulfide is oxidized in the troposphere, ultimately giving sulfate aerosols. According to the CLAW hypothesis,[152] there is a negative feedback mechanism that acts to stabilize climate: a rise in sea surface temperature stimulates emission of DMS to the atmosphere, which forms sulfate aerosol, opposing the original change. One product of DMS oxidation in the troposphere is carbonyl sulfide (COS), the most abundant sulfur-containing gas in the atmosphere. COS, together with SO_2, is photooxidized in the stratosphere to provide a nonvolcanic background concentration of sulfate to the stratospheric sulfate aerosol (Junge) layer discussed in Chapter 4.

Studies of the distribution of the abundant sulfur isotopes ^{32}S and ^{34}S have been central for unraveling the present sulfur cycles and their geological origins (see Section 1.3c).

b. Anthropogenic sulfur cycles

Sulfuric acid

Sulfuric acid is produced in the largest quantity of any single compound (see Table 8.7). Until the middle of the 1700s, sulfuric acid was produced by distilling hydrated iron(II) sulfate (green vitriol) obtained from the slow oxidation of pyrite in air (see Equation 7.31). At that time it was known that sulfuric acid could be made by oxidizing sulfur dioxide with nitric acid, but the use of fragile glass jars prevented large-scale production until the lead chamber was invented. Its introduction caused the price of sulfuric acid to fall, an important development in the industrial revolution (see Table 7.8). The first step was oxidation of FeS_2 to SO_2 in air using a special oven, the pyrite burner. Next, SO_2 was oxidized to H_2SO_4 in the lead chamber, with HNO_3 serving as catalyst. Atmospheric oxygen was the oxidant, and the NO_x gases produced were reoxidized using air, forming nitric acid that was recycled.

The contact method, in which SO_2 is oxidized to SO_3 with oxygen using a platinum catalyst, was discovered in 1831 and is technically superior to the chamber method. The details of a modern, environmentally benign industry are discussed next. Despite the fact that a cheap catalyst of V_2O_5 was invented early in the 20th century, the last lead chamber factory did not close until after World War II.

In modern production facilities for sulfuric acid, the first step is to burn elemental sulfur to sulfur dioxide at 1800 °C in a oxygen-lean atmosphere. Next comes an "afterburn" reactor at 700 °C. In all,

$$S(s) + O_2(g) \rightarrow SO_2(g) \qquad A_f^\circ = 300 \text{ kJ/mol} \qquad (7.37)$$

As seen, this process is thermodynamically favorable; nevertheless, the amount of oxygen is controlled in order to prevent the formation of NO_x. Some of the SO_2 product is cooled, compressed, and sold directly as liquid SO_2. However, the majority of SO_2 is oxidized with pure oxygen using the V_2O_5 catalyst in the exothermic, reversible process

$$SO_2(g) + \tfrac{1}{2}O_2(g) \rightleftharpoons SO_3(g) \qquad \Delta H^\circ = -98 \text{ kJ/mol} \qquad (7.38)$$

In agreement with Le Chatelier's principle, the yield is increased by using high pressure, low temperature, and continuous removal of the product. The catalyst makes the process profitable at temperatures as low as 500 °C. Sulfuric acid is obtained by reacting the sulfur trioxide with water. In the mid-1970s, a typical "converter" ran the equilibrium (Equation 7.38) three times, resulting in 99.7 % conversion efficiency. Since then, environmental controls have become more strict, and now the reaction is brought to equilibrium five times and the exhaust gas is treated with hydrogen peroxide:

$$SO_2(g) + H_2O_2(l) \rightarrow H_2SO_4(l) \qquad (7.39)$$

This method does not produce any by-products, and SO_2 cannot be detected in the exhaust of modern factories.

The raw material for this process is elemental sulfur. Superheated steam is used to melt the sulfur, which can then be pumped to the surface from deep-lying deposits (the Frasch process, 1891).[a] More recently, large amounts of sulfur have also been obtained by oxidation of hydrogen sulfide from natural gas (the Claus process; see Equation 8.44). World production of elemental sulfur in 2007 was 66.0 Tg; in the same year, U.S. production was 8.8 Tg, of which 8.2 Tg was recovered at petroleum refineries.[252]

Microbial leaching

As discussed in the introduction, chalcophilic metals are prepared from their sulfides, by roasting, but other methods of obtaining these metals from their sulfides have also

---

[a] S_8; $\theta_{fus} = 115.21$ °C; $\theta_{vap} = 444.60$ °C.

been attempted. Although chemical leaching with chlorine, nitric acid, or sulfuric acid has been used to oxidize sulfides, these methods turn out to be difficult to implement in an environmentally acceptable way.

Microbial leaching can extract metals from sulfides on an industrial scale. We now discuss the extraction of copper as one of several examples of ongoing efforts to create environmentally acceptable production methods. The bacterium *Thiobacillus ferrooxidans* obtains energy by oxidizing iron(II) to iron(III). When acting on pyrite, the net result is an oxidation by atmospheric oxygen:

$$4\,FeS_2(s) \; + \; 15\,O_2(g) \; + \; 2\,H_2O(l) \quad \rightarrow$$
$$4\,Fe^{3+}(aq) \; + \; 8\,SO_4{}^{2-}(aq) \; + \; 4\,H^+(aq) \quad (7.40)$$

(see Equation 1.25). The production of sulfuric acid according to Equation 7.31 results in an acidic environment in which *T. ferrooxidans* can live by catalyzing the oxidation of iron(II):

$$2\,Fe^{2+} \; + \; \tfrac{1}{2}O_2 \; + \; 2\,H^+ \quad \rightarrow \quad 2\,Fe^{3+} \; + \; H_2O \qquad (7.41)$$

As seen, this reaction decreases the acidity, and when the pH rises above approximately 2.5, an intensely yellow mineral called jarosite, $KFe_3(SO_4)_2(OH)_6$, precipitates,[a] halting the process. In order to prevent this, sulfuric acid is added, and iron(III) instead reacts with pyrite:

$$FeS_2 \; + \; 14\,Fe^{3+} \; + \; 8\,H_2O \quad \rightarrow \quad 15\,Fe^{2+} \; + \; 2\,SO_4{}^{2-} \; + \; 16\,H^+ \quad (7.42)$$

yielding iron(II) that is used by the bacteria through Equation 7.41.

The overall process is used in large scale to produce copper and molybdenum. About 15 % of copper production in the United States is from ores with a copper content below 0.4 %, which is too small for the usual processes to be feasible. In one facility a valley is used as an ore depot, Bingham Canyon in Utah, with a capacity of 3.6 Pg of ore. The bottom of the valley has been lined with an impermeable layer of plastic, and 0.01 M sulfuric acid is sprayed onto the ore from above. The solution contains dissolved oxygen and percolates downward through the ore, feeding the bacterial process. The ore naturally contains 10^6 g^{-1} of *T. ferrooxidans* per gram of pyrite. Using chalcopyrite as an example:[b]

$$4\,CuFeS_2(s) \; + \; 17\,O_2(aq) \; + \; 2\,H_2SO_4(aq) \quad \rightarrow$$
$$4\,CuSO_4(aq) \; + \; 2\,Fe_2(SO_4)_3(s) \; + \; 2\,H_2O(l) \quad (7.43)$$

The solution exiting the process contains about 2 g/l of Cu, which is recovered by electrolysis or by reduction with waste iron:

$$3\,Cu^{2+} \; + \; 2\,Fe \quad \rightarrow \quad 3\,Cu \; + \; 2\,Fe^{3+} \qquad (7.44)$$

There are a number of technical and environmental issues that must be addressed in such a large-scale facility. Whenever possible, natural solutions are preferred. For example, free sulfur occurring in the pyrite can block the bacteria's access to material. A sulfur-oxidizing bacterium, *T. thiooxidans*, is used to remove this sulfur.

---

[a] Minerals where K^+ is replaced by Na^+ or H_3O^+ are common in "acid mine drainage."
[b] The iron(III) sulfate represents several less soluble mixed Fe(II)-Fe(III) hydroxido sulfato complexes ("hydroxysulfates") as well as jarosites.

The chemistry can be written in a straightforward way by considering the bacteria to act as catalysts. Current research attempts to identify the active proteins and the genes that code for them.

7.6 Chlorine

Up until the 1960s it was believed that covalently bound chlorine was very rare in biomolecules, but over the following 30 years, more than 1,500 organochlorine compounds were identified.[171]

The anthropogenic contributions to the chlorine cycle follow the general biological path. As detailed in Chapter 4, there is a flux of anthropogenic chlorinated species to the atmosphere, and also in these cases chlorine returns as HCl. The environmental aspects of this chemistry are discussed in greater detail later in the book. Table 7.9 outlines some developments in the history of chlorine of relevance to environmental chemistry.

Table 7.9 Summary of the history of chlorine	
≈ 450 BC	Herodotos writes about salt.
≈ 950 AD	Icelandic sagas mention using salt to preserve meat.
1774	$Cl_2(g)$ prepared by C. W. Scheele; see Equation 8.48.
1785	Chemical bleaching ($KOH + Cl_2$); C.-L. Berthollet.
1787	Technical production of HCl from NaCl; N. Leblanc.
1851	Diaphragm cell for production of Cl_2; see Figure 8.7.
1863	The Alkali Act (UK) prohibited air pollution; see Section 8.3c.
1909	Gastric juice is shown to contain HCl.
1928	Freon, CCl_2F_2, a nontoxic, nonflammable gas for refrigeration; T. Midgley.
1950	NMR signals for ^{35}Cl and ^{37}Cl.
1974	Cl atoms catalyze stratospheric ozone destruction; R. S. Stolarsky.
	CFCs provide the Cl atoms; M. J. Molina, F. S. Rowland.
1987	The Montreal Protocol; see Table 10.5.

The discovery of HCl in gastric juice showed for the first time that the element chlorine is not only "inorganic" but also plays a role in life processes.

The geochemical cycle of chlorine is in fact the cycle of common salt, NaCl, as shown in Figure 7.8. It differs from the previous cycle because here the oceans hold the largest reservoir of the element, and the role of the Wilson cycles is uncertain. However, there are two steady-state equilibria with quite different lifetimes: the lifetime of sea salt in the atmosphere is very short, 0.4 d, and the lifetime of salt in the cryosphere is as long as the lifetime of this sphere itself.

Biota contribute with fluxes to the atmosphere, giving rise to a steady-state concentration of chlorine-bearing species. The most abundant, chloromethane (methyl chloride), has the mass fraction $w_{CH_3Cl} \approx 0.6 \times 10^{-9}$.[a] Eventually, such compounds are broken down to the end product HCl, which is returned to the Earth's surface by rain.

---

[a] CH_3Cl is formed during the degradation of methoxy groups in lignin.

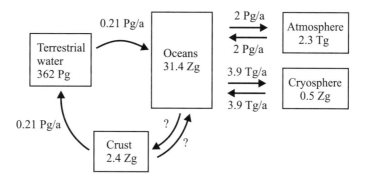

Figure 7.8 *The geochemical cycle of chlorine*

Notes: (1) There is a small flux of sea salt aerosols from atmosphere to land. (2) Biological activity gives rise to important cycles of chlorine in the troposphere; they are discussed in Chapter 4. (3) Sea salt aerosol + NO_x + light gives $Cl^\bullet$. (4) Most stratospheric chlorine, >90 %, is anthropogenic.

Setting up the natural global chlorine cycle

Figure 7.8 has been constructed using the tables within this book and some additional information.[169]

1. *The crust.* Table 3.2: $m_{Cl} = 23.6 \times 10^{24} \, \text{g} \times 126 \times 10^{-6} = 2.97 \, \text{Zg}$
2. *The oceans.* Tables 3.12 and 3.13:

$$m_{Cl} = 1.625 \times 10^{24} \, \text{g} \times 19.35 \times 10^{-3} = 31.4 \, \text{Zg}$$

3. *Groundwater.* Table 3.12; average chloride concentration 40 mg/l:

$$m_{Cl} = 9.1 \times 10^{21} \, \text{g} \times 40 \times 10^{-6} = 0.36 \, \text{Eg}$$

4. *Surface waters (lakes and rivers).* Tables 3.12 and 3.14:

$$m_{Cl} = 280 \times 10^{18} \, \text{g} \times 5.75 \times 10^{-6} = 1.61 \, \text{Pg}$$

5. *Cryosphere.* Tables 3.12; when ice is formed, ions such as sulfate and nitrate are expelled but chloride is retained to some extent, the average concentration being 17 mg/l:

$$m_{Cl} = 29 \times 10^{24} \, \text{g} \times 17 \times 10^{-6} = 0.49 \, \text{Pg}$$

6. *Atmosphere.* Table 3.2; $w_{CH_3Cl} = 620 \times 10^{-12}$; $w_{HCl} = 40 \times 10^{-12}$:

$$m_{Cl} = 5.2 \times 10^{21} \, \text{g} \times 620 \times 10^{-12} \times (35.5/50.5)$$
$$+ \; 5.2 \times 10^{21} \, \text{g} \times 40 \times 10^{-12} \times (35.5/36.5) = 2.3 \, \text{Tg}$$

7. *Flux of freshwater to oceans.* Tables 3.12 and 3.14:

$$\dot{m}_{Cl} = 36 \, \text{Eg/a} \times 5.75 \times 10^{-6} = 0.21 \, \text{Pg/a}$$

8. *Flux: crust to freshwater.*

Steady state requires this value to be equal to that of 7.

9. *Flux of sea salt aerosol between oceans and troposphere.*[a]

$$\dot{m}_{Cl} = 2.0 \, \text{Pg/a}$$

10. *Flux of water between oceans and cryosphere.*

Annual variation of water bound in the cryosphere: 230 Pg

$$\dot{m}_{Cl} = 230 \times 10^{15} \, \text{g} \times 17 \times 10^{-6} = 3.9 \, \text{Tg/a}$$

In 2007, world production of salt was 250.0 Tg. In the same year, U.S. production was 43.8 Tg.[252]

7.7 Aluminium and silicon

An experiment published in 1989 showed that trout living in natural water, carefully freed from silicates, died from damage to gill epithelia and loss of osmoregulatory capacity.[145] The explanation is that Al(III) forms strong complexes with oxygen donors (e.g., in the tissue), but it also has a strong affinity for silicate, which normally prevents Al from entering the gills. At that time it was known that the effects of "acid rain" include dissolution of aluminium hydroxides, but the mechanism of aluminium toxicity was not known.

Over the following 20 years, these hydroxyaluminosilicates have been characterized,[112] and they seem to be protoimogolite and protoallophane, eventually forming imogolite and allophane (see Table 6.4).

---

[a] IPCC: 3300 Tg/a NaCl.

8 The chemicals industry

8.1 Introduction

The chemicals industry manufactures commodities on a large scale.[a] Table 8.1 shows the commodities that, on a worldwide basis, are produced in the greatest amounts.[b] Some industries use significant amounts of recycled material; this is the reason why annual production of pig iron is much less than that of steel.[c] Production on this scale is very sensitive to the costs of raw materials, transport, and energy, and even small changes in synthetic efficiency can shift production patterns and have large socioeconomic impacts.

Environmental protection accounts for an increasing part of industrial costs, and we present here a survey of the chemicals industry because an understanding of why and how our society runs such a huge enterprise is a prerequisite for decision making on environmental issues.

Most production relies on energy whose origin is fossil carbon in a low oxidation state (see Table 8.3). Conversion of this fuel to energy requires a reduction of atmospheric O_2 to water;[d] other reductions are also of industrial significance, such as the reduction of Fe^{III} in Fe_2O_3 to Fe.

The raw materials of commodities are biota and minerals, and various raw materials enter chemical processes in different ways. Biota constitute human food and are also the source of the most important natural polymer, cellulose. Food generation and processing are large-scale activities that, together with the industrial production of fertilizers (Section 7.3), pesticides (Section 9.1), and plastics (Table 8.2) require environmental protection schemes. Minerals are normally quarried from heterogeneous rock that eventually undergoes chemical treatment. Accordingly, they require a high degree of mechanical processing (blasting, crushing, grinding) at the place of origin, followed by transport to the chemical processing plant. In all cases, environmental protection demands extensive chemical knowledge.

---

[a] Commodities include primary feedstock such as bulk chemicals, wood, and certain agricultural products.

[b] In order to give the reader a feeling of the order of magnitudes, consider some U.S. data from 2007: the production of cement[252] was 96.4 Tg, that is, 3.05 Mg/s, corresponding to 23.4 Mg/s of concrete; this is 2.4 Mg/a of concrete per capita. (Use: 1 a $\approx$ 31,556,952 s; $m_{concrete}/m_{cement} \approx 7.67$; population $\approx$ 305×10^6.) For comparison, the U.S. production of municipal solid waste (2007; household garbage, of which one-third was paper) was 254 Tg/a, corresponding to 830 kg/a per capita. Waste quantities per capita for Western Europe were 400–450 kg/a, and for India (2002), 35 kg/a.

[c] The iron industry is discussed in Section 8.2.

[d] A combustion is a redox reaction; the reductor is called fuel.

Table 8.1 World production of energy and commodities				
Material	Mass, 2007	Mass, 1994	Largest producer, 2007	Fraction of total
Energy[1]	11.4 Pg	8.4 Pg		
Wood[2]	2.5 Pg	2.4 Pg	USA	0.56
Cement	2.6 Pg	1.2 Pg	China	0.50
Raw steel	1.3 Pg	0.7 Pg	China	0.37
Pig iron	0.9 Pg	0.5 Pg	China	0.49
Paper	370 Tg [3]	270 Tg	Asia	0.40
Plastic	260 Tg	100 Tg	Europe	0.25
Salt, NaCl	257 Tg	190 Tg	China	0.23
Sugar	167 Tg	117 Tg	Brazil	0.20
Sulfuric acid	165 Tg [4]	156 Tg		
Ammonia	124 Tg [5]	94 Tg		
Ethene	113 Tg [6]			
Chlorine	63 Tg [7]	47 Tg	See Table 8.16	
Textiles	44 Tg [3]	40 Tg		
Aluminium	38 Tg	20 Tg	China	0.32

[1] Oil equivalent; energy equivalent 480 EJ in 2007;[a] see Table 8.3.
[2] Wood production assuming a density of 0.7 g/ml.
[3] 2005; [4] 2001; [5] 2006; [6] 2004; [7] 2008.
Refs. 222, 225, 226, 238, 252.

Table 8.2 Details of world production of plastics, 2007				
Polyethylene	Polypropylene	PVC	Others	World
80 Tg	54 Tg	35 Tg	91 Tg	260 Tg

a. Energy

The worldwide consumption of energy in 2008 was estimated to be the equivalent of 11.4 Pg of oil (480 EJ) (Table 8.3), and it grew at an average rate of 2.3 % per year from 1995 to 2005 but dropped from 2008 to 2009. Most of the world's energy, almost 90 %, comes from fossil fuels.

The main uses of energy are for transport (40 %), manufacturing (33 %), and residential buildings (11 %, not transport). In addition, manufacturing processes use carbon as a general reducing agent and as the starting material for other materials.

[a] 480 EJ/a = 15 TW.

Table 8.3 Global energy consumption 2008 by fuel[222]			
Category	C / Tg a^{-1}	P_1 / EJ a^{-1}	$10^2 P_2$
Oil[1]	3892.3	163.48	34.1
Coal[2]	3303.7	138.76	28.9
Natural gas	2726.1	114.50	23.9
Hydroelectricity	717.5	30.14	6.3
Nuclear power	619.7	26.03	5.4
Wind power[3]	116.7	4.90	1.0
Fuel ethanol	35.6	1.50	0.3
Geothermal electricity[3]	11.4	0.46	0.1
Photovoltaic energy[3]	7.5	0.32	0.1
Total	11430.5	480.09	100.1

World yearly energy consumption fell by 1.1 % in 2009.

C Consumption, oil equivalent.

P_1 Energy production calculated by assuming 42 kJ/g of oil equivalent.

P_2 Energy production, fraction of P_1.

[1] Oil includes crude oil and biodiesel; fuel ethanol is listed separately.

[2] Coal includes commercially sold solid fuels, that is, bituminous, anthracite, lignite, and brown coal.

[3] Based on installed capacity.

b. A survey of the chemicals industry

The examples of Table 8.1 show the diverse nature of the chemicals industry. Nevertheless, one may identify some general chemical principles and use them to organize the presentation.

Resources

The chemicals industry's feedstocks are energy, minerals, cellulose, and water, and its tools include access to bases, acids, oxidizing and reducing agents, and catalysts. An additional requirement is knowledge and an organization able to design and operate chemical plants. Processing is organized by putting together *unit operations*, that is, industrial processes, to perform large-scale chemical operations such as heating, cooling, distillation, agitation, filtration, and transportation known from laboratory work. Measures that ensure environmental protection are indispensible for the design and operation of a chemical plant.

Bases

Quicklime, CaO, and slaked lime, $Ca(OH)_2$, together called *lime*, are by far the most important strong bases.[a] They are prepared from chalk by limeburning (see Equations 7.3 and 7.4) and used directly as cheap general bases. The strong base

[a] World production[252] (2006): 271 Tg of lime, that is, quicklime, CaO, and hydrated (slaked) lime, $Ca(OH)_2$.

NaOH is produced in the *chlor-alkali industry* (Section 8.3). Production of the weak base Na_2CO_3 (soda) is discussed in Section 8.3b. Also, the weak base NH_3 is available in large quantities (see Table 8.1).

Acids

Sulfuric acid is the most important mineral acid (see Table 8.1); the chemistry of its manufacture was discussed in Section 7.5. It is a general-purpose acid, and its high boiling point (338 °C) and strength make it useful for preparing volatile and/or weaker acids from their salts. An example is the classical salt-cake process (Equation 8.41) for producing hydrochloric acid.

World production of HCl is in excess of 20 Tg/a (2005), which shows its industrial importance. A significant contribution comes from the heavy organic-chemicals industry; see the later discussion of Equation 8.4.

Oxidizing agents

Atmospheric oxygen is the ultimate oxidant. Its origin is biological: oxygen is formed during photosynthesis by reduction of carbon dioxide and carbonates (see Section 1.4c).

Liquid oxygen ($\theta_{vap} = 90$ K) is produced in large quantities together with liquid nitrogen ($\theta_{vap} = 77$ K) from the distillation of liquefied atmospheric air (see Table 8.7). Its main use is for manufacturing iron and steel (see Section 8.2b).

Reducing agents

The ultimate reducing agents are carbon, crude oil, and natural gas. The present high oxidation state of the Earth's surface is a result of microbiological activity: virtually

---

Table 8.4 Survey of chemical industries

Resources:
   Minerals including fossil material; cellulose; water; energy.
Heavy industry:
   The products listed in Tables 8.1, 8.6, and 8.7.
Inorganic chemicals industry:
   Aluminium; borates; cyanide derivatives; silicates, titanium dioxide;
   phosphorus derivatives; peroxides (H_2O_2 and derivatives); chlorine and
   chlorination; halogens (F_2, Br_2, I_2 and derivatives);
   industrial gases (O_2, N_2, Ne, Ar, H_2, He);
   water treatment;
   nuclear fuel and waste.
Organic chemicals industry:
   The top 23 manufactured organic chemicals are listed in Table 8.7;
   processed chemicals include polymers; dyes; soaps and detergents;
   pesticides; explosives; aromas and fragrances.
Biotechnological industry:
   Fermentation; pharmaceuticals; enzymes.
Agriculture and food industries:
   Dairy, livestock, sugar, beverages.

all matter that was originally in a reduced state has been oxidized. Accordingly, the only alternative to carbon and its derivatives is to use energy to generate reducing agents. That leaves only nonfossil energy, that is, nuclear power or solar energy, directly or indirectly.

A survey of the chemicals industry

The survey shown in Table 8.4 divides chemical industrial products into classes with some common features. Obviously, no sharp distinctions exist between the entries "heavy industry" and the groupings for "inorganic" and "organic" chemicals. In fact, industries such as the chlor-alkali industry, sulfuric acid production, and the production of fertilizers cannot be placed into a single category.

Many of the 23 most produced organic chemicals are used as feedstock for further processing; several examples are given later.

c. The agriculture and food industries

Production

Table 8.5 gives an overview of the agricultural sector. Wood used in construction, especially when in contact with soil and groundwater, can be impregnated with biocides. The dairy industry and livestock production are large-scale enterprises, but they are not proper "chemical industries." The same conclusion applies to rice and grain production (the figure for wheat is given in the table; other types of grain are of the same order of magnitude), but all such production requires fertilizers and pesticides. The manufacture of sugar and beer takes place in factories, where it is easier to prevent adverse effects on the environment.

Fertilizers

It has been recognized since the middle of the 1800s that sustained agricultural production requires that elements removed from soil be replaced. The present demand

Table 8.5 World production of some agricultural products [225, 248]	
Material	Production rate, 2007
Wood[1]	2.5 Pg a^{-1}
Milk	676 Tg a^{-1}
Rice	398 Tg a^{-1} [3]
Meat	242 Tg a^{-1} [2]
Sugar	167 Tg a^{-1}
Beer	155 Tg a^{-1} [3]
Wine	31 Tg a^{-1} [3]
Wheat	610 Gg a^{-1}
Coffee	7 Gg a^{-1}

[1] Wood production assuming a density of 0.7 g/ml.
[2] 2002; [3] 2004.

Table 8.6 Production of NPK fertilizer in the United States, 1994	
Compound	Production rate, 1994
Ammonia	16.55 Tg a^{-1}
Ammonium nitrate	3.35 Tg a^{-1}
Ammonium sulfate	2.67 Tg a^{-1}
Urea	5.28 Tg a^{-1}
Ammonium hydrogenphosphate	15.80 Tg a^{-1}
Ammonium dihydrogenphosphate	3.29 Tg a^{-1}
Phosphate rocks, such as apatites	42.93 Tg a^{-1}
Potassium chloride	1.87 Tg a^{-1}

Compare with the examples at the end of this section.

for food requires more fertilizer than can be provided by manure, and fields cannot lie fallow every other year. Therefore, industrially produced fertilizers are necessary. Different kinds of nitrogen-phosphorus-potassium (NPK) compounds in common use are listed in Table 8.6.

It is important to point out that fertilizers do not contain any compounds that do not occur naturally. Soils may be naturally too acidic because of CO_2, and in some areas because of H_2SO_4 and HNO_3. In such cases, $CaCO_3$ has been used as a neutralizing base for centuries (see Equation 6.28).[a] For a basic soil, titration may be carried out using $H_2PO_4{}^-$ or H_3PO_4. Many soils require additional magnesium, boron, manganese, or selenium. Such treatments must be given periodically, but without excess: for example, too much (1) fixed nitrogen causes nitrate problems, (2) calcium reduces the availability of trace metals (ion exchange), (3) dihydrogen phosphate releases aluminium, (4) sewage sludge introduces unusual and unwanted elements.

Pesticides and pest control (integrated pest management) are discussed in Section 9.1.

Examples: Commercial measures of fertilizer potency

1. The NPK value is used to rate fertilizers; it is equal to the mass fractions of N, P_2O_5, and K_2O in percent, NPK $= 100 \times w(N)$-$100 \times w(P_2O_5)$-$100 \times w(K_2O)$; zeros are allowed.[b]
   a. A fertilizer with the formula $KH_2PO_4 \cdot (NH_4)_2HPO_4$ has $w(N) = 0.104$, $w(P_2O_5) = 2.293 \times w(P) = 0.529$ and $w(K_2O) = 1.205 \times w(K) = 0.176$; thus NPK = 10-53-18.
   b. Triple superphosphate has NPK = 0-56-0.
   c. K_2SO_4 has NPK = 0-0-45-18S; this example shows how other elements are noted.

---

[a] In Western Europe, sulfur deposition may reach 20 kg ha^{-1} a^{-1} of S, causing a leaching of 125 kg ha^{-1} a^{-1} of $CaCO_3$ (ha = hectare).
[b] $w(P) \, M(P_2O_5) = 2 \, w(P_2O_5) \, M(P)$; $w(K) \, M(K_2O) = 2 \, w(K_2O) \, M(K)$.

2. BPL (bone phosphate of lime) has been used to rate phosphate rocks; it is equal to the mass fraction of $Ca_3(PO_4)_2$, that is, $w_{BPL} = w(Ca_3(PO_4)_2) = 5.013\ w(P)$. Today, the mass fraction $w(P_2O_5) = 2.293 \cdot w(P)$ has gained acceptance.

d. Chemical production

Production statistics for the major compounds produced by the chemicals industry (see Table 8.4) in the United States are shown in Table 8.7. This is a famous compilation[184] that has not been updated since 1994. Annual production has increased since then, but the relative amounts are virtually unchanged and are representative of world production.

The entries of Table 8.7 are not independent. The following examples show why, and also how general chemical principles bring some order to the very diverse reactions and products. The numbers refer to the table, and one of the most important applications is given in parentheses.

Ethene, #4

Ethene is produced in the petrochemical industry by cracking and is isolated by condensation and distillation.

(The following reaction equations are collectively referred to as Equation 8.1.)

$$C_2H_4 \hspace{10cm} (8.1)$$

$+ n\,C_2H_4 \rightarrow$ polyethylene (thermoplastic)

$+ H_2O \rightarrow$ ethanol $\hspace{2cm} \rightarrow$ butadiene (#36)(rubber)

$\hspace{5.2cm} \rightarrow$ chloroethane $\rightarrow$ ethylcellulose(thickener)[a]

$\hspace{5.2cm} \rightarrow$ ethanal $\rightarrow$ acetic acid

$\hspace{5.2cm} \rightarrow$ ethyl esters(solvents, softeners)

$\hspace{5.2cm} \rightarrow$ diethyl ether(solvent)

$\hspace{5.2cm} \rightarrow$ ethyl amines (pharmaceuticals)

$+ O_2 \rightarrow$ ethylene oxide $\hspace{0.6cm} \rightarrow$ ethylene glycol (#30)(antifreeze)

$\hspace{2.2cm}$ (#26) $\hspace{2cm} \rightarrow$ acrylonitrile (#39)(polymers)

$\hspace{5.2cm} \rightarrow$ ethylene glycol ethers(cellosolves)

$\hspace{5.2cm} \rightarrow$ polyethylene glycols(emulsifiers)

$\hspace{5.2cm} \rightarrow$ ethanolamines(surfactants)

$+ Cl_2 \rightarrow$ dichloroethane $\hspace{0.3cm} \rightarrow$ vinyl chloride (#16) $\rightarrow$ poly(vinyl chloride)

$\hspace{2.2cm}$ (#12) $\hspace{2cm} \rightarrow$ ethylenediamine(derivatives, e.g., edta[b])

$+ C_6H_6 \rightarrow$ ethylbenzene $\hspace{0.3cm} \rightarrow$ styrene (#20) $\rightarrow$ polystyrene(thermoplastic)

$\hspace{2.2cm}$ (#19)

Many of the organic chemical products of Table 8.7 are used in the polymers industry, for example, to produce polyesters and polyamides.

---

[a] Increases the viscosity of a liquid.
[b] Ethylenediaminetetraacetic acid, $(HOOCCH_2)_2NCH_2CH_2N(CH_2COOH)_2$.

	Table 8.7 Production of chemicals in the United States in 1994 [184]	Mass/Tg
1	Sulfuric acid, H_2SO_4	44.86
2	Nitrogen, N_2	30.64
3	Oxygen, O_2	22.53
4	Ethene (ethylene), $H_2C=CH_2$	22.01
5	Lime, CaO, and $Ca(OH)_2$	17.40
6	Ammonia, NH_3	17.20
7	Propene (propylene), $CH_3-CH=CH_2$	13.08
8	"Phosphoric acid," P_4O_{10}	12.79
9	Sodium hydroxide, NaOH	12.53
10	Chlorine, Cl_2	10.98
11	Sodium carbonate,[1] Na_2CO_3	9.33
12	Dichloroethane, ClH_2C-CH_2Cl	8.48
13	Nitric acid, HNO_3	8.01
14	Ammonium nitrate, NH_4NO_3	7.99
15	Urea, $CO(NH_2)_2$	7.32
16	Vinyl chloride, $CH_2=CHCl$	6.72
17	Benzene, C_6H_6	6.65
18	Methyl-*tert*-butylether,[2] $CH_3O-C(CH_3)_3$	6.29
19	Ethylbenzene, $C_6H_5-C_2H_5$	5.38
20	Styrene, $C_6H_5-CH=CH_2$	5.11
21	Carbon dioxide,[3] CO_2	4.98
22	Methanol, CH_3OH	4.90
23	*o*- and *m*-Xylene, $C_6H_4(CH_3)_2$	4.11
24	Terephthalic acid,[4] $C_6H_4(COOH)_2$	3.92
25	Formaldehyde,[5] HCHO	3.60
26	Ethylene oxide, C_2H_4O	3.08
27	Toluene, $C_6H_5CH_3$	3.06
28	Hydrogen chloride, HCl	3.04
29	*p*-Xylene, $C_6H_4 (CH_3)_2$	2.83
30	Ethylene glycol, $(CH_2OH)_2$	2.52
31	Cumene, $C_6H_5(CH_3)_2$	2.34
32	Ammonium sulfate, $(NH_4)_2SO_4$	2.30
33	Phenol,[6] C_6H_5OH	1.84
34	Acetic acid, CH_3COOH	1.73
35	Propylene oxide, C_3H_6O	1.68
36	Butadiene,[7] $CH_2=CH-CH=CH_2$	1.54
37	Carbon black, C	1.50
38	Potassium carbonate,[8] K_2CO_3	1.42
39	Acrylonitrile, $CH_2=CH-CN$	1.40
40	Vinyl acetate, $CH_3C(O)O-CH=CH_2$	1.37
41	Acetone, $(CH_3)_2CO$	1.26
42	Titanium dioxide, TiO_2	1.24
43	Aluminium sulfate, $Al_2(SO_4)_3$	1.04
44	Sodium silicate, Na_2SiO_3	0.966
45	Cyclohexane, C_6H_{12}	0.957

		Mass/Tg
46	Adipic acid, $HOOC-(CH_2)_4-COOH$	0.815
47	ε-Caprolactam,[9] $C_6H_{11}NO$	0.762
48	Bisphenol A, $HOC_6H_4-C(CH_3)_2-C_6H_4OH$	0.671
49	1-Butanol, $C_4H_9\,OH$	0.658
50	2-Propanol, $CH_3-CHOH-CH_3$	0.630
51	Sodium sulfate, Na_2SO_4	
52	Sodium, Na	
53	Phosphorus, P_4	
54	Potassium hydroxide, KOH	
55	Hydrogen peroxide, H_2O_2	
56	Sodium triphosphate, $Na_5P_3O_{10}$	
57	Sodium chlorate, $NaClO_3$	

The table does not include minerals such as apatite, salt, gypsum, or sulfur. Neither does it contain "petrochemicals" such as methane, ethane, propane, or butane.

[1] Includes natural and synthetic soda.
[2] 2-Methoxy-2-methylpropane.
[3] Only liquid and solid.
[4] Includes both the acid and its esters.
[5] Water solution with mass fraction of formaldehyde 0.37.
[6] Only the synthetic product.
[7] Only the raw chemical supplied for rubber production.
[8] Given as K_2O equivalent.
[9] Azepan-2-one, that is, lactam of 6-aminohexanoic acid.

Polyesters

In general, hydroxy acids are capable of forming linear polyesters by condensation, that is, by splitting off water. More important, polyesters may also be formed from divalent alcohols or phenols and dicarboxylic acids. As a first example, ethylene glycol (#30) and terephthalic acid (benzene-1,4-dicarboxylic acid (#24)) yield a fiber-forming polyester, which constitutes more than 50 % of the world's production of synthetic fiber:

$$H-[-O-CH_2CH_2-O-CO-C_6H_4-CO-]_n-OH \tag{8.2}$$

The former trade names Terylene (ICI, UK) and Dacron (DuPont, USA) have been replaced by the generic name *polyester*.

Divalent phenols may react with carbonyl chloride to yield a special class of polyesters called polycarbonates. The first step is the condensation of phenol (#33) with acetone (#41) to yield the divalent phenol, "bisphenol A" (#48):

$$2\,C_6H_5OH \,+\, CH_3COCH_3 \;\rightarrow\; HO-C_6H_4-C(CH_2)_2-C_6H_4-OH \,+\, H_2O \tag{8.3}$$

The systematic name is 4,4'-(propan-2,2-diyl)diphenol.[a] Next, carbonyl chloride (phosgene), $COCl_2$, which is the acid chloride of carbonic acid, reacts by releasing HCl to yield the polycarbonate:

$$n\, \text{bisphenol-A} \;+\; n\, \text{carbonyl chloride} \;\rightarrow$$

$$\text{H--[--O--C}_6\text{H}_4\text{--C(CH}_3)_2\text{--C}_6\text{H}_4\text{--O--CO--]}_n\text{--OH} \;+\; 2n\, \text{HCl} \qquad (8.4)$$

The trade name of polycarbonate is Lexan, and its main use is as a replacement for glass.

Polyamides

Linear polyamides are formed analogously to polyesters of aminocarboxylic acids or of diamines and dicarboxylic acids. Thus, ε-aminocaproic acid or its lactam, ε-caprolactam (#47), gives nylon 6:

$$n\, \text{H}_2\text{N--(CH}_2)_5\text{--COOH} \;\rightarrow\; \text{H--[--NH--(CH}_2)_5\text{--CO--]}_n\text{--OH} \;+\; (n-1)\text{H}_2\text{O}$$
$$(8.5)$$

Similarly, hexamethylenediamine (hexane-1,6-diamine) and adipic acid (hexanedioic acid) (#46) form nylon 6.6:

$$\text{H--[--NH--(CH}_2)_6\text{--NHCO--(CH}_2)_4\text{--CO--]}_n\text{--OH} \qquad (8.6)$$

and with sebacic acid (decanedioic acid), nylon 6.10 is obtained. Molten nylons can be squeezed through narrow nozzles to form threads. After stretching, they resemble silk but are stronger and chemically more stable.

8.2 Heavy industry

We have selected a few key industrial processes for closer examination. Even small changes in these processes can have large environmental (and economic) impacts. In the following, we will present a physical and chemical description of the manufacture of cement, steel, paper, and detergents, and chlorine production using electrolysis. Tables 8.3, 8.4, and 8.7 show that many important industries are not included in our discussion. In many cases, the environmental consequences of related processes resemble those of the ones that are presented. The basic chemistry of many of the other processes can be found in textbooks on organic chemistry, inorganic chemistry, and biochemistry.

a. Cement

Hydraulic cement was first invented by the ancient Egyptians; in fact, sections of the pyramids are built using concrete. Other civilizations, including the Greeks,

---

[a] Bisphenol B, 4,4'-(butan-2,2-diyl)diphenol, is much less used.

Babylonians, Romans, and Aztecs, had their own cements, which were mixtures of crushed brick, volcanic ash, and limestone. An important step in the development of modern cement was taken by John Smeaton in the mid-1700s as he prepared to build the Eddystone Lighthouse near Plymouth, England. Smeaton (re)discovered the role of clay in creating hydraulic cement; hydraulic cements harden in water.

During the construction of the Houses of Parliament in London in the mid-1800s it was discovered that so-called Portland cement was far superior to Roman cement. Joseph Aspin patented Portland cement in 1824, and later his son improved the technique to achieve higher temperatures of operation. At high temperature, the patented mixture of limestone ($CaCO_3$, that is, CaO; see Equation 7.3) and aluminosilicates sinters into so-called clinker, giving a stronger cement. The clinker is ground and mixed with water and gravel or sand to produce concrete. The strength of concrete can be further improved using iron or plastic reinforcement.

The rotary kiln (1899) allowed continuous production of cement. It consists of a cylinder rotating at a slanting angle. Finely ground precursors are fed into the upper end, and a flame made from air or oxygen combined with coal dust is blown through the lower end; sintering into clinker occurs at temperatures above 1450 °C. The product is called Portland cement because of its resemblance to limestone quarried on the Portland peninsula in Dorset, England.

The most important components of Portland cement are shown in Table 8.8. The two first have well-defined orthosilicate-type crystal structures with several polymorphs at high temperature. The fourth component can be seen as a solid solution in which Fe^{III} and Al^{III} can substitute isomorphically within certain limits. Common Portland cement has the empirical formula $Ca_{45}Al_5FeSi_{12}O_{78}$. A cubic meter of concrete contains roughly 300 kg cement, 150 kg water, 800 kg sand, and 1050 kg gravel. During hardening, the clinker is hydrolyzed, a process that forms amorphous fibers and cement paste. The properties of concrete (e.g., ability to harden, mechanical strength, resistance to weathering) depend on the raw materials. Small amounts of Na^I, K^I, Mg^{II}, and P^V are detrimental. The conditions in the oven and the way cement is mixed are also important.

Toward the end of the 20th century, the largest kilns could produce 7 Gg/d (81 kg/s) of cement, using 3 MJ/kg, implying a power consumption of 240 MW. This is the amount of power produced by a medium-sized power plant.

Table 8.8 Ranges of mass fraction, w_m, of the components of clinker in Portland cement

Clinker component	$10^2\, w_m$
Ca_2SiO_4	5 - 30
Ca_3SiO_5	50 - 80
$Ca_3Al_2O_6$	1 - 10
$Ca_4Al_2Fe_2O_{10}$	1 - 10

$Ca_3Al_2O_6$ is a cyclic aluminate; the well-defined anion $Al_6O_{18}^{18-}$ has alternating Al and O atoms in the ring.

Cement production is a major industry, and two fundamental environmental factors are apparent. The first is that the occurrence of relatively small concentrations of impurities (e.g., Cd^{II}) in the raw materials can have significant consequences for the local and even the global environment. The second is the opportunity for using the by-products of other processes as raw material for cement, with a corresponding savings of energy and material.

Exhaust gases from cement manufacture may be trapped and separated. For example, CO_2 can be put to many uses, and 25% of global industrial production of CO_2 is used by the bottling industry. Regardless of its use, virtually all CO_2 ends up in the atmosphere, and only a tiny fraction will react with the calcium hydroxide that is formed as cement hardens. Carbon dioxide emission from cement manufacture can be reduced by replacing the raw material with different recycled materials, such as slag from steel manufacture, fly ash from coal burning, silica waste from silicon and ferrosilicate manufacturing, and recycled concrete from the demolition of older structures. Another interesting example is "eco-cement" made by heating sludge from municipal waste-water treatment to 1000 to 1200 °C.[238]

Cement manufacture results in the formation of NO_x, a catalyst for the formation of air pollution (see Chapter 4), from atmospheric nitrogen and oxygen. The equilibrium concentration is a theoretical maximum that is not observed; typical NO_x concentrations in oven exhaust are around 200×10^{-6}. There are several reasons. The kinetics of NO_x formation, described in Chapter 4, is slow relative to the residence time of air in the oven, so the reaction does not have time to go to completion. Also, the equilibrium Equation 7.13 shifts to the left as the system cools, and the NO_x concentration observed does not correspond to the maximum temperature achieved by the system. In addition, the conditions in the oven can be changed to discourage NO_x formation. The proportion of oxygen in the flame can be reduced to about 10%. Further, the amount of time the system spends at 1450 °C is reduced by using a preburner to process the limestone at 900 °C, a temperature at which NO_x formation is not significant. Finally, residual NO_x can be reduced using ammonia before emission into the atmosphere.

Particle emissions from cement manufacture are also of environmental concern. Fine particles are generated during production; at several stages the material is finely ground. In addition, material that becomes volatile at high temperature condenses, forming particles, as the exhaust gas cools to ambient temperature. Dust can be removed from the exhaust using electrostatic filters and centrifugal traps, and process technology is being improved continually. Nonetheless, some dust is inevitably released into the environment, especially from plants without environmental controls. Of particular concern are Mn^{II} and Cd^{II} present in the raw material, limestone and clay, and the part, around 10%, of Cr^{III} from the lining of the oven that is oxidized to toxic Cr^{VI}. Other undesirable metals can be introduced if recycled material is fed into the system. One example is that slag from iron and steel production is often used, introducing Zn, Cu, Pb, and As into the cement, the exhaust gas, and the local environment.

Cement ovens can be used to destroy CFC gases. An example is the destruction of CFC-11 at 1450 °C:

$$CFCl_3 \; + \; 2\,H_2O \quad \rightarrow \quad 3\,HCl \; + \; HF \; + \; CO_2 \tag{8.7}$$

Both HCl and HF react with CaO at the temperature of the oven, producing the respective calcium salts. In small quantities, these salts do not affect the quality of the cement.

b. Coal and steel

The manufacture of iron

Iron is obtained from ores that contain iron oxides, especially hematite, Fe_2O_3, and magnetite, Fe_3O_4, together with the silicate rock. The iron oxides are reduced with carbon in a concerted process with oxygen:

$$2\,Fe_2O_3 + 6\,C + 3\,O_2 \quad \rightarrow \quad 4\,Fe + 6\,CO_2 \tag{8.8}$$

The mechanism of the reaction, which takes place at a range of temperatures (Figure 8.1), involves carbon monoxide:

$$2\,C + O_2 \quad \rightleftharpoons \quad 2\,CO \tag{8.9}$$

$$Fe_2O_3 + 3\,CO \quad \rightleftharpoons \quad 2\,Fe + 3\,CO_2 \tag{8.10}$$

Further, calcite (limestone) is added, and this reacts with SiO_2:

$$CaCO_3 + SiO_2 \quad \rightarrow \quad CaSiO_3 + CO_2 \tag{8.11}$$

Calcium silicate melts and is removed as slag. Without this reaction, the SiO_2 would not melt at the operating temperature.

This method of producing iron has been known since the Iron Age. At that time, iron was smelted by blowing air through a tall (about 3 m) oven filled with layers of wood, charcoal, or coal, and iron ore. Ancient shaft ovens resembled the top portion of the blast furnace shown in Figure 8.1. The technical problem was to keep the reaction temperature low enough that only the slag melted, flowing out of the column and leaving the iron intact as so-called sponge iron for later refinement.

A modern blast furnace is a large chemical reactor composed of a conical shaft resting on a base, also shaped like a cone (see Figure 8.1). The interior surface of the oven is made of refractory brick to withstand high temperatures, about 2000 °C. In operation, the furnace is continually fed with material from the top: crushed iron ore, coke, and limestone. Hot air is blown in from the bottom. The chemical reactions and temperature variations in the furnace are shown in the figure. Liquid iron and slag flow to the bottom of the furnace. The liquid slag is less dense and is drawn off at the top of the reservoir; liquid iron is drawn from the bottom. The raw iron product is called pig iron and has a carbon mass fraction w_C of 4.3 %. The gas exiting the top of the furnace has a high concentration of carbon monoxide and can be burned to produce heat for the stoves (Figure 8.2). Once a blast furnace is started, it will typically be run continuously for 4 to 10 years with only short stops for maintenance.

A blast furnace is surrounded by a substantial support system. Before the charge can enter the oven, the iron ore must be dried, crushed, and sintered using limestone. Raw coal is heated to drive off water, coal gas, and coal tar. The product, called coke, is porous and about 40 % less dense than the starting material, and it is enriched in

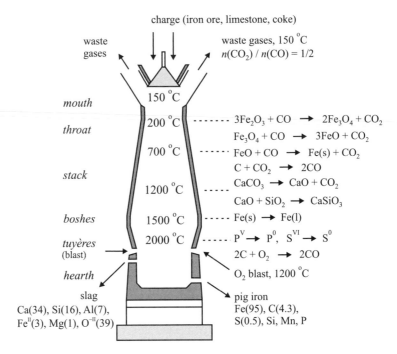

Figure 8.1   *The blast furnace*

Numbers in parentheses are mass fractions in percent. The furnace is lined with refractory materials such as chamotte (Al_2O_3; stack and boshes) and carbon (hearth) and cooled with air (stack and boshes) and water (hearth). The height of the furnace depends on the type of coal; with coke (see p. 288), it is 45–50 m.

carbon. If the blast furnace is operated using oxygen instead of air, then approximately 250 kg oxygen per Mg steel is required. Modern, more environmentally friendly processes require up to 750 kg per Mg steel, with the benefit of using less energy and reducing emissions to the atmosphere.

The molten iron is led into a pear-shaped container and treated with oxygen to reduce its carbon content. This process, called the basic oxygen process (BOP), also removes sulfur and phosphorus from the raw iron. When $w_C < 1.7\%$, the product is called steel and can be worked. The addition of elements such as chromium, vanadium, nickel, molybdenum, and/or tungsten to steel imparts special qualities, such as corrosion resistance and strength. The BOP uses about 70 kg of oxygen per Mg of steel; the oxygen is obtained by condensing and distilling atmospheric air.

Iron and steel

The phase diagram of the carbon-iron system is central to the production of iron and steel (Figure 8.3).

Iron is dimorphic, meaning that it exists in two forms. The form that is stable at ambient temperatures is called α-iron. This form has a body-centered cubic structure and is ferromagnetic. At the Curie temperature, 768 °C, the magnetic state changes to paramagnetic; this transition does not involve a change in the geometric structure

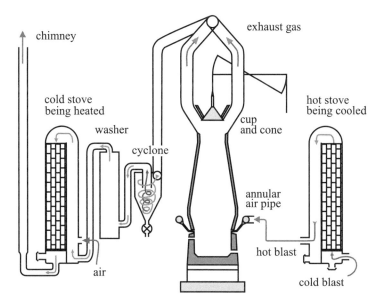

Figure 8.2 *Blast furnace and Cowper stoves*
The cold blast, air or O_2, is heated in the hot stove (≈ 1200 °C) and led through
the annular air pipe and the tuyères into the furnace. The exhaust gas, which can be
burned to produce heat, is purified by passing the cyclone (dust remover) and washer,
then mixed with air and burned to heat another stove. The flows of blast and exhaust
gas, respectively, are switched between two or more stoves.

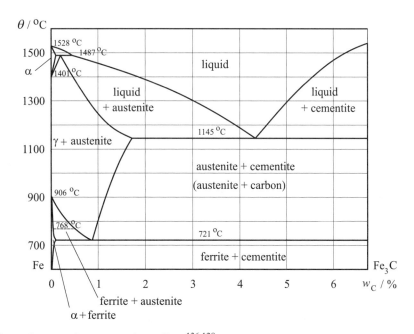

Figure 8.3 *Phase diagram for iron-carbon alloys*[136,128]
The abscissa shows the mass fraction w_C of carbon in iron. $w_C = 0.0669$ is
cementite, Fe_3C. Steel is an alloy with $w_C < 0.017$. Data at 1 bar.

of the solid. Between 906 and 1401 °C the stable phase is γ-iron with a face-centered cubic structure. Between 1401 °C and the melting point, α-iron is once again stable. These relationships are summarized as follows:

$$\text{Fe}(s, \alpha) \underset{\overset{906\,°C}{\rightleftharpoons}}{} \text{Fe}(s, \gamma) \underset{\overset{1401\,°C}{\rightleftharpoons}}{} \text{Fe}(s, \alpha) \underset{\overset{1528\,°C}{\rightleftharpoons}}{} \text{Fe}(1) \qquad (8.12)$$

Part of the iron-carbon phase diagram is shown in Figure 8.3. Iron with small carbon impurities forms ferrite (C in α-Fe) and austinite (C in γ-Fe), interstitial forms in which the C atoms are randomly distributed between the Fe atoms. The large-scale Fe structure is not affected by C impurities up to a certain level. Different hardnesses arise when austenite is cooled quickly. The higher the carbon content (up to the limit, 1.7%), the harder the steel. If the carbon content is increased beyond the limit, white iron carbide (white cast iron) is formed, a state called cementite with the composition Fe_3C. A high Si content deters the formation of Fe_3C, resulting in gray cast iron, darkened because of the presence of graphite. There is a eutectic point at $w_C = 4.3\%$ and $\theta = 1145$ °C, which is the reason that pig iron has precisely this carbon content.

The environmental impact of steel manufacture extends from iron ore and coal mines to the process itself, which consumes energy and produces NO_x, SO_2, and particle pollution, in addition to slag. Modern plants use advanced emission control systems to comply with environmental regulations.

Coal and coke

Coke is obtained by heating bituminous coal to as high as 1100 °C without oxygen. Volatile species are driven out, resulting in a porous solid with a high concentration of carbon. The liquid product, bitumen, has a high viscosity and can be used directly for paving roads. The volatile species in coal (besides water vapor) are coal gas, a variety of hydrocarbons, and coal tar. Coal gas is used for heating and as raw material for organic synthesis. The compositions of bituminous coal and coal gas are given

Table 8.9 Typical composition of bituminous coal	
Carbon	0.634
Sulfur	0.007
Phosphorus	0.001
Water	0.040
Volatile hydrocarbons[1]	0.314

Values presented as mass fraction.
[1] See Table 8.10.

in Tables 8.9 and 8.10. In 1974, 0.37 Pg of coke was produced, of which over 90% was used in steel production. In the same year, 3.59 Pg of carbon black was produced through combustion of oil or natural gas in oxygen-poor conditions. The main part of the carbon black (soot), about 3.3 Pg, was used by the rubber industry to manufacture tires. A typical automobile tire contains about 3 kg of elemental carbon in the form

Table 8.10 Coal gas	
H$_2$	0.579
CH$_4$	0.303
C$_2$H$_4$	0.033
CO	0.045
CO$_2$	0.018
N$_2$	0.020

Typical composition of coal gas; mass fraction. This gas has a heating value of 21.1 MJ/m^3 at STP.[a] See Table 8.9.

of soot. The rest was used in ink and dyes for paint, paper, and plastic. Active carbon, that is, graphite with a large surface area, 1000 to 2000 m^2/g, is used to purify sugar, industrial gases, and waste water. In 1979, 90 Gg of active carbon was produced in the United States.

The system Fe-C-O at 1500 K

As shown in Figure 8.1, the blast furnace reduces iron oxides, for example:

$$FeO(s) + CO(g) \quad \rightleftharpoons \quad Fe(s) + CO_2(g) \tag{8.13}$$

The following data apply at 1500 K:

$$C(s) + O_2(g) \quad \rightleftharpoons \quad CO_2(g) \qquad A^\circ = 396.26 \text{ kJ/mol} \tag{8.14a}$$

$$2\,C(s) + O_2(g) \quad \rightleftharpoons \quad 2\,CO(g) \qquad A^\circ = 487.48 \text{ kJ/mol} \tag{8.14b}$$

$$2\,Fe(s) + O_2(g) \quad \rightleftharpoons \quad 2\,FeO(s) \qquad A^\circ = 350.90 \text{ kJ/mol} \tag{8.14c}$$

Using this information, the following data for reaction 8.13 at 1500 K can be calculated:

$$A^\circ = -22.93 \text{ kJ/mol} \qquad K_p = 0.159 \tag{8.15}$$

Therefore, in order for FeO to be reduced, the ratio p_{CO}/p_{CO2} must be greater than 6.3 (at 1227 °C); otherwise, Fe will be oxidized by CO$_2$. Now, consider Figure 8.4. Three lines represent Equations 8.14. The general feature is that K_p increases with decreasing temperature; around 1000 K it is approximately 1.

c. Metals

Many metals are found as oxides in Nature, or their oxides may be prepared by simple means. Accordingly, reduction of oxides to produce pure metals is an essential manufacturing process. Carbon is the most important reduction agent (used for Zn, Cd, Sn, Bi, Mn, Fe, Co, Ni, Cu, Pb) and the physical chemistry of these processes will be discussed in depth below. Conversion of sulfides (Pb, Bi, Cu, Cd, As, Sb, Sn, Hg, Mo, Co, Ni)[b] to oxides and the ensuing environmental problems were discussed

---

[a] STP = standard temperature and pressure: $T = 273.15$ K, $p = 1.01325$ bar.
[b] See Figure 1.3, class III and IV.

in Chapter 7. Most large-scale industries cause unintended problems; as an example, we will discuss aluminium production, which yields large amounts of a (red) basic suspension of iron(III) oxide as by-product.

The affinity of a chemical reaction is determined from the expression (see Equation 2.169)

$$A = -\frac{\partial H}{\partial \xi} + T\frac{\partial S}{\partial \xi} \tag{8.16}$$

and thus it is favored by decreasing enthalpy (exothermic reaction) and increasing entropy. Tables of reaction enthalpies and entropies show that for reactions in the condensed phase (excepting allotropic phase changes), the inequality

$$\left|\frac{\partial H}{\partial \xi}\right| \gg T\left|\frac{\partial S}{\partial \xi}\right| \tag{8.17}$$

is a good approximation. Therefore,

$$A \approx -\frac{\partial H}{\partial \xi} \tag{8.18}$$

normally gives a useful estimate of the affinity of the reactants.

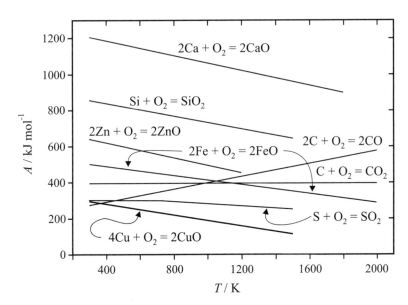

Figure 8.4 *Ellingham diagram*
Affinities A for the processes shown.[129]

The physical chemistry of the reduction of metal oxides with carbon is illustrated in Figure 8.4, which shows the affinity of formation of selected oxides from the pure elements as a function of temperature, a so-called Ellingham diagram.[202] In this case, we are dealing with a transition between a condensed phase (solid or liquid) and the gas phase, for which the reaction entropy typically makes a significant contribution to the affinity. In agreement with this, it is seen that the affinity normally decreases with increasing temperature, that the affinity is independent of temperature for the dioxides of sulfur and carbon,[a] and that the only example of an increase in Figure 8.4 is for the formation of carbon monoxide. Among the elements that form normal oxides in the temperature range shown, copper has the smallest affinity for oxygen and calcium the largest. The figure shows that copper, iron, and zinc can be reduced from their oxides by simply heating them in the presence of carbon, whereas calcium cannot be reduced from its oxide in this way.

Elements such as Al and Ti have an affinity for oxygen that lies between that of Si and Ca. Therefore, the metals cannot be obtained by direct reduction of the oxides with carbon. In such cases, cathodic reduction is a convenient synthetic path provided that a liquid electrolyte (molten salt) can be found. In the case of aluminium, one can lower the melting point of the oxide by addition of a flux, as discussed later. No such method has been found for titanium dioxide, which instead is converted to the low-melting titanium tetrachloride (liquid range -25 to $+136\,°C$) using carbon and chlorine.

This synthesis was discovered by H. C. Ørsted (1824) in the first synthesis of elemental aluminium. The industrial production of $TiCl_4$ from TiO_2 (rutile) is

$$TiO_2 + 2\,C + 2\,Cl_2 \quad \rightarrow \quad TiCl_4 + 2\,CO \qquad (8.19)$$

and the metal is obtained by electrolysis of liquid $TiCl_4$.

Ørsted, after synthesizing anhydrous $AlCl_3$ in a similar way, obtained aluminium via reduction with potassium amalgam:

$$4\,AlCl_3 + 3\,K(Hg) \quad \rightarrow \quad Al + 3\,KAlCl_4 + Hg(l) \qquad (8.20)$$

The formation of the tetrachloroaluminate is crucial. In fact, the process

$$AlCl_3 + 3\,K(Hg) \quad \nrightarrow \quad Al + 3\,KCl + Hg(l) \qquad (8.21)$$

with the amount ratio $n(AlCl_3)\,/\,n(K) = 1/3$ does *not* result in the formation of any free metal. F. Wöhler, who had visited Ørsted, tried in vain to make the metal using the ratio 1/3, but in 1828 he succeeded using the correct ratio, 4/3.

The tremendous growth in the aluminium industry is illustrated by the numbers in Table 8.11.

Table 8.11 World production of aluminium			
Year	1900	1960	2000
Production rate	5.7 Gg/a	4.7 Tg/a	24.3 Tg/a
See Table 8.1.			

Aluminium oxide (alumina, Al_2O_3, $\theta_{fus} = 2072\,°C$) is soluble in molten cryolite (Na_3AlF_6, $\theta_{fus} = 1020\,°C$), and the electrolysis is carried out at the melting point of the mixture, 940–980 °C.[b] The main aluminium ore, bauxite, consists of various

[a] For $S + O_2$ only up to the boiling point of sulfur, $T_{vap} = 717.8\,K$ ($\theta_{vap} = 444.6\,°C$), then the affinity decreases with increasing temperature.

[b] This was discovered by Hall and Héroult (1886) and the Hall-Héroult process has been in use since 1888.

hydrous oxides of Al^{III} and Fe^{III}. The ore is refined using Bayer's method (1887):[97] bauxite is crushed, milled, and treated with aqueous NaOH at $\approx 180\,^{\circ}C$:

$$Al_2O_3 + 2\,HO^- + 3\,H_2O \;\rightarrow\; 2\,Al(OH)_4^- \qquad (8.22)$$

Iron(III) oxide, which is not amphoteric, is removed by filtration. The precipitate, suspended in the basic mother liquor, is called "red mud," and its disposal is a serious problem. The trihydrate precipitates when the filtrate is cooled, and, after calcination at 1200 °C, it is mixed with cryolite and electrolyzed in a carbon-lined steel reactor. Cryolite, Na_3AlF_6, is a very rare mineral,[98] and therefore, synthetic cryolite is manufactured by the reaction

$$Al(OH)_3 + 6\,HF + 3\,NaOH \;\rightarrow\; Na_3AlF_6 + 6\,H_2O \qquad (8.23)$$

d. Pulp and paper

Cellulose is one of the most important raw materials for the chemical industry in terms of annual production. Dry wood contains around 40 % cellulose and 15 - 40 % lignin. Cellulose (Figure 8.5) is the most common organic compound in the biosphere, followed by chitin and lignin (Figure 8.6).

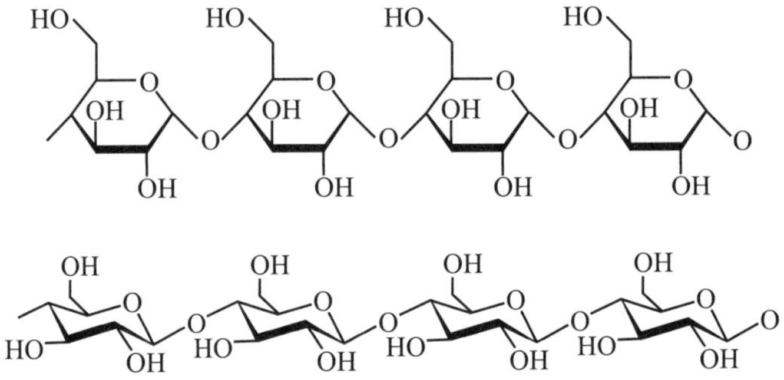

Figure 8.5 *Section of starch (top) and cellulose (bottom)*
Upper panel: Section of the starch polymer: poly(maltose), $C_6H_{10}O_5$; Haworth representation. Maltose is the dimer 4-O-α-D-glucopyranosyl-D-glucose.
Lower panel: Section of the cellulose polymer: poly(cellobiose), $C_6H_{10}O_5$; chair representation. Cellobiose is the dimer 4-O-β-D-glucopyranosyl-D-glucose.

Paper is manufactured from pulp, which is a suspension of cellulose fibers in water, and most paper pulp is made from wood. Cellulose is a naturally occurring long-chain polymer of glucose as shown in Figure 8.5. Lignin is a macromolecular

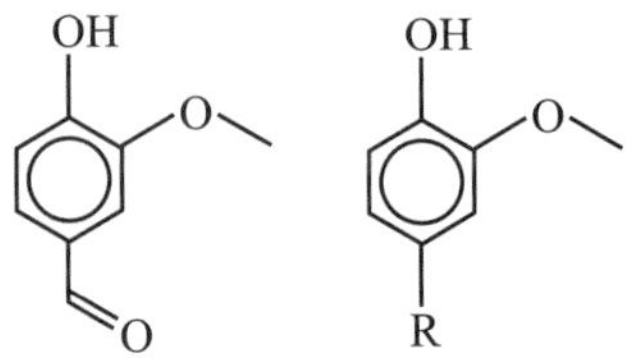

Figure 8.6 *Vanilla (left) and a subsection of lignin (right)*

chemical mainly composed of 4-hydroxy-3-methoxyphenylpropyl units (Figure 8.6). The detailed structure of lignin is strongly dependent on the type of wood.

Pulp can be produced by separating the cellulose fibers from lignin mechanically or chemically. One method is to use large mechanical grinders, sometimes with the help of steam. The yield is high because all of the wood is used, but the resulting paper is not strong, and it becomes brittle with time. The chemical processes begin with a slurry of wood chips in water. Chemicals (see later discussion) are added to dissolve lignin, releasing the cellulose fibers, which are removed by filtration. The chemical processes produce longer cellulose fibers, resulting in stronger paper; however, the yield of paper is lower, and it is more expensive.

There are two main chemical processes for producing paper, the sulfite-cellulose method and the kraft method. In the first of these, the wood suspension is treated with magnesium or calcium hydrogensulfite, obtained by reacting SO_2 and water with the carbonates. During heating the lignin is dissolved as a sulfonate, allowing the cellulose to be removed by filtration. The lignin solution has several uses. It can be used to produce ethanol via fermentation. Yeast can be used to convert it into animal feed. Finally, vanilla can be produced through oxidation with O_2 in basic solution. Vanilla is an important aroma used in the food, cosmetics, and pharmaceutical industries.

In the kraft method, sodium hydroxide and sodium sulfide react with the wood slurry at 170 °C and 8 bar. This method results in a very strong paper.[a] The chemical process is hydrolysis of lignin by base, and a main function of the sulfide is to produce hydroxide in a buffering reaction,

$$S^{2-} + H_2O \rightarrow HS^- + HO^- \tag{8.24}$$

After removal of the cellulose fibers by filtration, water is evaporated from the solution, and the solid residue burned to produce energy. Before burning, chemicals such as turpentine and dimethyl sulfide can be obtained. After combustion, sodium sulfide can be obtained from the sulfate formed in the reaction mixture:

$$Na_2SO_4 + 2C \rightarrow Na_2S + 2CO_2 \tag{8.25}$$

The resulting Na_2S is a valuable compound.

Until recently, most paper was bleached using chlorine. Chlorine bleaches by adding to double bonds, thereby removing conjugate bonds that absorb visible light:

$$RCH{=}CHR' + Cl_2 \rightarrow RCHCl{-}CHClR' \tag{8.26}$$

This method leads to the formation of polychlorinated biphenyls (PCBs), benzenes, dioxins, and dibenzofurans, which are powerful environmental toxins. The chlorine bleaching method has therefore been replaced by chlorine dioxide bleach. ClO_2 forms hypochlorous acid, HOCl, in aqueous solution. HOCl adds to double bonds in a reaction analogous to Equation 8.26, resulting in chlorohydrins. In Scandinavia, hydrogen peroxide has replaced chlorine, with the ensuing formation of diols. The use of hypochlorous acid or hydrogen peroxide does not result in the formation of the aforementioned hazardous chlorinated species.

---

[a] Ger. *kraft* = strength.

The production of paper pulp demands large amounts of raw materials (Table 8.12). In all, 4% of the world's production of Cl_2, 16% of NaOH, and 4% of Na_2CO_3 are used in making pulp (2008). The paper industry has historically produced large amounts of pollution, both locally around mills and regionally, as shown in Table 8.13, which considers the environmental effects of 1.5% of the world's production of paper in 1986. In the intervening years, new solutions have been introduced; in particular, atmospheric emissions have been strongly reduced by the collection and burning of the exhaust gases in recovery boilers.

Table 8.12 Paper industry		
	Mass ^{1}/Gg	Mass ^{2}/Gg
NaOH	2350	9600
Cl_2	975	2500
$NaClO_3$	885	
O_2	430	
Na_2CO_3	180	
H_2O_2	140	
Other (e.g., $CaCO_3$, TiO_2)	515	

^{1} Paper industry's use of chemicals in the United States, 1993.
^{2} Paper industry's use of chemicals worldwide, 2008.

Table 8.13 Emissions from the production of 3.7 Tg of paper, 1986	
Compound	mass/Tg
Emitted into the hydrosphere	
w_{BOD} see Section 3.2b	0.730
Suspended solids	0.347
(In addition, 5.8×10^6 m^3 of toxic discharge)	
Emitted into the atmosphere	
Dust	0.100
Sulfur dioxide, SO_2	0.080
Carbon monoxide, CO	0.060
Nitric oxide, NO	0.015
Hydrocarbons	0.012

The data are based on a Canadian facility.

8.3 The inorganic chemicals industry

The entry of Table 8.4 that covers the most diverse group of industries and environmental problems is the inorganic chemicals industry.[96, 97] In the present section, we have chosen as the main subject the chlor-alkali industry, because it illustrates essential points of discovery, production, and use of connected and related chemical species. In general the steps are (a) realizing that a chemical species is useful, (b) inventing of appropriate methods of production, (c) considering the impact of other industries on pricing and production, and (d) solving environmental problems (frequently in that order!).

The rise of the chlorine industry in the beginning of the 19th century was in part due to increasing demands for chemicals by the textile industry. Manufacturers found that they could produce better clothing by first removing plant waxes from cotton using base, bleaching it with chlorine, and finally utilizing the changed surface structure of the fibers for dying.

Before this time textiles were treated with potassium carbonate (potash) and bleached in sunlight for months. However, with increasing demand, the sources of potash (oak- or beechwood) were not sufficient. In the last part of the 18th century, the Leblanc process (1792) for producing sodium carbonate (soda ash) was developed to meet demand from the glass industry, and this base was subsequently used in the textile industry as well. About that time Scheele (1774) isolated chlorine and discovered that it could be used as a bleaching agent. Chlorine quickly found widespread use, and the methods of production are discussed in Section 8.3d.

The chlor-alkali industry produces chlorine, sodium hydroxide,[a] and sodium carbonate.[b] During the 20th century, chlorine was mainly produced by electrolysis of a saturated aqueous solution of sodium chloride,[c]

$$NaCl(aq) + H_2O(l) \xrightarrow{\text{electrolysis}} \tfrac{1}{2}Cl_2(g) + \tfrac{1}{2}H_2(g) + NaOH(aq) \quad (8.27)$$

Sometimes, for example, around the year 2000, the demand for chlorine is greater than for sodium hydroxide, and so the latter is converted to sodium carbonate (particularly in Europe and Japan):

$$2\,NaOH(aq) + CO_2(g) \rightarrow Na_2CO_3(aq) + H_2O(l) \quad (8.28)$$

At other times, such as about 1950, the demand for sodium hydroxide is larger, and additional sodium hydroxide is produced from sodium carbonate by means of a method known since antiquity:

$$Ca(OH)_2(aq) + Na_2CO_3(aq) \rightarrow CaCO_3(s) + 2\,NaOH(aq) \quad (8.29)$$

Calcium hydroxide was produced by limeburning (see Section 7.2d).

The electrolytic process (Equation 8.27) was discovered in 1800, but the invention of the dynamo (1872) was necessary before electrolysis became a useful industrial technique. Around 1890, continuous processes were developed: the diaphragm method and the mercury method, both described next. Since 2000, the membrane method has grown in popularity for environmental reasons.

---

[a] Trade name/technical name: caustic soda.
[b] Trade name/technical name: soda ash.
[c] Technical name of this solution: brine.

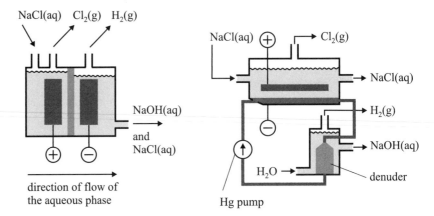

Figure 8.7   *Diaphragm cell (left) and mercury cell (right)*
Typical area of each electrode of a diaphragm cell: 55 m^2 (the areas that face one another).
Typical anode area of a mercury cell: 35 m^2.

a. The electrolytic cell

The diaphragm cell

A sketch of the cell is shown in the left-hand side of Figure 8.7. The anode compartment is separated from the cathode compartment by an asbestos diaphragm. A weakly acidic (HCl) aqueous solution of saturated sodium chloride (mass fraction $w_{NaCl} = 0.26$) is pumped toward a titanium anode, where chloride is discharged and converted into gaseous chlorine:

$$Cl^-(aq) \xrightarrow{\ 85\,°C,\,pH\,5\ } {}^1\!/_2\,Cl_2(g) + e^-(Ti) \tag{8.30}$$

The solution flows constantly through the diaphragm, and water is reduced to gaseous hydrogen and hydroxide at the steel cathode:

$$e^-(steel) + H_2O(l) \ \rightarrow \ {}^1\!/_2\,H_2(g) + HO^-(aq) \tag{8.31}$$

The solution that leaves the cathode compartment contains sodium hydroxide, $w_{NaOH} = 0.11$, but also rather a lot of sodium chloride, $w_{NaCl} = 0.11$. It is evaporated, typically using steam, and NaCl precipitates. The final liquid product has $w_{NaOH} = 0.50$ and $w_{NaCl} = 0.01$.

    The flow prevents hydroxide from migrating backward into the anode compartment, where it would give rise to undesirable side reactions forming hypochlorite and chlorate:

$$Cl_2 + 2\,HO^- \ \rightarrow \ ClO^- + Cl^- + H_2O \tag{8.32}$$

$$3\,Cl_2 + 6\,HO^- \ \rightarrow \ ClO_3^- + 5\,Cl^- + 3\,H_2O \tag{8.33}$$

The typical current density for a diaphragm cell is 2.7 kA/m². Such a cell (see Figure 8.7), working at 150 kA, will produce 4.4 Mg/d of chlorine, assuming a current efficiency of 96 %.[a]

The mercury cell

The mercury cell is sketched in the right-hand side of Figure 8.7. The cell has one compartment for both electrodes and is fed with saturated sodium chloride. The oxidation at the titanium anode is given by Equation 8.30.

However, in contrast to the diaphragm cell, the process at the mercury cathode is the reduction of sodium(I):

$$e^-(Hg) \ + \ Na^+(aq) \quad \rightarrow \quad Na(Hg) \qquad (8.34)$$

and a solution of sodium in mercury, an amalgam, is formed. The amalgam remains liquid and is transferred into a separate vessel, the denuder, where sodium reacts with water in a process catalyzed by graphite:

$$Na(Hg) \ + \ H_2O(l) \quad \xrightarrow{\text{graphite}} \quad {}^1\!/_2 H_2(g) \ + \ NaOH(aq) \qquad (8.35)$$

The advantage of this method is that the hydroxide contains no chloride, and no heat is necessary to produce a concentrated solution, $w_{NaOH} = 0.50$. The disadvantage is that emission of mercury cannot be avoided, and therefore the process is being phased out.

The typical current density in the mercury cell is 11.5 kA/m². This means that the cell (Figure 8.7), working at 410 kA, will produce 12.1 Mg/d of chlorine, assuming a current efficiency of 96 %.

The membrane cell

The mercury cell is being phased out because of the invention of the ion-exchange membrane (see Section 9.5c) and its use in the membrane cell. Figure 8.8 shows a cell using a cation-exchange membrane. The anode functions as discussed previously. The difference is that a solution depleted of sodium chloride flows out of the anode compartment for recharging and recycling. The membrane allows cations to flow into the cathode compartment. As shown, sodium hydroxide is concentrated from $w_{NaOH} = 0.30$ or less to $w_{NaOH} = 0.40$ at the cathode, often a steel mesh, placed near the membrane.

Ideally the membrane would not allow anions to pass, but imperfections allow some hydroxide to pass and react according to Equations 8.32 and 8.33.

The membrane cell has been improved in recent years and may now work at a current density of 4.0 kA/m², that is, 6.5 Mg/d of chlorine.[b] This is much better than the diaphragm cell, but far from the output of the mercury cell.

---

[a] $I = 2700 \times 55$ A $\approx 1.5 \times 10^5$ C/s; $t = 84000$ s/d; $F = 96485$ C/mol; $M(Cl) = 35.45$ g/mol; $\Delta m / \Delta t = I t M \, 0.96 / F \approx 4.4$ Mg/d.

[b] Assuming a membrane cell with the same electrode area and efficiency as for the mercury cell, Figure 8.7.

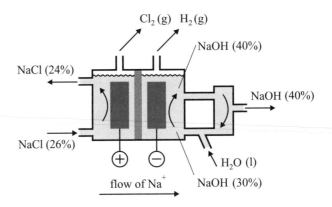

Figure 8.8 *Membrane cell*
The cation-exchange membrane is placed between the electrodes.

Electrode potentials and energy conditions

The decomposition potentials may be estimated using electrode potentials, as calculated by using the Nernst equation. Relevant standard electrode potentials are listed in Table 8.14. As discussed in Section 5.3a, the reversible decomposition potential of water is 1.23 V at all pH values, but in the present cases the applied potential is not simply the difference of the standard electrode potentials.

Table 8.14 Selected standard electrode potentials

$2H^+(aq) + 2e^- = H_2(g)$	$E^\circ = 0\,V$
$O_2(g) + 4H^+(aq) + 4e^- = 2H_2O(l)$	$E^\circ = +1.23\,V$
$Na^+(aq) + e^- = Na(s)$	$E^\circ = -2.71\,V$
$Na^+(aq) + e^- = Na(Hg)$	$E^\circ = -1.80\,V$
$Cl_2(g) + 2e^- = 2Cl^-(aq)$	$E^\circ = +1.36\,V$

In the diaphragm cell, the potential at the anode is approximately

$$E_{\text{anode}}/V \approx 1.36 - 0.071 \times \lg(5.3) = 1.31 \tag{8.36}$$

because $\theta \approx 85\,^\circ C$ and $c(Cl^-) \approx 5.3$ M. The potential at the cathode is approximately

$$E_{\text{cathode}}/V \approx -0.065 \times 14.5 = -0.93 \tag{8.37}$$

because $\theta \approx 50\,^\circ C$ and $c(HO^-) \approx 3.3$ M. This rough estimate gives a decomposition potential slightly above 2.2 V, compared with the measured value of 2.15 V. The overpotential is ≈ 0.33 V (from both electrodes), and cell resistance requires ≈ 1.00 V (the sum of the resistances of the diaphragm, solution, and electrodes) giving a working potential of 3.48 V. The product $FE = 96485 \times 3.48\,J/mol = 336\,kJ/mol$ is a measure of the consumption of energy per amount of substance.[a] The membrane cell

---

[a] That is, 2.63 kWh/kg of chlorine. Suppose that 500 V is available for a factory, allowing 143 diaphragm cells (of 55 m², see Figure 8.7 and footnote a, p. 297) to be connected in series. This plant will yield approximately 230 Gg/a (≈ 4.4 Mg/d × 143 × 365 d/a) of chlorine.

has the same working potential and therefore requires the same energy per amount of substance produced.

The main reason why the mercury cell works is the large overpotential for the formation of hydrogen. The decomposition potential of the cell is 3.05 V, and the total working potential is 4.32 V. The product $FE = 417$ kJ/mol is a larger amount of energy per amount of substance produced relative to the other cells, but as mentioned, the advantage is the rather direct synthesis of pure sodium hydroxide.

b. Sodium hydroxide

Sodium hydroxide (NaOH, caustic soda) is an important general strong base. Global production and principal applications are given in Table 8.15. Since antiquity, it has been prepared from soda (see Equation 8.29), but more exotic methods have been suggested for industrial use, such as

$$2\,NaCl \;+\; PbO \;+\; H_2O \;\;\rightarrow\;\; 2\,NaOH \;+\; PbCl_2 \tag{8.38}$$

(Scheele) because lead(II) oxide is easily prepared, and $PbCl_2$ is sparingly soluble.

Table 8.15 Use of sodium hydroxide, 2005	w_m
Organic chemicals	0.19
Pulp and paper	0.16
Inorganic chemicals	0.12
Alumina	0.10
Detergents	0.08
Textile industry	0.08
Sodium hypochlorite, NaOCl (bleach)	0.03
Water treatment	0.02
Other uses[1]	0.22

Mass fraction of sodium hydroxide production (caustic soda, total mass $\approx$ 60 Tg) used in the product or used making the product.
[1] Includes food industry, $w_m \approx 0.01$.

c. Sodium carbonate

In antiquity and through medieval times, sodium carbonate was obtained from the Egyptian trona flats[a] and similar deposits around the world. It was used in the production of glass and soap and as a common weak base. During the 19th and 20th centuries, synthetic industrial production took over, and today one-fourth of the world's production has a natural source and three-fourths is synthetic, with a total production of more than 40 Tg/a.

[a] Trona: $2Na_2CO_3 \cdot NaHCO_3 \cdot 2H_2O$.

Synthetic industrial production in the United States ceased in 1955 in favor of natural sources. This is due to trona from the Green River Basin of Wyoming (1938); reserves are estimated at more than 60 Pg. Trona is calcinated (i.e., heated) to yield carbonate, and the carbon dioxide product is used for other purposes (in particular beverages):

$$2\,Na_2CO_3 \cdot NaHCO_3 \cdot 2\,H_2O(s) \quad \rightarrow$$
$$3\,Na_2CO_3(s) + CO_2(g) + H_2O(g) \qquad (8.39)$$

Large-scale synthetic production of sodium carbonate is a classic example of an industry that created environmental problems and – after pressure from society – was driven to invent new and cleaner methods.

The basic problem is to convert naturally occurring salt and limestone into soda ash:[a]

$$2\,NaCl + CaCO_3 \quad \rightarrow \quad Na_2CO_3 + CaCl_2 \qquad (8.40)$$

As described later, this problem was first solved by Leblanc (1792), who utilized sulfuric acid.

The Leblanc process

The first step is to convert sodium chloride into sodium sulfate:

$$2\,NaCl(s) + H_2SO_4(l) \xrightarrow{600\,^\circ C} Na_2SO_4(s) + 2\,HCl(g) \qquad (8.41)$$

Direct treatment of sodium chloride with sulfuric acid yields the hydrogensulfate, and the sulfate is obtained after heating to red-hot. The product is known as *salt cake*.

The sulfate is then mixed with coal and chalk and heated:

$$Na_2SO_4(s) + 4\,C(s) + CaCO_3(s) \quad \rightarrow$$
$$Na_2CO_3(s) + CaS(s) + 4\,CO(g) \qquad (8.42)$$

The reaction mixture, called *black ash* because it is colored by unreacted carbon, is treated with water to dissolve sodium carbonate, leaving calcium sulfide as a by-product.

Environmental issues

The first Leblanc factory in England was put into operation in 1814. In the beginning, gaseous hydrogen chloride (Equation 8.41) was emitted directly into the atmosphere because the exhaust gas is rather dilute and conversion into a useful aqueous hydrochloric acid is demanding. Obviously, the exhaust is destructive to the environment and intolerable to neighbors. Many years later it was utilized in the Weldon process (see Equation 8.48).

---

[a] The chalk cliffs of Dover are stable with respect to seawater!

Further, black piles (colored by carbon) of solid calcium sulfide reacted with carbon dioxide in the air to give calcium carbonate and gaseous hydrogen sulfide,

$$CaS + H_2O + CO_2 \rightarrow CaCO_3 + H_2S \qquad (8.43)$$

which polluted the atmosphere around the factories. In addition, the sulfides were slowly oxidized by atmospheric oxygen, giving sulfuric acid, which eventually ended up in the hydrosphere.

This uncontrolled pollution of air and water caused the British Parliament to pass environmental legislation. The Alkali Act (1863) restricted the release of hydrogen chloride produced by the process of Equation 8.41. Later amendments restricted the exhaust of hydrogen sulfide through the process of Equation 8.43.

Very pure sulfur can be prepared by controlled application of the process Equation 8.43 followed by oxidation of hydrogen sulfide by atmospheric oxygen:

$$2\,H_2S(g) + O_2(g) \rightarrow 2\,H_2O(l) + \left(\tfrac{1}{4}\right)S_8(s) \qquad (8.44)$$

This is known as the Claus process (1882) and is still widely used. The process is cost effective because the sulfur is used to make sulfuric acid. Today, the main source of hydrogen sulfide is sulfur in crude oil, which must be removed before use.

The Solvay process

The Solvay process,[a] developed in the years 1863–1869, is more economical and less polluting than the Leblanc process, and the last Leblanc factory closed in 1920. The Solvay process is based on the fact that sodium hydrogencarbonate precipitates from a concentrated aqueous solution of sodium chloride and ammonium hydrogen-carbonate:

$$Na^+ + HCO_3^- \rightarrow NaHCO_3(s) \qquad (8.45)$$

Calcination of the hydrogencarbonate gives the carbonate:

$$2\,NaHCO_3(s) \xrightarrow{150\,°C} Na_2CO_3(s) + H_2O(g) + CO_2(g) \qquad (8.46)$$

together with carbon dioxide. The key to the method is that carbon dioxide is obtained from limeburning (see Equation 7.3), which also supplies the base, $Ca(OH)_2$, that liberates ammonia used in reaction 8.45 for recycling:

$$NH_4^+ + HO^- \rightarrow NH_3(g) + H_2O(l) \qquad (8.47)$$

The Solvay process executes the reaction, Equation 8.40, in an elegant but energy-demanding manner.

Applications

World production of sodium carbonate was 43 Tg, of which three-fourths was man-ufactured synthetically (2007). Total annual production in the United States was 11 Tg (using natural sources), of which 5 Tg was exported.[252] Within the United

[a] Or: The Ammonium Soda Process.

States, sodium carbonate was used for glass (50%), chemicals (29%), soap and detergents (9%), pulp and paper (2%), flue gas desulfurization (2%), water treatment (1%), and miscellaneous (7%).

d. Chlorine

During the first part of the 19th century, chlorine, prepared using Scheele's method (1774), was mainly used to bleach textiles. Scheele was the first to describe the oxidation of hydrochloric acid with the mineral pyrolusite, MnO_2:

$$4\,HCl(aq) + MnO_2(s) \;\rightarrow\; Cl_2(g) + MnCl_2(aq) + 2\,H_2O(l) \quad (8.48)$$

This method was expensive and remained so even after Weldon (1866) discovered that Mn^{II} can be oxidized in basic solution with oxygen into Mn^{III}-Mn^{IV} oxides (see Equations 3.10a and 3.10b).

The direct oxidation of hydrogen chloride with oxygen was discovered by Deacon (1868):

$$2\,HCl(g) + \tfrac{1}{2}O_2(g) \;\xrightarrow{\;CuCl_2,\,450\,^\circ C\;}\; Cl_2(g) + H_2O(g) \quad (8.49)$$

The Deacon process is superior to the Weldon process and is still in use (2008) because chlor-alkali production of chlorine does not satisfy demand.

The sources of hydrogen chloride include large amounts formed as a by-product of vinyl chloride production (see Equation 8.1). In more detail:

$$CH_2{=}CH_2 \;\xrightarrow{\;Cl_2\;}\; ClH_2C{-}CH_2Cl \;\xrightarrow{\;500\,^\circ C\;}\; CH_2{=}CHCl + HCl \quad (8.50)$$

Some HCl from the pyrolysis is reused on-site through oxychlorination of ethene, a reaction catalyzed in the same way as the Deacon process:

$$CH_2{=}CH_2 + 2\,HCl + \tfrac{1}{2}O_2 \;\rightarrow\; ClH_2C{-}CH_2Cl + H_2O \quad (8.51)$$

Besides the Deacon process, HCl is converted into chlorine by the "Kel-Chlor" process, which uses nitrogen oxides as catalysts:

$$2\,HCl + NO_2 \;\rightarrow\; Cl_2 + NO + H_2O \quad (8.52)$$

$$NO + \tfrac{1}{2}O_2 \;\rightarrow\; NO_2 \quad (8.53)$$

Table 8.16 Distribution of world production of chlorine, 2008				
China	North America	EU	Asia-Pacific	Others[1]
0.26	0.23	0.20	0.19	0.12

Mass fraction, w_m; total mass: ≈ 63 Tg.[255]
[1] Others include South America, the Levant, Africa, and Eastern Europe.

Table 8.16 shows the main figures for world production of chlorine in 2008. The importance of this commodity chemical is reflected in the rate of growth of production.

In the years 1975-1995, production increased by a factor of 1.02 a^{-1}, corresponding to a doubling every 35 years; since then, the growth factor has increased to 1.04 a^{-1}, corresponding to a doubling every 18 years.

Applications[164]

World production of chlorine was 63 Tg in 2008 (see Table 8.1). More than half of the products sold by the chemical industry and approximately 90 % of all pharmaceuticals use chlorine. Table 8.17 shows the major groups, classified according to mass fraction. We now focus on environmental chemical aspects of the production summarized in Table 8.17.

Table 8.17 Application of chlorine derivatives, 2002	w_m
Vinyl chloride, $CH_2{=}CHCl$	0.34
Phosgene, $COCl_2$	0.09
Hydrogen chloride, HCl	0.08
Solvents[1]	0.08
Propylene oxide[2]	0.07
Water treatment[3]	0.05
Allyl chloride, 3-chloroprop-1-ene	0.04
Pulp and paper	0.04
Other organic chemicals[4]	0.07
Other inorganic chemicals[5]	0.14

Mass fraction of chlorine (total mass ≈ 60 Tg) in the product or used making the product.
[1] Chloro congeners[a] of methane and ethane.
[2] Made from propylene chlorohydrin, that is, 2-chloropropane-1-ol and 1-chloropropane-2-ol.
[3] Only the use of elemental chlorine.
[4] Includes derivatives of chloroacetate and chloro congeners of benzene.
[5] Discussed in the text.

Poly(vinyl chloride)

Polyaddition of vinyl chloride gives poly(vinyl chloride), PVC[b]:

$$n\,CH_2{=}CHCl(g) \quad \rightarrow \quad [-CH(Cl)CH_2-]_n(s) \qquad (8.54)$$

which is a solid ($\theta_{fus} \approx 100\text{--}250\ ^\circ C$) at room temperature. PVC is a thermoplastic, meaning that it becomes elastic above its glass transition temperature, about $80\ ^\circ C$. PVC possesses several valuable properties: it is cheap, the mass fraction of chlorine being 57 %, and its production requires relatively little petroleum. It is a relatively inflammable construction material with a flammability only slightly higher than that of concrete; it is impermeable to gases; it is easy to use in processing and it is compatible with many additives such that its properties can be modified to a great

[a] See Section 9.2.
[b] Poly(1-chloroethane-1,2-diyl); acronym: PVC.

extent; it is durable and can be reused in many ways, from drinking bottles to drain pipes; and finally, it can be recycled many times.

More than half of all PVC is used in the building sector, including items such as sewer pipes, door and window frames, electrical insulation, and floor coverings. The rest is used for (a) consumer goods, including articles for home and office, cars, credit cards, and toys; (b) packaging, including food, cosmetics, and pharmaceuticals; and (c) medical equipment.

Additives, including plasticizers and stabilizers, may harm the environment. Plasticizers modify the mechanical properties of the polymer, making it more flexible. Diesters of phthalic acid (benzene-1,2-dicarboxylic acid) with long chain alcohols are most common, an example being bis(2-ethylhexyl)phthalate, $C_6H_4[C(O)CH_2CH(C_2H_5)C_4H_9]_2$. Stabilizers protect the polymer against thermal and photolytic degradation: tetrabutyltin, $Sn(C_4H_9)_4$, has been widely used as a heat stabilizer, and pentaerythritol, $C(CH_2OH)_4$, is an example of an antioxidant.

8.4 The biotechnology industry

Humans have used microbiological processes for thousands of years. In the classical fermentation process, yeast is used to produce beer, wine, and bread. Molds are used to produce citric acid and gluconic acid (oxidized glucose), and bacteria convert milk into dairy products such as cheese and yogurt, in addition to acetic acid from wine.

More recently, genetically engineered organisms have been used to produce a wide range of natural products. For example, fermentation is used to produce antibiotics, and bacteria produce proteins used for antiviral vaccines, human insulin, and enzymes for washing and tanning.

The nascent field of environmental biotechnology explores naturally occurring bacteria that have a rich and as yet not fully described metabolic diversity with the goal of identifying those that can be used in industrial and environmental processes. In several cases it has been possible to isolate the genes that code for proteins that degrade toxic waste and purify waste water. Examples include the genes coding for biodegradation of 2,4,5-T (see Equation 9.11) and atrazine (see Equation 9.9).

8.5 Sustainable synthetic chemistry: Green chemistry

Green chemistry seeks to reduce and prevent pollution by promoting products and chemical processes that minimize the use and generation of hazardous substances. This includes discouraging the production and use of potentially dangerous substances, even ones that have no direct toxicity to humans.

Green chemistry includes consideration of the following:[239]

1. Feedstock. Use of renewable raw materials, such as agricultural products and wastes from other processes, is preferred to use of feedstocks obtained from fossil fuels or mining.
2. Synthesis. Design of processes such that final products contain most of the input material. This is termed *atom economy*. It implies that protection groups or other

temporary modifications of the reactants should be avoided because they are later discarded, and that efficient catalysts should be used whenever possible.[a]

3. Waste. Restrict the use of solvents; avoid toxic and nondegradable solvents.

As seen, green chemistry typically addresses only the production process. In contrast, life cycle analysis includes the entire product life "from cradle to grave."

Examples of green chemistry include catalysts designed for asymmetric syntheses, the use of solvents in a supercritical state, hydrogen peroxide used in clean oxidations, and synthesis in engineered microorganisms.

Catalyst design

A main goal of green chemistry is the development of new catalysts for industrial syntheses. Iron catalysts used in the production of ammonia and the V_2O_5 catalysts used to make sulfuric acid have been discussed previously. In the two following examples, the importance of catalysts that promote stereospecific breaking and formation of carbon-carbon bonds is stressed.[227] The importance of the subject is demonstrated by the list of Nobel prizes given in this area: Grignard, Sabatier, Diels, Alder, Ziegler, Natta, Brown, Wittig, all known from undergraduate courses in organic chemistry.

1. Asymmetric catalysis

In 2001, Knowles, Noyori, and Sharpless were awarded the Nobel prize for their research on catalytic asymmetric synthesis. Here, a chiral catalyst is used to prepare one of the enantiomers preferentially, by addition to a carbon-carbon double bond. In general there are two modes of addition: a reductive one, hydrogenation, and an oxidative one, epoxidation.

Consider first asymmetric hydrogenation. It was found that the rare amino acid L-dopa,[b] (S)-2-amino-3-(3,4-dihydroxyphenyl)propanoic acid (see Equation 8.57), is useful in treating Parkinson's disease. The reactions comprising the first industrial process (Hoffmann-LaRoche) are shown in the following reaction equations. Vanillin was chosen as the starting point because it is a cheap by-product of paper production (see Figure 8.6):

vanillin   N-benzoyl glycine

$$H_2O \; + \; \text{(structure A)} \qquad (8.55)$$

A

---

[a] The apt phrase is: benign by design.

[b] The acronym is derived from the obsolete name 3-(3,4-dioxyphenyl)alanine.

Compound **A** was hydrogenated with H_2, using a Pd/C catalyst, to give a racemic mixture of **B**, dopa with two protection groups:

$$ \text{(8.56)} $$

B

Resolution of the stereoisomers was effected by precipitation of one of the pair of **B** acids with the base D-(+)$_D$-1-phenetylamine, $C_6H_5CH(NH_2)CH_3$. The diastereomer precipitate was freed from the base and deblocked with acid to give L-dopa:

$$ \text{(8.57)} $$

(S)-2-amino-3-(3,4-dihydroxyphenyl)propanoic acid

One of the two product enantiomers **B** is not valuable even though it is resolved, so at least half of the yield is lost. It is clear that the crucial step is the hydrogenation, and systematic research was initiated in order to create an asymmetric hydrogenation catalyst. The rhodium(I) complex of the ligand called R, R-dipamp was found to be best:[a]

$$ \text{(8.58)} $$

(1R,2R)-bis[(2-methoxyphenyl)phenylphosphino]ethane

The Monsanto L-dopa process uses [Rh(R, R-dipamp)cod]$^+$ BF$_4^-$ as the catalyst; L-dopa produced in this way has an $ee = 0.95$.[b] This example clearly fulfills green chemistry's goal of using natural precursors and selective catalysts, and of reducing waste.

The second mode of addition is the asymmetric epoxidation. The example of this approach starts with the female gypsy moth, *Lymantria dispar*, which produces the

[a] pamp is phenyl-2-anisylmethylphosphine.
[b] In the Monsanto process, the phenol and the amine of **A** are protected with acetyl, $CH_3CO–$. cod is cycloocta-1,5-diene.

pheromone (7R,8S)-dispalure:

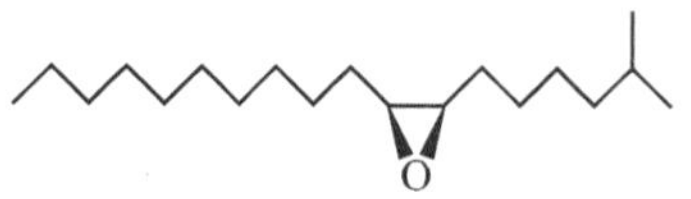

$$ (8.59) $$

(7R,8S)-7,8-epoxy-2-methyl octadecane

Asymmetric epoxidation has been used successfully on an industrial scale to produce this pheromone. The oxidation agent is *tert*-butyl hydroperoxide,[a] and the catalyst the enantiomerically pure diethyltartrato complex of titanium(IV). The ligand is the diethyl ester of D-tartaric acid, called D-det:

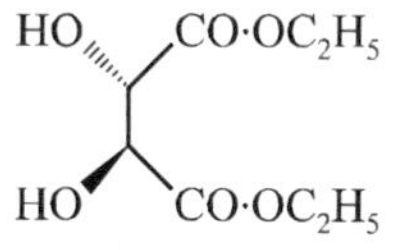

$$ (8.60) $$

diethyl (2S,3S)-2,3-dihydroxybutanedioate

The following oxidation scheme presents the generic form of the reaction

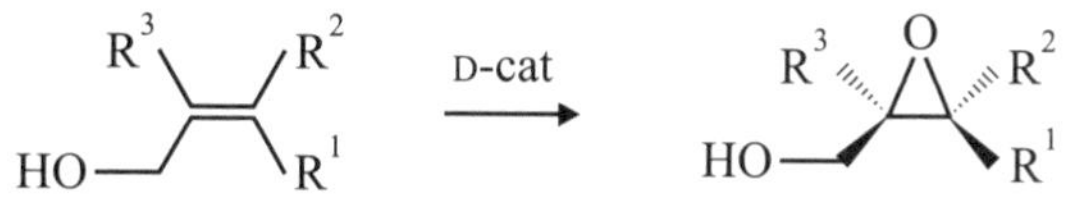

$$ (8.61) $$

where D-cat denotes the catalyst just described. The enantiomeric L-cat promotes the mirror image of the product.

2. Metathetic catalysis

The 2005 Nobel Prize in Chemistry was awarded to Chauvin, Grubbs, and Schrock for their research on metathesis in organic synthesis.[b] A metathetical reaction is an exchange reaction, such as AB + CD → AC + BD, where parts of two molecules are swapped. Consider the carbene (alkylidene) exchange between olefins (alkenes):

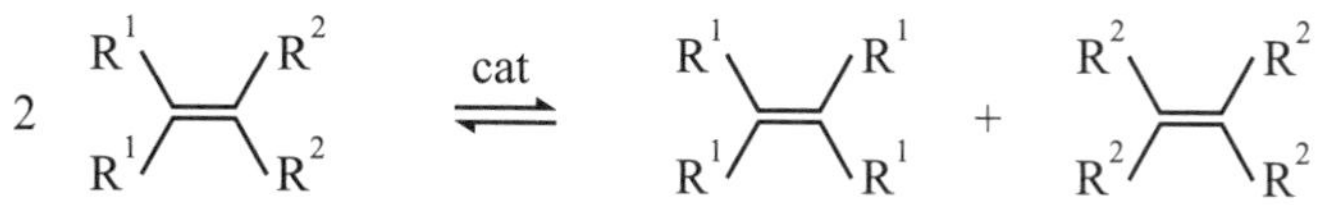

$$ (8.62) $$

Metathesis catalysts are metal-alkylidene complexes where the metal is from the second or third transition series: Ta, Mo, W, Ru. This mechanism is extremely versatile and gives rise to a wealth of synthetic transformations,[203] including straightforward exchange (cross metathesis) (Equation 8.62), closure of large rings, formation of dienes from cyclic and acyclic olefins, and polymerization of cyclic olefins and acyclic dienes.

  As an example, consider Grubbs' catalysts shown on the left-hand side of Equation 8.63. In the so-called first-generation catalyst, the ligand X is tris(cyclohexyl)

---

[a] That is, (1,1-dimethylethyl)dioxidane.
[b] Gk. μετα ≙ beyond; θεση ≙ position.

phosphane, P(Cy)$_3$. Here the two chlorido ligands are almost trans to each other, and the benzylidene ligand is coplanar with the RuCl$_2$ moiety.[a] In the second-generation catalyst, the X ligand is derived from 1,3-bis(mesityl)imidazolidine, which is shown to the right of Equation 8.63.[b]

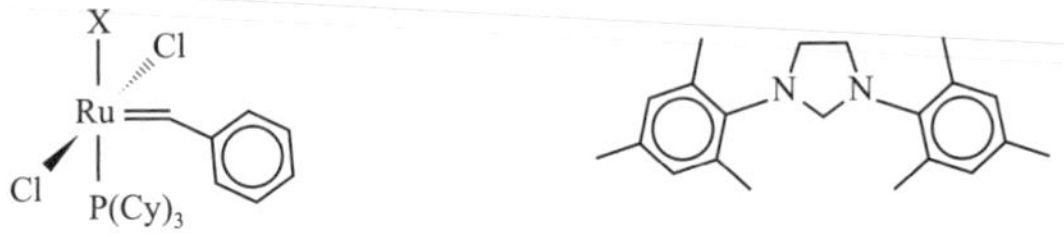

$$(8.63)$$

The catalyst's function is due to its ability to maintain two double bonds during the reaction. Grubbs' second catalyst has been used to manufacture the pheromone 11-tetradecyl acetate that is excreted by several species of Lepidoptera, an insect group that includes butterflies and moths. The following synthesis targets the omnivorous leaf roller moth (*Platynota stultana*), which is a pest on some deciduous plants, including grapes, kiwis, and nuts. The pheromone is prepared by cross metathesis:

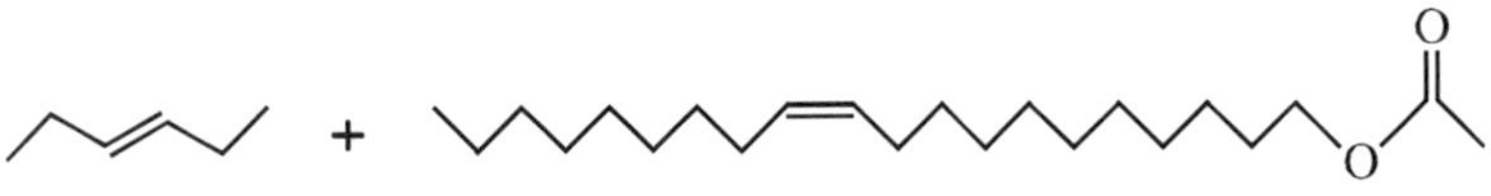

3-hexene   11-eicosenyl acetate

$$(8.64)$$

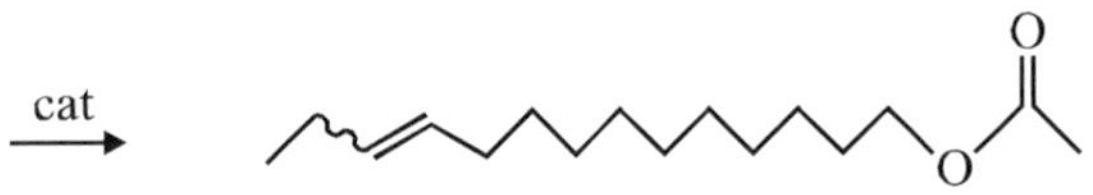

11-tetradecanyl acetate ($n_E / n_Z = 82{:}18$)

The synthetic product has the same *E/Z* ratio as in the natural product.

3. General comments and examples

The three modern examples of catalyst design just discussed are very different in scope. Methods using rare metals are only practical for minor specialty chemicals. In contrast, as the catalyst for asymmetric oxidation is based on titanium, a common element, one can foresee widespread application of this synthetic path.

Supercritical solvents

Supercritical fluids are fluids at a temperature and pressure above their critical point. Their properties are intermediate between those of the liquid phase and the gas phase, and they can be tuned using temperature and pressure. These fluids provide a new class of solvents that have been successfully applied in various areas, including chemical synthesis, analysis, and extraction.[41] Previously, we have mentioned CO$_2$(sc) and

---

[a] The systematic name of Grubbs' first-generation catalyst is benzylidenedichloridobis[tris(cyclo-hexyl)phosphane]ruthenium(II).

[b] The systematic name of Grubbs' second-generation catalyst is benzylidene[1,3-bis(2,4,6-trimethyl phenyl)imidazolidin-2-ylidene]dichlorido[tris(cyclohexyl)phosphane]ruthenium(II).

$H_2O(sc)$, but other fluids are in use, such as $CHF_3(sc)$ and mixtures of CO_2 and CH_3CN.[a] All supercritical fluids are completely miscible with each other; that is, any mixture of fluids is a single phase if the critical point of the mixture is exceeded.

Supercritical solvents can be expected to replace halogenated solvents that are harmful to groundwater and the ozone layer.

Waste water

Although diluted waste water is often difficult to purify by chemical means, it can be treated using biodegradation. For example, waste water from a factory making nylon 6 (see Equation 8.5) was treated in a sludge blanket reactor[b] for an extended period in order to remove unreacted monomer. After some time a microorganism (a variety of *Pseudomonas aeruginosa*) was isolated from the sludge. This organism was able to remove caprolactam in a short time with simultaneous reduction in w_{COD}.[187]

Air pollution

A chemical plant making carbonyl chloride, $COCl_2$, used as a feedstock for the production of Kevlar, was emitting tetrachloromethane, CCl_4.[c] $COCl_2$ is made from carbon monoxide and chlorine, and the CCl_4 resulted from the reaction of chlorine with the carbon catalyst. A new, oxidatively stable catalyst was invented that reduced the unwanted emission by 95%.

[a] CO_2 $q_c = 31.0\,°C$ $p_c = 73.8\ bar$
    H_2O $q_c = 373.99\,°C$ $p_c = 220.64\ bar$
    CHF_3 $q_c = 25.6\,°C$ $p_c = 48.4\ bar$
[b] Water purification and recycling is discussed in Section 9.5.
[c] DuPont's Chambers Works plant, NJ, USA, 1997.

Environmental impact of selected chemicals

Industrial chemicals may be broadly classified as commodity chemicals (see Table 8.1) and specialty chemicals (Table 9.1). The two classes are not exact, but they may be characterized as *bulk chemicals* (produced in large volume) and *fine chemicals* (produced in low volume), respectively.

The most important specialty chemicals[97] are listed in Table 9.1. The list is in decreasing order of mass fraction per year, around 2009. An assessment of the actual values is difficult because (1) the delimitation of each group depends on the source of the data, and (2) whereas most data are well known for the United States and the European Union, they are less precise for developing countries. A global production rate of 1 Tg/a corresponds to an average consumption rate of 150 g/a per human being.

The environmental problems that have driven the development of environmental chemistry are associated with the production of goods that we would not want to do without. Industrial production has been built up sequentially: each advance builds on what has come before. One may therefore expect that environmental problems, documented with help from analytical chemistry and spectroscopy, can be solved using the same empirical process. Many compounds either have been phased out or are the subject of regulation. One example is the chlorofluorocarbons described in Chapter 4; another is the pesticides listed in Table 9.2. By the end of the 20th century, useful alternatives had been found for tetraethyl lead, an antiknock additive in gasoline, and its use has been banned (an example of the toxicity of lead can be found in Section 1.3c). The field of environmental chemistry is changing as its focus moves from one pollutant to the next.

9.1 Pesticides

Pesticides are substances that are used to kill pests, such as bacteria, weeds, fungi, insects, mites, nematodes, rodents, and viruses. The common nomenclature is simple: the suffix *-cide* is added to the pest, giving rise to the generic name *pesticide* and the specific names *bactericide, herbicide, fungicide, insecticide, miticide, nematicide, rodenticide,* and *viricide,* respectively.[a] This chapter presents the chemistry of some insecticides, herbicides, and fungicides.

---

[a] International Standard ISO 1750, *Pesticides and other agrochemicals – Common names,* 1981.
International Standard ISO 257: *Pesticides and other agrochemicals – Principles for the selection of common names,* 3rd Ed., June 15, 2004.
International Standard ISO 765: *Pesticides considered not to require common names,* 1976.

Table 9.1 World consumption of major specialty chemicals, 2009	
Chemical	Mass fraction, $100\ w_m$
Pesticides	11.9
Specialty polymers	8.3
Electronic chemicals	6.4
Surfactants	5.9
Construction chemicals	5.8
Industrial and institutional cleaners	5.3
Flavors and fragrances	5.0
Printing inks	4.6
Specialty coatings	4.4
Food additives	3.9
Water-soluble polymers	3.7
Specialty paper chemicals	3.6
Oil field chemicals	3.0
Adhesives and sealants	3.0
Plastic additives	2.6
Water management chemicals	2.6
Cosmetic chemicals	2.4
Catalysts	2.3
Textile chemicals	1.9
Other	12.9

The data are extracted from ref. 246. The production rate is around 20 Tg/a (estimated from the world consumption of pesticides[251] in 2007).

Table 9.2 Insecticides that are forbidden or restricted in the United States[a]			
Aramite	1955	Chlordecone	1978
Aminotriazole	1959	Hexachlorocyclohexane	1978
Dialkyl mercury	1970	2,4,5-T, Silvex, endrin	1979
DDT, DDD	1971	Pyriminil, nitrofene	1980
Strychnine, sodium fluoroacetate	1972	Aldicarb	1981
Aldrin, dieldrin	1974	Toxaphene	1983
Heptachlor	1976	Camphechlor	1986
Strobane, mirex, leptophos	1976	Dicofol, dinoseb	1986
Chloranil, chlordimeform	1977	Alachlor, chlordane	1987
		Parathion	1991

Data compiled from the Internet and the EPA (U.S. Environmental Protection Agency); see also ref. 254.

Humanity cannot accept large losses of crops due to rats or mold; on the other hand, experience shows that the use of pesticides is associated with risks to human health and the environment. For more than 40 years, use of pesticides has been reduced by combining them with alternative pest controls, including mechanical, biological, and

---

[a] RUP = Restricted Use Pesticide. Note that there are many examples of chemicals that are forbidden in the United States or the European Union that are produced and used in developing countries.

genetic management. Such integrated pest management (IPM) is now commonplace in modern agriculture. Improved analytical methods mean that it is possible to follow the environmental fate of pesticides in detail. Scientific progress within chemistry, biochemistry, biology, and environmental chemistry is necessary to develop the next generation of pesticides, as will be discussed later.

a. Insecticides

The first insecticides were simply the most poisonous inorganic chemicals that could be found: for example lead arsenate, which was applied to plants in powder form. Cryolite and fluorosilicates were also used. Moth larvae were fought using 1,4-dichlorobenzene, or simply by washing the moth-eaten textiles with ligroin. Later, chlorine-containing compounds were found to be promising, and after some initial work, a method was found for chemically fastening them to textiles. However, the discovery of DDT marked a milestone in the development of insecticides.

DDT

Although DDT[a] (1,1,1-trichloro-2,2-bis(4-chlorophenyl)ethane; see Equation 9.2) was discovered in 1874, its potency as an insecticide was not discovered until 1940, at which time large-scale production was begun in the Geigy factories in Switzerland. DDT was used against lice, preventing typhus fever, and it was a true balm for soldiers and prisoners of war in World War II. Later it was mainly used against mosquitos, flies, and insects in barns with livestock.

DDT is synthesized by condensing chloral (2,2,2-trichloroethanal) with chlorobenzene. Chlorine is added to a mixture of ethanol and water, yielding the geminal diol:

$$C_2H_5OH + 4\,Cl_2 + H_2O \rightarrow CCl_3 \cdot CH(OH)_2 + 5\,HCl(aq) \quad (9.1)$$

Distillation with sulfuric acid removes water and converts the gem-diol into chloral, which is condensed in situ with chlorobenzene to produce a mixture, whose main component is p,p'-DDT:

p, p'-DDT

The mixture typically contains 15 to 20 % p,p'-DDT and up to 12 other species. DDT was banned in the United States in 1971 because it is a lipophilic species that accumulates in living organisms (bioaccumulation) and becomes concentrated in the

[a] The acronym comes from an erroneous name of the compound $(p\text{-}ClC_6H_4)_2CHCCl_3$: **d**ichloro-**d**iphenyl-**t**richloroethane.

upper levels of the food chain (biomagnification). In bioaccumulation, the rate of uptake of a given compound into the organism greatly exceeds the rate of excretion, and in some cases this leads to biomagnification at higher trophic levels.

Diels-Alder products

Eventually, as DDT-resistant insects evolved, many other chlorinated compounds were investigated. The most well-known is aldrin, which is synthesized by a Diels-Alder reaction from hexachlorocyclopentadiene and bicyclo[2.2.1]hepta-2,5-diene:

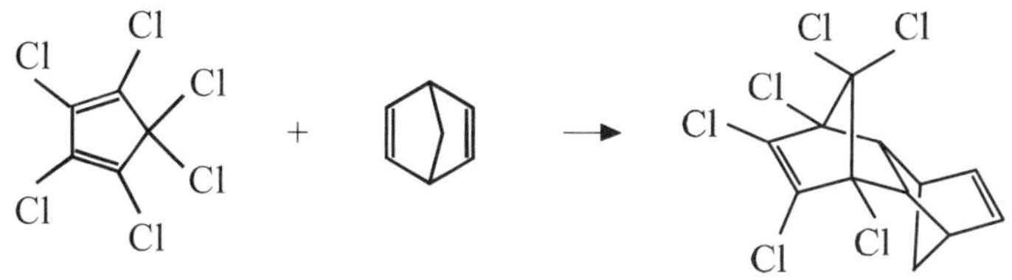

$$(9.3)$$

aldrin

Aldrin[a] is rather effective against grasshoppers, which periodically swarm, devastating agricultural production in entire regions.[b,c] Formally, aldrin is a naphthalene derivative, and a number of important insecticides have been made by addition to its nonchlorinated double bond; for example, dieldrin is produced through epoxidation:

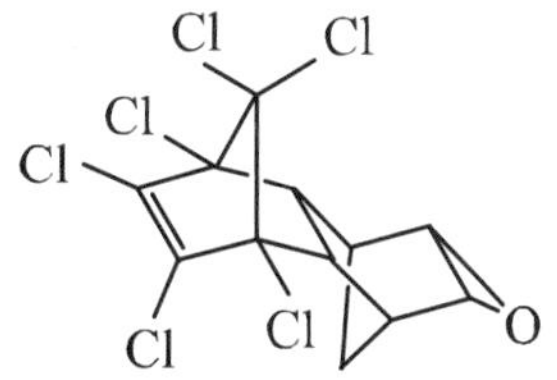

$$(9.4)$$

dieldrin

Aldrin and dieldrin bioaccumulate and cause cancer in mammals. Although these compounds were banned in 1974 and have been phased out, traces are still found in the environment.

The cyclical sulfite endosulfane is still in common use:[d]

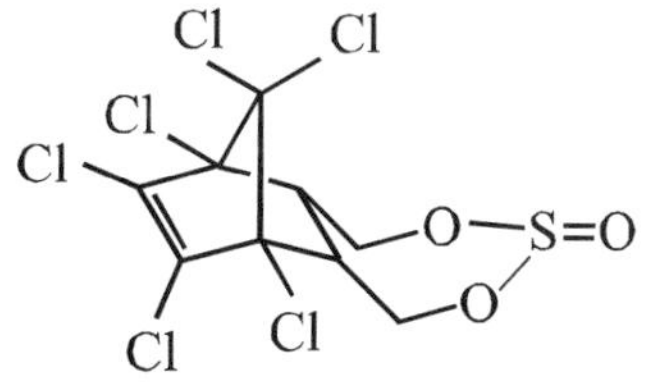

$$(9.5)$$

endosulfane

[a] 1,2,3,4,10,10-Hexachloro-1,4,4a,5,8,8a-hexahydro-1,4:5,8-dimethanonaphthalene.
[b] The name *aldrin* originates from the name of Alder, *dieldrin* from Diels-Alder.
[c] See the 8th Egyptian plague described in Exodus 10:4–6.
[d] endo = internal ring; sulfane = sulfonate (medical nomenclature), not a derivative of hydrogen sulfide = sulfane.

Endosulfane can easily hydrolyze to give SO_2 and a diol that is significantly less toxic. However, the diol is still hazardous, and workers in corn fields should take precautions to avoid exposure via the skin. Its effects include permanent memory loss.

In the Diels-Alder reaction, a double bond adds 1,4 to a conjugated diene so that the product is a stable six-membered ring, the reaction being a cycloaddition. This type of synthesis is easy to perform, rapid, cheap, and very versatile (see Equation 9.3, where the double bond belongs to the bicycloheptadiene). As the original Diels-Alder products were banned, new species were tested, and in particular, effective optically active products were invented, as discussed later in Section 9.2.

Thiophosphates

A mixture of tetraethyldiphosphate and oligophosphate esters was among the first nonchlorine insecticides to be used. However, because of its toxicity to mammals, it was soon replaced by the esters of mono- and dithiophosphoric acid. The most widely known of these is parathion, a tri-*O*-ester of thiophosphoric acid:

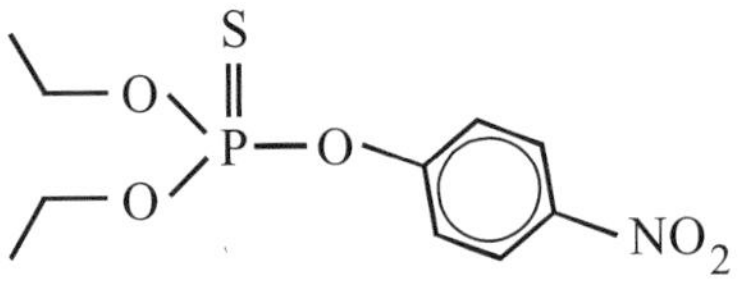

$$\tag{9.6}$$

parathion

Starting in 1991, parathion has been phased out because of its toxicity to large mammals. However, malathion, a malonic ester derivative of *O,O*-dimethyldithiophosphoric acid, is in general use as an insecticide:

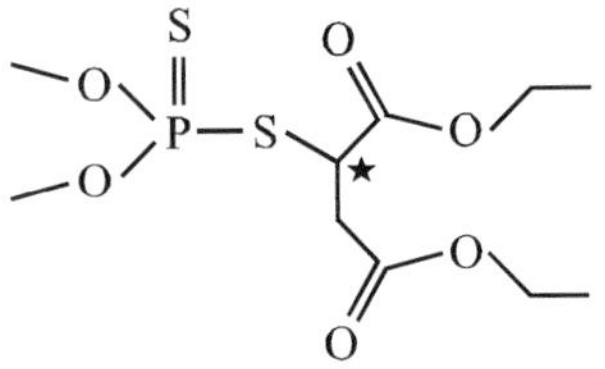

$$\tag{9.7}$$

Malathion;[a] the asymmetric center is marked with an asterisk

Malathion is an effective pest control agent with low toxicity toward humans, and it has been widely used through aerial dispersion to control mosquitos and the West Nile virus. Thus, malathion was sprayed over Pasadena, California, into the 1990s to fight the Mediterranean fruit fly, and over Winnipeg, Manitoba, in 2005 to fight mosquitos spreading the West Nile virus. A disadvantage is that human metabolism

---

[a] Dithiophosphoric acid-*S*-[1,4-diethoxy-1,4-dioxobutane-2-yl]-*O,O'*-dimethyl ester.

converts malathion to malaoxon, whereby its toxicity increases by a factor of more than 50.

$$malaoxon \qquad\qquad isomalathion \qquad (9.8)$$

At present, isomalathion arises as an unwanted by-product in malathion production, and the mixture of isomers is significantly more toxic to humans than pure malathion. The two asymmetric centers give rise to four isomers, which have been chromatographically resolved by enantioselective HPLC; the toxicity of the individual isomers has not been examined.

Green chemistry

As knowledge of the chemistry of natural products has increased, it has become possible to control the population of damaging insects without spreading more or less toxic xenobiotic compounds in nature. For example, several beetles, such as the ambrosia beetle, *Trypodendron lineatum*, and the European spruce bark beetle, *Ips typographus*,[248b] are serious pests in coniferous forests in Europe and North America. From 1975 to 1985, about 5 hm^3 of timber were destroyed in Scandinavia as the bark beetles dug holes beneath the trees' bark. Male beetles dig out 4 to 6 mating chambers per dm^2 and attract females by releasing a mixture of three sex pheromones. Pheromones are used as signal compounds by insects and vertebrates. After mating, two other pheromones are sent out that repel additional females. Chemists at the University of Oslo succeeded in synthesizing a crucial pheromone,[208] and it is now produced commercially. Pheromone traps are placed in key areas to remove females before they can mate. A few survive, of course, and the overall result is that the population is held at a sustainable, low level without being exterminated, and without unwanted secondary effects to the ecosystem.

b. Herbicides[132]

Herbicides are divided into *generations*, a loosely defined concept meaning "something better and different in action."

The first generation comprises the so-called universal herbicides that kill all plant life. Many inorganic compounds have been used in this way, the most widespread being sodium chloride, sulfuric acid, and iron(II) sulfate.

Sym-triazines

The next generation of herbicide is made up of derivatives of 1,3,5-triazine: Simazine (1956) is 6-chloro-N-N'-diethyl-1,3,5-triazine-2,4-diamine. Atrazine (1958) is like simazine, but one ethyl group has been changed into isopropyl:

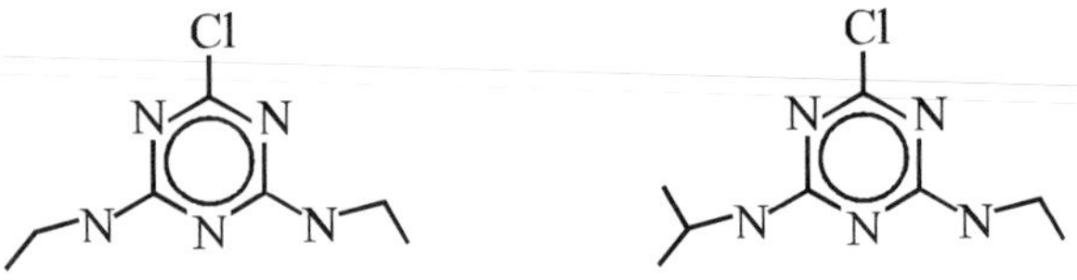

simazine		atrazine

(9.9)

The triazines are universal herbicides that work by inhibiting photosynthesis. Remarkably, atrazine does not have an effect on C_4 plants such as corn and sugarcane,[a] and therefore it is widely used on these crops in the United States. In the European Union, these compounds have been almost completely phased out (since 1996), in part because they have been found in groundwater following excessive use, but also because these crops are not grown as widely in Europe.

Atrazine is retained by humus (by both van der Waals and Coulomb forces) and is mineralized over the course of about a year. Bacterial degradation of atrazine produces NH_4Cl, NH_4HCO_3, and H_2O. The individual intermediates have been identified, and the enzymes responsible for the breakdown have been isolated. Already in the first step, chlorine is broken off and atrazine is transformed into a source of fixed nitrogen that can be used by plants.

Chlorsulfurone, a benzenesulfonamide formed from urea and a 1,3,5-triazine, is an example of the third generation of herbicides.

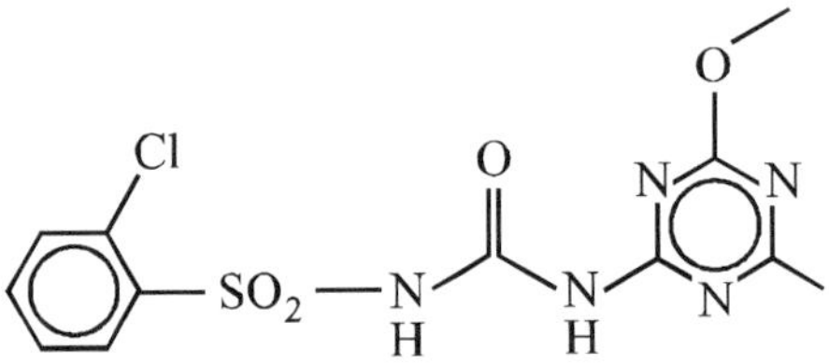

(9.10)

chlorsulfurone

This compound acts specifically on dicots and is very efficient; in practice 4 g ha^{-1} is sufficient.[b]

Phenoxyacetic acids

The second generation of herbicides includes derivatives of phenoxy acetic acid. The most important are 2,4-D (2,4-dichlorophenoxyacetic acid), 2,4,5-T

---

[a] See the discussion in Section 3.4b.
[b] See footnote a page 136.

(2,4,5-trichlorophenoxyacetic acid), and MCPA (4-chloro-2-methylphenoxyacetic acid):

$$\text{2,4-D} \qquad\qquad \text{2,4,5-T} \qquad\qquad \text{MCPA} \tag{9.11}$$

These herbicides act as growth hormones in plants, causing the plants to grow so quickly that they die. They are most effective on dicots and can therefore be used to control weeds in fields of monocots, for example, cornfields and lawns. The herbicide *Agent Orange* dispersed from aircraft by the United States during the Viet Nam War was a 1:1 mixture of 2,4-D and 2,4,5-T, mixed with diesel fuel. The purpose was to defoliate trees and deny the enemy cover, and in that respect it was a success: forests and farmland were destroyed. However, in the synthesis of 2,4,5-T, substantial amounts of the dioxin TCDD (see Equation 9.23), is generated as a by-product, and this is very toxic to mammals; see Section 9.2.

The three herbicides of Equation 9.11 were eventually replaced by the optically active propionic acid derivatives, (*R*)-2-(2,4-dichlorophenoxy)propanoic acid, called dichlorprop-P, and (*R*)-2-(4-chloro-2-methylphenoxy)propanoic acid, called mechlorprop-P:

$$\text{dichlorprop-p} \qquad\qquad\qquad \text{mechlorprop-p} \tag{9.12}$$

Originally the two herbicides were applied as racemic mixtures, but recent advances in asymmetric syntheses allow syntheses of the enantiomeric compounds on a commercial scale.

Glyphosate

The herbicide glyphosate, *N*-(phosphonomethyl)glycine (see Equation 9.13), was introduced as a universal herbicide in 1971. It is sold as a dilute aqueous solution under the trade name Roundup[a] and is one of the herbicides that is least toxic to bacteria, insects, fish, birds, and mammals.

$$
\text{HO} - \underset{\underset{\text{O}^-}{|}}{\overset{\overset{\text{O}}{\|}}{\text{P}}} \diagup \overset{+}{\text{NH}}_2 \diagup \overset{\text{OH}}{\diagdown} \text{O}
\tag{9.13}
$$

As can be seen from the structure, glyphosate resembles an amino acid in that it is an amphoion (also called a zwitterion) and is therefore crystalline with a high melting point. The protonated form is a tetravalent acid (pK_a values of approximately 0.3, 2.2, 5.4, and 10.1) with an isoelectric point at a pH of 1.1. It is good at forming complexes and is easily adsorbed by components of soil such as mineral surfaces and humus. Its primary degradation products are acetic acid and aminomethylphosphonic acid, called AMPA, which is used as a marker for glyphosate in the environment.

Glyphosate functions by inhibiting the enzyme EPSP-synthase and thus preventing the plant from producing essential aromatic amino acids, Tyr, Phe, and Trp,[b] as well as the vitamins E, K, and folic acid, and also lignin. Glyphosate is clearly catastrophic for wide segments of a plant's biochemistry. However, other organisms that do not use this enzyme are unaffected by glyphosate. A great deal of work has gone into the development of seeds that tolerate glyphosate, and soybeans and rapeseed have been developed with this property.

To demonstrate the extent of the literature concerning a key herbicide such as glyphosate, consider a literature survey covering the peak period of research on this species (1981 to 1997), which found 720 scientific papers and technical reports in addition to a large monograph.[23]

c. Fungicides

Fungicides are an integral part of modern food production.[132] As examples, we discuss the protection of seed and fruits.

The most widely known prophylactic agent against parasitic fungi on plants is probably "Bordeaux wash," made from 1 % copper sulfate and 3 % lime in water. The result is a basic suspension of copper hydroxide that was previously used throughout southern Europe to protect grapevines, with peak use from 1880 to 1950. Its effect is due to copper(II) ions (also toxic to humans) that kill fungal spores and prevent their reproduction.

Alkyl and aryl mercury compounds of the form R-Hg-Cl were formerly used to protect seeds, but this has been discontinued to prevent dispersion of mercury, and mercury poisoning should this seed accidentally be used for food.

---

[a] The Monsanto Company.
[b] See Appendix A1.9.

During the latter half of the 20th century, much effort was invested in producing biologically active compounds that interfere with specific enzymatic processes. An early example (1966) is oxycarboxin, which prevents respiration in fungi:

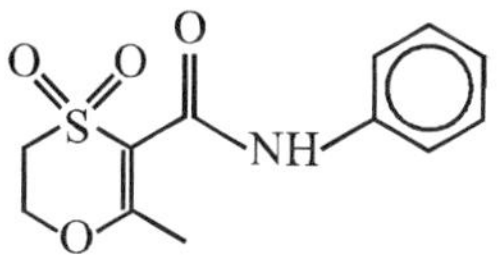

(9.14)

oxycarboxin =
5,6-dihydro-2-methyl-4,4-dioxo-N-phenyl-1,4-oxathiin-3-carboxamide

Dithiocarbamates

A series of salts of N,N-dimethyldithiocarbamic acid,

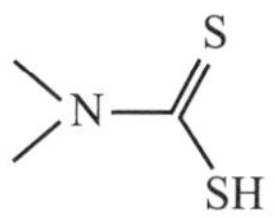

(9.15)

have been used to fight fungi (mildew) on fruit trees. Examples include the zinc salt (called ziram) and the iron(III) salt (called ferbam), as well as the derivative tetramethylthiuramdisulfide[a] (thiram):

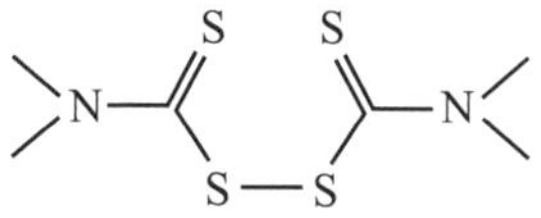

(9.16)

and salts of ethylene-N,N'-bis(dithiocarbamic acid),

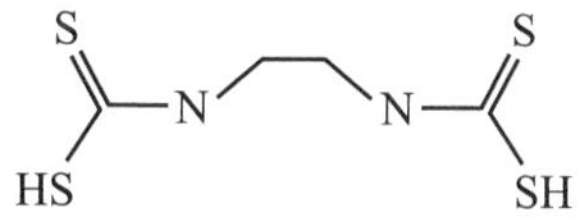

(9.17)

notably the disodium salt (nabam), the zinc salt (zineb), and the manganese(II) salt (maneb). This class of biocides has now been banned, in the United States as well as in the European Union (2009).

Strobilurins[140]

Strobilurin fungicides occur naturally and were originally isolated from mushrooms; the name is derived from the mushroom genus *Strobilurus*. Strobilurin fungicides have been in the trade since 1996.

The first strobilurin to be isolated was strobilurin A (Equation 9.18) with $R_1=R_2=$H, and all natural strobilurins retain this moiety.[b] Modification of the sites R_1 and R_2 leads to a series of strobilurins used against different crop diseases.

[a] Tetraethylthiuramdisulfide is the medicine Antabuse, used to treat alcoholism.
[b] Strobilurin A ($R_1 = R_2 =$ H), $C_{16}H_{18}O_3$, (2E,3E,5E)-2-(methoxymethylene)-3-methyl-6-phenyl-hexa-3,5-dienoic acid methylester.

Structurally, a common feature is β-methoxyacrylic acid, shown in Equation 9.18,[a] and in some synthetic strobilurin fungicides, QSAR-equivalent groups have been used with success.[b]

Strobilurins [footnote b, p.319] β-methoxyacrylic acid (R = H)

$$(9.18)$$

These compounds work by preventing electron transfer reactions in the fungal mitochondria, disrupting energy production and eventually killing the fungus. This mode of action is novel, as is the discovery that fungicides are produced by fungi.

The primary use of strobilurins is in agriculture, but recent applications also include control of mold and mildew in pulp (paper industry), cooling towers, and heat exchangers. Strobilurins are nontoxic to plants and mammals but weakly harmful to fish and algae; however, because of their rapid degradation in soil, this class of fungicides is generally considered nonharmful.

d. Enantiomeric xenobiotics

In the beginning of the 19th century it was discovered that quartz crystals rotate the plane of linearly polarized light (Biot, 1812), and in the following years chemical species such as sugars and terpenes were found to have the same property, whether in the solid, liquid, or gas phase. This phenomenon, called *optical activity*, is due to the existence of enantiomerism, which on the molecular level was shown to be an essential element of life.

Enantiomerism is basic to enzymatic processes, which almost invariably utilize only one enantiomer of a racemic mixture. The discovery of enantiomers, which can be detected using chromatographic techniques, has allowed the production of more potent herbicides that minimize undesirable side effects.

In this section we first outline the hierarchy of isomers, then discuss some concepts related to optical activity, and finally address the subject of enantiomeric xenobiotics.

Enantiomers

An empirical formula is a statement of the atomic composition of a chemical species. An isomer is one of several species that have the same empirical formula but different constitutions, that is, different interatomic distances and bond angles. In this

---

[a] (*E*)-3-Methoxy-prop-2-enoic acid.
[b] QSAR = quantitative structure-activity relationship.

discussion, the distinction between a chemical species and a molecular entity is temporarily suspended, and all concepts apply to both categories. Therefore, one may say that different isomers are nonsuperposable molecular entities of the same atomic composition. Isomers are divided into two groups: enantiomers and stereoisomers. An enantiomer is one of a pair of isomers that are mirror images of each other.[a] Stereoisomers are non-enantiomers. In the following discussion, we shall use the generic labels L and D for the two enantiomers of a pair.

The term *chiral species* refers to a subgroup of enantiomers for which the terms *left* and *right* have a defined significance; this includes molecular helices and certain other species, such as left- and right-handed crystals of quartz.

Additional concepts: (1) A mixture of equal amounts of the two enantiomers of a pair is called a racemic mixture. (2) Diastereomers are stereoisomers that contain enantiomeric parts. *Example*: A racemic pair of species B_L and B_D reacts with the enantiomer C_L to form the diastereomers $B_L C_L$ and $B_D C_L$.

Experimentally, enantiomers are detected by means of the way they interact with polarized electromagnetic radiation.

Polarized light

Photons with the angular frequency ω carry the energy, $E = \hbar\omega$, the linear momentum, $p = \hbar\omega/c_0$, and the angular momentum, $j = \pm\hbar$.[b,33] Energy and linear momentum depend on the frequency, and the spectrum of energies is discussed in Chapter 10. In contrast, the angular momentum of a photon can take just two values: photons with $j = +\hbar$ are left circularly polarized, and photons with $j = -\hbar$ are right circularly polarized. A light beam whose photons are all in a state with $j = +\hbar$ is called left circularly polarized light, and similarly, a right circularly polarized light beam has all its photons in a state with $j = -\hbar$. The spatial trace of the electric field vector of a beam of photons with $j = +\hbar$ is a left-handed helix. If the proportion of the two kinds of photons is the same, then a light beam is said to be linearly polarized. Natural light is randomly polarized, and an optical device whose input is natural light and whose output is polarized light is called a polarizer.[c]

Interaction between light and matter

An absorption process consists of the transfer of energy, linear momentum, and angular momentum from the beam of photons to the target, a chemical species B. At the resonant frequency ω_a, the *absorbance A* is given by the Lambert-Beer law,

$$A(\omega_a) = \varepsilon_B(\omega_a)\, c_B\, l \qquad (9.19)$$

---

[a] Gk. $\varepsilon\nu\alpha\nu\tau\iota\sigma\varsigma \,\hat{=}\,$ = opposite; $\mu\varepsilon\rho\sigma\varsigma \,\hat{=}\,$ part.

[b] $\omega = 2\pi v$; in vacuum $v = c_0/\lambda$. Values of the fundamental constants are given in Appendix A1.8.

[c] Polarizers were used by the Vikings at sea. Light scattering through the atmosphere becomes linearly polarized, and the direction of polarization (the direction to the Sun) can be detected by viewing the sky through a crystal of the mineral cordierite (dichroite); this even works under an overcast sky. The principle can be verified using polarized sunglasses. A recent 3D film technique (used in the motion picture *Avatar*, 2009) uses oppositely circularly polarized light for each eye.

where ε is the *absorption coefficient*, c the molar concentration, and l the length of the light path. The absorption coefficient is a measure of the linear momentum transferred from the light wave to the target molecule.

One enantiomer of a pair will interact differently with the two kinds of circularly polarized light, and the difference between the absorbance of left (A_+) and right (A_-) circularly polarized light, $\Delta A = A_+ - A_-$, is also given by a Lambert-Beer law,

$$\Delta A(\omega_a) = \Delta \varepsilon_B(\omega_a) c_B l \tag{9.20}$$

where $\Delta \varepsilon$ is called the *circular dichroism*. Circular dichroism is a measure of the net angular momentum imparted to the target. A chemical species that at some frequency of the impingent light shows circular dichroism is called an *optically active* species.[a]

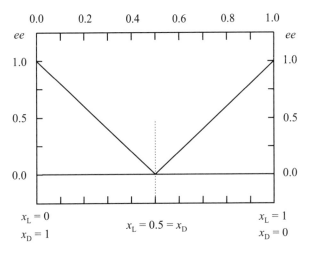

Figure 9.1   *Enantiomeric excess, ee*
Enantiomeric (or enantiomer) excess (Equation 9.21) as a function of the mole fraction of the enantiomers.

Enantiomeric excess, ee

The circular dichroism of two enantiomers has opposite sign for all wavelengths. It turns out that for all applications for environmental studies relation, Equation 9.20, is linear in concentration. Therefore, one defines the *enantiomeric excess, ee*, as the absolute value of the difference of the mole fractions of the L and the D isomers in a mixture:

$$ee = |x_L - x_D| \tag{9.21}$$

where x denotes mole fraction, and $x_L + x_D = 1$. A racemic mixture of a pair of enantiomers has $ee = 0$. Figure 9.1 illustrates these concepts. The observed ΔA_{obs} of a mixture of enantiomers is a direct measure of the enantiomeric excess: $ee = \Delta A_{obs}/\Delta A_{max}$, where ΔA_{max} is the value of ΔA for the pure enantiomer, obtained under the equivalent experimental conditions. The reason this works is that the circular

---

[a] Sometimes optical activity is detected using refraction. Outside regions of absorption, the different velocities of the two circularly polarized waves may be measured as the rotation of the plane of linearly polarized light; this is called *optical rotatory power*.[126]

dichroism of enantiomers cancels such that Equation 9.21 simply shows what is left to be detected.

Enantiomeric xenobiotics

Advances in enantioselective analytical methods have given rise to a change of the focus of environmental studies: studies that do not include the different effects of diastereomers and enantiomers are simply not complete and can be misleading. The parallel to the pharmaceutical industry is obvious. Here several cases are known where the enantiomers of medications have different effects; one is curative, and the other without effect[a] or even harmful.[b] Of course, production has been changed to produce only the curative enantiomer. Analogous requirements for the production of pesticides have been discussed (see Equation 9.12), but they are expected to be more widespread in the future.

A list of more than seventy enantiomeric xenobiotics was published a decade ago.[43] The minor part includes the nonchlorine compounds shown earlier, and the greater part consists of the organochlorine compounds to be discussed in the next section.

9.2 Organochlorine compounds

Most of the chemical species called *persistent organic pollutants* are organochlorine compounds. They were produced over the past 70 years and were used mainly as insecticides (against lice, termites, or mosquitos). They have also found use as additives in building materials, as electrical insulators, and as intermediates in the production of other pesticides. Obviously they are released into the environment, but the problem is that some bioaccumulate before they decompose, and often they are hazardous to higher animals and humans. International conventions have restricted the use of some organochlorine compounds and prohibited the production of others (see Table 9.2 and ref. 254), but new ones are still being developed. The present section presents selected examples illustrating the chemistry and the general principles of characterization, synthesis, and analytical detection. The examples include compounds (HCHs, PCCHs, bromocyclene) whose production has been terminated as of 2010.[254]

a. Dioxins and (polychloro)biphenyls

Dioxin is a common designation for chloro derivatives of dibenzo[1,4]dioxin:

$$(9.22)$$

Dibenzo[1,4]dioxin, oxanthrene

---

[a] Ibuprofen, (*S*)-2-[4-(2-methylpropyl)phenyl]propanoic acid; the enantiomer, *R*, has no curative effect.
[b] Thalidomide, (*R*)-2-(2,6-dioxopiperidine-3-yl)-1*H*-isoindole-1,3(2*H*)-dione; the enantiomer is harmful.

One may substitute from 1 to 8 hydrogen atoms with chlorine atoms leading to a total of 75 (polychloro)dibenzo[1,4]dioxins, collectively designated by the acronym PCDD. At one time it was thought that all 75 species might be formed simultaneously by the incineration of chlorine-containing organic waste. This is the reason for calling a species of this set a *congener*, meaning a member of the set of all chlorine derivatives that can be generated from the same carbon skeleton.[a]

It turns out that 17 of the congeners are quite toxic to humans, with the congener TCDD being the most potent,

$$(9.23)$$

2,3,7,8-Tetrachlorodibenzo[1,4]dioxin, TCDD. $\theta_{fus} = 305\,^\circ C$; $\theta_{vap} = 446\,^\circ C$; the octanol-water partition ratio $K_c^{ow} = 10^{6.8}$, sec Equation 2.210

Toxic effects for this kind of poison are quantified according to their damage to test animals. TCDD at a dose of $1\ \mu g\ kg^{-1}\ d^{-1}$ causes internal hemorrhaging and kidney failure in male guinea pigs. The results of such measurements may be converted into toxic equivalents; for example, one finds that 1,2,3,7,8-pentachlorodibenzodioxin possesses half the toxicity of TCDD. PCDDs are generally formed in the gas phase at temperatures below $\approx 850\ ^\circ C$, not only in incineration plants but also by forest fires and in general by heating organic material containing HCl or NaCl. However, they may also be formed accidentally at much lower temperatures, and in the liquid phase. The synthesis of 2,4,5-T (Equation 9.11) is an example: it is made via hydrolysis of 1,2,4,5-tetrachlorobenzene to 2,4,5-trichlorophenolate in a basic solution of glycol and xylene at $\approx 150\ ^\circ C$:

$$(9.24)$$

1,2,4,5-Tetrachlorobenzene   2,4,5-Trichlorophenolate

followed by condensation with sodium monochloroacetate. Here 2,4,5-trichlorophenolate forms TCDD (see Equation 9.23) by an unavoidable side reaction.

Dioxins are hydrophobic species with high values of the octanol-water partition ratio (see Equation 9.23). Accordingly, when present in industrial aqueous effluent, dioxins may be absorbed in suspended solid particles. Therefore, they can be removed with traditional technology such as flocculation and filtration after addition of an appropriate flocculant.[b]

The congeners of dibenzofuran and dibenzothiophene are important classes of compounds with properties close to the dioxins:

[a] Lat. *con* $\hat{=}$ *cum* = with; Gk. $\gamma\varepsilon\nu\nu\alpha\omega$ $\hat{=}$ form.
[b] The use of flocculation is discussed in Section 9.5c.

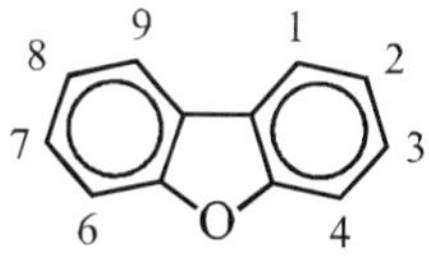

Dibenzofuran 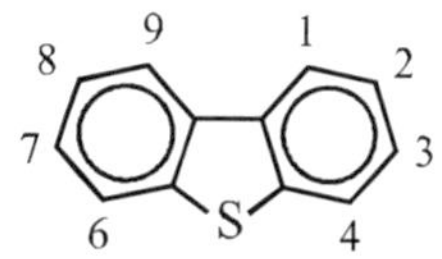   Dibenzothiophene

(9.25)

where the heteroatoms have number 5. Each class consists of 135 congeners, (poly-chloro)dibenzofurans, PCDF, and (polychloro)dibenzothiophenes, PCDT.

Based on the biphenyl skeleton,

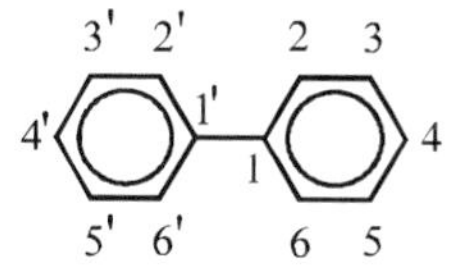

Biphenyl

(9.26)

a total of 209 congeners, (polychloro)biphenyls, PCBs, may be formed. Of these, 130 are anthropogenic. As an example, mixtures of PCBs containing the liquid 3,3',4,4'-tetrachlorobiphenyl,

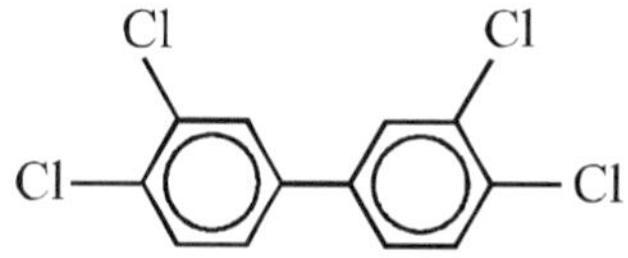

3,3',4,4'-Tetrachlorobiphenyl

(9.27)

have been used to insulate high-voltage transformers and capacitors.[a] During incineration, the following PCDF may be formed:

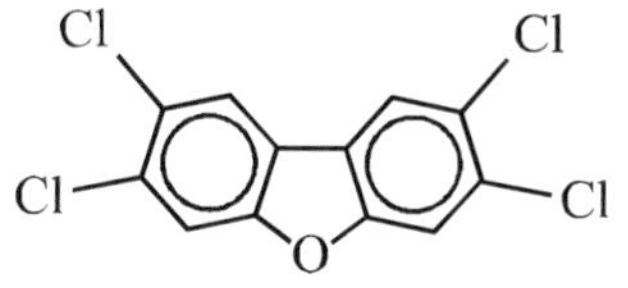

2,3,7,8-Tetrachlorodibenzofuran

(9.28)

in a typical gas-phase reaction: hydroxyl radicals attack the 6,6' positions of the PCB, which forms a radical that reacts with oxygen to give the PCDF. In order to prevent such reactions and instead burn it to CO_2, H_2O, and HCl, incineration must take place at $1100\,°C$ for at least 2 s.[198] Note: Burning clean wood produces dioxins and (polychloro)dibenzofurans because they are formed in all processes (at elevated temperature) where organic molecules and bound chlorine or chloride are present. Wood contains up to 2 ‰ chlorine, and dioxins have been detected everywhere on the surface of the Earth. Their origin is natural as well as anthropogenic.

The PCBs include interesting examples of hindered rotation (rather than an asymmetric carbon center) as the cause of enantiomerism. The two aryls of each of the

---

[a] PCBs have been phased out but were widely used in buildings as plasticizers in caulks, sealants, and paints.

congeners depicted in Equation 9.29 cannot rotate freely because of the chlorines at 2,2' and 6,6'. The mirror image of pentachlorobiphenyl is an identical molecule but the hexachlorobiphenyl shown, is one of a pair of enantiomers.

$$(9.29)$$

2,2',3,6,6'-Pentachlorobiphenyl   2,2',3,3',6,6'-Hexachlorobiphenyl

b. Hexachlorocyclohexane, HCH, and pentachlorocyclohexene, PCCH

HCHs and PCCHs are powerful insecticides that have been studied in great detail in part because of the observation of enantioselective microbial transformations.

Nine isomers of 1,2,3,4,5,6-hexachlorocyclohexane exist, but only four have been in general use as insecticides. They are called α-HCH, β-HCH, and γ-HCH, and α-HCH comprises an enantiomeric pair denoted $(-)_D$-α-HCH and $(+)_D$-α-HCH.[a] The absolute configuration of the α-HCH isomers has been determined and is shown in Equation 9.30.[196,b] Note that in these drawings, a line signifies a hydrogen atom, not a methyl group.

$$(9.30)$$

$(-)_D$-α-HCH; $(1R,2S,3S,4R,5R,6R)$ $(+)_D$-α-HCH; $(1S,2R,3R,4S,5S,6S)$

The structures of the β-HCH and γ-HCH (lindane) are:

$$(9.31)$$

β-HCH; $(1\alpha,2\beta,3\alpha,4\beta,5\alpha,6\beta)$ γ-HCH; $(1\alpha,2\alpha,3\beta,4\alpha,5\alpha,6\beta)$

---

[a] The $(-)_D$-isomer (the $(+)_D$-isomer) has a negative (positive) angle of optical rotation at the sodium D-line, 589 nm.

[b] The R,S nomenclature used here uses an extension of the Cahn, Ingold, Prelog rules, the so-called CIP priority.[194]

where the descriptors α and β denote "on the same side" and "on the opposite side," respectively, of the (ideal) plane of the six carbons.

A racemic mixture of HCH isomers is easily prepared by the addition of chlorine to benzene under conditions where free radicals are present. HCH is a cheap insecticide. Under weakly basic conditions, HCH undergoes *anti*-elimination of HCl, forming 1,3,4,5,6-pentachlorocyclohexene, PCCH. As seen from Equations 9.30 and 9.31, β-HCH has no Cl trans to an H, and it was found that the rate of conversion for this isomer is 7000 times slower than that of the slowest of the other isomers, which vary by a factor of 3.[63] Figure 9.2 shows the PCCHs formed from α-HCHs and γ-HCH.

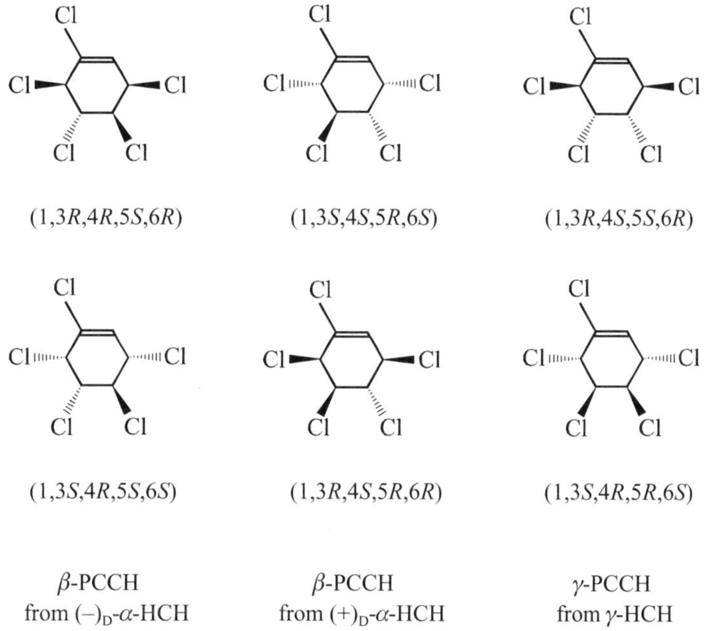

(1,3R,4R,5S,6R)	(1,3S,4S,5R,6S)	(1,3R,4S,5S,6R)

(1,3S,4R,5S,6S)	(1,3R,4S,5R,6R)	(1,3S,4R,5R,6S)

| β-PCCH | β-PCCH | γ-PCCH |
| from $(-)_D$-α-HCH | from $(+)_D$-α-HCH | from γ-HCH |

Figure 9.2   *1,3,4,5,6-Pentachlorocyclohexenes*
Hydrogen atoms have been omitted for clarity.[a] The PCCHs are obtained by *anti*-elimination (an E2-reaction) of hydrogen chloride from the HCHs indicated. The γ-PCCHs are enantiomers, but the two β-PCCHs derived from each of the two α-HCHs are diastereomers. The enantiomers of the two isomers derived from $(-)_D$-α-HCH are the two isomers derived from $(+)_D$-α-HCH.

The enzyme dehydrochlorinase LinA, extracted from a soil bacterium,[b] is able to anti-eliminate only one of the two topologically different HCl-pairs of γ-HCH to give the pure enantiomer (1,3R,4S,5S,6R)-PCCH. This is just one of many examples demonstrating a natural preference for a specific enantiomer.

Similar results are observed in Nature: an enantiomeric excess, *ee* ≈ 0.1, of $(-)_D$-α-HCH was observed at 21 stations in the North Sea and the Baltic Sea, indicating a preferential microbial transformation of $(+)_D$-α-HCH.[43]

---

[a] This kind of drawing is called a *Mills depiction*.
[b] *Sphingomonas paucimobilis* UT26.[211]

Although HCHs have been phased out virtually everywhere, lindane is still produced for medicinal use: a mixture of malathion (Equation 9.7) and lindane (Equation
9.31) is used to treat head and body lice on humans.

c. Bromocyclenes

The great versatility of the Diels-Alder syntheses has lead to a series of chlorine
derivatives analogous to aldrin (Equation 9.3) and dieldrin (Equation 9.4). The insecticide bromocyclene has a low mammalian toxicity and is used to treat domestic
animals.

$$\tag{9.32}$$

(R)-Bromocyclene[a]   (S)-Bromocyclene

Bromocyclene is the adduct of hexachlorocyclopentadiene and 3-bromoprop-1-ene
(allyl bromide) and is synthesized as a racemic mixture. Although its absolute configuration has not been determined, field studies show that $(+)_D$-bromocyclene is taken
up by rainbow trout in German rivers, leaving a surplus of $(-)_D$-bromocyclene in the
aqueous phase.[43]

9.3 Metal compounds

Unlike organic pollutants, metals do not decay, and thus toxic metals pose a unique
problem for remediation. We will give a few examples of pollutants of environmental
significance from groups 12, 14, and 15 of the periodic table. They illustrate that some
metals and their compounds are inherently toxic (Hg), some are virtually harmless
except for a few compounds with metal-carbon bonds (Sn) when found in specific
environments, and some are not human-made pollutants but rather a part of Nature
that a local population cannot easily avoid (As).

Mercury (group 12)

Mercury was first described by Aristotle (350 BC), who wrote that the mineral
cinnabar, HgS, found in Almaden (Spain), was roasted in air to give the liquid metal,

$$HgS(s) + O_2(g) \rightarrow Hg(l) + SO_2(g) \tag{9.33}$$

Work at the mines was done by prisoners of war who did not survive for more than a
few months. Mercury makes alloys with most metals, except iron, but including gold.

---

[a] 5-Bromomethyl-1,2,3,4,7,7-hexachlorobicyclo[2.2.1]hept-2-ene.

The technique by which a thin layer of gold is placed on a surface by evaporating Hg from Au(Hg) is known as fire gilding; also here, the prisoners given this job did not survive for long.

Mercuric nitrate was formerly used to treat animal skins in the production of quality felt for men's hats. Cases of mercury poisoning were widespread, leading to the phrase "mad as a hatter" and the character the Mad Hatter in Lewis Carroll's novel *Alice in Wonderland*. There is no doubt that all authorities should know about the toxicity of mercury and its compounds.

Methylmercury, $H_3C-Hg^+X^-$, where X^- is Cl^-, HO^- or NO_3^-, is a potent biocide that has been used to protect seeds. For half a century, a large industrial facility emitted wastewater containing methylmercury directly into in Minamata Bay, Japan, resulting in a major disposal incident (1965). Thousands died, and thousands more were injured for life. The Minamata incident alerted society to the hazards of mercury pollution.

Tin (group 14)

Tin has been known since ancient times. Around 3000 BC, the Stone Age was succeeded by the Bronze Age. Bronze is an alloy of copper and tin, which at that time was obtained by adding cassiterite, SnO_2, to copper ore before processing (see Section 7.5). The free metal was probably produced in Egypt in 1500 BC (by reduction of the dioxide with carbon), and from 700 BC it was common to wrap mummies in tinfoil.

Tin and its compounds are generally considered to be nontoxic (in contrast to lead; see Section 1.3c). The metal is used in various alloys, and the dioxide is used in ceramic materials as an opacifier, and also for surface treatment of bottles where it increases the strength of the glass. Organotin compounds, such as R_2SnX_2, where R is an alkyl residue and X is laurate or maleate, are used as stabilizers for PVC plastics and drinking bottles, resulting in material with an excellent clarity.

In the early 1960s, it was discovered that "tributyltin" (TBT), Bu_3SnOH, is a very potent agricultural biocide. The advantage is that this type of compound hydrolyzes to give harmless compounds (butanol and tin dioxide) that are nontoxic to mammals. However, when it was used as a marine antifouling agent, TBT was found to disrupt the hormonal systems of gastropods (snails and slugs living in the ocean), leading to imposex, a condition in which individuals develop reproductive organs of the opposite sex. TBT was responsible for severe losses in the European oyster industry, and in 2004 it was banned by the International Maritime Organization.

Arsenic (group 15)

Since ancient times, diarsenic trioxide, As_2O_3, has been prepared from the minerals arsenopyrite, FeAsS, and the two sulfides realgar, As_4S_4, and orpiment, As_2S_3, by heating them in air. The poisonous character of arsenic compounds was known: the sulfides were used as a rodenticide,[a] and the slaves who worked in the quarries only

---

[a] Ru. *Мышьяк* ≙ mouse poison is the Russian name of the element arsenic; compare with Shakespeare's use of the term *ratsbane* (rat death) for diarsenic trioxide.

lived for a short time. The famous "Mithridates" antidote' was said to contain As_2O_3 because a moderate daily intake may prevent assassination by arsenic.[a]

Arsenic occurs naturally in groundwater as a mixture of As^{III} and As^V. The relevant data are (25 °C):

$$\text{arsenic acid, } H_3AsO_4 \qquad pK_1 = 2.26, \qquad pK_2 = 6.76, \qquad pK_3 = 11.29$$
$$\text{arsenous acid, } H_3AsO_3 \quad pK_1 = 9.29 \tag{9.34}$$
$$H_3AsO_4 + 2H^+ + 2e^- = H_3AsO_3 + H_2O \qquad E° = 0.560\,V$$

showing that near pH 7 and at a redox potential near 0 V, the dominant species is $H_2AsO_4^-$, $HAsO_4^{2-}$, or H_3AsO_3, depending on the specific circumstances.[b] The World Health Organization (WHO) recommends a threshold mass fraction of $w_{As} < 10^{-8}$ arsenic in potable water. They estimate that 57×10^6 people are drinking groundwater with an As content above this value. The problem is acute in Bangladesh, where there is a massive epidemic of "black foot disease" caused by arsenic poisoning (2009). The disease leads to cancer and death.

9.4 Detergents

The region close to the boundary between two phases is called an interface. The interface is an inhomogeneous zone with properties significantly different from, but related to, those of the two phases.[c] The structure and chemistry of interfaces is an important research area (1) because important chemical processes take place in this region and (2) because modern experimental techniques have made surface studies possible at the molecular level. Examples include the heterogeneous chemistry of the stratosphere, cloud droplet activation, adsorption processes in soil, and the absolutely central role that cell walls and membranes play in biological systems. Instruments used to study interfaces include updated versions of Langmuir-Blodgett methods known since the 1930s and modern spectroscopic techniques combined with electrochemistry using micro electrodes; see footnote a, page 245. General references include 1, 7, 65, 91, 161.

Surface tension is a property of liquid interfaces. Laplace's equation 2.99, which was discussed for water, is one manifestation of this phenomenon. Generally, for a solution, a solute changes the surface tension of the solvent, and, as we shall see, it may be raised or lowered. We shall use the following result: if the concentration of the solute is increased in the surface region then the surface tension is lowered (see Equation 9.45).

---

[a] Mithridates, 120–60 BC, king of Pontus. The myth has been used by several authors and composers including A. Dumas, D. L. Sayers, and W. A. Mozart.

[b] It may be of help to draw an E-pH diagram; see Section 5.3a.

[c] When this zone is the specific object of a study, it is called an interphase or a surface phase.

A solute that lowers the surface tension of the solvent is called a *surface active agent*, a *surfactant*.[a] A surfactant having cleaning properties in dilute aqueous solution is called a *detergent*. Detergents are divided into soaps and syndets: a *soap* is a salt of a fatty acid, and a *syndet* is a *syn*thetic *det*ergent, that is, by definition, a detergent other than soap.

Syndets were first produced in large quantities in Germany during World War I, but in the United States, mass production of syndets did not exceed that of soap until 1952. Around 2000, the production of detergents in the United States and the European Union was 4 Tg, of which only 10 % was soap. Detergents have caused major environmental problems due to discharge directly into rivers and lakes. This practice has been discontinued, but builders (see later discussion) that contain phosphates and complexing agents still cause problems in the hydrosphere.

a. Soaps

Soaps are salts of fatty acids containing more than eight carbon atoms, that is, salts of aliphatic monocarboxylic acids, whether saturated or unsaturated. They are obtained by basic hydrolysis of animal or vegetable fats and oils.[b]

Soaps have been known and used for washing and laundering since antiquity.[c] They were prepared in soaperies by boiling waste fat with soap-boiler's lye (a concentrated solution of NaOH or KOH), which was prepared using the reaction of Equation 8.29 with soda (Na_2CO_3) or potash (K_2CO_3). An example is the saponification of common candle wax giving sodium octadecanoate (sodium stearate):

$$CH_2 \cdot O \cdot CO \cdot C_{17}H_{35}$$
$$CH \cdot O \cdot CO \cdot C_{17}H_{35} \quad + \quad 3\,NaOH \quad \longrightarrow$$
$$CH_2 \cdot O \cdot CO \cdot C_{17}H_{35}$$

$$\text{(9.35)}$$

$$C_3H_5(OH)_3 \; + \; 3\,Na^+ \; + \; 3$$

Afterward, sodium chloride was added to the reaction mixture, yielding two phases: an upper one with the soap and a lower one with glycerol (used for skin care), salt, and surplus hydroxide.

Salts of saturated higher fatty acids (C_{18}, stearate, and C_{16}, palmitate) give hard soaps, whereas unsaturated acids (C_{18}, for example, oleic acid) and saturated lower fatty acids (C_{10}, C_{12}, and C_{14}) give softer soaps. However, soaps possess some disadvantages: First, the solubility of salts of divalent cations is low. In particular, hard water that contains $Ca(HCO_3)_2$ precipitates "lime soap," wasting soap and

---

[a] Surfactants include soaps, detergents, syndets, emulsifiers, and foaming agents.
[b] Lipids are substances of biological origin that are soluble in nonpolar solvents. They are classified as saponifiable lipids (e.g., glycerides and phospholipids) and nonsaponifiable lipids (e.g., steroids). Fats and oils are glycerides.
[c] Jeremiah 2:22, about 600 BC.

forming hard deposits.[a] Further, optimal rinsing of laundry in hard water requires elevated temperatures, even to the point of boiling, which together with the basicity of the soap swells textile fibers and bleaches colors.

The cleaning action of soap is due to its amphiphilic molecular structure. This means that each molecule possesses a lipophilic part and a hydrophilic part (see the molecule depicted on the right-hand side of Equation 9.35). The mechanism is twofold. First, soap decreases the surface tension of the solution, and the lowered surface tension allows the soap to penetrate the interface between the dirt particle and the surface, freeing it. Then a micelle is formed: the lipophilic part of the soap molecule surrounds the particle, and the hydrophilic part keeps it dissolved as a colloid anion while it is carried away. Chemical modifications of a soap may accentuate certain properties such as its ability to form a foam or to act as a moisturizing or emulsifying agent.

Salts of the fatty acids with Mg^{II}, Zn^{II}, Al^{III}, Mn^{II}, Sn^{II}, Pb^{II}, and Co^{II} are surfactants that have technical uses as additives for lubricating oils, paints, and lacquers; for impregnation and smoothing of textiles; and for coating medicinal tablets.

b. Syndets

Syndets have amphiphilic molecular structures, analogous to the alkyl carboxylates, and examples are shown in Table 9.3. The lipophilic part is an aliphatic chain, often derived from a fatty acid. The hydrophilic group may be anionic (sulfonates, sulfate esters), cationic (quaternary ammonium), zwitterionic[b] (betaines) or non-ionic (N-oxides, polyethers, polyamides). The great advantage of syndets is that they do not precipitate calcium ions.

World production of syndets in 1996, 15 Tg, was distributed in the following way: anionic syndets (50 %) were mainly used for washing textiles; cationic syndets (10 %) were used for disinfection and for softeners that prevent static electricity in artificial textiles; nonionic syndets (40 %) in liquid syndets were used for laundry and dishwashing because they emulsify well and are poor at generating foam.

The first syndets were prepared from linear alkanes with 12 to 18 carbon atoms through sulfonyl chlorides,

$$RH + SO_2Cl_2 \rightarrow R \cdot SO_2Cl + HCl \tag{9.36}$$

which were then converted into the sulfonate

$$R \cdot SO_2Cl + 2\,NaOH \rightarrow R \cdot SO_2ONa + NaCl + H_2O \tag{9.37}$$

This approach is necessary because alkanes cannot be sulfonated directly. In contrast, aromatic systems react easily with sulfuric acid, and this is essentially why detergents are synthesized from alkanes attached to a benzene group. As an example, dodecylbenzene

$$C_{12}H_{25} \cdot C_6H_5 + HOSO_3H \rightarrow C_{12}H_{25} \cdot C_6H_4 \cdot SO_3H + H_2O \tag{9.38}$$

[a] Hardness of water; see Section 3.2b.
[b] See Section 5.1c.

Table 9.3 Synthetic detergents		
Linear alkylbenzene sulfonates	$C_{12}H_{25} \cdot C_6H_4 \cdot SO_3^-$	
Alkane sulfonates	$C_nH_{2n+1} \cdot SO_3^-$	$n = 12\text{--}18$
Fatty acid esters of hydroxyethane sulfonate	$R \cdot COO \cdot CH_2CH_2 \cdot SO_3^-$	
Fatty acid amides of the N-methyltaurine anion	$R \cdot CO \cdot N(CH_3) \cdot CH_2CH_2 \cdot SO_3^-$	
Primary and secondary alkylsulfates	$C_nH_{2n+1} \cdot OSO_3^-$	$n = 12\text{--}18$
Anions of sulfuric acid monoesters of monoglycerides	$R \cdot COO \cdot CH_2 \cdot CHOH \cdot CH_2 \cdot OSO_3^-$	
Alkyl(benzyl)dimethylammonium	$(C_nH_{2n+1})(C_6H_5CH_2)(CH_3)_2N^+$	$n = 8\text{--}18$
Fatty alkyl betaines	$(C_nH_{2n+1})N^+(CH_3)_2CH_2COO^-$	$n = 10\text{--}20$
Fatty amine oxides	$(C_nH_{2n+1})(CH_3)_2N^+\text{-}O^-$	$n = 10\text{--}20$
Esters of fatty acids with poly(ethylene oxide)	$R \cdot COO \cdot (CH_2CH_2O \cdot)_nH$	$n = 4\text{--}8$
Amides of fatty acids with poly(ethylene oxide)	$R \cdot CONH \cdot (CH_2CH_2O \cdot)_nH$	$n = 4\text{--}8$
Ethers of fatty alcohols with poly(ethylene oxide)	$R \cdot CH_2 \cdot O \cdot (CH_2CH_2O \cdot)_nH$	$n = 4\text{--}8$
Secondary fatty amines with poly(ethylene oxide)	$R \cdot CH_2 \cdot NH \cdot (CH_2CH_2O \cdot)_nH$	$n = 4\text{--}8$
Anisols with poly(ethylene oxide)	$R \cdot C_6H_4(\cdot OCH_2CH_2)_nOH$	$n = 4\text{--}8$

Examples of anionic, cationic, and nonionic syndets. R indicates the aliphatic part of a fatty acid, R·COOH. The functional groups are discussed in the text.

gives dodecylbenzene sulfonic acid, whose sodium salt is the most common linear alkylbenzene sulfonate, generally termed LAS.[a] In the past, dodecylbenzene was prepared by direct combination of dodecene (a tetramer of propene) with benzene. This produced a mixture of isomers, including branched aliphatic chains, which are very slowly decomposed in Nature. Today, LASs are prepared exclusively from linear carbon chains.

Primary alkane sulfonates are esters prepared from long-chain alcohols (C_{12} - C_{18}, formed by reduction of esters of fatty acids) and sulfuric acid; they are used in shampoos, toothpaste, and cremes. However, in this century production of secondary alkane sulfonates is increasing because the price of 1-decene, "C_{12} α-olefin," has decreased. Presently (2009), the olefin market, including alkenes with C_6 - C_{30}, is of the order of magnitude 3 Tg/a.

The polyethers of Table 9.3 are soluble in water because of the formation of hydrogen bonds; thus, their solubility increases with decreasing temperature. They all have a terminal alcohol group that may form an ester with sulfuric acid, increasing their foaming ability.

Builders

Early industrial methods producing syndets were based on sulfonates (see Equation 9.38), which were treated with sodium hydroxide in order to neutralize sulfonic acid and a surplus of sulfuric acid. The resulting sodium sulfate acted as a filler, making the product easier to handle, so it was not removed. This was called a *builder*. The name is still used, but now it refers to chemical species that supplement the action of syndets.

[a] This particular LAS is sometimes termed LABSA, from laurylbenzenesulfonate, an obsolete name.

Various phosphates are frequently used as builders, including sodium salts of triphosphate, tetraphosphate, and poly(metaphosphate). These compounds function as pH buffers and complexing agents by removing Ca^{2+} and Mg^{2+} (changing the hardness of the water) and Fe^{III} (reducing deposition of rust on the textiles). However, when released into the hydrosphere phosphates cause algal blooms, leading to eutrophication (see Section 9.5a). For example, over the course of a few years around 1960, the concentration of phosphate in Bodensee (a lake between Germany and Switzerland) increased by a factor of 15. New builders were invented to prevent eutrophication, and phosphate-free syndets are now used in the United States, Western Europe, Japan, and South Korea, but not yet in Southern and Eastern Europe, Africa, Asia, and Central and South America.

Carbon-based builders are mainly soda, trona, and sodium hydrogencarbonate (see Section 8.3c). In addition, nitrilo triacetate, $N(CH_2COO^-)_3$, is frequently used because it forms stable complexes with divalent and trivalent metal ions and is decomposed rather quickly in Nature. Builders based on silicon include various zeolites. They are not soluble in water but act as ion exchangers, switching Na^+ with $Ca^{2+}(aq)$ but not with the larger aqua ion, $Mg^{2+}(aq)$. Global consumption of builders in 2001 included sodium carbonate, 2.7 Tg; sodium triphosphate, 1.5 Tg; and zeolites, 1.3 Tg.

Some syndets are less efficient at forming micelles, which may cause the dirt particles to reprecipitate. Carboxymethylcellulose is added to prevent this. It encapsulates the particles and acts as a surfactant, changing the surface properties of the textile. Carboxymethylcellulose is prepared by reacting cellulose with chloroacetate in basic solution, yielding ethers of hydroxy acetate (glycolate), $R \cdot O \cdot CH_2 \cdot COO^-$. One glucose can react with at most three chloroacetates (see Figure 8.5), but the normal product contains only 0.9 chloroacetates per glucose.

c. The Gibbs isotherm

This section presents the classic (1879) thermodynamic explanation for the connection between surface concentration and surface tension: The definition (Equation 9.39) leads to Equations 9.44 and 9.45, implying that the surface tension of a solvent depends on the difference between the concentration of the solute near the surface and in the interior of the solvent.[67]

The surface tension of a liquid, γ, is the ratio between the infinitesimal work dG needed to change the area of the surface by an infinitesimal area dA_s and this area (see Equation 2.100):[a]

$$\gamma = \left(\frac{\partial G}{\partial A_s}\right)_{T,p,n_B} = \left(\frac{\partial U}{\partial A_s}\right)_{S,V,n_B} \tag{9.39}$$

The surface tension is measured at constant temperature, pressure, and amount of substance, as shown in the middle term. The right-hand term is more convenient for

[a] The SI-unit is $J\,m^{-2} = N\,m^{-1}$.

calculations, and it will be used in the following; it is related to the middle term in the same way as Equation 2.4 is related to Equation 2.6.

Consider a liquid phase α, including its interface with another phase, which is not part of the system to be studied. A change of state of the system will fulfill the fundamental equation 2.4, including a term accounting for the surface tension (Equation 9.39):

$$dU^\alpha = T^\alpha \, dS^\alpha - p^\alpha \, dV^\alpha + \gamma^\alpha \, dA_s^\alpha + \sum_B \mu_B^\alpha \, dn_B^\alpha \qquad (9.40)$$

This gives rise to a Gibbs-Duhem equation,

$$0 = S^\alpha \, dT^\alpha - V^\alpha \, dp^\alpha + A_s^\alpha \, d\gamma^\alpha + \sum_B n_B^\alpha \, d\mu_B^\alpha \qquad (9.41)$$

The system is now divided into the surface phase σ and the bulk phase α. At constant T and p, Equation 9.41 is

$$0 = A_s \, d\gamma + \sum_B n_B^\sigma \, d\mu_B \qquad \text{constant } T, p \qquad (9.42)$$

for the surface phase, and from Equation 2.118,

$$0 = \sum_B n_B^\alpha \, d\mu_B \qquad \text{constant } T, p \qquad (9.43)$$

for the bulk phase. Any one of the μ_B's could be eliminated from these two equations. Consider a binary mixture of solvent A and solute B, and eliminate μ_A to obtain

$$\frac{d\gamma}{d\mu_B} = \frac{n_A^\sigma}{A_s} \left(\frac{n_B^\alpha}{n_A^\alpha} - \frac{n_B^\sigma}{n_A^\sigma} \right) \qquad \text{constant } T, p \qquad (9.44)$$

This is called the Gibbs isotherm.

Assume an ideal solution, that is, $\mu_B = RT \ln m_B^\alpha$, and introduce the molar mass M_A of species A to change Equation 9.44 into

$$\frac{d\gamma}{d \ln m_B^\alpha} = \frac{RT M_A n_A^\sigma}{A_s} \left(m_B^\alpha - m_B^\sigma \right) \qquad (9.45)$$

where m_B^α and m_B^σ are the molalities of B in phases α and σ, respectively. Equation 9.45 shows that γ is lowered by the positive adsorption of a solute into the surface, $d\gamma/d \ln m_B^\alpha < 0$ for $m_B^\sigma > m_B^\alpha$. For example, the surface tension of water is lowered by the addition a small quantity of detergent. Conversely, $d\gamma/d \ln m_B^\alpha > 0$ for $m_B^\sigma < m_B^\alpha$, so the surface tension is raised by the negative adsorption of a solute from a surface. For example, the surface tension of water is raised by the addition a small quantity of sodium chloride.

9.5 Water treatment

In this section we discuss two separate aspects of society's use of water: domestic and industrial. The reason is that the pre- and post-treatment are very different in these two cases.

The technical requirements for treatment of water for domestic use are simple. Often filtering and prophylactic treatment are sufficient. In the European Union, potable water may contain 50–500 mg/kg of dissolved salts; an example of river water with 100 mg/kg is given in Table 3.14. After use, the composition of sewage (feces, paper, detergents) requires more elaborate treatment for environmental protection, including prevention of eutrophication, and control of sludge from sewage treatment plants.

The most exacting industrial processes (certain boilers, production of electronics) specify a maximum impurity of 20 µg/kg, implying that the concentration of salts must be reduced by a factor of up to 25,000. Industrial waste water is normally rather simple to monitor and rinse because the methods of production in a single plant are limited and well known. However, water from the food industry often has a high level of organic waste, which varies significantly throughout the year, in quantity and quality.[205]

a. Domestic water

When the first settlements were formed, the sanitary problems caused by gathering many people in a small area quickly became obvious. Sanitation was gradually developed and reached a zenith in ancient Rome, where each citizen (the population was more than 10^6) had access to 600–1000 l/d of water, and sewers from all of the city led into the river Tiber. The aqueducts leading water from the mountains to Rome are famous architectural wonders. However, in Ostia, a harbor city located downstream, unsanitary conditions arose that today would be classified as severe.

Sanitary conditions vary and were particularly bad in 19th-century Europe. Legislation became necessary, such as the Public Health Act (UK, 1848) and the Acts of Technical and Medicinal Hygienics (Denmark, 1853). Legislation of this type plays a fundamental role in modern society, and a range of controls, targeting water quality and pollution, have been developed. Pollution is a subject of immense public and political interest, and the science-based advising of governmental institutions is an important application of environmental chemistry.

Purification and disinfection

Purification of domestic water normally consists of two steps. First, water is purified using appropriate techniques, as discussed in Section 9.5c. The next step is prophylactic disinfection to kill microorganisms that may be harmful.[a] A general chemical method would be to use a powerful oxidant, and the most commonly used are chlorine and chloroazane (chloramine), which hydrolyze, giving the active species hypochlorous acid/hypochlorite ($pK_a \approx 7.5$):

$$Cl_2 + H_2O \;\rightleftharpoons\; HOCl + H^+ + Cl^- \tag{9.46}$$

$$NH_2Cl + 2\,H_2O \;\rightarrow\; NH_4^+ + HOCl + HO^- \tag{9.47}$$

---

[a] An example: Cholera is a widespread bacterial disease spread by untreated water.[61] Cholera epidemics are feared because dehydration and death can occur within hours of infection.

The standard electrode potential of hypochlorous acid at pH 7 is given by

$$HOCl + 2\,e^- = Cl^- + HO^- \qquad\qquad E^{\circ\prime} = 1.07\,V \qquad (9.48)$$

Sometimes dioxochloride (chlorine dioxide) is used:

$$2\,ClO_2 + H_2O \rightarrow ClO_2^- + ClO_3^- + 2\,H^+ \qquad\qquad (9.49)$$

where the active species at pH 7 are chlorite and chlorate:

$$ClO_2^- + 2\,H_2O + 4\,e^- = Cl^- + 4\,HO^- \qquad E^{\circ\prime} = 1.17\,V \quad (9.50)$$

$$ClO_3^- + 3\,H_2O + 6\,e^- = Cl^- + 6\,HO^- \qquad E^{\circ\prime} = 1.03\,V \quad (9.51)$$

Although ozone is more expensive, it is nonetheless used at low concentrations in many places:

$$O_3 + H_2O + 2\,e^- = O_2 + 2\,HO^- \qquad\qquad E^{\circ\prime} = 1.66\,V \quad (9.52)$$

Use of ozone avoids the undesirable taste and smell of chlorinated water, and residual ozone decomposes to oxygen soon after treatment, $O_3 \rightarrow \frac{3}{2}\,O_2$.

We now consider the treatment of sewage effluent and agricultural runoff carrying fertilizers, the major topics being eutrophication and the physical quantities used in monitoring wastewater.

Eutrophication

Eutrophication was briefly discussed in Section 7.4. When a surplus of nutrients, in particular dihydrogenphosphate–hydrogenphosphate, has been discharged to fresh-water and marine environments, a series of events may take place. First photosynthesis is increased, in principle

$$CO_2(aq) + H_2O(l) \rightarrow CH_2O(biota) + O_2(g) \qquad\qquad (9.53)$$

manifested as an increased growth of algae and cyanobacteria. This is called an *algal bloom*. These may also occur naturally, but untreated waste water frequently causes algal blooms. This biomass is transformed by other organisms through respiration, the reciprocal process,

$$CH_2O(biota) + O_2(aq) \rightarrow CO_2(aq) + H_2O(l) \qquad\qquad (9.54)$$

depleting the water of dissolved oxygen. The normal level of oxygen, ≈ 10 mg/l, is maintained in the surface, but below that, deficiency of oxygen may cause organisms requiring an oxygenated environment to die.

The dead biomass, including algae, settles at the bottom and various microorganisms continue the decay according to decreasing redox potential, as shown in Table 9.4.

The person equivalent, PE

The EU has implemented several directives aimed at the treatment of urban waste water. The basic physical quantity used is the *person equivalent* of water,

$$q_m(PE, water) = 200\,kg/d \approx 0.2\,m^3\,d^{-1} \qquad\qquad (9.55)$$

Table 9.4 Electrode potentials of waters close to organisms under decay

		E	>	+0.7 V	Oxygen reduction
+0.7 V	>	E	>	+0.2 V	Nitrate reduction
+0.2 V	>	E	>	−0.3 V	Sulfate reduction
−0.3 V	>	E			Methanogenesis

Approximate limits of the potential of an inert electrode at pH 7 together with the characteristic bacterial respiration process; see Table 3.29 and Figure 5.9.

which is a mass flux. The goal is to specify the size and efficiency of wastewater treatment plants on the basis of the volume of sewage produced by the underlying population. Experimentally, the equivalent fluxes of BOD, N, and P are:

$$q_m(\text{PE,BOD}) = 60\,\text{g/d}, \quad q_m(\text{PE,N}) = 12\,\text{g/d}, \quad q_m(\text{PE, P}) = 4\,\text{g/d} \quad (9.56)$$

which correspond to the following mass fractions:

$$w_{\text{BOD}}(\text{PE}) = 300\,\text{mg/kg}, \quad w_\text{N}(\text{PE}) = 60\,\text{mg/kg}, \quad w_\text{P}(\text{PE}) = 20\,\text{mg/kg} \quad (9.57)$$

These figures are determined as discussed in Section 3.2b.

The European Union directives fix the maximum allowed size of these quantities from an urban sewage treatment plant to the recipient:

$$w_{\text{BOD}}(\text{PE}) = 15\,\text{mg/kg}, \quad w_\text{N}(\text{PE}) = 8.0\,\text{mg/kg}, \quad w_\text{P}(\text{PE}) = 1.5\,\text{mg/kg} \quad (9.58)$$

for a population equivalent of 6,000 $q_m(\text{PE, water})$.

Example

Consider a case in which 1000 persons live in a farming area, to which a waterworks delivers an average of 150 m^3/d. The wastewater treatment plant receives 250 m^3/d, including drainage water. The seepage, mainly rainwater, is therefore 100 m^3/d, and the nominal number of users 250/0.2 = 1250.

In rural districts with a low population density, one may rinse sewage using a septic tank combined with a plant root zone. Certain plant roots (willow, cattail) effectively remove nutrients such as BOD, P, and N and actively supply the soil with oxygen, around 6 mg/kg. With a lifetime $\tau = 15$ d for water in the root zone treatment system (see Equation 2.48), a reduction of

$$w_{\text{BOD}}(\text{PE}):75\%, \quad w_\text{N}(\text{PE}):40\%, \quad w_\text{P}(\text{PE}):35\% \quad (9.59)$$

may be achieved. In the European Union, this is accepted for at most 300 $q_m(\text{PE, water})$.

b. Industrial water

The required purity of industrial water depends on its use. Generally water is produced and used in vast amounts, and its purification per se is a demanding industrial process.[a] The chemical problems include prevention of scaling and corrosion, which are discussed in the next section.

As an example, consider boiler feedwater for power plants. High-pressure steam requires pure water. The higher the pressure, the better the economic efficiency, and the higher the requirement for water purity. The conditions (boiler pressure, purity, treatment) were (25 - 35 bar, 300 - 5 mg/kg, softening) before World War II, and after the war, in 1955 (100 bar, 0.1 mg/kg, ion exchange). Today, the conditions of critical water are approached: (210 bar, 0.02 mg/kg, ion exchange).

c. General methods

Distillation and reverse osmosis are frequently used for water purification. For example, consider the Yuma Desalting Plant in Arizona, which uses reverse osmosis. Irrigation waters from the Colorado River have percolated through the soil, leaching out saline components. This water cannot be pumped back into the river because the high level of salt would disturb users downstream in Mexico. The desalination plant has a capacity of 273 Gg/d and extracts 325 Mg/d of salt, which is deposited in the desert.

A number of other methods are discussed in the following paragraphs.

Filtration
River water, which may contain up to 5 g/kg of silt (e.g., the Nile), is often used directly for irrigation. For other uses, it should be filtered, often using a sand bed.

Flocculation
Colloidal clay particles may be removed by flocculation (coprecipitation) with hydroxides of aluminum(III) or iron(III) that are produced by precipitation of the aquaion:

$$M^{3+}(aq) + 3\,H_2O \quad \rightarrow \quad M(OH)_3(s) + 3\,H^+(aq) \qquad (9.60)$$

where M = Al or Fe (see Tables 5.15 and 5.16). The amount used is ≈ 0.1 mmol/l, and the acid produced is normally neutralized by the natural alkalinity (HCO_3^-) of the raw water. The hydroxides form positively charged flocs that attract and carry down negatively charged colloidal clay particles and anions of humic and fulvic acids.

Potassium alum, $KAl(SO_4)_2 \cdot 12H_2O$ (potash alum), is the most widely used flocculation agent. It is produced by leaching alumina from bauxite, which is then reacted with potassium sulfate.

Iron is frequently added as chlorinated iron(II) sulfate:

$$Fe^{2+} + \tfrac{1}{2}\,Cl_2 + 3\,H_2O \quad \rightarrow \quad Fe(OH)_3 + 3\,H^+ + Cl^- \qquad (9.61)$$

---

[a] The level of Fe, Cu, Ni in analytical-grade HCl is 1.0 mg/kg. The total impurity level, 20 μg/kg (0.02×10^{-6}), of boiler feedwater corresponds to a purity of 99.999 998 %.

This method is preferred because oxidation of iron(II) using the natural dissolved oxygen,

$$2\,Fe^{2+} + \tfrac{1}{2}\,O_2 + 5\,H_2O \;\rightarrow\; 2\,Fe(OH)_3 + 4\,H^+ \tag{9.62}$$

is difficult to control. Iron(III) sulfate, applied according to Equation 9.60, is of increasing importance.

Softening

Hard water (see Section 3.2b and Section 5.2c) contains calcium and magnesium ions, and the process by which they are removed is called *softening*.

The addition of calcium hydroxide (slaked lime, $Ca(OH)_2$) is a chemical method for softening water. Slaked lime removes temporary hardness that would otherwise precipitate and form scale when the water is boiled (see Equation 3.9):

$$Ca(HCO_3)_2(aq) + Ca(OH)_2(aq) \;\rightarrow\; 2\,CaCO_3(s) + 2\,H_2O(l) \tag{9.63}$$

Sodium carbonate (soda, Na_2CO_3) is used to remove permanent hardness caused by $CaSO_4(aq)$ and $MgCl_2(aq)$, which do not precipitate simply by boiling:

$$CaSO_4(aq) + Na_2CO_3(aq) \;\rightarrow\; CaCO_3(s) + Na_2SO_4(aq) \tag{9.64a}$$

$$MgCl_2(aq) + 2\,Na_2CO_3(aq) + 2\,H_2O(l) \;\rightarrow\;$$
$$Mg(HCO_3)_2(s) + 2\,NaCl(aq) + 2\,NaOH(aq) \tag{9.64b}$$

where the magnesium hydrogencarbonate is a mixture of basic magnesium carbonates of varying composition. The problem of sludge disposal has been solved by replacing lime softening with ion exchange.

Ion exchange

An ion exchange substrate is composed of an insoluble porous organic polymer species called *resin*, to which chosen functional groups have been attached. Cation-exchange substrates used for water softening are normally made of small balls of polystyrene resin with sulfonic acid groups. The beads are packed into tubes and charged with sodium ions by passing sodium chloride through the tube:

$$RSO_3H(s) + Na^+(aq) \;\leftrightharpoons\; RSO_3Na(s) + H^+(aq) \tag{9.65}$$

where R denotes the resin.[a] Next the hard water (containing calcium and magnesium hydrogencarbonate) is passed through the tube:

$$2RSO_3Na(s) + Ca(HCO_3)_2(aq) \;\leftrightharpoons\; (RSO_3)_2Ca(s) + 2\,NaHCO_3(aq) \tag{9.66}$$

Before use, the ion-exchange substrate must be swelled, that is, hydrated. In operation, the exchange capacity is approximately 2 mol/l resin, and about 70 % of this

---

[a] Previously, we have used the symbol $\equiv R^-$ for a general cation-exchange resin and, analogously, $\equiv R^+$ for an anion-exchange resin. However, functional groups in artificial resins are chemically well defined, as explained later.

can be used before the water's hardness rises above $w_{CaO} = 10^{-6}$. The ion-exchange resin is regenerated using NaCl:

$$(RSO_3)_2Ca(s) + 2\,Na^+(aq) \;\rightleftharpoons\; 2\,RSO_3Na(s) + Ca^{2+}(aq) \quad (9.67)$$

The cation-exchange substrate acts as a strong acid in solid form in every respect.

Carboxylate can also serve as the functional group in cation exchange. However, because it is a weak acid, it is protonated in weak acid solutions and is therefore less useful for exchanging cations from salts of strong acids. Even so, it is well suited to removing carbonates quantitatively:

$$2\,RCOOH(s) + Ca(HCO_3)_2(aq) \;\rightleftharpoons\;$$
$$(RCOO)_2Ca(s) + 2\,CO_2(aq) + H_2O \qquad (9.68)$$

because the carbon dioxide can be degassed.

Anion exchange uses an analogous mechanism that is based on amine derivatives. $RN(CH_3)_2$ is a weakly basic anion-exchange substrate that forms a hydrochloride with hydrochloric acid:

$$RN(CH_3)_2(s) + HCl(aq) \;\rightleftharpoons\; RN(CH_3)_2HCl(s) \qquad (9.69)$$

whereas the quaternary ammonium ion is a strongly basic anion-exchange substrate:

$$RN(CH_3)_3OH(s) + HCl(aq) \;\rightleftharpoons\; RN(CH_3)_3Cl(s) + H_2O \quad (9.70)$$

Modern ion-exchange facilities for water purification use a mixed bed: strongly acidic and strongly basic ion-exchange substrates are merged so that water is deionized as it passes through the system, in principle:

$$RSO_3H + RN(CH_3)_3OH + NaCl \;\rightarrow\;$$
$$RSO_3Na + RN(CH_3)_3Cl + H_2O \qquad (9.71)$$

The resins are regenerated after separation by means of a flow of water utilizing the different density of exchangers.[a] This is achieved in the same physical unit using an appropriate network of pipes. The development of modern power plants using high-pressure boilers could not have taken place without this technique, which may reduce the salt content to 0.02 mg/kg.

Electrodialysis

Another form of deionization, electrodialysis, is made possible as a result of the development of mechanically stable ion exchange membranes. A cell consists of a cation- and an anion-exchange membrane. Figure 9.3 shows a system in which many such membrane cells are joined in series, a stack. A plant may consist of several stacks in parallel. The process is a mass transfer: sulfate ions move to the right until they are stopped by a cation-exchange membrane; similarly, sodium ions move to the left until they are stopped by an anion-exchange membrane. Typically, a stack is operated at 250 V, 50 A, and to obtain sufficient conductivity, a processing temperature of

---

[a] The cation exchanger has the greater density.

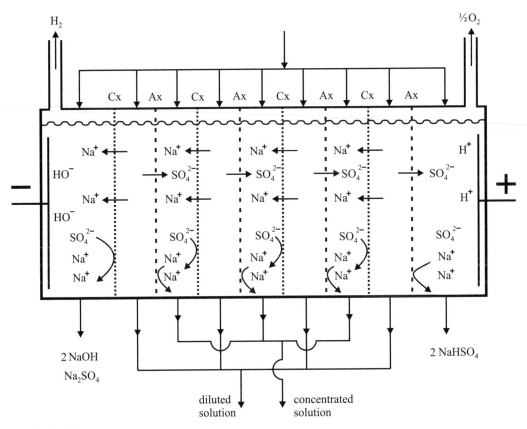

Figure 9.3 *Electrodialysis cells*[96]

An electrodialysis stack is composed of several hundred cells, of which seven are shown. The cathode (–) is placed to the far left and the anode (+) to the far right. Cation-exchange membranes Cx and anion-exchange membranes Ax are placed alternately between the electrodes; the spacing between each membrane is about 1 mm. The plant is fed continuously from above, and solutions are drawn from the individual cells and combined. A Na_2SO_4 solution is used as an example.

30–45 °C is necessary. At concentrations around 1.5 mmol kg^{-1}, the process ceases because the electric resistance is too high.

In the figure, the cells are fed continuously from above with a Na_2SO_4 solution; the solution is removed from the individual cells from below. Na_2SO_4 and H_2SO_4 run off from the anode volume, and $O_2(g)$ escapes from the top; Na_2SO_4 and NaOH run off from the cathode volume, and $H_2(g)$ escapes from the top. Water from cells with less salt than the original solution is combined for subsequent purification and use. The solution from cells that contain more salt is also combined. In the deionizing process this is a waste, but the technique is also used to recover salts of valuable acids: a sodium salt made as a dilute aqueous solution can be concentrated through electrodialysis.

The chemistry of climate change

Diverse and detailed data show that climate[a] has always been in a state of change. Climate is driven by radiation energy from the Sun, and for this reason we begin by presenting the basic physics of thermal radiation. Next we will look at the origin of climate and describe some of the many factors that are responsible for natural climate variation. Finally, we will examine natural and anthropogenic factors that are known to have driven climate change since 1750 AD, and some possible climate futures will be discussed. One of the main challenges for the next generation of chemists will be to solve the issues described here that link chemistry, energy, and environment.

Evidence for long-term climate variation can be found in the glacial ice of Antarctica and Greenland, and in deep-sea sediments containing shells from plankton. These records are interpreted by examining changes in the abundance of naturally occurring stable isotopes, for example, oxygen isotopes in water in the ice or in the carbonates of sediments, and sulfur isotopes in pyrite and calcium sulfate in shallow-water sediments. Long-term climate changes are correlated with changes in the Sun-Earth geometry as a consequence of the interaction of the Earth with the Sun and its planets.

The period since 1750 is short indeed compared to geological timescales. However, climate has been measured by various means throughout this period, and changes have been observed: on average, surface temperature, ocean water temperature, sea level, and precipitation have all increased. The important issue is to what degree these changes are of natural and/or anthropogenic origin. Human activities have changed the composition of the atmosphere, and therefore, much research has been aimed at establishing their impact on climate. The 1995 Nobel prize–winning atmospheric chemist Paul Crutzen regards anthropogenic effects as being so significant as to constitute a new geological era. He advanced the word *Anthropocene* to describe the most recent period in the Earth's history.

Modern understanding of climate begins with the observation that there is a balance between incoming solar energy (mainly UV-Vis radiation) and the energy that is radiated back toward space (mainly IR radiation). Physical quantities that cause this balance to change and thereby affect the global average temperature are called *forcings*. The forcings of infrared active gases are easy to establish. Linking forcings to climate change is not as easy, as we will discuss.

---

[a] Climate is the time average of weather.

10.1 The physics of thermal radiation

The following discussion presents concepts that are directly relevant to the Sun-Earth system.

It is readily seen that physicochemical systems emit radiation. In physical terms, they emit radiant energy whose quantity and quality depend on the temperature of the surface. For example, the rate at which an incandescent lamp emits radiation roughly increases with the fourth power of the temperature of the filament. The quality of the radiation also changes with temperature: a solid starts glowing at 525 °C and gets whiter at higher temperatures; the bright red color of hot iron sets in at around 1300 °C. Radiation whose quantity and quality depend only on the temperature of the emitting system is called *thermal radiation*. A characteristic property of thermal radiation is that its prism-dispersed spectrum is continuous.

When radiation hits the surface of a system, a fraction may be transmitted (i.e., it passes through the system), reflected (scattered without dissipation of energy), or absorbed (interacts with the system which is heated). Examples:[a] aluminium foil reflects 98 % and absorbs 2 %, soot (carbon black) reflects 5 % and absorbs 95 %, and glass reflects 4 % and transmits 96 % of the energy received.

Such observations will be made quantitative. First, quantities for the energy emitted or received by physicochemical systems will be established. Next, flux densities for the radiant energy of a so-called blackbody will be given as functions of temperature and wavelength. Finally, the Sun-Earth system will be discussed.

a. Quantitative expressions

The source

The rate at which a system emits radiant energy Q is called the *radiant power P*:

$$P = dQ/dt \tag{10.1}$$

It is seen that P is a flux of energy. The flux density is called the *radiant excitance* and denoted M,

$$M = dP/dA_{source} \tag{10.2}$$

where A is the surface area of the emitting system. M is a sum of contributions of radiation from all wavelengths, and one may write

$$M = \int_0^\infty M_\lambda \, d\lambda \tag{10.3}$$

[a] For visible radiation with the incoming ray perpendicular to the surface; room temperature.

where M_λ, called the *spectral excitance*, is the excitance within a narrow wavelength interval $d\lambda$ centered at λ (see Figure 10.1). Working in the opposite direction gives the expressions for M_λ,

$$M_\lambda = \frac{dM}{d\lambda} = \frac{d^2P}{dA\,d\lambda} \tag{10.4}$$

which have the SI unit W m^{-3}.

The receptor

We now turn our attention to the system that receives radiant energy and define the *irradiance* $I(\theta)$ as the radiant power dP that is received at the surface dA:

$$I(\theta) = \frac{dP}{dA}\cos\theta \tag{10.5}$$

where θ is the angle between the ray (from a distant source[a]) and the surface normal. We shall use the abbreviation I for $I(\theta = 0)$.

We define the concepts discussed in the introduction as follows:

$$\text{transmittance } \tau = I_{\text{trs}}/I \tag{10.6a}$$
$$\text{reflectance } \rho = I_{\text{rfl}}/I \tag{10.6b}$$
$$\text{absorptance } \alpha = I_{\text{abs}}/I \tag{10.6c}$$

with the condition $\alpha + \rho + \tau = 1$. In analogy to Equation 10.4, the *spectral irradiance* I_λ is defined,

$$I_\lambda = dI/d\lambda \tag{10.7}$$

in order to discuss the quantities of Equations 10.6 as functions of the wavelength of the incoming radiation.

A system that absorbs all incident radiation, that is, one having $\alpha = 1$, is called a blackbody. No such system actually occurs in Nature, but some come close, such as carbon black. Nevertheless, as will be seen presently, an ideal blackbody is a useful concept in the physics of radiation.

b. Radiation theory

In contrast to most textbooks on optics, this section does not detail the historical development of the subject. Instead, it is more convenient to derive the pertinent expressions from Planck's radiation law (see Equation 10.9). We begin by introducing Kirchhoff's law.

---

[a] Because we restrict the applications to the Sun-Earth system, we do not require the adaptation of solid angles. To a good approximation, rays from the Sun are parallel.

Kirchhoff's radiation law (1859)

For a closed system in thermal equilibrium, the radiant excitance is equal to the absorbed irradiance.

This statement is in accordance with the discussions of Section 2.3. In the mid-19th century, Kirchhoff's empirical conclusion was found to be one of the foundations of thermodynamics.

Restating,

$$M = \alpha I \tag{10.8}$$

where α is the fraction of the irradiance that is converted into heat and the ensuing thermal radiation. Kirchhoff also noted that a system with $\alpha = 1$ was "a completely blackbody."[33]

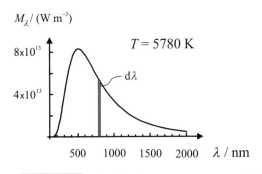
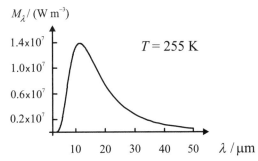

Figure 10.1 *Blackbody emission spectra of the Sun and the Earth*
Left panel: Spectrum of the Sun, $T_\odot = 5780$ K: $\lambda_{max} = 501$ nm $= 0.501$ μm; $M_\odot = 63.29$ MW m^{-2}; $P_\odot = 3.853 \times 10^{26}$ W.
Right panel: Spectrum of the Earth, $T_\oplus = 255$ K: $\lambda_{max} = 11.4$ μm; $M_\oplus = 239$ W m^{-2}; $P_\oplus = 1.22 \times 10^{17}$ W.

Planck's radiation law (1900)

The determination of the formula for spectral excitance was called "Kirchhoff's challenge." It was a significant problem for theoretical physicists for 40 years, until Planck solved it.[a]

According to Planck, the spectral excitance of a blackbody is given by

$$M_\lambda = \frac{2\pi h c_0^2}{\lambda^5} \frac{1}{\exp[h c_0/(\lambda k_B T)] - 1} \tag{10.9}$$

where all symbols have their standard meaning.[126] Figure 10.1 shows the spectrum of a blackbody at two different temperatures.

---

[a] Oddly enough, his derivation contained an error, and it took several more years before it was corrected by Bose and Einstein.[33] The correct expression is given in Equation 10.9.

Wien's displacement law (1893)

Differentiation of the spectral radiance (Equation 10.9) with respect to wavelength gives the position of the maximum. Wien's experiments demonstrated the inverse-proportional relationship between the temperature T and the maximum emission wavelength λ_{max}:

$$T \, \lambda_{max} = 0.00028978 \text{ K m} \tag{10.10}$$

The higher the temperature, the shorter the wavelength of the maximum emission.

The Stefan-Boltzmann law (S 1879; B 1884)

Integration of the spectral radiance (Equation 10.9) over wavelength according to Equation 10.3 gives the radiant excitance of a blackbody:

$$M_{bb} = \frac{2\pi^5}{15} \frac{k_B^4}{h^3 c_0^2} T^4 \overset{\text{def}}{=} \sigma_{SB} \, T^4 \tag{10.11}$$

which is proportional to the fourth power of temperature. The constant $\sigma_{SB} = 5.670400 \times 10^{-8}$ W m^{-2} K^{-4} is called the Stefan-Boltzmann constant.

Kirchhoff's law revisited

The radiant excitance, M, from a phase is always smaller than that from an ideal blackbody at the same temperature,

$$M = e \, M_{bb} \tag{10.12}$$

Here e is the emittance, and for real systems $e < 1$. This means that Equations 10.9, 10.10, and 10.11 cannot be used directly.[a] Nevertheless, Equation 10.11 is used in discussions of thermal radiation in the Sun-Earth system.

We can now return to Equation 10.8 and observe that

$$\alpha I = e \, M_{bb} \tag{10.13}$$

Actually, for phases with $\alpha > 0$, one finds that $M_{bb} = I$, so for these cases $e = \alpha$.

The Lambert-Beer law (L 1760; B 1852)

Consider a phase having $\alpha > 0$. This means that the irradiance decreases as the radiation passes through the phase:

$$dI = -I(\kappa - \beta) c_B dl \quad \text{or} \quad d\ln I = -(\kappa - \beta) c_B \, dl \tag{10.14}$$

Lambert observed the linear dependence of intensity on path length l, and Beer observed the linear dependence of intensity on concentration c_B of the absorbing species B. The constant κ is the molar Napierian absorption coefficient, and β

---

[a] In real systems, the power of the temperature in the Stefan-Boltzmann law (Equation 10.11) may be different from 4; for example, it is 5 for platinum.

describes the effect of blackbody emission by the sample. At room temperature, thermal radiation β is very small, and we have

$$\kappa c_B = \frac{\kappa}{N_A}(c_B N_A) = \sigma_{net} C_B \qquad (10.15)$$

where σ_{net} is the so-called net absorption cross section, N_A the Avogadro constant, and C_B the number concentration (see Table A1.7). It is a matter of convenience which of the two expressions is used. Integration of the right-hand side of Equation 10.14, putting $I = I_0$ at $l = 0$, that is, where the light enters the phase, and conversion into decadic logarithms leads to the familiar Lambert-Beer law,

$$A_\lambda \overset{def}{=} -\lg\tau_\lambda = \varepsilon_\lambda c_B l \qquad (10.16)$$

where $\varepsilon = \kappa \lg e$. Thermal radiation and luminescence have been neglected in this expression. This equation defines the *absorbance* A_λ at wavelength λ and connects it to the (molar decadic) *absorption coefficient* ε_λ, the concentration c_B, and the path length l of the radiation in the phase. At the outset, Equation 10.16 was an empirical relationship, but today the absorption coefficient can be described using fundamental quantum chemical quantities.

c. Application to the Sun-Earth system

The Sun
Observations of the continuous part of the solar spectrum show that the Sun acts as a blackbody to a very good approximation. Use of Wien's law gives the surface temperature $T_\odot = 5780$ K (see Figure 10.1). Subsequent application of the Stefan-Boltzmann law gives the radiant excitance, $M_\odot \approx 63.29$ MW m^{-2}. The radius of the Sun is $r_\odot = 0.6960$ Gm, and therefore its total radiant power is $P_\odot = M_\odot 4\pi r_\odot^2 \approx 385.3$ YW.

The Earth
The power (rate of energy flow) passing through an imaginary spherical shell around the Sun is constant, independent of the radius of the shell. Therefore, the irradiance at the distance r_S of 1 astronomical unit,[a] $r_S = 1$ au ≈ 149.6 Gm, is[144]

$$I_S = \frac{P_\odot}{4\pi r_S^2} = 1370\,\text{W m}^{-2} \qquad (10.17)$$

The constant I_S is called the *solar constant*. Using I_S and assuming thermal equilibrium, one can assess the blackbody temperature of the Earth as shown in Equation 10.19. In geophysical literature, the solar irradiance outside the atmosphere on a plane parallel to the surface (see Equation 10.5) is called *insolation*.[b] The 24 h mean

---

[a] The astronomical unit, au $= 1.49\,597\,870\,691 \times 10^{11}$ m, is a unit of length approximately equal to the mean Earth-Sun distance, r_S.[126]
[b] The word is formed from "incident solar radiation."

insolation as a function of date and geographical latitude is called the insolation formula; it is a major subject of Section 10.2.

Now, let the Earth be the system to be studied. The radiant power intersecting the Earth is the product of the cross section $\pi r_\oplus{}^2$ and the solar constant I_S. At radiative equilibrium, the radiant flux density $J_\oplus$ is this quantity divided by the area of the radiating surface, $4\pi r_\oplus{}^2$:

$$J_\oplus = I_S \frac{\pi r_\oplus^2}{4\pi r_\oplus^2} = \frac{I_S}{4} = 343\,\text{W m}^{-2} \qquad (10.18)$$

The subscript $\oplus$ indicates that this intensive quantity pertains to the entire Earth.

For the Earth system, Equations 10.6 give rise to the relation $\rho_\oplus + \alpha_\oplus = 1$ because the transmittance vanishes; here $\rho_\oplus$ is called the *albedo*.[a] The albedo comprises a sum of contributions due to reflection from the atmosphere and the surface, that is, scattering by molecules and aerosols, snow, ice, roads and buildings, and so forth; satellite observations give the average value $\rho_\oplus \approx 0.30$. Accordingly, at radiative equilibrium, the fraction $\alpha_\oplus J_\oplus = 240\,\text{W m}^{-2}$ is the total radiant energy flux density absorbed by the Earth. Use of Equations 10.8 and 10.11 gives the blackbody radiant excitance, $M_\oplus$, and the blackbody temperature, $T_\oplus$, respectively, of the Earth as seen from space:

$$M_\oplus = \alpha_\oplus J_\oplus = 240\,\text{W m}^{-2}, \qquad T_\oplus = \sqrt[4]{\frac{\alpha_\oplus J_\oplus}{\sigma_{SB}}} = 255\,\text{K} \qquad (10.19)$$

In contrast, the average surface temperature of the Earth is $T_E \approx 287\,\text{K}$ (see Figure 10.8) corresponding to a blackbody radiant excitance of $M_E = 385\,\text{W/m}^2$. The fact that the ratio $\mu = M_\oplus/M_E = (T_\oplus/T_E)^4 \approx 0.76$ is less than 1 is due to insulation of the surface by the atmosphere. This phenomenon is called the *greenhouse effect*.

The greenhouse effect

Changes in absorptance $\alpha_\oplus$ and albedo $\rho_\oplus$ will induce changes in $M_\oplus$ and, as a result, changes in $T_\oplus$. Consider an increase of $\alpha_\oplus$ by 1 %. The logarithmic derivative[b] of the right-hand side of Equation 10.19 gives

$$\text{d}\,T_\oplus = \frac{T_\oplus}{4}\frac{\text{d}\alpha_\oplus}{\alpha_\oplus} = 0.6\,\text{K} \qquad (10.20)$$

Similarly, if the albedo $\rho_\oplus$ is increased by 1 %, then

$$\text{d}\,T_\oplus = -\frac{T_\oplus}{4}\frac{\text{d}\rho_\oplus}{\rho_\oplus}\frac{\rho_\oplus}{\alpha_\oplus} = -0.3\,\text{K} \qquad (10.21)$$

The greenhouse effect is concerned with the spectral absorptance, $\alpha_{\oplus,\lambda}$. The explanation is as follows. Solar irradiance, mainly at visible wavelengths, reaches the surface. Part of this energy is reradiated back through the atmosphere, and an infrared active molecule may absorb some of the infrared radiation. When the molecule reemits

[a] Lat. *albus* $\hat{=}$ white; the term refers to all wavelengths.
[b] See Section 2.1b, footnote *b*.

this energy, it does so isotropically, that is, as much energy goes down as up. Thus, part of the reemitted radiation warms the surface.

Radiation from the Sun has its maximum spectral excitance at the wavelength $\lambda = 0.5$ μm (see Figure 10.1). Radiation with $\lambda < 0.3$ μm, some 3 % of the energy, is absorbed by atmospheric ozone (see Figure 4.1), and a small portion with $\lambda > 1$ μm is also absorbed, mainly by water. The radiant excitance, M_E, from the surface of the Earth is similar to that shown in the right panel of Figure 10.1 and occurs at $\lambda > 2$ μm. The IR spectra of the most important of the atmospheric absorbers, the so-called greenhouse gases, are shown in Figure 10.2. These gases absorb different bands of thermal radiation; accordingly, the spectrum of the Earth as seen from space is the blackbody radiation M_E minus some radiation at the wavelengths of the IR-absorption bands.

Radiant forcing and global warming potential

The greenhouse effect, $\mu = M_\oplus/M_E$, depends on the infrared absorptance of certain gases. This is a function of the concentrations of the gas, which is determined by the balance between emission rate and atmospheric lifetime.

In order to describe the atmosphere as a whole, it is convenient to imagine a cylinder extending from the surface of the Earth to space. As an example, in Section 4.2a the overhead ozone was given as the amount of ozone in a so-called column, per cross section.

A change in radiant excitance $M_\oplus$ due to a change in the amount of a species B in the atmosphere is called the *radiant forcing* of B and denoted dF_B.[a] It is defined by

$$dF_B = -\left(\frac{\partial M_\oplus}{\partial n_B}\right)_{\forall n_C \neq n_B} dn_B \tag{10.22}$$

where the amount of other species C, n_C, and other thermodynamic variables are kept constant. The minus sign accounts for the convention that dF is considered positive for decreasing $M_\oplus$. The partial molar radiant excitance is called the *molar radiant efficiency*,

$$A_B \overset{\text{def}}{=} -\left(\frac{\partial M_\oplus}{\partial n_B}\right)_{\forall n_C \neq n_B} \tag{10.23}$$

Note that A_B depends (parametrically) on all other species present (as always for partial molar quantities of real mixtures). The *specific radiant efficiency* a_B, defined by

$$a_B \overset{\text{def}}{=} -\left(\frac{\partial M_\oplus}{\partial n_B}\right)_{\forall C \neq B} \tag{10.24}$$

is often used in discussions of climate. The radiant forcings of some gases are given later in Table 10.3.

---

[a] The approved notation for radiant forcing is dM_B,[126] but IPCC uses dF_B, and the term *radiative forcing*, for this quantity.[230]

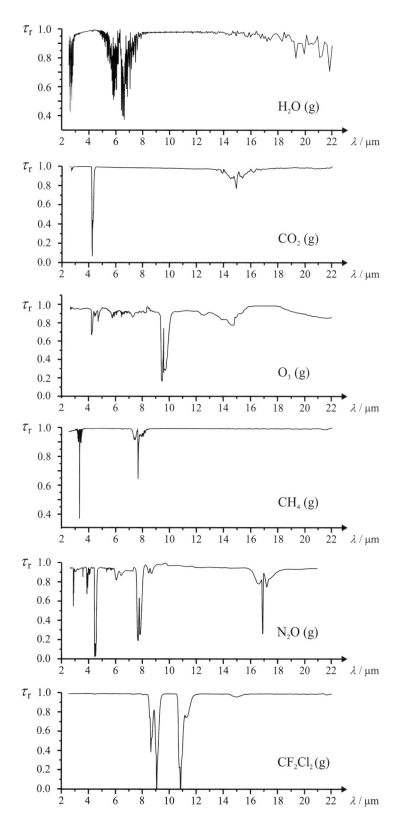

Figure 10.2 *IR spectra of some greenhouse gases*[243]

Abscissa: wavelength, Ordinate: relative transmittance.

The partial pressures are 0.03–0.26 bar diluted to 1 bar with $N_2(g)$.

The peak at 4.2 μm in the O_3 spectrum is due to a trace impurity of CO_2.

For a finite change of Δn_B of a species B, the radiant forcing can be obtained as the product

$$\Delta F_B = A_B \, \Delta n_B \qquad (10.25)$$

In the assessment of the relative effect of greenhouse gases, it is important to consider the radiant forcing over a specific time interval Δt starting at t_0:

$$\Delta F_B(\Delta t) = \int_{t_0}^{t_0 + \Delta t} A_B \, n_B(t) \, dt = A_B \int_{t_0}^{t_0 + \Delta t} n_B(t) \, dt \qquad (10.26)$$

where the last term follows because radiant efficiency is a thermodynamic property that does not depend on time. The detailed behavior of amount fluxes of open systems has been discussed previously (see Section 2.2b). The expression in Equation 10.26 may be thought of as the radiant forcing after a time Δt from the injection of an infinitesimal amount of species B into the atmosphere. It decays according to

$$n_B(t) = n_B(t_0) \exp\left(-\frac{t}{\tau_B}\right) \qquad (10.27)$$

where τ_B is the lifetime of B (see Equation 2.48).

The *global warming potential*, ψ_{GWP}, of a species B is defined by[230]

$$\psi_{GWP}(B, \Delta t) \overset{def}{=} \frac{\Delta F_B(\Delta t)}{\Delta F_{CO_2}(\Delta t)} = \frac{\displaystyle\int_{t_0}^{t_0 + \Delta t} a_B \, m_B(t) \, dt}{\displaystyle\int_{t_0}^{t_0 + \Delta t} a_{CO_2} m_{CO_2}(t) \, dt} \qquad (10.28)$$

where the time horizon Δt is a key parameter. Global warming potentials for various greenhouse gases are discussed later in this chapter.

Water is the most abundant greenhouse gas. However, it is not an independent forcing agent, because its lifetime in the atmosphere is very short, ≈ 10 d (see Section 3.2). The amount of water in the atmosphere depends on the ocean surface temperature and weather: the water vapor concentration is a dependent variable, and therefore, most climate models treat water dynamically, that is, temperature changes cause changes in the hydrological cycle, which in turn affects climate.

10.2 Astronomical forcing

The Sun is by far the most important source of energy to the surface of the Earth, and so thermal radiation from the Sun is the obvious starting point in an examination of climate. The Sun's radiant power (the solar constant) is almost constant with time, varying by about 0.1 % over the course of the 11-year solar cycle. Table 10.1 shows that contributions by geothermal, fossil-fuel, and tidal sources are less than the deviation of solar power.

Table 10.1 Sources of power to the surface of the Earth		
Solar radiation[1]	174600 TW	99.977 %
Geothermal energy	23 TW	0.013 %
Fossil fuel combustion (2005)	14 TW	0.008 %
Tidal energy	3 TW	0.002 %

[1] $I_S \pi r_{\oplus}^2 = 174.6$ PW.

Historical changes in global climate have been triggered by changes in insolation (Croll, 1864; Milankovitch, 1920).[5] These variations are caused by slow changes in the geometry of the Earth's orbit in response to predictable changes in the gravitational field experienced by the Earth. This field is mainly due to the Sun (Kepler, 1609), with weak perturbations from the Moon and the seven other planets. Moreover, climate is a complex phenomenon with intercoupled feedbacks. Some, such as the biosphere and ice-albedo feedbacks, are discussed later in this chapter. One behavior of chaotic systems is that they can move from one quasisteady state to another: for example, climate has shifted between glacial and interglacial periods. Variations in the Earth's orbit around the Sun have been most important perturbation triggering these shifts throughout Earth's history. Notable exceptions include intense volcanic activity, meteors, and, most recently, human activity.

In Section 10.2a, we clarify and discuss the present-day insolation as a function of geographical latitude and time of year. Section 10.2b contains an astronomical explanation of the forcing periods, and Section 10.2c deals with astronomical forcing as supported by studies of ice cores and core samples of deep sea sediments.

a. The insolation formula

Although the quantitative expression for insolation, the insolation formula, is the basis for understanding long-term climate variation, we have not found an accessible account for use by chemists. Therefore, in Appendix 5 we have defined pertinent astronomical concepts and quantities and derived the relevant formula. In the present section some figures from the appendix will be used.

The insolation formula is an expression for the 24-h mean insolation $I_{\oplus}$ as a function of the calendar date λ and the geographical latitude φ:

$$I_{\oplus}(\lambda, \varphi; e, \varepsilon, \omega) = I_S(a) \, \chi(\lambda; e, \omega) \, \psi(\lambda, \varphi; \varepsilon) \qquad (10.29)$$

The parameters $a, e, \varepsilon, \omega$ pertain to the orbit and are explained later.

The solar constant I_S is a function of the length of the semimajor axis a, cf. Figure A5.2, but a constant value close to that of Equation 10.17 is used in most climate models.

The second factor χ describes the yearly variation of the irradiance. It is a function of the calendar date λ and contains parametric information about the Earth's orbit: the eccentricity e and the longitude of the perihelion ω,

$$\chi(\lambda; e, \omega) = \left(\frac{1 + e\cos(\lambda - \omega)}{1 - e^2} \right)^2 \qquad (10.30)$$

The calendar date is a measure of the date: $\lambda = 0$ at vernal equinox (March 21), $\lambda = \pi/2$ at summer solstice (June 21), $\lambda = \pi$ at fall equinox (September 21), $\lambda = 3\pi/2$ at winter solstice (December 21); see Figure A5.2B. Other dates are approximate because they do not coincide exactly with the true anomaly.[a]

The quantities λ and ω are measured at the ecliptic with the vernal equinox γ as zero point (see Figure A5.2). The *vernal equinox* is defined as the point (at the equator) where the declination δ of the Sun is zero (see Figures A5.1 and A5.3), changing from negative to positive values. The position of γ changes through time as shown in Figures A5.3 and A5.4; this is called the *moving equinox*. In the astronomical climate theory, the longitude of the perihelion is referred to the moving equinox and denoted $\hat{\omega}$.

The third factor ψ represents the distribution of the irradiance as a function of the calendar date λ and the geographical latitude φ. It contains the obliquity of the ecliptic ε indirectly,

$$\psi(\lambda, \varphi; \varepsilon) \ = \ (\sin h_0 \cos \varphi \cos \delta \ + \ h_0 \sin \varphi \sin \delta)/\pi \qquad (10.31\text{a})$$

where δ is the declination of the Sun,

$$\delta \ = \ \arcsin(\sin \varepsilon \sin \lambda) \qquad (10.31\text{b})$$

and h_0 is the hour angle at sunrise and sunset,

$$h_0 \ = \ \text{Re}\,[\arccos(-\tan \varphi \tan \delta)] \qquad (10.31\text{c})$$

To clarify the significance of h_0: the length of daylight dl is $dl = 2\,h_0$.[b] Examples: Assume $\delta > 0$. If $\varphi > 0$, then $dl > 12$ h (summer in the Northern Hemisphere); if $\varphi \geq \pi/2 - \delta$, then $dl = 24$ h (midnight sun); if $-\varphi > 0$, then $dl < 12$ h (winter in the Southern Hemisphere); if $-\varphi \geq \pi/2 - \delta$, then $dl = 0$ (polar night); if $\varphi = 0$, then $dl = 12$ h (equator, length of day throughout the year). Because the present orbital parameters are known:

$$\text{The eccentricity, } e \ = \ 0.01670 \qquad (10.32\text{a})$$
$$\text{The obliquity, } \varepsilon \ = \ 23.433° \qquad (10.32\text{b})$$
$$\text{The longitude of the perihelion, } \omega \ = \ 282.8° \qquad (10.32\text{c})$$

one can calculate the 24 h mean insolation at any date λ and any geographical latitude φ.

Figure 10.3, left side, shows the insolation $I_\oplus$ (see Equation 10.29), for four values of the calendar date and all values of the geographical latitude, $-90° \leq \varphi \leq +90°$. Note that the series of functions $0° \leq \lambda \leq 180°$ and the series of functions $180° \leq \lambda \leq 360°$ are not symmetric with respect to the equator, $\varphi = 0°$. For example, the polar values at the solstices are $I_\oplus$ $(90°, +90°; e, \varepsilon, \omega) = 527$ Wm^{-2} (summer solstice) and $I_\oplus$ $(270°, -90°; e, \varepsilon, \omega) = 562$ Wm^{-2} (winter solstice), reflecting the variation in the Sun-Earth distance over a year.

---

[a] The term *true anomaly* emphasizes that the quantity is measured at the ecliptic; a related but different concept is *eccentric anomaly*. Similarly, longitudes are here only measured at the ecliptic; therefore we may omit the word *true* in *true longitude*.

[b] For example, for $\delta = 0$, $h_0 = \pi/2 \ \Rightarrow \ dl = (24\,h_0/\pi)$ h.

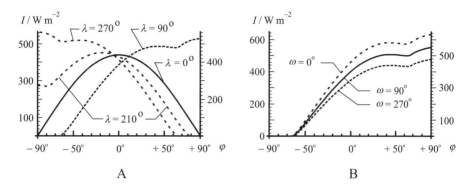

Figure 10.3 *Examples of the insolation $I_\oplus$*
The value of $I_\oplus (\lambda, \varphi; e, \varepsilon, \omega)$ (see Equation 10.29) as a function of the geographical latitude φ at four calendar dates λ (left panel) for various values of the longitude of the perihelion ω at summer solstice, $\lambda = 90°$ (right panel). In each case the remaining orbital parameters have their present values (see Equations 10.32).

The position of the longitude of the perihelion ω relative to the vernal equinox γ is a critical parameter in astronomical climate theories. The right side of Figure 10.3 compares three characteristic positions of ω at summer solstice and shows that the value of this parameter has a profound influence on the yearly distribution of heat and thereby climate. Note that the chosen value the parameter $\omega = 270°$ is close to the present value of ω; see Equation 10.32c.

The integrated insolation is given by

$$I_S \frac{1}{4} \int\limits_{\lambda=0}^{2\pi} \int\limits_{\varphi=-\pi/2}^{+\pi/2} \chi(\lambda; e, \omega)\, \psi(\lambda, \varphi; \varepsilon)\, \mathrm{d}\lambda\, \mathrm{d}\varphi = I_S \frac{2 + e^2}{2(1 - e^2)^2} \quad (10.33)$$

which is the variation of the solar constant due to the eccentricity e.

Calculation of the Milankovitch cycles consists of an examination of the long-term variations of the orbital parameters e, ε, and ω using celestial mechanics. Some results are presented in Section 10.2b.

b. Time dependence of insolation

Milankovitch[5] (1920) carried out the first in-depth calculation of the time dependence of insolation. Celestial mechanics deals with a many-body problem (Sun, Earth, Moon, seven other planets) solved using Newtonian mechanics (in the Lagrange formalism), and this was a considerable undertaking considering the numerical methods and technical means available at that time. Present research attempts to correlate the theoretical findings with geological and archaeological results, and this is discussed in the following sections.

The generic insolation formula (Equation 10.29) is

$$I_\oplus(\lambda, \varphi; e, \varepsilon, \hat{\omega}) = I_S \left(\frac{1 + e\cos(\lambda - \hat{\omega})}{1 - e^2} \right)^2 \psi(\lambda, \varphi; \varepsilon) \quad (10.34)$$

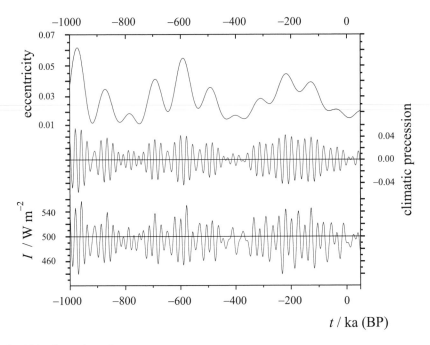

Figure 10.4 *The Milankovitch cycles*
Each curve includes five time-dependent Fourier terms.[143]
Upper curve: The eccentricity (Equation 10.37 and Table 10.2).
Middle curve: The climatic precession (Equation 10.38 and Table 10.2).
Lower curve: The insolation at 65° N and summer solstice (Equation 10.35).
Over the past million years the eccentricity has ranged between 0.01 and 0.06 and
the climatic precession between -0.045 and $+0.045$. The obliquity has oscillated
rather regularly between 22.2° and 24.3°.

where the longitude of the perihelion $\hat{\omega}$ is referred to the moving vernal equinox.
Some special cases are particularly interesting: the insolation at the equator[144] where
the last function is $\psi = \pi^{-1} \cos \delta$, and the insolation at the solstices, $\lambda = \pi/2$ and
$\lambda = 3\pi/2 = -\pi/2$. The latter is given by:

$$I_{\oplus}(\pm\pi/2, \varphi; e, \varepsilon, \hat{\omega}) = I_S \left(\frac{1 \pm e \sin(\lambda - \hat{\omega})}{1 - e^2} \right)^2 \psi(\pm\pi/2, \varphi; \varepsilon) \quad (10.35)$$

Milankovitch was interested in the correlation of the onset and the duration of
ice ages with his calculated insolations. He considered climate at the geographical latitude $\varphi = 65°$ N at summer solstice as particularly important in this respect
(Figure 10.4, lower curve). Often in the early calculations e^2 and higher powers of e
were neglected;[a] this approximation gives

$$I_{\oplus}(\pm\pi/2; e, \hat{\omega}) = I_S(1 \pm 2e \sin \hat{\omega})\psi(\pm\pi/2, \varphi; \varepsilon) \quad (10.36)$$

The important quantity $e \sin \hat{\omega}$ is called the *climatic precession*.

[a] Even at $e = 0.07$, the error of this approximation is only 5 ‰.

Table 10.2 A few Fourier coefficients and periods for the eccentricity e, the obliquity of the ecliptic ε, and the climatic precession, $e \sin \hat{\omega}$

	$a_0/2$	a_1	c_1	a_2	c_2	
e	0.0281	0.0113	404 ka	0.008	95 ka	
ε	23.264°	−0.547°	41 ka	−0.251°	40 ka	
	b_1	s_1	b_2	s_2	b_3	s_3
$e \sin \hat{\omega}$	0.0190	23.7 ka	0.0163	22.4 ka	0.0130	19.0 ka

More recent results include the following.[143] The eccentricity $e(t)$ and the obliquity of the ecliptic $\varepsilon(t)$ as functions of time t can be expanded as a Fourier cosine series

$$f(t) = \frac{a_0}{2} + \sum_{i=1}^{\infty} a_i \cos(2\pi t/c_i + p_i) \tag{10.37}$$

in terms of the amplitude a_i, the period c_i, and a phase p_i. $a_0/2$ is a constant that, together with the phases, defines the boundary conditions. Although the number of calculated terms exceeds 45, Table 10.2 shows the two largest terms; the zero of time is taken as 1950 AD. The eccentricity over the past 1 Ma is shown in Figure 10.4.[a] The climatic precession $e \sin \hat{\omega}(t)$ can be given in an analogous way as a Fourier sine series

$$f(t) = \sum_{i=1}^{\infty} b_i \sin(2\pi t/s_i + q_i) \tag{10.38}$$

Data are given in Table 10.2, and the results over the past 1 Ma are shown in Figure 10.4.

The Milankovitch cycles explain changes in the sunlight reaching the Earth – for example, overhead angle, the intensity of the seasons, and similar external factors:

1. The most important term in the series expansion of eccentricity has a period of 404 ka. The next four terms have periods between 95 ka and 131 ka; therefore, in a low-resolution spectrum of insolation, one sees peaks with period ≈ 100 ka as is observed in Figure 10.4.
2. The spectrum of obliquity is simple and is dominated by components with periods near 41 ka. This spectrum is not shown in Figure 10.4, but the enlarged insolation curve of Figure 10.6 clearly shows a peak frequency of slightly more than 41 ka.
3. The climatic precession $e \sin \hat{\omega}$ is composed of several components with periods as shown in Table 10.2. A low-resolution spectrum has a period near 22 ka, and this was the precession observed by Hipparchos (140 BC) by comparison with older data; see Figure A5.3, legend B.

The astronomical theory of climate is concerned with the effects of variations in the Earth's orbit on climate. The calculations begun by Milankovitch can now be carried out with great precision, and the main objective is to establish links between

---

[a] Each curve in Figure 10.4 includes five time-dependent Fourier terms with phases (see ref. 143).

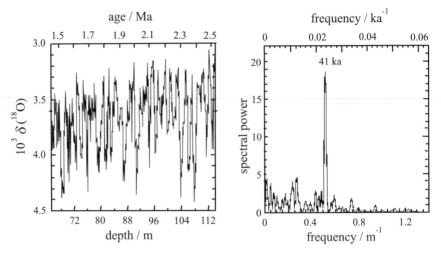

Figure 10.5   *Deep sea-floor drilling*
Left panel: $\delta(^{18}O)$ as a function of the depth of the sediment.
Right panel: The Fourier transform shows a peak at 0.52 m^{-1}. To place a time scale on the data, a sedimentation rate of 45.6 m Ma^{-1} was assumed.
Adapted from ref. 195.

the changing seasonal and latitudinal distribution of solar irradiation and climate. The global insolation integrated over a year changes by only a tiny amount as a result of 1, 2, and 3, which is not enough, in itself, to explain the ice ages. However, the Milankowitch cycles are able to trigger a powerful positive feedback mechanism in the Earth system, the ice-albedo feedback. During the winter at $65°$ N in a period of high annual Milankowitch variability, glaciers will grow. A white surface reflects sunlight, preventing the glacier from melting as much in the summer as it grew in the winter. Thus, even under conditions of equal average annual temperature, the growth of glaciers is promoted when annual climate variation is high and suppressed when it is low. The feedback mechanism of ice growth leading to increased reflectivity for solar radiation, leading to a cooler climate is linked to the onset of ice ages. Such correlation is shown in Figure 10.6.

In the absence of anthropogenic influences, the Milankovitch cycles predict that the long-term cooling trend that began around 6 ka BP will continue for the next 23 ka[177] (see Figure 10.4).

c. Climate recorded in sediments and glacial ice

The left panel of Figure 10.5 shows the record of the oxygen isotope ratio $\delta(^{18}O)$ in ocean sediments as a function of depth, a proxy for temperature (see Equation 1.15). The right panel of Figure 10.5 shows the Fourier transform of the isotope record with a clean sharp peak at a frequency of 41 ka, matching the Milankovitch obliquity

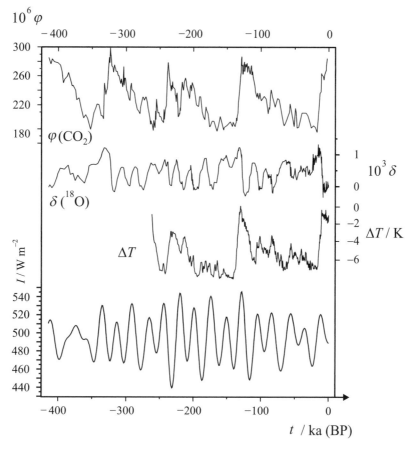

Figure 10.6 *Vostok ice core record*
From top: concentration of atmospheric CO_2, volume fraction φ; isotopic
enrichment δ (^{18}O); temperature record ΔT; insolation I at summer solstice, 65° N.
Zero of time taken as 1950 AD. Based on ref. 200; see ref. 249b.

signal. Figures 10.6 and 10.7 show a longer reconstruction of Earth's climate based
on the Vostok ice core from Antarctica.

Three-fourths of Earth's freshwater is locked in glaciers in polar regions, including
the ice caps of Antarctica and Greenland (see Table 3.12). Snow falls onto the tops
of these massive glaciers, and at a certain depth the pressure of the snow cap fuses
the crystals into solid ice. Air in the snow is trapped in bubbles, roughly 10 % of the
ice by volume. By drilling into the glacier, it is possible to retrieve ancient air, and
analysis of this air shows the changes in greenhouse gas concentrations through time.
The time record of carbon dioxide is shown in Figure 10.6; similar data for methane
and nitrous oxide also exist. In addition, the isotopic composition of the water itself
records temperature because the isotopic phase equilibrium is temperature dependent
(see Equation 1.15). The $\delta(^{18}O)$ and $\delta(D)$ records are used to reconstruct the climate
of the Southern Hemisphere (Antarctica) and the Northern Hemisphere (Greenland).
Figure 10.6 shows a clear correlation between the Milankovitch forcing and the

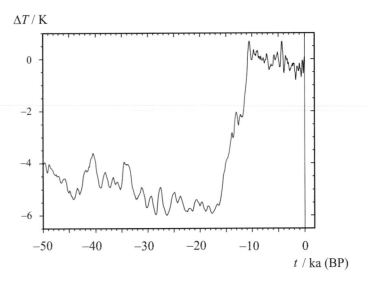

ΔT / K

t / ka (BP)

Figure 10.7 *Vostok ice core record: Temperature*
As discovered by Dansgaard,[157] stable isotopes in atmospheric water vapor are fractionated by evaporation and condensation. As the isotopic composition of the ocean is constant, the isotopic variation trapped in precipitation in ice caps are proxy for hemisphere temperature.
Based on ref. 200; see ref. 249b.

record of glacial and interglacial periods. The ice core drilled at the Vostok station on Antarctica reaches back 420 ka BP; drilling stopped just short of the enormous Lake Vostok beneath the ice. The core records four glacial cycles. The EPICA project[a] has yielded the longest core on record – by drilling 3190 m, ice has been retrieved from 800 ka BP; this core is still being analyzed (2010). Figure 10.7 shows the temperature of the most recent glacial and interglacial period recorded in the Vostok core.

10.3 Modern climate

Climate is "average weather," commonly defined using a 30-year period. Figure 10.8 shows the thermometer record of surface temperature for the last 130 years: climate is changing. Increases are also seen in total rainfall, sea level, and ocean temperature.

Figure 10.8 shows two main periods of warming, the first from 1910 to 1940 and the second starting in the mid-1970s. The 15 warmest years since the invention of the thermometer have occurred since 1990. The climate has warmed by about 0.7 K in the period shown. The rate of increase over the past 30 years has been 0.2 K per decade. Although the global average surface temperature might have been slightly warmer around 11 ka BP than at present (see Figure 10.7), if the warming

---

[a] European Project for Ice Coring in Antarctica.

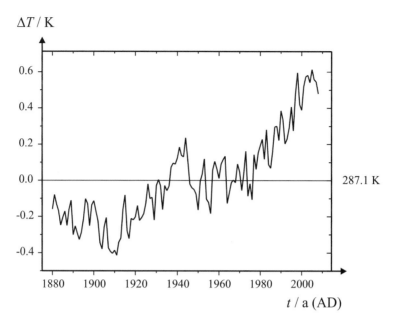

Figure 10.8 *Global annual mean surface temperature 1880 to 2008*
Instrument record of global mean surface temperature.[249a] The zero of the temperature
axis is the global mean temperature (13.9 °C) in the period 1901-2000. In this period,
the land surface mean temperature was 8.5 °C, and the sea surface mean temperature,
16.5 °C.

trend continues, Earth will soon have the warmest climate of any time for the past
1 million years.

a. Causes of climate change

The natural greenhouse effect clearly has a large effect on climate. The key greenhouse
gases, in order of importance, are H_2O, CO_2, CH_4, O_3, N_2O, and the CFCs. During
the past 250 years, the concentrations of the CO_2, CH_4, and N_2O gases have increased
exponentially (Table 10.3).

Similar figures for the change in concentration of CFC-11 are (year: mole fraction
y) 1950: $y = 0$; 1960: $y = 8 \times 10^{-12}$; 1965: $y = 25 \times 10^{-12}$; 1970: $y = 60 \times 10^{-12}$; 1990:
$y = 270 \times 10^{-12}$; 2000: $y = 260 \times 10^{-12}$. The value of y declined after 1990 as a result
of the Montreal Protocol. As discussed in Chapter 4, the concentration of tropospheric
ozone has more than doubled in the Northern Hemisphere since preindustrial times.
The question is how much of an effect the increased concentrations of the greenhouse
gases have had on the climate (see Figure 10.8).

b. Energy flux densities in the atmosphere

The global radiation budget may be summarized as follows. Of the solar radiation that
is not reflected (i.e., 240 W m^{-2}), 68 W m^{-2} are absorbed within in the atmosphere,

Table 10.3 Atmospheric concentrations and radiant forcings of some gases				
Year (AD)	1750	1800	1900	2000
$10^6 \, x(CO_2)$	280	280	300	340
$\Delta F(CO_2) \, / \, W \, m^{-2}$	0.1	0.1	0.4	1.46
$10^9 \, x(CH_4)$	700	750	900	1745
$\Delta F(CH_4) \, / \, W \, m^{-2}$	0.0	0.0	0.1	0.48
$10^9 \, x(N_2O)$	270	271	280	314
$\Delta F(N_2O) \, / \, W \, m^{-2}$	0.0	0.0	0.06	0.15

for example, by stratospheric ozone and other gas-phase molecules, and by particles such as soot and dust; 170 W m^{-2} are absorbed by the surface of the planet. Less than 10 % of the radiation emitted by the surface goes directly to space; most of it is absorbed by the atmosphere. There is about twice as much energy received by the surface from the atmosphere as directly from the Sun. Most of the planet's emission to space occurs from the atmosphere and not from the surface.

c. Radiant forcing

The addition of a greenhouse gas inhibits radiation from escaping to space, and this trapped energy may cause the surface temperature to increase. The problem can be divided into two parts. The first, the radiant forcing, which is relatively easy to calculate, is the effect that the addition of a gas or change in some other factor has on the radiation balance. The second part is to convert a radiant forcing into a climate forcing. The Intergovernmental Panel on Climate Change (IPCC) has evaluated the radiant forcings caused by the concentration changes shown in Table 10.3. The total radiant forcing of the long-lived greenhouse gases is $\Delta F_{tot} = 2.64$ W m^{-2}, which is a ≈ 1 % change relative to the part of the Sun's irradiance that reaches the surface.

There are many factors in addition to the greenhouse gases that cause radiant forcing. The IPCC has summarized all known changes in radiant forcing from 1750 to the present,[231] as shown in Table 10.4.

As seen, many factors have been identified. The causes of the changes in the concentrations of the long-lived greenhouse gases CO_2, CH_4, CFCs, and N_2O have been discussed in Chapter 4, and the changes themselves were described in this section. In addition, Table 10.4 introduces some new factors that we now briefly discuss.

1. Tropospheric and stratospheric ozone
As described in Chapter 4, the tropospheric ozone concentration has more than doubled since preindustrial times. Air pollution has intensified, especially in the Northern Hemisphere, because of increases in NO_x, carbon monoxide, and hydrocarbons. In contrast, also as described in Chapter 4, stratospheric ozone concentrations have decreased because of increased chlorine and water concentrations in the stratosphere. Ozone is a greenhouse gas, so the decrease of stratospheric ozone has led to cooling

Table 10.4 Present-day global average radiant forcings relative to 1750		
Species	$\Delta F / \text{W m}^{-2}$	LOSU
CO_2	+1.66 [+1.49 - +1.83]	H
CH_4	+0.48 [+0.43 - +0.53]	H
CFC's	+0.34 [+0.31 - +0.37]	H
N_2O	+0.16 [+0.14 - +0.18]	H
Tropospheric ozone	+0.35 [+0.25 - +0.65]	M
Stratospheric ozone	−0.05 [−0.15 - +0.05]	M
Stratospheric H_2O from CH_4	+0.07 [+0.02 - +0.12]	L
Albedo – land use change	−0.2 [−0.4 - −0.0]	M-L
Albedo – soot on snow	+0.1 [+0.0 - +0.2]	M-L
Direct aerosol	−0.5 [−0.9 - −0.1]	M-L
Indirect aerosol (cloud albedo)	−0.7 [−1.8 - −0.3]	L
Contrails (aviation induced)	+0.01 [+0.003 - +0.03]	L
Total net anthropogenic	+1.6 [+0.6 - +2.4]	
Solar irradiance	+0.12 [+0.06 - +0.30]	L

Anthropogenic forcing compared with the change of the natural solar irradiance. The limits [. . .] on the estimates may not be symmetrical and the sum cannot be obtained by simple addition.[231]
Level of Scientific Understanding (LOSU) is H for high, M for medium, L for low.

and the increase of tropospheric ozone has led to warming. An additional effect is that the decreased amount of ozone in the stratosphere has led to increased UV-radiation in the troposphere, and therefore enhanced hydroxyl concentrations. This has slightly decreased the concentration of methane (and other reactive gases), giving a cooling effect.

2. Stratospheric water vapor

In order to enter the stratosphere, air must pass through the cold tropopause. Condensation and precipitation therefore limit the amount of water that can enter the stratosphere. Nonetheless, stratospheric water vapor concentrations have increased by 50 % in the last half of the 20th century. Part of the increase is due to the increase in atmospheric methane, which is oxidized in the stratosphere to produce water (and CO_2). The rest is linked to water vapor transport and may be due to greater strength of tropical convective storms. In addition, precipitation may be lifted through the tropopause. Ice crystals would normally fall out of the atmosphere, but in an updraft the crystals can be blown upward, later releasing the water as vapor. Overall, the increase in stratospheric water vapor, from methane and transport, has caused a positive radiative forcing.

3. Albedo – land use change

Equation 10.19 shows the important role of albedo. The average albedo of the earth, $\rho_{\oplus} \approx 0.30$, is made up of many components. The ocean's albedo, typically less than

0.2, is highly variable, depending on factors such as the solar zenith angle, wind speed, and chlorophyll concentration.[181] In contrast, fresh snow can have a reflectivity up to 0.90, and the albedo of clouds varies from 0.4 to 0.8. Human activity has increased the albedo as forests with a low albedo (conifer forest 0.1) have been replaced with fields (bare soil 0.17), grasslands (0.25), and concrete (0.55) that are more reflective. Desertification also plays a role, as desert sand has an albedo around 0.40. As seen in Table 10.4, land use change has had a cooling effect.

4. Albedo – soot on snow

Soot is concentrated at the surface when snow and ice sublime. The normally high albedo of snow (0.8 to 0.9) and ocean ice (0.5 to 0.7) can be changed dramatically by the addition of a small amount of black carbon soot. Charcoal with an albedo of 0.04 is one of the darkest substances known. The soot-on-snow effect is associated with positive feedback, as the absorptance of the surface increases as a consequence of surface warming.[174]

5. Direct aerosol

Particles in the atmosphere scatter radiation, increasing the albedo and cooling the surface. The main component is sulfate aerosol due to the use of coal and diesel oil containing sulfur. Figure 10.8 shows that the global mean surface temperature did not increase in the years after World War II, a period of rapid industrialization. Higher resolution data show that the industrial regions of the planet – the Great Lakes region and East Coast of North America, Europe, and Asia – actually cooled during this period, whereas other regions showed warming, giving an overall neutral trend. Soot, fossil fuel organic carbon and black carbon, biomass burning, nitrate, and mineral dust also contribute to the direct aerosol effect.

6. Indirect aerosol

Air pollution increases the number of cloud condensation nuclei, seed particles on which cloud droplets grow. Imagine a cloud with a certain mass fraction of liquid water. In a clean environment with few nuclei, this water would be divided among relatively few, large droplets. In a polluted environment, the same amount of liquid water would be partitioned among many more, but smaller, droplets. The smaller droplets are better at scattering light. Thus, clouds forming in polluted regions are "whiter" than they would otherwise be, increasing the planetary albedo and giving a cooling effect.

7. Contrails

The white condensation trails (contrails) from high-altitude airplanes are clouds that occur in the cold upper troposphere or lower stratosphere. The warming effect of the additional water vapor outweighs the increased albedo, giving on balance a warming effect.

8. Solar irradiance

Since 1750, the Sun has moved out of a shorter period of lower activity called the Maunder Minimum into a phase of relatively higher activity. This small increase in solar output results in warming.

In addition, from time to time there is explosive volcanic activity of sufficient strength to affect the global climate. For example, the eruption of Mount Pinatubo in 1991 cooled the planet by about 0.5 K. After 1 - 2 years, the enhanced stratospheric sulfate aerosols generated by the volcanic gases had decayed such that the climate attained its preeruption state. Volcanic activity is episodic and has not affected the trend of climate from 1750 to present.

d. Global warming potential

The global warming potential ψ_{GWP} is used to compare the climate effects of different gases. It is defined as the radiant forcing ΔF_B induced by the addition of an amount of a gas B over some length of time Δt relative to the same amount of a reference compound, usually carbon dioxide (see Equation 10.28).

Carbon dioxide is an inefficient greenhouse gas relative to many other gases. The reason is that it has a high atmospheric concentration and relatively few infrared absorption bands. Most of its atmospheric absorptions are saturated, so the radiation budget is relatively insensitive to changes in concentration. As seen in Figure 10.2, there is an "infrared window" from 8 to 13 μm where H_2O, CO_2, CH_4, and N_2O do not absorb, but where the surface emits (see Figure 10.1). Therefore, gases that absorb in this region such as O_3 tend to have large ψ_{GWP}'s. Because the frequency of the carbon-fluorine stretching vibration falls in this window, chlorofluoro carbons are powerful greenhouse gases. In fact, the Montreal Protocol on substances that deplete the ozone layer, including CFCs (Chapter 4), has reduced greenhouse forcing in the atmosphere five times more than the Kyoto Treaty limiting greenhouse gases.

Equation 10.28 involves the atmospheric lifetime of a gas, but for CO_2 there is not a single exponential lifetime but rather the decay function (Equation 7.1), with an approximate lifetime of 1.1 a. The ψ_{GWP}'s of some greenhouse gases are listed in Table 10.5.

e. Climate sensitivity

The *climate sensitivity parameter* λ links radiant forcing to temperature change, using Equation 10.39:

$$\Delta T = \lambda \Delta F \tag{10.39}$$

Using the simple one-layer atmospheric model introduced earlier, λ is calculated to be 0.3 K m^2 W^{-1}. Substituting in the radiant forcing due to long-lived greenhouse gases ($\Delta F_{tot} = 2.64$ W m^{-2}, Table 10.4) yields a temperature change of 0.8 K, in agreement with the observed temperature change. However, the climate response is much more complicated than this simple model, and estimates of the climate

Table 10.5 Global warming potentials ψ_{GWP}

Code	Formula	$10^{12} x(1750)$	$10^{12} x(1998)$	τ / a	ψ_{GWP}
	CO_2	– See Table 10.3 –		1.1	1
	CH_4	700,000	1,745,000	12	23
	N_2O	270,000	314,000	114	296
	CF_4	40	80	<50,000	5,700
	C_2F_6	–	3.0	10,000	11,900
	SF_6	–	4.2	3,200	22,200
HFC-23	CHF_3	–	14	260	12,000
HFC-134a	CF_3CH_2F	–	7.5	13.8	1,300
HFC-152a	CH_3CHF_2	–	0.5	1.4	120
CFC-11	$CFCl_3$	–	268	45	4,600
CFC-12	CF_2Cl_2	–	533	100	10,600
CFC-13	CF_3Cl	–	4	640	14,000
CFC-113	$CF_2Cl\,CFCl_2$	–	84	85	6,000
CFC-114	$CF_2Cl\,CF_2Cl$	–	15	300	9,800
CFC-115	CF_3CF_2Cl	–	7	1,700	7,200
	CCl_4	–	102	35	1,800
	CH_3CCl_3	–	69	4.8	140
HCFC-22	CHF_2Cl	–	132	11.9	1,700
HCFC-141b	CH_3CFCl_2	–	10	9.3	700
HCFC-142b	CH_3CF_2Cl	–	11	19	2,400
Halon-1211	CF_2ClBr	–	3.8	11	1,300
Halon-1301	CF_3Br	–	2.5	65	6,900
Halon-2402	$CF_2Br\,CF_2Br$	–	0.45	<20	

x mole fraction. τ lifetime. The ψ_{GWP}'s are calculated using a 100-year time horizon, relative to CO_2. Greenhouse gases included in the Montreal Protocol and its amendments are the gases including and below CFC-11. See ref. 86.

sensitivity parameter appearing in the literature vary from 0.3 to 1.4 K m^2 W^{-1}. Positive feedbacks in the earth system (see later discussion) mean that the climate is more sensitive than indicated by a simple algebraic model. Thus, there is "room" in the budget for the cooling effect of aerosols (Table 10.4).

The one-layer atmospheric model considers the instantaneous radiative equilibrium. In contrast, there are many elements in the climate system that change with time, producing feedbacks on climate. The long-term effect of increased radiative forcing will be larger than the short-term effect. One example is that more than 80% of the additional energy in the climate system is going into the ocean, which will take hundreds of years to achieve a steady state. Some climate feedbacks are related to the albedo. Cloud distributions will change in response to higher sea surface temperatures, and the distributions of biomes will change in response to climate change. Deserts will continue to grow, and the distributions of forests and grasslands will change, changing the albedo. As temperature rises, ice will continue to melt, reducing the albedo and giving continued temperature rise. Rising temperatures will also affect the greenhouse gases themselves. The most obvious effect, as mentioned, is the water vapor feedback, in which increased sea surface and atmospheric temperatures lead to greater atmospheric water vapor concentrations. In addition, as the climate warms, soils are likely to give off CO_2. Also, climate change will change the chemistry of the atmosphere. Models show that although perturbations such as the injection of a pulse of CH_4 or CO show some short-lived chemical effects, these pulses also produce long-term changes in greenhouse gases. For example, a pulse of H_2 will consume OH, which is therefore not available to remove CH_4; thus, H_2 (and other reducing gases) acts as an indirect greenhouse gas. Another example is that a pulse of N_2O, according to the chemistry described in Chapter 4, will result in more NO_x in the stratosphere. This will break down ozone catalytically and the reduced ozone shield will transmit more UV light into the troposphere, generating more OH. This OH will reduce the lifetime of methane, mitigating the global warming caused by the additional N_2O. The effect is significant; on a mole fraction basis, the ratio of CH_4 change per N_2O change is -36%. Sophisticated computer models are used to predict the future of climate.

f. Climate change

In this chapter we have shown that climate science can give a good account of variations occurring over at least the past 1 million years, and it is clear that there is a large natural greenhouse effect of around 32 K (see Equation 10.19). In addition, we know that human activity has caused an increase in the concentrations of greenhouse gases in the atmosphere (see Table 10.3), and as seen in Table 10.4, these changes in atmospheric composition have caused a radiative forcing of 1.6 W m^{-2}, relative to a solar input of 240 W m^{-2}. Table 10.6 lists climate changes that have been observed. However, the complexity of the Earth system means that the future course of climate cannot be known with certainty. Models and emission scenarios are necessary in order to predict future climate.

Table 10.6 Climate changes that have been observed

- Sea surface temperatures have increased 0.4 to 0.8 °C since the late 19th century.
- Sea level increase of approximately 1.5 mm/a over the 20th century, total increase 10 to 20 cm.
- Global ocean heat content to 300 m depth increased 0.04 K/daa since the 1950s.[1]
- Massive retreat of mountain glaciers.
- Two-week decrease in ice duration in lakes and rivers at mid to high latitude.
- Arctic sea ice: summer thickness decrease of 40%, 10 to 15% decrease in extent.
- Antarctic sea ice: no significant change since 1978.
- 20% increase in stratospheric water vapor since 1980.
- 2% increase in clouds over land during 20th century.
- 2% increase in clouds over ocean since 1952.
- 5 to 10% increase in Northern Hemisphere mid- to high-latitude precipitation since 1900, much of this in extreme events.

[1] daa = 10 a = decayear.
From IPCC technical summary.[231]

In the beginning, climate models considered meteorological factors, and with time they have advanced to include changes in the biosphere, land surface, ocean, and atmosphere, as well as interactions with human society. Independent climate models have been developed at research centers around the world and are run on the world's most advanced computers. They show clearly that natural variation alone is unable to explain the climate variation seen over the past 100 years (see Table 10.6). In particular, the period since 1970 can only be explained by taking both natural and anthropogenic factors into account. The recent advances in understanding are reflected in the successively stronger terms used by the United Nations expert panel on climate change. In 1990 the IPCC wrote that "growth in greenhouse gas emissions may lead to significant increase in global temperature." In 1996, "the balance of evidence suggests a discernable human influence on global climate." And, in 2001, "greenhouse gases produced mainly by burning of fossil fuels are likely to have contributed substantially to observed warming over the past 50 years." In 2007, the IPCC stated that there is "very high confidence that the globally averaged net effect of human activities since 1750 has been one of warming." According to models, these effects will be even more clearly visible in the next 10 years with a global average temperature increase relative to the postwar baseline[a] of 0.4 K in 2010, 0.8 K in 2020, and 2.3 K in 2050.

However, the change in the global average temperature does not tell the whole story. The observed climate changes have been larger over land than sea, and they have also been larger near the poles than at the equator. These trends are predicted by the climate models and are predicted to continue. Projected changes by 2050 include increased temperatures of 3 to 4 K over land, compared to 1 to 2 K over the oceans. Rainfall patterns will also change – overall, the hydrological cycle is expected to intensify because of increased sea surface temperature leading to increased precipitation and flooding, especially at high latitudes. The dry atmospheric subsidence regions outside of the tropics are predicted to have reduced rainfall, leading to further drought and desertification.

Given the risk of disruption to society, there is a significant effort to find solutions to the problem. Reduction of greenhouse gas emissions is undoubtedly the "Plan A" solution, a path that is recommended by political leaders around the world. This plan would involve only using coal in combination with carbon capture and storage, using oil and gas only for transportation because of its demand for high energy-density fuel, removing CO_2 from the atmosphere using agriculture and forestry, increasing the use of renewable energy (wind, sun, waves, biomass) and nonfossil alternative energy (nuclear), and not least using energy more effectively. Methods of CO_2 abatement are associated with different costs; for example, improving building insulation "costs" about -160 € Mg^{-1} of CO_2 equivalent[b] and low-energy lightbulbs cost -75 € Mg^{-1} of CO_2 equivalent. Clearly these are "no regret" options because they reduce CO_2 emissions by saving energy, and save money at the same time. Other methods have positive costs: for example, solar power costs 20 € Mg^{-1}, and carbon capture and storage 30 € Mg^{-1}. Ethanol fuel from sugarcane costs -25 € Mg^{-1} and ethanol from

[a] 287.1 K, see Figure 10.8.
[b] € Mg^{-1} = Euro per ton.

cellulose around 5 € Mg^{-1}. CO_2 emissions are currently (2010) trading at around 12 € Mg^{-1} on the European market.

The reasons why more progress has not been made on reducing greenhouse gas emissions are varied. The climate threat is diffuse and slow (but steady), whereas business and political planning cycles are short. Although energy is the key to the economy, different countries have different energy use per GDP;[a] the ratio of primary energy use to GDP is twice as high in the United States as in the European Union. Power plants take decades to build and last for a generation or more, and buildings can last for hundreds of years. In addition, the problem can seem large and our ability to act expensive. Because of the greenhouse gases that have already been released, even if all emissions were stopped today, the temperature would continue to rise: CO_2 lingers in the atmosphere and associated reservoirs for hundreds of years.

In addition to stabilizing greenhouse gas emissions, adaptation will be necessary. For example, there are plans to protect Venice from rising sea levels, and many regions are planning infrastructure in response to the anticipated future climate. Beyond to "Plan B" (adaptation), "Plan C" would seek to actively intervene in the climate system, for example, shelling the stratosphere with sulfur to cool the earth with sulfuric acid aerosol, in analogy to the known cooling effect of large volcanic eruptions.

From 1980 to 2005, fossil fuel CO_2 emissions and population have both grown by about 30 %. During this same period, the per capita gross world product increased by about 70 %. Thus, even though CO_2 emissions have increased, the energy intensity of the economy has decreased significantly. At the same time, the world has become more efficient in the amount of energy produced per amount of CO_2 emitted. The conclusion is that the largest single factor contributing to rising greenhouse gas emissions is population growth see Figure 1, page 2.[201]

---

[a] The gross domestic product (GDP) is the amount of goods and services produced per year in a country.

Appendix 1

Appendix 1 provides a summary of internationally accepted quantities, units, and symbols from physical chemistry relevant to this book.[126] The CODATA[a] recommended values for fundamental physical constants may be found in the same reference, or on the Internet.[119]

A1.1 Symbols of the elements

Table A1.1 The periodic table of the elements

	1	2	3	4	5	6	7	8	9	10	11	12	13	14	15	16	17	18
1	H																	He
2	Li	Be											B	C	N	O	F	Ne
3	Na	Mg											Al	Si	P	S	Cl	Ar
4	K	Ca	Sc	Ti	V	Cr	Mn	Fe	Co	Ni	Cu	Zn	Ga	Ge	As	Se	Br	Kr
5	Rb	Sr	Y	Zr	Nb	Mo	Tc	Ru	Rh	Pd	Ag	Cd	In	Sn	Sb	Te	I	Xe
6	Cs	Ba	La	Hf	Ta	W	Re	Os	Ir	Pt	Au	Hg	Tl	Pb	Bi	Po	At	Rn
7	Fr	Ra	Ac	Rf	Db	Sg	Bh	Hs	Mt	Ds	Rg	Cn						
6				Ce	Pr	Nd	Pm	Sm	Eu	Gd	Tb	Dy	Ho	Er	Tm	Yb	Lu	
7				Th	Pa	U	Np	Pu	Am	Cm	Bk	Cf	Es	Fm	Md	No	Lr	

The periodic table of the chemical elements and the designations of groups (columns) and periods (rows) shown here follow the IUPAC recommendations.[234]

The symbol of an element may be surrounded by indices; in such cases, their positions and meanings are as follows:[126]

Left superscript	Mass number
Left subscript	Atomic number, Z
Right superscript	Charge number; oxidation number; excitation symbol
Right subscript	Number of atoms per entity

Examples: $^{79}Br^+$, $^{79}Br^-$, $_{28}Ni^0$, Ni^{2+}, $^{16}O^{-II}$, O^{2-}, $^{36}Cl^{\bullet}$, $^{32}S_8$

[a] CODATA = Committee on Data for Science and Technology, an International Council for Science (ICSU) committee, http://www.codata.org (2010).

A1.2 Atomic weights of the elements

Table A1.2 Atomic weights of the elements abridged to four significant digits[215]			
Atomic number	Element name	Symbol	Atomic weight
1	Hydrogen	H	[1.007; 1.009]
2	Helium	He	4.003
3	Lithium	Li	[6.938; 6.997]
4	Beryllium	Be	9.012
5	Boron	B	[10.80; 10.83]
6	Carbon	C	[12.00; 12.02]
7	Nitrogen	N	[14.00; 14.01]
8	Oxygen	O	[15.99; 16.00]
9	Fluorine	F	19.00
10	Neon	Ne	20.18
11	Sodium	Na	22.99
12	Magnesium	Mg	24.31
13	Aluminium	Al	26.98
14	Silicon	Si	[28.08; 28.09]
15	Phosphorus	P	30.97
16	Sulfur	S	[32.05; 32.08]
17	Chlorine	Cl	[35.44; 35.46]
18	Argon	Ar	39.95
19	Potassium	K	39.10
20	Calcium	Ca	40.08[1]
21	Scandium	Sc	44.96
22	Titanium	Ti	47.87
23	Vanadium	V	50.94
24	Chromium	Cr	52.00
25	Manganese	Mn	54.94
26	Iron	Fe	55.85
27	Cobalt	Co	58.93
28	Nickel	Ni	58.69
29	Copper	Cu	63.55
30	Zinc	Zn	65.38(2)
31	Gallium	Ga	69.72
32	Germanium	Ge	72.63
33	Arsenic	As	74.92
34	Selenium	Se	78.96(3)
35	Bromine	Br	79.90
36	Krypton	Kr	83.80[1]
37	Rubidium	Rb	85.47[1]
38	Strontium	Sr	87.62[1]
39	Yttrium	Y	88.91
40	Zirconium	Zr	91.22[1]
41	Niobium	Nb	92.91

Atomic number	Element name	Symbol	Atomic weight
42	Molybdenum	Mo	95.96(2)[1]
43	Technetium[2]	Tc	
44	Ruthenium	Ru	101.1[1]
45	Rhodium	Rh	102.9
46	Palladium	Pd	106.4[1]
47	Silver	Ag	107.9[1]
48	Cadmium	Cd	112.4[1]
49	Indium	In	114.8
50	Tin	Sn	118.7[1]
51	Antimony	Sb	121.8[1]
52	Tellurium	Te	127.6[1]
53	Iodine	I	126.9
54	Xenon	Xe	131.3[1]
55	Cesium	Cs	132.9
56	Barium	Ba	137.3
57	Lanthanum	La	138.9
58	Cerium	Ce	140.1[1]
59	Praseodymium	Pr	140.9
60	Neodymium	Nd	144.2[1]
61	Promethium[2]	Pm	
62	Samarium	Sm	150.4[1]
63	Europium	Eu	152.0[1]
64	Gadolinium	Gd	157.3[1]
65	Terbium	Tb	158.9
66	Dysprosium	Dy	162.5[1]
67	Holmium	Ho	164.9
68	Erbium	Er	167.3[1]
69	Thulium	Tm	168.9
70	Ytterbium	Yb	173.1[1]
71	Lutetium	Lu	175.0
72	Hafnium	Hf	178.5
73	Tantalum	Ta	180.9
74	Tungsten	W	183.8
75	Rhenium	Re	186.2
76	Osmium	Os	190.2
77	Iridium	Ir	192.2
78	Platinum	Pt	195.1
79	Gold	Au	197.0
80	Mercury	Hg	200.6
81	Thallium	Tl	[204.3; 204.4]
82	Lead	Pb	207.2
83	Bismuth	Bi	209.0[1]
84	Polonium[2]	Po	
85	Astatine[2]	At	
86	Radon[2]	Rn	

Table caption: **Table A1.2** (*continued*)

Table A1.2 (*continued*)			
Atomic number	Element name	Symbol	Atomic weight
87	Francium[2]	Fr	
88	Radium[2]	Ra	
89	Actinium[2]	Ac	
90	Thorium[2]	Th	232.0
91	Protactinium[2]	Pa	231.0
92	Uranium[2]	U	238.0[1]

The atomic weight of an element B is the relative molar mass $M_{r, B} = M_B/M_u$ where $M_u = 1 \text{ g mol}^{-1}$.

The atomic weights of many elements are not invariant, but depend on the origin and treatment of the material. The standard values of $M_{r, B}$ and the uncertainties (in parentheses, following the last significant figure to which they are attributed) apply to elements of natural terrestrial origin. The last significant figure of each tabulated value is considered reliable to ± 1 except when a larger single-digit uncertainty is inserted in parentheses following the atomic weight. For 10 of these elements, an atomic-weight interval is given with the symbol $[a; b]$ to denote the range of atomic-weight values in normal materials; thus, $a \leq M_{r, B} \leq b$. The symbols a and b denote the lower and upper bounds of the interval.

[1] Values may differ in some naturally occurring samples because of variation in the abundances of the element's stable isotopes.

[2] The element has no stable isotopes.

A1.3 The international system of units, SI

Table A1.3a Names and symbols for the SI base units		
Physical quantity	Name of SI unit	Symbol for SI unit
l (length)	metre	m
m (mass)	kilogram	kg
t (time)	second	s
I (electric current)	ampere	A
T (thermodynamic temperature)	kelvin	K
n (amount of substance)	mole	mol
I_v (luminous intensity)	candela	cd

Table A1.3b SI prefixes					
Factor	Prefix	Symbol	Factor	Prefix	Symbol
10^{-1}	deci	d	10^{1}	deca	da
10^{-2}	centi	c	10^{2}	hecto	h
10^{-3}	milli	m	10^{3}	kilo	k
10^{-6}	micro	μ	10^{6}	mega	M
10^{-9}	nano	n	10^{9}	giga	G
10^{-12}	pico	p	10^{12}	tera	T
10^{-15}	femto	f	10^{15}	peta	P
10^{-18}	atto	a	10^{18}	exa	E
10^{-21}	zepto	z	10^{21}	zetta	Z
10^{-24}	yocto	y	10^{24}	yotta	Y

A1.4 Nonstandard units and suffixes

Table A1.4a Some frequently used non-SI units			
Physical quantity	Name of unit	Symbol for unit	Value in SI units
Area	hectare	ha	$= 1\ hm^2$
Volume	liter	l	$= 1\ dm^3$
Mass	ton	t	$= 1\ Mg$
Mass	unified atomic mass unit	u, Da [1]	$= m_a(^{12}C)/12$
Temperature[2]	degree Celsius	°C	$= 1\ K$
Time	year	a	$\approx 31\ 559\ 952\ s$

Note that the unit a is used with the prefixes k, M, and G, but *never* with h. The unit ha is only used for hectare, $1\ ha = 10^4\ m^2$.

[1] $1\ Da = 1.660538762 \times 10^{-27}$ kg.

[2] Celsius temperature θ is related to thermodynamic temperature T by the equation $\theta\ /\ °C = T\ /\ K - 273.15$.

Table A1.4b Fractions occasionally used in chemistry		
Name	Symbol	Value
percent	%	10^{-2}
permil	‰	10^{-3}
part per million	ppm	10^{-6}
part per billion	ppb	10^{-9}
part per trillion	ppt	10^{-12}
part per quadrillion	ppq	10^{-15}

Note: These multiples of the unit one are not part of the SI, and ISO[a] does not recommend their use.

[a] ISO = International Organization for Standardization; see http://www.iso.org/ (2010).

A1.5 Transport properties

Name	Symbol	Definition	SI unit
General physical quantity	X		(varies)
Flux (of quantity X)[1]	$\dot{X}, q_X$	$\dot{X}, dX/dt$	(varies)
Flux density (of quantity X)	J_X	$J_X = \dot{X}/A$	(varies)
Mass flux	$\dot{m}, q_m$	$\dot{m} = dm/dt$	kg s^{-1}
Mass flux density	J_m	$J_m = \dot{m}/A$	kg m^{-2} s^{-1}
Amount flux[2]	$\dot{n}, q_n$	$\dot{n} = dn/dt$	mol s^{-1}
Amount flux density[3]	J_n	$J_n = \dot{n}/A$	mol m^{-2} s^{-1}
Energy	E		J
Energy flux	$\dot{E}, q_E$	$\dot{E} = dE/dt$	W
Energy flux density	J_E	$J_E = \dot{E}/A$	W m^{-2}
Diffusion coefficient[4]	D_B	$J_{n_B} = -D_B \, \mathbf{grad} \, c_B$	m^2 s^{-1}

[1] Any flux or flux density can be defined as a vector quantity, for example, $q_X = (dX/dt) \, e$, where e is a unit vector in the direction of the flow.

[2] Abbreviation for "amount-of-substance flux."

[3] Abbreviation for "amount-of-substance flux density."

[4] c is the chemical concentration (*i.e.* amount-of-substance concentration).

A1.6 Electricity

Table A1.6 Selected electrical quantities with SI units[80,126]

Name	Symbol	SI unit
Electric current	I	A
Electric current density[1]	$\mathbf{j}$	A m^{-2}
Electric potential difference	$\Delta\varphi$	$V = J \, C^{-1}$
Electric field strength[1]	$\mathbf{E}$	V m^{-1}
Dielectric displacement[1]	$\mathbf{D}$	C m^{-2}
Permittivity[2]	ε	F m^{-1}
Electric resistance	R	$\Omega = V \, A^{-1}$
Resistivity	ρ	Ω m
Conductivity	κ	S m$^{-1} = \Omega^{-1}$ m^{-1}
Molar conductivity	Λ	S m^2 mol^{-1}
Ionic conductivity	λ	S m^2 mol^{-1}
Migration speed	v	m s^{-1}
Electric mobility[3]	u	m^2 V^{-1} s^{-1}
Transport number	t	1
Zeta potential	ζ	V

[1] $\mathbf{E}$, $\mathbf{D}$, and $\mathbf{j}$ are vector quantities; recall that $\mathbf{D} = \varepsilon \, \mathbf{E}$ and $\mathbf{j} = \kappa \, \mathbf{E}$.

[2] The unit F $= C \, V^{-1}$.

[3] The unit is (m s^{-1})/(V m^{-1}); see Equation 5.92.

A1.7 General chemistry

Table A1.7 General chemistry			
Name	Symbol	Definition	SI unit[1]
Number of entities	N		1
Amount (of substance)[2]	n	$n_B = N_B/N_A$	mol
Avogadro constant	N_A		mol^{-1}
Molar mass	M	$M_B = m/n_B$	$kg\ mol^{-1}$
Molar volume	V_m	$V_{m,B} = V/n_B$	$m^3\ mol^{-1}$
Mass fraction	w	$w_B = m_B/\Sigma\ m_i$	1
Mole fraction[3]	x, y	$x_B = n_B/\Sigma\ n_i$	1
Pressure	p		Pa
Partial pressure	p_B	$p_B = y_B\ p$	Pa
Mass density[4]	ρ	$\rho = m/V$	$kg\ m^{-3}$
Specific volume	v	$v = V/m$	$m^3\ kg^{-1}$
Surface density	ρ_A	$\rho_A = m/A$	$kg\ m^{-2}$
Specific surface area	a	$a = A/m$	$m^2\ kg^{-1}$
Number concentration[5]	C	$C_B = N_B/V$	m^{-3}
Amount concentration[6]	c	$c_B = n_B/V$	$mol\ m^{-3}$
Specific amount[7]	b	$b_B = n_B/m$	$mol\ kg^{-1}$
Surface concentration	Γ	$\Gamma_B = n_B/A$	$mol\ m^{-2}$
Stoichiometric coefficient[8]	v		1
Extent of reaction	ζ	$n_B = n_{B,0} + v_B\ \zeta$	mol

Notes: (*a*) The elementary entity, B, to which the unit "mol" refers must never be part of the unit. Expressions such as mol_c (mol of charge) or mol(+) (mol of positive charge) are not defined in the SI. (*b*) All expressions refer to a single phase.

[1]   Given here in terms of base units. Practical units are constructed using the prefixes found in Table A1.3. For example, commonly used units of molar mass are g/mol (tables), mg/mmol (laboratory), or Eg/Emol (atmosphere).

[2]   Also called *chemical amount.*

[3]   That is, amount fraction or number fraction; for a gaseous mixture, y is used.

[4]   Also called *mass concentration.*

[5]   Also called *number density.*

[6]   Amount of substance concentration, normally just called *concentration*. In this book we use [B] to represent actual concentrations and c_B to represent stoichiometric concentrations. If $V_m = V/\Sigma_A\ n_A$ is the molar volume of a mixture, then $x_B = c_B\ V_m$.

[7]   In the special case of $b_B = n_B/m_A$ where B is a solute and A the solvent, the name is *molality.*

[8]   The stoichiometric coefficient is defined through the reaction equation. It is negative for reactants and positive for products. A symbolic way to write a general chemical equation is $0 = \Sigma\ v_i\ B_i$.

A1.8 Fundamental constants

Quantity	Symbol	Value
Avogadro constant	N_A	$6.022\ 141\ 79(30) \times 10^{23}\ \mathrm{mol^{-1}}$
Boltzmann constant	k_B	$1.380\ 650\ 4(24) \times 10^{-23}\ \mathrm{J\ K^{-1}}$
Faraday constant	F	$9.648\ 533\ 99(24) \times 10^4\ \mathrm{C\ mol^{-1}}$
Planck constant	h	$6.626\ 068\ 96(33) \times 10^{-34}\ \mathrm{J\ s}$
	h	$h/2\pi \approx 1.055 \times 10^{-34}\ \mathrm{J\ s}$
Molar gas constant	R	$8.314\ 472(15)\ \mathrm{J\ K^{-1}\ mol^{-1}}$
Speed of light in vacuum	c_0	$299\ 792\ 458\ \mathrm{m\ s^{-1}}$ (defined)
Magnetic constant[1]	μ_0	$4\pi \times 10^{-7}\ \mathrm{H\ m^{-1}}$ (defined)
Electric constant[1]	$\varepsilon_0 = 1/\mu_0 c_0{}^2$	$8.854187817\ldots \times 10^{-12}\ \mathrm{F\ m^{-1}}$
Standard atmosphere		$101\ 325\ \mathrm{Pa}$ (defined)
Newtonian constant of gravitation	G	$6.674\ 28(67) \times 10^{-11}\ \mathrm{m^3\ kg^{-1}\ s^{-2}}$
Standard acceleration of gravity	g_0	$9.806\ 65\ \mathrm{m\ s^{-2}}$ (defined)

[1] $\mathrm{H = Wb/A};\ \mathrm{H/m = N/A^2};\ \mathrm{F = C/V}.\ \mathrm{H = henry},\ \mathrm{Wb = weber},\ \mathrm{F = farad}.$

A1.9 α-Amino acids of proteins

Trivial name	Symbol	Systematic name
Alanine	Ala	2-Aminopropanoic acid
Arginine	Arg	2-Amino-5-guanidinopentanoic acid
Asparagine	Asn	2-Amino-3-carbamoylpropanoic acid
Aspartic acid	Asp	2-Aminobutanedioic acid
Cysteine	Cys	2-Amino-3-mercaptopropanoic acid
Glutamine	Gln	2-Amino-4-carbamoylbutanoic acid
Glutamic acid	Glu	2-Aminopentanedioic acid
Glycine	Gly	Aminoethanoic acid
Histidine	His	2-Amino-3-(1H-imidazol-4-yl)propanoic acid
Isoleucine	Ile	2-Amino-3-methylpentanoic acid
Leucine	Leu	2-Amino-4-methylpentanoic acid
Lysine	Lys	2,6-Diaminohexanoic acid
Methionine	Met	2-Amino-4-(methylthio)butanoic acid
Phenylalanine	Phe	2-Amino-3-phenylpropanoic acid
Proline	Pro	Pyrrolidine-2-carboxylic acid
Serine	Ser	2-Amino-3-hydroxypropanoic acid
Threonine	Thr	2-Amino-3-hydroxybutanoic acid
Tryptophan	Trp	2-Amino-3-(1H-indol-3-yl)propanoic acid
Tyrosine	Tyr	2-Amino-3-(4-hydroxyphenyl)propanoic acid
Valine	Val	2-Amino-3-methylbutanoic acid

See ref. 178.

A1.10 The Greek alphabet

Greek letter		Greek name	Phonetic equivalent	Greek letter			Greek name	Phonetic equivalent
A	α	Alpha	a	N	ν		Nu	n
B	β	Beta	b	Ξ	ξ		Xi	x
Γ	γ	Gamma	c	O	o		Omicron	ŏ
Δ	δ	Delta	d	Π	π		Pi	p
E	ε	Epsilon	ĕ	P	ρ		Rho	r
Z	ζ	Zeta	x	Σ	σ	ς	Sigma	s
H	η	Eta	ë	T	τ		Tau	t
Θ	θ ϑ	Theta	th	Y	υ		Upsilon	u
I	ι	Iota	i	Φ	φ		Phi	ph
K	κ	Kappa	k	X	χ		Chi	ch
Λ	λ	Lambda	l	Ψ	ψ		Psi	ps
M	μ	Mu	m	Ω	ω		Omega	ö

Appendix 2

This appendix contains the algebraic relations necessary for ordinary analysis of acid-base and complex equilibria in solution. We give two computer program examples written in Mathematica.

A2.1 Polyvalent acids

The purpose of this section is twofold:

1. To write down general forms of the quantities used in Section 5.1d,[197] and
2. to present an implementation, written in Mathematica.

The acid *dissociation constant* of an n-valent weak acid is of the general form (see Equation 5.27)

$$H_{n-i}A^{-i} \rightleftharpoons H^+ + H_{n-i-1}A^{-i-1} \qquad K_{ai} = \frac{[H^+][H_{n-i-1}A^{-i-1}]}{[H_{n-i}A^{-i}]}$$

$$\text{for } i = 0, 1, \ldots, n-1 \tag{A2.1}$$

with the consecutive equilibrium constants K_{ai}. The cumulative constants are the products $\kappa_1 = K_{a1}$, $\kappa_2 = K_{a1}K_{a2}$, and so forth, used in the formulas; they have the general form

$$\kappa_i = \prod_{j=1}^{i} K_{a,j} \tag{A2.2}$$

The stoichiometric concentration of the acid is the general form of Equation 5.28:

$$c = \sum_{i=0}^{n} [H_{n-i}A^{i-}]$$

$$= [H_nA]\left(1 + \sum_{i=1}^{n} \frac{\kappa_i}{[H^+]^i}\right) \overset{\text{def}}{=} [H_nA]\sigma_0 \tag{A2.3}$$

where the subindex 0 distinguishes this sum from the sums σ_1 and σ_2,

$$\sigma_1 \overset{\text{def}}{=} \sum_{i=1}^{n} \frac{i\,\kappa_i}{[H^+]^i} \qquad \sigma_2 \overset{\text{def}}{=} \sum_{i=1}^{n} \frac{i^2\kappa_i}{[H^+]^i} \tag{A2.4}$$

to be used later.

The degree of formation of a generic species α_{n-i} is

$$\alpha_{n-i} = \frac{[H_{n-i}A^{i-}]}{c} = \frac{\kappa_{n-i}[H^+]^{i-n}}{\sigma_0} \quad i = 0, 1, \ldots, n \quad (A2.5)$$

and the αs are subject to the condition

$$\sum_{i=0}^{n} \alpha_i = 1 \quad (A2.6)$$

The alkalinity is

$$
\begin{aligned}
A &= [HO^-] - [H^+] + [H_{n-1}A^-] + \cdots + i[H_{n-i}A^{i-}] + \cdots + n[A^{n-}] \\
&= [HO^-] - [H^+] + c(\alpha_{n-1} + 2\alpha_{n-2} + \cdots + (n-1)\alpha_1 + n\alpha_0) \\
&= [HO^-] - [H^+] + cd \quad (A2.7)
\end{aligned}
$$

which also defines the *dissociation function d*, which in turn may be expressed in terms of the sums σ_0 and σ_1:

$$A = [HO^-] - [H^+] + c\frac{\sigma_1}{\sigma_0} \quad (A2.8)$$

The formal expression of the hydron condition (Equation 5.34) is unchanged:

$$H([H^+]) = K_w/[H^+] - [H^+] + cd + c_b - c_a = 0 \quad (A2.9)$$

Here K_w is the ion product of water, c_b and c_a are the concentrations of strong base and acid, respectively, and $d = \sigma_1/\sigma_0$.

The buffer value is defined by

$$
\begin{aligned}
\beta &= \frac{dA}{d\,pH} = \left[c\left(\frac{\sigma_2}{\sigma_0} - d^2\right) + [H^+] + [HO^-]\right] \ln 10 \\
&= \left[c\frac{\sigma_0\sigma_2 - \sigma_1^2}{\sigma_0^2} + [H^+] + [HO^-]\right] \ln 10 \quad (A2.10)
\end{aligned}
$$

The quantity that is the "acid analogue" of the alkalinity is the concentration of titrable hydron,

$$
\begin{aligned}
c_H &= [H^+] - [HO^-] + n[H_nA] + \cdots + (n-i)[H_{n-i}A^{i-}] + \cdots + [HA^{(n-1)-}] \\
&= [H^+] - [HO^-] + c(n\alpha_n + (n-1)\alpha_{n-1} + \cdots + 2\alpha_2 + \alpha_1) \\
&= [H^+] - [HO^-] + c\bar{n} \quad (A2.11)
\end{aligned}
$$

which defines $\bar{n}$. $\bar{n}$ is called the formation function of the acid-base system; it is simply the mole fraction of bound hydron with respect to the all species of the acid. $\bar{n}$ is implicitly given here and explicitly defined by Equation A2.25.

Program example 1
This program calculates the pH, alkalinity, buffer value, and formation function of a polyvalent acid.

Declarations

$K_w := 10^{-pK_w}$ (* ion product of water/(mol/l)^2 *)

$H := 10^{-pH}$ (* hydron concentration/(mol/l) *)

$HO := K_w/H$ (* hydroxide concentration/(mol/l) *)

$K_a := 10^{-pK_a}$ (* acidity constants/(mol/l) *)

$\kappa[j_-] := \prod_{i=1}^{j} K_a[\![i]\!]$ (* cumulative acidity constants $((mol/l)^j$ *)

$\sigma_0 := 1 + \sum_{i=1}^{n} \dfrac{\kappa[i]}{H^i}$ (* defined sum 0 *)

$\sigma_1 := \sum_{i=1}^{n} \dfrac{i\,\kappa[i]}{H^i}$ (* defined sum 1 *)

$\sigma_2 := \sum_{i=1}^{n} \dfrac{i^2\kappa[i]}{H^i}$ (* defined sum 2 *)

$d := \dfrac{\sigma_1}{\sigma_0}$ (* dissociation function *)

$hc := HO - H + cd - c_b + c_a$ (* hydron condition/(mol/l) *)

$A := HO - H + cd$ (* alkalinity/(mol/l) *)

$\beta := \left(HO + H + c\left(\dfrac{\sigma_2}{\sigma_0} - d^2\right)\right) Log[10]$ (* buffer value/(mol/l) *)

Initializations

$pK_w = 14.;$ (* ion product of water/M^2 at 25 °C *)

$n = 2;$ (* n-valent acid *)

$pK_a = \{6.0, 9.5, 10.3\};$ (* list of acidity exponents, the first n values are used *)

$c = 0.002;$ (* unit: mol/l *)

$c_a = 0.; c_b = c;$ (* unit: mol/l *)

Calculation of pH

$pH = pH_{min} = -1.3;$ $pH_{max} = pK_w + 1.3;$ $pH_{dif} = 10^{-6};$ $hc_{min} = hc;$

While $[Abs[(pH_{max} - pH_{min})] > pH_{dif},$ $pH = (pH_{max} + pH_{min})/2;$

$\quad$ If $[hc\, hc_{min} < 0, pH_{max} = pH, pH_{min} = pH; hc_{min} = hc]]$

Results

Print["Acidity, pH = ", NumberForm[p H, 4]];
Print["Alkalinity = ", ScientificForm[A, 4] "mol/l"];
Print["Buffer value, β = ", ScientificForm[β, 4] "mol/l"];
Print["Value of formation function, n = ", NumberForm[$n - d$, 4]];

Acidity, pH = 7.747
Alkalinity = 2. $\times 10^{-3}$ mol/l
Buffer value, $\beta = 1.595 \times 10^{-4}$ mol/l
Value of formation function, $n = 1$.

Program example 2
This program generates plots of speciation, alkalinity, buffer value, and formation function of a polyvalent acid as a function of pH. This program was used to make the figures in the text.

Declarations

(* The declarations are as in program example 1, except that the hydron condition
   hc is removed and replaced by a declaration of the fractions α_i *)

$$\alpha[i] := \frac{\kappa[n - i]\, H^{i-n}}{\sigma_0} \qquad \text{(* mole fraction of species } i \text{ *)}$$

$\kappa[0] = 1;$ (* for evaluation of $\alpha[n]$ *)

Initializations

$pK_w = 14.;$ (* ion product of water/M^2 at 25 °C *)
pHmin = 0;   pHmax = pK_w;   (* limits of pH step *)
$n = 2;$ (* n-valent acid *)
$pK_a = \{6.0, 9.5, 10.3\}$ (* list of acidity exponents, the first n values
                                    are used *)

$c = 0.002;$ (* unit: mol/l *)

Speciation curves, alkalinity, buffer value, and acid formation curve

Plot [Evaluate[Table[α [i] , { i, 0, n }]], { pH, pHmin, pHmax }]
Plot [A, { pH, pHmin, pHmax }]
Plot [β, { pH, pHmin, pHmax }]
Plot [$n - d$, { pH, pHmin, pHmax }]

A2.2 Mononuclear complexes

The purpose of this section is to derive algebraic expressions for the formation of mononuclear complexes. As inferred from Section 5.2, complex formation in most natural waters (oceans, freshwater, soil solution) is simple and does not require a complex algebraic description. On the other hand, for a deeper understanding of Nature and for the correct use of tables of constants found in the literature, a collection of expressions using generally accepted nomenclature is indispensable.

Mononuclear complexes ML_i ($i = 1, 2, \ldots, n$) are chemical species consisting of a central entity M that can combine with up to n ligands L in a stepwise fashion. M is not restricted to metal ions but includes bases that take up hydrons, and L may be chelate ligands (bi-, tridentate, etc.).[a] Thus, we have n equilibria with associated consecutive *formation constants* K_i and *cumulative constants* β_i:

$$M + L \rightleftharpoons ML \qquad K_1 = \frac{[ML]}{[M][L]} \qquad \beta_1 = K_1 \qquad \text{(A2.12a)}$$

$$ML + L \rightleftharpoons ML_2 \qquad K_2 = \frac{[ML_2]}{[ML][L]} \qquad \beta_2 = K_1 K_2 \qquad \text{(A2.12b)}$$

$$\ldots \qquad\qquad \ldots$$

$$ML_{n-1} + L \rightleftharpoons ML_n \qquad K_n = \frac{[ML_n]}{[ML_{n-1}][L]} \qquad \beta_n = K_1 K_2 \cdots K_n \qquad \text{(A2.12c)}$$

The cumulative constants fulfill the general expression:

$$M + iL \rightleftharpoons ML_i \qquad \beta_i = \frac{[ML_i]}{[M][L]^i} = \prod_{j=1}^{i} K_j \qquad \text{(A2.13)}$$

The stoichiometric concentrations of the central entity M and ligand L are given by

$$\begin{aligned} c_M &= [M] + [ML] + [ML_2] + \cdots + [ML_n] \\ &= [M]\left(1 + [L]\beta_1 + [L]^2\beta_2 + \cdots + [L]^n\beta_n\right) \end{aligned} \qquad \text{(A2.14)}$$

and

$$\begin{aligned} c_L &= [L] + [ML] + 2[ML_2] + \cdots + n[ML_n] \\ &= [L] + [M]\left([L]\beta_1 + 2[L]^2\beta_2 + \cdots + n[L]^n\beta_n\right) \end{aligned} \qquad \text{(A2.15)}$$

The degree of formation of each species is a mole fraction, defined by

$$\alpha_0 = \frac{[M]}{c_M} \quad \cdots \quad \alpha_i = \frac{[ML_i]}{c_M} \quad \cdots \quad \alpha_n = \frac{[ML_n]}{c_M} \qquad \text{(A2.16)}$$

subject to the condition

$$\sum_{i=0}^{n} \alpha_i = 1 \qquad \text{(A2.17)}$$

---

[a] Gk. $\chi\eta\lambda\eta \;\hat{=}\;$ crab claw; Lat. dens $\hat{=}$ tooth.

As was the case with the polyvalent acids, the algebraic expressions become compact by defining the three sums s_0, s_1, and s_2:

$$s_0 \stackrel{\text{def}}{=} 1 + \sum_{i=1}^{n} [L]^i \beta_i = \sum_{i=0}^{n} [L]^i \beta_i \tag{A2.18}$$

$$s_1 \stackrel{\text{def}}{=} \sum_{i=0}^{n} i\,[L]^i \beta_i = s_0 \sum_{i=0}^{n} i\,\alpha_i \tag{A2.19}$$

$$s_2 \stackrel{\text{def}}{=} \sum_{i=0}^{n} i^2 [L]^i \beta_i = s_0 \sum_{i=0}^{n} i^2 \alpha_i \tag{A2.20}$$

Here, the extension of the sum s_0 to include the zeroth term requires the definition

$$\beta_0 \stackrel{\text{def}}{=} 1 \tag{A2.21}$$

Note that these sums are functions of only one parameter, the free ligand concentration, [L]. The stoichiometric concentrations are now

$$c_{\text{M}} = [\text{M}]\,s_0 \tag{A2.22}$$

$$c_{\text{L}} = [\text{L}] + [\text{M}]\,s_1 \tag{A2.23}$$

and the degrees of formation are

$$\alpha_0 = \frac{1}{s_0} \quad \cdots \quad \alpha_i = \frac{[L]^i \beta_i}{s_0} \quad \cdots \quad \alpha_n = \frac{[L]^n \beta_n}{s_0} \tag{A2.24}$$

Next, we define the average number of coordinated ligands:

$$\bar{n} = \frac{[\text{ML}] + 2\,[\text{ML}_2] + \cdots + n\,[\text{ML}_n]}{c_{\text{M}}} = \frac{c_{\text{L}} - [\text{L}]}{c_{\text{M}}}$$

$$= \alpha_1 + 2\alpha_2 + \cdots + n\alpha_n = \sum_{i=0}^{n} i\,\alpha_i = \frac{s_1}{s_0} \tag{A2.25}$$

which is called the *formation function*. Note that an experimental determination of n only requires measurement of the concentration of the free ligand [L],[a] because c_{M}, c_{L}, and the dilution are known at the outset.

The dissociation function

$$d = n\alpha_0 + (n-1)\,\alpha_1 + \cdots + 2\,\alpha_{n-2} + \alpha_{n-1} = n - \bar{n} \tag{A2.26}$$

which is very useful for interpretations based on dissociation constants, is of less importance here.

Analogously to the definition of pH, pL is defined as

$$\text{pL} \stackrel{\text{def}}{=} -\lg \frac{[\text{L}]}{\text{M}} \tag{A2.27}$$

---

[a] *Free ligand* means "not bound to the central entity."

which includes the specific unit of concentration, $M = mol/l$. The derivatives of the formation function $\bar{n}$ with respect to pL,

$$\frac{d\bar{n}}{d\,pL} = \left(\bar{n}^2 - \frac{s_2}{s_0}\right) \ln 10 \tag{A2.28}$$

and of the dissociation function d with respect to pL,

$$\frac{d\,d}{d\,pL} = \left(\frac{s_2}{s_0} - d^2\right) \ln 10 \tag{A2.29}$$

give quantities that are proportional to buffer values. When applied to acid-base systems in aqueous solutions (pL $=$ pH), the latter expression is expanded to β as given in Equation A2.10.

Finally, we discuss the derivative of α_i with respect to pL. The logarithmic derivative is a compact and convenient expression:

$$\frac{d\lg\alpha_i}{d\,pL} = \bar{n} - i \tag{A2.30}$$

which is zero for $\bar{n} = i$.

a. Polynuclear complexes

An environmental chemist doing experimental work will often need to treat polynuclear complexes that have simultaneous acid-base properties quantitatively. One example is vanadium(V) (see Table 5.8).

Consider first the reaction

$$p\,M + q\,L \rightleftharpoons M_pL_q \tag{A2.31}$$

between the metal ion M and the ligand L forming a series of complexes M_pL_q with varying values of $p > 1$ and $q > 0$. The general expression for the cumulative constants of formation is

$$\beta_{pq} = \frac{[M_p L_q]}{[M]^p [L]^q} \tag{A2.32}$$

and the stoichiometric concentrations of metal c_M and ligand c_L may be given as

$$c_M = \sum_{p,q} p\,\beta_{pq} [M]^p [L]^q \qquad q = 0,\ldots \tag{A2.33}$$

$$c_L = [L] + \sum_{p,q} q\,\beta_{pq} [M]^p [L]^q \qquad q = 1,\ldots \tag{A2.34}$$

Because $p \geq 2$, c_M is *not* linear in the concentration [M]. This is an important reason why this kind of investigation cannot be carried out and interpreted without modern spectroscopic equipment and computers. Some results are given in Tables 5.15 and 5.16.

When the complexes also have acid-base properties, one must use expressions of the type

$$pH^+ + qM + rL \rightleftharpoons H_pM_qL_r \tag{A2.35}$$

$$\beta_{pqr} = \frac{[H_pM_qL_r]}{[H^+]^p[M]^q[L]^r} \tag{A2.36}$$

An example was discussed in connection with Table 5.8.

Appendix 3

A3.1 The activity of electrolytes

a. The Debye-Hückel limiting law

The primitive model for the activity of electrolytes considers Coulomb's law for rigid-sphere ions in a homogeneous dielectric fluid.

Coulomb's law gives the force F between two point charges Q_1 and Q_2 separated by a distance r as[35]

$$F = \frac{Q_1 Q_2}{4\pi\varepsilon r^2} \tag{A3.1}$$

where $\varepsilon = \varepsilon_0\varepsilon_r$ is the permittivity of the fluid.[a] The relative permittivity ε_r was formerly called the *dielectric constant*. The superposition principle applies to electrostatic forces between point charges: the electrostatic force between a pair of point charges is unaltered by the addition of further point charges. The electrostatic potential $\varphi(r)$ at position r due to the charge Q is given by

$$\varphi(r) = \frac{Q}{4\pi\varepsilon r} \tag{A3.2}$$

Consider a positive ion, and let the potential at distance r from the ion be $\varphi(r)$. An ion with charge $\pm ze$ placed at r has the potential energy $\pm z e \varphi(r)$. Far from the center, the potential vanishes, i.e., $\varphi(r) \to 0$. Let the number concentration at this location be denoted C. According to the Boltzmann equation, the concentration of positive and negative ions at r from the ion is

$$C_+ = C e^{-ze\varphi/k_B T} \quad \text{and} \quad C_- = C e^{+ze\varphi/k_B T} \tag{A3.3}$$

respectively. The total charge density $\rho(r)$ is a sum of all contributions,

$$\rho(r) = e\sum_i C_i z_i\, e^{-z_i e\varphi(r)/k_B T} \tag{A3.4}$$

---

[a] Fundamental constants are given in Appendix A1.8.

However, if $ze\varphi$ is small relative to $k_B T$, an approximation can be made, keeping only the first term in a series expansion of the exponential:

$$\rho \approx e \sum_i C_i z_i \left(1 - \frac{z_i e \varphi}{k_B T} \right)$$

$$= e \sum_i C_i z_i - \frac{e^2 \varphi}{k_B T} \sum_i C_1 z_i^2 \tag{A3.5}$$

Because the solution is electrically neutral, that is, $\Sigma\, C_i\, z_i = 0$,

$$\rho = -\frac{e^2 \varphi}{k_B T} \sum_i C_i z_i^2 \tag{A3.6}$$

Coulomb's law implies Poisson's equation, meaning that the charge density determines the electric potential through

$$\nabla^2 \varphi = -\frac{\rho}{\varepsilon} \tag{A3.7}$$

Inserting and expanding yields

$$\frac{\partial^2 \varphi}{\partial r^2} + \frac{2}{r}\frac{\partial \varphi}{\partial r} = \frac{e^2 \varphi}{\varepsilon k_B T} \sum_i C_i z_i^2 = \kappa^2 \varphi \tag{A3.8}$$

where the following substitution has been made:

$$\kappa = \sqrt{\frac{e^2}{\varepsilon k_B T} \sum_i C_i z_i^2} \tag{A3.9}$$

The quantity κ^{-1} is called the Debye length and is a characteristic distance of the ion atmosphere. In the present discussion, κ is simply a convenient definition.[a] The general solution of the differential equation is

$$\varphi(r) = \frac{Ae^{-\kappa r}}{r} + \frac{Be^{\kappa r}}{r} \tag{A3.10}$$

Here $B = 0$ because φ must be equal to zero for large values of r. In the limit of $r = 0$, Equation A3.2 applies, leading to $A = (z_i e)/(4\pi\varepsilon)$ and

$$\varphi(r) = \frac{z_i e}{4\pi\varepsilon r} e^{-\kappa r} \tag{A3.11}$$

If the central ion were alone, the potential at r would be $(z_i e)/(4\pi\varepsilon r)$. The difference

$$\frac{z_i e}{4\pi\varepsilon r} e^{-\kappa r} - \frac{z_i e}{4\pi\varepsilon r} = -\frac{z_i e}{4\pi\varepsilon} \left(\frac{1 - e^{-\kappa r}}{r} \right) \tag{A3.12}$$

---

[a] In this appendix, the symbol κ is only used as shown; it does not refer to conductivity. At $\theta = 20\ °C$, the relative permittivity of water is $\varepsilon_r = 80.02^{129}$; then a 0.05 M aqueous solution of a 1-1 electrolyte has $\kappa^{-1} = 1.36$ nm.

expresses the effect of the ionic atmosphere. A series expansion of the exponential gives

$$\varphi_i = \frac{z_i e \kappa}{4 \pi \varepsilon} \tag{A3.13}$$

The energy of a charge Q in a potential φ is $Q\varphi/2$.[35] Therefore, the energy of the central ion due to the ionic atmosphere is

$$\frac{1}{2} z_i e \varphi_i = -\frac{z_i^2 e^2 \kappa}{8 \pi \varepsilon} \tag{A3.14}$$

Accordingly, this contribution to the molar energy is

$$\frac{1}{2} z_i F \varphi_i = -\frac{z_i^2 e^2 \kappa N_A}{8 \pi \varepsilon} \tag{A3.15}$$

using the relation $F = e N_A$.

Assuming an infinitely dilute solution with no electrostatic interaction between the ions, its behavior will be ideal, and the chemical potential of an ion i is

$$\mu_i = \mu_i^\circ + RT \ln \frac{c_i}{c^\circ} \tag{A3.16}$$

In the nonideal case, the activity is no longer equal to the concentration but is given by (see Equation 2.145)

$$a_i = \frac{c_i}{c^\circ} \gamma_i \tag{A3.17}$$

such that the chemical potential is

$$\mu_i = \mu_i^\circ + RT \ln a_i = \mu_i^\circ + RT \ln \frac{c_i}{c^\circ} + RT \ln \gamma_i \tag{A3.18}$$

Assuming that deviation from ideality is due to the ionic atmosphere,

$$RT \ln \gamma_i = -\frac{z_i^2 e^2 \kappa N_A}{8 \pi \varepsilon} \tag{A3.19}$$

Moving from entities to molar quantities using $C_i = c_i N_A$ and $k_B = R/N_A$, and defining the ionic strength by $I = \frac{1}{2} \Sigma c_i z_i^2$,[a] gives

$$- \lg \gamma_i = \frac{e F^2}{(\varepsilon RT)^{3/2} 4\sqrt{2\pi} \ln 10} z_i^2 \sqrt{I} = A z_i^2 \sqrt{I} \tag{A3.20}$$

This is the Debye-Hückel limiting law, which works well in dilute solutions. The value of the constant A is defined to include the choice of standard state. At 25 °C, the permittivity of water is $\varepsilon = 78.54 \, \varepsilon_0$, and with $c^\circ = 1$ M, one finds the value $A = 0.509$ M$^{-1/2}$. Using SI units, $c^\circ = 1$ mol m^{-3}, the value of A is 1.610×10^{-2} m$^{3/2}$ mol$^{-1/2}$. Expression A3.20 applies to ionic strengths less than

---

[a] IUPAC-NIST[166] recommends the symbol I_c for this quantity and $I_m = 1/2 \, \Sigma \, m_i z_i^2$ for the ionic strength on a molality scale.

about 0.01 M. For concentrations up to about 0.1 M, the following expression may be used:

$$-\lg \gamma_i = \frac{A z_i^2 \sqrt{I}}{1 + B \sqrt{I}} \qquad (A3.21)$$

with $B = 1.0$ M$^{-1/2}$ (Güntelberg, 1926).[a,b]

This equation is one of several empirical expressions based on the theoretically substantiated relationship[82]

$$-\lg \gamma_i = \frac{A z_i^2 \sqrt{I}}{1 + \kappa a} \qquad (A3.22)$$

where κ^{-1} is the Debye length and a a measure of the ion size. The numerator accounts for long-distance Coulomb forces (the limiting law) and the denominator for short-range repulsive interactions.

An extensive body of work on the estimation of activity coefficients can be found in the literature.[170,210] Reference 235 is of particular interest in environmental chemistry: software for "The Adjustment, Estimation and Uses of Equilibrium Constants in Aqueous Solution" is accessible from this site, the programs being of the same structure as those of ref. 133.

---

[a] Note: If experimental data have been extrapolated to give stability constants valid at $I_c = 0$ M using the SIT model (specific ionic interaction theory), then $B = 1.5$ M$^{-1/2}$ at 25 °C.[166]

[b] These expressions are of no use at high ionic strengths and charges. As an example, for 1 M SmCl$_3$ $\lg \gamma_\pm > 0.08$ and $\gamma_\pm > 1.2$.

Appendix 4

A4.1 Convection

Consider an isolated system consisting of a fluid phase without external fields. Thermal forcing of the system will give rise to flows that will continue until thermal equilibrium has been achieved. This kind of movement is called *convection*, and we will use the present study as an example of the equations developed in Section 2.3.

A fluid phase in a gravitational field may be at mechanical rest without being in thermal equilibrium. This means that the conditions for mechanical stability (see Equation 2.77) may be fulfilled despite the presence of temperature gradients. The aim of this section is to determine the requirements for the absence of convection in such systems.

Consider an element of a fluid phase α at an altitude z, with the molar volume $V_m^\alpha(p, S)$. Here p and S are the equilibrium values of pressure and entropy, respectively, at that altitude. Suppose now that this element is moved adiabatically upward by an amount $+dz > 0$. The molar volume of the element will thereby change to $V_m^\alpha(p + dp, S)$. On the other hand, the element that is crowded out at this altitude, $z + dz$, has $V_m^\alpha(p + dp, S + dS)$. Mechanical stability requires that the former element be more dense so that it falls back; this condition is

$$V_m^\alpha(p + dp, \ S + dS) \ - \ V_m^\alpha(p + dp, \ S) \ > \ 0 \qquad (A4.1)$$

By considering the derivative of $V_m^\alpha(p, S)$ with respect to altitude,

$$\frac{dV_m^\alpha(p, S)}{dz} = \left(\frac{\partial V_m^\alpha}{\partial p}\right)_S \frac{dp}{dz} + \left(\frac{\partial V_m^\alpha}{\partial S}\right)_p \frac{dS}{dz} \qquad (A4.2)$$

it is seen that Equation A4.1 is equivalent to the statement

$$\left(\frac{\partial V_m^\alpha}{\partial S}\right)_p \frac{dS}{dz} > 0 \qquad (A4.3)$$

Here we make use of one of Maxwell's equations,[67]

$$\left(\frac{\partial V_m^\alpha}{\partial S}\right)_p = \frac{T}{C_{p,m}^\alpha} \left(\frac{\partial V_m^\alpha}{\partial T}\right)_p \qquad (A4.4)$$

where $C_{p,m}^\alpha$ denotes the molar heat capacity at constant pressure. This is a positive quantity, as is the temperature; therefore, the condition for absence of convection is

$$\left(\frac{\partial V_m^\alpha}{\partial T}\right)_p \frac{dS}{dz} > 0 \qquad (A4.5)$$

The first factor is related to the cubic expansion coefficient, $\alpha = (1/V)\,(\partial V/\partial T)_p$,[a] which is positive for most substances, but not, for example, for water between $0\ ^\circ\mathrm{C}$ and $4\ ^\circ\mathrm{C}$.

The atmosphere has $\alpha > 0$, and Equation A4.5 reduces to

$$\frac{\mathrm{d}\,S}{\mathrm{d}\,z} > 0 \qquad (\mathrm{A4.6})$$

That is, *the absence of convection in the atmosphere requires that entropy increases with increasing altitude*. The temperature gradient $d\,T/d\,z$ (see Equation 2.62) may now be derived from Equation A4.6: the molar entropy is $S_{\mathrm{m}} = S_{\mathrm{m}}\,(T, p, y_{\mathrm{B}}, \dots)$, where the phase label $\alpha = $ gas has been omitted and $y_{\mathrm{B}}, \dots$ denotes the mole fractions of all chemical species including water.

$$\frac{\mathrm{d}\,S_{\mathrm{m}}}{\mathrm{d}\,z} = \left(\frac{\partial\,S_{\mathrm{m}}}{\partial\,T}\right)_{p,y_{\mathrm{B}}}\frac{\mathrm{d}\,T}{\mathrm{d}\,z} + \left(\frac{\partial\,S_{\mathrm{m}}}{\partial\,p}\right)_{T,y_{\mathrm{B}}}\frac{\mathrm{d}\,p}{\mathrm{d}\,z} + \left(\frac{\partial\,S_{\mathrm{m}}}{\partial\,y_{\mathrm{aq}}}\right)_{T,p,y_{\mathrm{air}}}\frac{\mathrm{d}\,y_{\mathrm{aq}}}{\mathrm{d}\,z} > 0$$

$$(\mathrm{A4.7})$$

We define "air" as the gases of the dry atmosphere (see Table 3.18) and "aq" means water. The sole approximation is that $\mathrm{d}y_{\mathrm{B}\neq\mathrm{aq}} = 0$, that is, only the effect of the change in the amount of water in the gas phase is included in the calculation. From thermodynamics:

$$\left(\frac{\partial\,S_{\mathrm{m}}}{\partial\,T}\right)_{p,y_{\mathrm{B}}} = \frac{C_{p,\mathrm{m}}}{T} \qquad (\mathrm{A4.8a})$$

$$\left(\frac{\partial\,S_{\mathrm{m}}}{\partial\,T}\right)_{p,y_{\mathrm{B}}} = -\left(\frac{\partial\,V_{\mathrm{m}}}{\partial\,T}\right)_{p,y_{\mathrm{B}}} = -\frac{R}{P} \qquad (\mathrm{A4.8b})$$

Here the last term applies to a perfect gas mixture. In addition,

$$\left(\frac{\partial\,S_{m}}{\partial\,y_{\mathrm{aq}}}\right)_{T,p,y_{\mathrm{air}}} = \frac{1}{T}\left(\frac{\partial\,H_{m}}{\partial\,y_{\mathrm{aq}}}\right)_{T,p,y_{\mathrm{air}}} = -\frac{\Delta_{\mathrm{vap}}H_{\mathrm{m,aq}}}{T} \qquad (\mathrm{A4.8c})$$

where the last term is the heat of vaporization or sublimation of water at temperature T. Next, $d\,T/d\,z$ is isolated to derive the condition for the absence of convection:

$$\frac{\mathrm{d}\,T}{\mathrm{d}\,z} > \frac{R\,T}{p\,C_{p,\mathrm{m}}}\frac{\mathrm{d}\,p}{\mathrm{d}\,z} + \frac{\Delta_{\mathrm{vap}}H_{\mathrm{m,aq}}}{C_{p,\mathrm{m}}}\frac{\mathrm{d}\,y_{\mathrm{aq}}}{\mathrm{d}\,z} \qquad (\mathrm{A4.9})$$

Euler's hydrostatic equation 2.78 expresses the equilibrium condition for a fluid phase at mechanical rest in a homogeneous field of gravity:

$$\frac{\mathrm{d}\,p}{\mathrm{d}\,z} = -g\rho = -g_{\mathrm{n}}\frac{M_{\mathrm{air}}}{V_{\mathrm{m,air}}} \qquad (\mathrm{A4.10})$$

where g_{n} is the standard acceleration of gravity. Inserting this gives the condition for stability of the atmosphere with respect to convection,

$$\frac{\mathrm{d}\,T}{\mathrm{d}\,z} > -\frac{g_{\mathrm{n}}}{c_p} + \frac{\Delta_{\mathrm{vap}}\,H_{\mathrm{m,aq}}}{C_{p,\mathrm{m}}}\frac{\mathrm{d}\,y_{\mathrm{aq}}}{\mathrm{d}\,z} \qquad (\mathrm{A4.11})$$

where $c_p = C_{p,\mathrm{m}}/M_{\mathrm{air}}$ is the specific heat capacity of air at constant pressure.

[a] As a physical quantity, α is written using italic type; the phase α is written using Roman type.

The negative gradient of the temperature profile is called the lapse rate Γ, $\Gamma = -dT/dz$. The derived lapse rate characterizes a homogeneous air parcel: if an actual temperature gradient, dT/dz is less than $-\Gamma$, then the atmosphere is unstable, and convection will occur until stability is restored. However, for all values $dT/dz > -\Gamma$ the atmosphere is stable.

Example

For a dry atmosphere, only the first term of Equation A4.11 applies:

$$\frac{dT}{dz} = -\frac{9.80665 \text{ m s}^{-2}}{1.007 \text{ J g}^{-1} \text{ K}^{-1}} \approx -9.7 \text{ K/km} \qquad (A4.12)$$

where g_n is the defined value and the value of c_p applies to air at $p = 1$ bar and $T = 200–300$ K.[a] However, the atmosphere is not always dry. An amount of $H_2O(g)$ liberates heat when it is transferred into a condensed phase, and this increases the temperature gradient. We choose $T \approx 263$ K, $\Delta_{vap}H_m \approx 52$ kJ mol^{-1} (see Table 3.16), and

$$\frac{dT}{dz} \approx -9.7 \text{ K/km} + 1.8 \text{ K} \frac{d\, p_{aq}/\text{mbar}}{dz} \qquad (A4.13)$$

If an air parcel at 1 bar and 263 K and saturated with water vapor ($p_{aq}{}^* \approx$ 2.6 mbar, see Table 3.15), is moved upward 1 km, its lapse rate is 5.1 K/km, that is, if $dT/dz > -5.1$K/m, then the atmosphere is stable. Thus, the upward convection of wet air, following a lapse rate of 5.1 K/km, transports heat into an otherwise dry atmosphere following a lapse rate of 9.7 K/km. This is a significant source of heat to the middle and upper troposphere. Further examples are shown in Figure 3.4.

[a] If $C_{p,m} \approx (7/2) R$ for a diatomic gas is assumed, then the use of M_{air} from Table 3.18 leads to the value $c_p \approx 1.005$ J/(K g).

Appendix 5

A5.1 Parameters of the insolation formula, Equation 10.29

This appendix defines astronomical quantities pertaining to Earth's orbit.[a] Some of the present figures are referred to in Section 10.2a.

The insolation formula was given in general terms in Section 10.2a:

$$I_\oplus(\lambda, \varphi; e, \varepsilon, \omega) = I_S(a)\, \chi(\lambda; e, \omega)\, \psi(\lambda, \varphi; \varepsilon) \tag{10.29}$$

and the purpose of this appendix is to present the derivation of the expression

$$I_\oplus(\lambda, \varphi; e, \varepsilon, \widehat{\omega}) = I_S \left(\frac{1 + e\cos(\lambda - \widehat{\omega})}{1 - e^2} \right)^2 \psi(\lambda, \varphi; \varepsilon) \tag{10.34}$$

where

$$\psi(\lambda, \varphi; \varepsilon) = (\sin h_0 \cos \varphi \cos \delta + h_0 \sin \varphi \sin \delta)/\pi \tag{10.31a}$$

We start by considering the ψ term, because this will introduce two of the geocentric coordinate systems used to describe celestial phenomena Figure A5.1.

Then, we present the ecliptic, which is the yearly orbit of the Earth (Figure A5.2). The daily rotation gives rise to a precession similar to that of a spinning top, causing movement of the vernal equinox relative to the perihelion, as shown. This phenomenon constitutes the major part of the *luni-solar precession*, whose other sources are perturbations from the Sun and the Moon. This is further clarified in Figure A5.3.

An additional cause of the movement of the vernal equinox is the *planetary precession*, whose driving forces are perturbations from the other planets of our solar system. This gives rise to changes of the obliquity of the ecliptic ε, which is the angle between the planes of the ecliptic and the equator, respectively (see Figure A5.3).

The combined effect of all perturbations is called the *general precession*. The main terms are shown in Figure A5.4. In this appendix we discuss only the main components of the equations of motion of the Earth; this is adequate for a broad understanding of the astronomical climate theory.

a. The 24-h mean insolation at a geographical latitude φ

Figure A5.1 is explained in the figure caption. Application of the spherical cosine relations to the shaded triangle yields

$$\cos \Theta = \cos(\pi/2 - \varphi)\cos(\pi/2 - \delta) + \sin(\pi/2 - \varphi)\sin(\pi/2 - \delta)\cos h$$
$$= \sin \varphi \sin \delta + \cos \varphi \cos \delta \cos h \tag{A5.14}$$

---

[a] A thorough understanding of this appendix is not necessary for the applications of Section 10.2b.

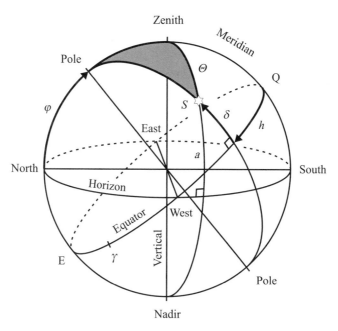

Figure A5.1 *The celestial sky as seen from the outside with the celestial coordinates used to calculate insolation.*

The sky as observed from the Earth is called the celestial sphere; we see only half of it, limited by the horizon. The point above the vertical is the zenith, and the opposite point is the nadir; the corners of the Earth are marked on the horizon. The sky moves from east to west, but the poles are fixed. The straight line connecting the poles stands normal to the celestial equator EQ. The polar altitude φ has the same value as the geographical latitude. The altitude a of a star S is the complement of the zenith distance Θ; $a = 0$ when a star rises or sets. The declination δ of a star is measured on a great circle through the star and the poles from the equator with positive values on the Northern Hemisphere. The hour angle h of a star is measured at the equator; its zero point is on the meridian at the point Q: $h = -\pi/2$ at the east point, and $h = +\pi/2$ at the west point.

Calculation of the insolation with the Sun at S requires calculation of the cosine of the zenith distance, $\cos\Theta$; see Equation 10.5.

Let γ denote the vernal equinox and S the Sun. The drawing represents the Sun near summer solstice, that is; the declination δ of the Sun is $\delta \approx +\varepsilon$.

The mean value of $\cos\Theta$ from sunrise $-h_0$ to sunset $+h_0$ is given by the integral

$$\frac{1}{2\pi}\int_{-h_0}^{+h_0}\cos\Theta\,\mathrm{d}h = \frac{1}{\pi}\left(h_0\sin\varphi\sin\delta + \sin h_0\cos\varphi\cos\delta\right)\qquad(A5.15)$$

which is Equation 10.31a. It is noted that Equation 10.31c follows from Equation A5.14 because at both sunrise and sunset, the altitude a of the Sun is $a = 0$ and its zenith distance is $\Theta = \pi/2$.

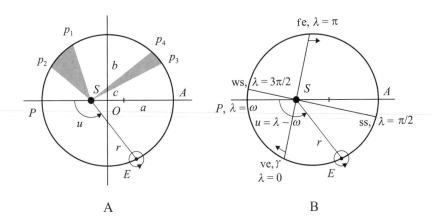

A   B

Figure A5.2   *The ecliptic as seen from celestial north*
Panel A: The orbit of the Earth E around the Sun S is called the ecliptic. According to Kepler's first law, the ecliptic is an ellipse with a focus located at S. O is the center of the ellipse, A the aphelion, P the perihelion, a the semimajor axis, b the semiminor axis, c the distance from the center to either focus, and r the radius vector of the Earth. The eccentricity e is given by $e = c/a$; on the drawing, $e \approx 0.25$ for clarity, but in reality $e = 0.0167$. The angle u is called the *true anomaly;*[a] for $u = \pi/2$ and $3\pi/2$, $r = a(1 - e^2) = b^2/a$. In general, $r = a(1 - e^2)/(1 + e \cos u)$ as in Equation 10.29. The areas of the shaded sectors are equal illustrating Kepler's second law: the planet uses the same time passing the two distances $p_1 - p_2$ and $p_3 - p_4$. The circle around E indicates the daily rotation; see $\omega_\oplus$, Equation 2.82.
Panel B: The calendar dates λ are marked on the ecliptic. The λs have the same values as the true longitude[b] of the Sun. At the vernal equinox, a geocentric observer sees the Sun in the constellation *Pisces*, whereas a heliocentric observer sees the Earth in the constellation *Virgo*, which is a change of phase of π. With this position in mind, we have set the zero point for λ as the vernal equinox "ve," with the symbol γ. The values of λ at summer solstice "ss," fall equinox "fe," winter solstice "ws," and the perihelion "P," $\lambda = \omega = 283°$, then follow. The true anomaly [c] has the perihelion as the zero point, $u = \lambda - \omega$. The daily rotation of the Earth is indicated. The luni-solar precession gives rise to the movement of the equinoctial line approximately $0.0139°$ a^{-1} in the direction shown.

b. The ecliptic

The ecliptic approximates an ellipse as shown in Figure A5.2A. The Sun is positioned at a focus; the perihelion is the point of the orbit where the Earth is closest to the Sun; at the aphelion it is farthest away.

The shape of the function $I_\oplus(\lambda, \varphi; e, \varepsilon, \omega)$, Equation 10.29, is a direct consequence of the elliptic orbit as explained in the legend of Figure A5.2A. The extreme values of the insolation are $I_S(1 - e)^{-2}\psi(\lambda_p, \varphi; \varepsilon)$ (at perihelion, the anomaly is

---

[a] See Footnote *a* page 354.

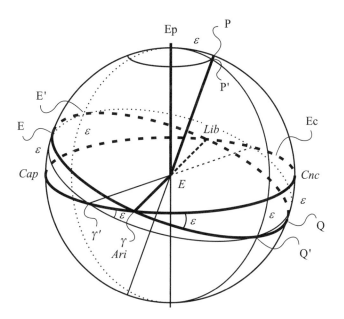

Figure A5.3   *The luni-solar precession*

The figure shows the celestial sphere with the Earth *E* in the center.

1. The heavy lines may be considered as a reference frame, which is taken to be 140 BC. During a year the Sun moves in the ecliptic Ec through the constellations of the Zodiac, of which four are shown, *Aries* (vernal equinox), *Cancer* (summer solstice), *Libra* (fall equinox), and *Capricorn* (winter solstice). Ep marks a pole of the ecliptic. The equator EQ with north pole P leans at an angle ε from the ecliptic; this angle is called the *obliquity of the ecliptic*. The vernal equinox γ is the point where the declination of the Sun goes from negative to positive; the extremal declinations are $\pm\,\varepsilon$.

2. Around 140 BC, the Greek astronomer Hipparchos discovered the movement of the vernal equinox in the direction shown in Figure A5.2B. Since then it has moved about 33° away from γ in the constellation *Aries* to the present position γ' in the constellation *Pisces*, that is, more than one full constellation (30°) in the Zodiac. The new equator is marked E'Q'. In the approximation depicted here, the obliquity of the ecliptic ε has not changed.

$u = 0;$[a] this occurs on January 4,[b] $\lambda_p = 282.8$ °) and $I_S\,(1 + e)^{-2}\,\psi(\lambda_a, \varphi; \varepsilon)$ (at aphelion, the anomaly is $u = \pi$; this occurs on July 4, where $\lambda_a = 102.8$ °), and for the present values of e and ω,

$$I_\oplus(\lambda_{\mathrm{p\,or\,a}}, \varphi; e, \varepsilon, \omega) = I_S(1.000 \pm 0.034)\,\psi(\lambda_{\mathrm{p\,or\,a}}, \varphi; \varepsilon) \qquad \text{(A5.16)}$$

Thus, the annual variation in total irradiance exceeds 6.5 %.

---

[a] See Footnote *a* page 354.
[b] All dates refer to the horizon of the Copenhagen Observatory, 55.686825° N, 12.575814° E.

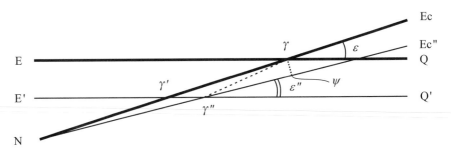

Figure A5.4 *The major precessions*

The figure is a continuation of figure A5.3 showing the main components of the general precession. The heavy lines represent an arc of the equator EQ and the ecliptic Ec, respectively; the vernal equinox γ at this time is marked. At a later time, owing to the luni-solar precession, the equator has moved to the position E'Q' and the vernal equinox to γ'. At this time, the obliquity of the ecliptic has also changed, from ε to ε''. Here N is the end point of the diameter around which the ecliptic moves. The general precession, arc $\gamma\gamma''$, represents the movement of the vernal equinox in the interval of time under consideration. The distance arc Nψ is marked as being equal to the arc Nγ. If one assumes that γ'' is the position of the present vernal equinox, then the arc $\gamma''\psi$ is the correction that must be added to all longitudes referred to the fixed vernal equinox γ. The values of the the quantities discussed herein are[a]

arc $\gamma\gamma'$	luni-solar precession	$50''.3758$ a^{-1}	(2000 AD)
arc $\gamma'\gamma''$	planetary precession	$0''.1059$ a^{-1}	(2000 AD)
arc $\gamma\gamma''$	general precesion	$50''.2787$ a^{-1}	(2000 AD)
arc Nγ''	present value $\approx 5.5°$	$33''$ a^{-1}	(2000 AD)

The definition of the calendar date, denoted λ, is explained in the legend of Figure A5.2B; note the phase difference of π between λ and the longitude of the Sun.

---

[a] The reason why arc Nγ'' moves less than $\approx 50''$ a^{-1} is that N moves $\approx 17''$ a^{-1} in the same direction as the vernal equinox.

References

1. Textbooks and Monographs

1. Adamson, A. W.; Gast, A. P. *Physical Chemistry of Surfaces*, J. Wiley & Sons, New York 1997.

2. Adriano, D. C., Ed. *Biogeochemistry of Trace Metals*, Lewis Publisher, Boca Raton 1992.

3. Apello, C. A. J.; Postma, D. *Geochemistry, Groundwater and Pollution*, A. A. Balkema, Rotterdam 1993; 2nd print 1994.

4. Berry, R. S.; Rice, S. A.; Ross, J. *Physical Chemistry*, 2nd Ed., Oxford University Press, Oxford 2000.

5. Berger, A.; Imbrie, J.; Hays, J.; Kukla, G.; Saltzman, B., Eds. *Milankovitch and Climate*, D. Reidel, Dordrecht 1984.

6. Berzelius, J. J. *Lehrbuch der Chemie Vol. 1*, 5th Ed., Arnoldische Buchhandlung, Leipzig 1856.

    a. idem *Lärbok i kemien*, 2nd Ed., Stockholm 1817.

7. Billing, G. D. *Dynamics of Molecule Surface Interactions*, J. Wiley & Sons, New York 2000.

8. Bitton, G.; Gerba, C. P., Eds. *Groundwater Pollution Microbiology*, J. Wiley & Sons, New York 1984.

9. Bjerrum, J. *Metal Ammine Complexes in Aqueous Solution*, P. Haase & Son, Copenhagen 1941. Reprint 1957.

10. Borggaard, O. K. *Soil Chemistry in a Pedological Context*, 4th Ed., DSR Press, Copenhagen 2000.

11. Brady, N. C. *The Nature and Properties of Soils*, 10th Ed., Macmillan Publishing Company, New York 1990.

12. Brasseur, G. P.; Orlando, J. J.; Tyndall, G. S.; Eds. *Atmospheric Chemistry and Global Change*, Oxford University Press, New York 1999.

13. Bridges, E. M. *World Soils*, 2nd Ed., Cambridge University Press, Cambridge 1988.

14. Brown, S. S. *History of IUPAC 1988–1999*, IUPAC 2001.

15. Bucher, K.; Frey, M. *Petrogenesis of Metamorphic Rocks*, 6th Ed., Springer Verlag, Heidelberg 1994.

16. Condie, K. C. *Plate Tectonics and Crustal Evolution*, 4th Ed., Butterworth-Heinemann, Oxford 1997.

17. Crichton, R. R. *Biological Inorganic Chemistry*, Elsevier, Amsterdam 2008.

18. Darnell, J. E.; Lodish, H.; Baltimore, D. *Molecular Cell Biology*, Scientific American Books, New York 1986.

19. Deer, W. A.; Howie, R. A.; Zussman, J. *An Introduction to the Rock Forming Minerals*, 1st Ed., Longman, Essex 1966; 15th Impression 1985.

20. Dzombak, D. A.; Morel, F. M. M. *Surface Complexation Modelling*, J. Wiley & Sons, New York 1990.

21. Fennell, R. *History of IUPAC 1919–1987*, Blackwell Science, Oxford 1993.

22. Franks, F. *Water*, The Royal Society of Chemistry, London 1983.

23. Franz, J. E.; Mao, M. K.; Sikorski, J. A. *Glyphosate: A Unique Global Herbicide*, ACS Monograph 189, The American Chemical Society 1997.

24. Fraústo da Silva, J. J. R.; Williams, R. J. P. *The Biological Chemistry of the Elements*, Clarendon Press, Oxford 1991.

25. Garrels, R. M.; Christ, C. L. *Solutions, Minerals and Equilibria*, Harper & Row, New York 1965.

26. Glasstone, S. *Textbook of Physical Chemistry*, 7th Ed., D. van Nostrand, New York 1946.

27. Greenwood, N. N.; Earnshaw, A. *Chemistry of the Elements*, Pergamon Press, Oxford 1984.

28. Hall, A. *Igneous Petrology*, Longman, Essex 1987.

29. Hargis, L. G. *Analytical Chemistry, Principles and Techniques*, Prentice-Hall International Editors, New Jersey 1988.

30. Harned, H. S.; Owen, B. B. *The Physical Chemistry of Electrolytic Solutions*, 3rd Ed., Reinhold, New York 1958.

31. Harrison, R. M., Ed. *Pollution: Causes, Effects & Control*, 2nd Ed., Royal Society of Chemistry, Cambridge 1990; 2nd Reprint 1993.

32. Hay, W. W., Ed. *Studies in Paleo-Oceanography 20*, 1974.

33. Hecht, E. *Optics*, 4th Ed. Addison-Wesley, San Francisco 2002.

34. Henderson, P. *Inorganic Geochemistry*, Pergamon Press, Oxford 1982.

35. Hinchliffe, A.; Mumm, R. W. *Molecular Electromagnetism*, John Wiley & Sons, Chichester 1985.

36. Hinshelwood, C. N. *The Structure of Physical Chemistry*, Oxford University Press, Oxford 1951.

37. Huheev, J. E. *Inorganic Chemistry. Principles of Structure and Reactivity*, 2nd Ed., Harper & Row, New York 1978.
    a. *idem*, pp. 162–163.

38. Irgolic, K. J.; Martell, A. E., Eds. *Environmental Inorganic Chemistry*, VCH Publishers, Deerfield Park 1985.

39. Jacob, D. J. *Introduction to Atmospheric Chemistry*, Princeton University Press, Princeton 1999.

40. Jensen, K. A. *Almen Kemi I–III*, Jul. Gjellerups Forlag, København 1957 (I), 1959 (II), 1964 (III).

41. Jessop, P. G.; Leitner, W., Eds. *Chemical Synthesis Using Supercritical Fluids*, Wiley-VCH, Weinheim 1999.

42. Kaim, B. G.; Schwederski, B. *Bioanorganische Chemie*, B. G. Teubner, Stuttgart 1991.

43. Kallenborn, R.; Hühnerfuss, H. *Chiral Environmental Pollutants*, Springer, Heidelberg 2001.

44. Kaplan, W. *Advanced Calculus*, Addison-Wesley, Reading, Massachusetts 1952; 5th print 1959.

45. Kerridge, J. F.; Matthews, M. S., Eds. *Meteorites and the Early Solar System*, University of Arizona Press, Tucson 1988.

46. Kondepudi, D.; Prigogine, I. *Modern Thermodynamics*, Wiley, Chichester 1998.

47. Krauskopf, K. B. *Introduction to Geochemistry*, 2nd Ed., McGraw-Hill, Tokyo 1979.

48. Krouse, H. R., Ed. *Stable Isotopes: Natural and Anthropogenic Sulphur in the Environment*, SCOPE 43, 1991.

49. Kuznetsov, A. M.; Ulstrup, J. *Electron Transfer in Chemistry and Biology*, J. Wiley & Sons, Chichester 1999.

50. Landau, L. D.; Lifschitz, E. M. *Lehrbuch der theoretischen Physik I, Mechanik*, 9th Ed., Akademie-Verlag, Berlin 1979.

51. Landau, L. D.; Lifschitz, E. M. *Lehrbuch der theoretischen Physik VI, Hydrodynamik*, 3rd Ed., Akademie-Verlag, Berlin 1978.

52. Lass, H. *Vector and Tensor Analysis*, McGraw-Hill, New York 1950.

53. Leeder, M. R. *Sedimentology*, 1st Ed., Chapman & Hall, London 1982; Reprint 1992.

54. Lehninger, A. L. *Biochemistry*, 2nd Ed., Worth Publishers, New York 1975.

55. Letcher, T. M., Ed. *Thermodynamics, Solubility and Environmental Issues*, Elsevier, Amsterdam 2007.

56. Lewis, G. N.; Randall, M. *Thermodynamics*, Revised by K. S. Pitzer and L. Brewer, 2nd Ed., McGraw-Hill, New York 1961.

57. Liebig, J. *Chemiske Breve*, P. G. Philipsens Forlag, Kjøbenhavn 1854.

58. Likens, G. E., Ed. *Some Perspectives of the Major Biogeochemical Cycles*, SCOPE 17, 1981.

59. Lippard, S. J.; Berg, J. M. *Principles of Bioinorganic Chemistry*, University Science Books, California 1994.

60. Macalady, D. L., Ed. *Perspectives in Environmental Chemistry*, Oxford University Press, New York 1998.

61. Madigan, M. T.; Martinko, J. M.; Parker, J. *Brock: Biology of Microorganisms*, 8th Ed., Prentice Hall, New Jersey 1997.

62. Malone, T. F.; Roederer, J. G., Eds. *Global Change*, ICSU Press, Cambridge 1985.

63. March, J. *Advanced Organic Chemistry: Reactions, Mechanisms, and Structure*, 2nd Ed., McGraw-Hill Kogakusha, Tokyo 1977.

64. Margenau, H.; Murphy, G. M. *The Mathematics of Physics and Chemistry*, 2nd Ed., D. van Nostrand, New Jersey 1956.

65. Masel, R. I. *Principles of Adsorption and Reaction on Solid Surfaces*, J. Wiley & Sons, New York 1996.

66. Mason, S. F. *Chemical Evolution; Origin of the Elements, Molecules, and Living Systems*, Clarendon Press, Oxford 1991.

67. McGlashan, M. L. *Chemical Thermodynamics*, Academic Press, London 1979.

68. Morel, F. M. M.; Hering, J. G. *Principles and Applications of Aquatic Chemistry*, J. Wiley & Sons, New York 1993.

69. Munson, B. R.; Young, D. F.; Okiishi, T. H. *Fundamentals of Fluid Mechanics*, J. Wiley & Sons, New York 1990.

70. Müller, A.; Krebs, B., Eds. *Sulfur; Its Significance for Chemistry, for the Geo-, Bio- and Cosmosphere and Technology*, Elsevier, Amsterdam 1984.

71. Nernst, W. *Theoretische Chemie*, 5. Aufl., Verlag von Ferdinand Enke, Stuttgart, 1907.

72. Noe-Nygaard, A. *Vulkaner*, Gyldendal, København 1979.

73. O'Neill, P. *Environmental Chemistry*, 2nd Ed., Chapman & Hall, London 1993.

73a. Partington, J. R. *A Short History of Chemistry*, 3rd Ed., Macmillan & Co., London 1960.

73b. Partington, J. R. *A History of Chemistry*, Macmillan & Co., London 1961 (Vol. 2), 1962 (Vol. 3), 1964 (Vol. 4).

74. Pauling, L. *The Nature of the Chemical Bond*, 3rd Ed., Cornell University Press, New York 1960.

75. Petersen, L. *Grundtræk af Jordbundslæren*, 4th Ed. Jordbrugsforlaget, København 1994.

76. Pourbaix, M. J. N. *Thermodynamics of Dilute Aqueous Solutions*, E. Arnold & Co., London 1949.

77. Price, M. *Introducing Groundwater*, 2nd Ed., Chapman & Hall, London 1996.

78. Prigogine, I. *Introduction to Thermodynamics of Irreversible Processes*, 3rd Ed., J. Wiley & Sons, New York 1967.

79. Ramdohr, P.; Strunz, H. *Klockmanns Lehrbuch der Mineralogie*, 16th Ed., Ferdinand Enke Verlag, Stuttgart 1978.

80. Reitz, J. R.; Milford, F. J.; Christy, R. W. *Foundations of Electromagnetic Theory*, 3rd Ed., Addison-Wesley, Reading, Massachusetts 1979.

81. Richter, C. F. *Elementary Seismology*, W. H. Freeman and Co., San Francisco 1958.

82. Robinson, R. A.; Stokes, R. H. *Electrolyte Solutions*, 2nd Ed., Butterworths, London 1959.

83. Schlesinger, W. H. *Biogeochemistry: An Analysis of Global Change*, Academic Press, San Diego 1991.

84. Schröder, D. *Soils – Facts and Concepts*, International Potash Institute, Bern 1984.

85. Schwarzenbach, R. P.; Gschwend, P. M.; Imboden, D. M. *Environmental Organic Chemistry*, 2nd Ed., Wiley-Interscience, New Jersey 2003.

86. Seinfeld, J. H.; Pandis, S. N. *Atmospheric Chemistry and Physics, from Air Pollution to Climate Change*, 2nd Ed., Wiley & Sons, New Jersey 2006.

87. Siegel, H.; Siegel, A. Eds. *Vanadium and Its Role in Life,* Vol. 31 in *Metal Ions in Biological Systems*. Marcel Dekker, Basel 1995.

88. Skoog, D. A.; Holler, F. J.; Nieman, T. A. *Principles of Instrumental Analysis*, 5th Ed., Saunders College Publishing, Philadelphia 1998.

89. Sposito, G. *The Chemistry of Soils*, Oxford University Press, New York & Oxford 1989.

90. Sposito, G. *The Surface Chemistry of Soils*, Oxford University Press, Oxford 1984.

91. Stumm, W. *Chemistry of the Solid-Water Interface*, J. Wiley & Sons, New York 1992.

92. Stumm, W.; Morgan, J. J. *Aquatic Chemistry*, 2nd Ed., J. Wiley & Sons, New York 1981.

93. Svensson, B. H.; Söderlund, R., Eds. *Nitrogen, Phosphorus and Sulphur – Global Cycles*, SCOPE 7, 1976.

94. Tan, K. H. *Principles of Soil Chemistry*, 3rd Ed.; Marcel Dekker, New York 1998.

95. Thompson, D., Ed. *Insights into Speciality Inorganic Chemicals*, The Royal Society of Chemistry, London 1995.

96. Thompson, R., Ed. *The Modern Inorganic Chemicals Industry*, The Chemical Society, London 1977.

97. Thompson, R., Ed. *Speciality Inorganic Chemicals*, The Royal Society of Chemistry, London 1981.

98. Topp, N.-H. *Kryolitindustriens Historie 1847–1990*, Kryolitselskabet Øresund, København 1990.

99. van Elsas, J. D.; Trevors, J. T.; Wellington, E. M. H., Eds. *Modern Soil Microbiology*, Marcel Dekker, New York 1997.

100. Violante, A.; Huang, P. M.; Gadd, G. M., Eds. *Biophysico-Chemical Processes of Heavy Metals and Metalloids in Soil Environments*, J. Wiley & Sons, New Jersey 2008.

101. Wayne, R. P. *Chemistry of Atmospheres*, 2nd Ed., Oxford University Press, Oxford 1999.

102. Weinberg, S. *The First Three Minutes*, André Deutsch, London 1977.

103. Wiberg, N. *Holleman-Wiberg: Lehrbuch der Anorganischen Chemie*, 102nd Ed., Walter de Gruyter, Berlin 2007.

104. Wilkins, R. G. *Kinetics and Mechanism of Reactions of Transition Metal Complexes*, 2nd Ed., VCH, Weinheim 1991.

105. Winther, L.; Henze, M.; Linde, J. J.; Jensen, H. T. *Spildevandsteknik*, 3rd Ed., Polyteknisk Forlag, Lyngby 2004.

106. Wood, M. *Soil Biology*, Blakie & Son, London 1989.

107. Yen, T. F. *Environmental Chemistry*, Prentice Hall, New Jersey 1999.

2. Selected Chapters

108. Ahrland, S. *Inorganic chemistry of the ocean*, in: Ref. 38, pp. 65–88.

109. Bolin, B; Rosswall, T.; Richey, J. E.; Freney, J. R.; Ivanov, M. V.; Rodhe, H. *C, N, P, and S cycles: Major reservoirs and fluxes*, in: Ref. 253, Chap. 2.

110. Bouwer, H. *Elements of soil science and groundwater hydrology*, in: Ref. 8, pp. 9–38.

111. Cronin, J. R. et al. *Organic matter in carbonaceous chondrites, planetary satellites, asteroids, and comets*, in: Ref. 45, pp. 819–857.

112. Exley, C. *The solubility of hydroxyaluminosilicates and the biological availability of aluminium*, in: Ref. 55, Chap. 17.

113. Freyer, H.-D. *Variations in the atmospheric CO_2 content*, in: *The Global Carbon Cycle*, SCOPE 13, 1979. Chap. 3. http://www.icsu-scope.org/downloadpubs/scope13/.

114. Granat, L.; Rodhe, H.; Hallberg, R. O. *The global sulphur cycle*, in: Ref. 93, pp. 89–134.

115. Ivanov, M. V. *The global biogeochemical sulphur cycle*, in: Ref. 58, pp. 61–78.
116. Lafon, G. M.; Mackenzie, F. T. *Early evolution of the oceans – a weathering model*, in: Ref. 32, pp. 205–218.
117. Pierrou, U. *The global phosphorus cycle*, in: Ref. 93, pp. 75–88.
118. Rosswall, T. *The biogeochemical nitrogen cycle*, in: Ref. 58, pp. 25–49.
119. Söderlund, R.; Svensson, B. H. *The global nitrogen cycle*, in: Ref. 93, pp. 23–73.
120. Thode, H. G. *Sulphur isotopes in Nature and the environment: An overview*, in: Ref. 48, pp. 1–26.

3. Tables and Handbooks

121. Fegley, B. Jr. in *Global Earth Physics: A Handbook of Physical Constants (AGU Reference Shelf 1)*, Ahrens, T. J. Ed., American Geophysical Union, Washington DC 1995, pp. 320–345.
122. International Union of Geological Sciences (IUGS) *Igneous Rocks: A Classification and Glossary of Terms*, Le Maitre, R. W., Ed.; 2nd Ed., Cambridge University Press, Cambridge 2002.
123. International Union of Pure and Applied Chemistry (IUPAC) *Compendium of Chemical Terminology*, 2nd Ed., Blackwell Science, Oxford 1997 ("The Gold Book"). http://old.iupac.org/goldbook/.
124. International Union of Pure and Applied Chemistry (IUPAC) *Nomenclature of Inorganic Chemistry*, RSC Publishing, Cambridge 2005 ("The Red Book").
    a. p. 42 and Table VI, p. 260.
125. International Union of Pure and Applied Chemistry (IUPAC) *Nomenclature of Organic Compounds*, Blackwell Scientific Publications, Oxford 1993 ("The Blue Book").
126. International Union of Pure and Applied Chemistry (IUPAC) *Quantities, Units and Symbols in Physical Chemistry*, 3rd Ed., RSC Publishing, Cambridge 2007 ("The Green Book").
127. International Union of Pure and Applied Chemistry (IUPAC) *Stability Constants of Metal-Ion Complexes*, compiled by E. Högfeldt, Pergamon Press, Oxford 1982.
128. Landoldt-Börnstein *Physikalisch-Chemische Tabellen*, Julius Springer, Berlin. 5. Aufl. 1936; 6. Aufl. 1955; Neue Serie 1976.
129. Lide, D. R., Ed. *Handbook of Chemistry and Physics*, 77th Ed., CRC Press, Boca Raton, 1996.
130. Milazzo, G.; Caroli, S. *Tables of Standard Electrode Potentials*, Wiley & Sons, New York 1978.
131. Moore, C. E. *Atomic Energy Levels*, National Bureau of Standards, Washington, D.C., 1949. Vol. I: H–V.
132. Ohkawa, H.; Miyagawa, H.; Lee, P. W., Eds. *Ullmann's Agrochemicals*, Wiley-VCH, Weinheim 2007.
133. Powell, K. J. *SolEq*, academic software, 1999. CD-ROM, http://acadasoft.co.uk.
134. Sillén, L. G.; Martell, A. E. *Stability Constants of Metal-Ion Complexes*, The Chemical Society, London.
    a. Special Publication No. 17, *Inorganic Ligands*, 1964.
    b. Special Publication No. 25, *Organic Ligands*, 1971.

135. Smith, R. M.; Martell, A. E. *Critical Stability Constants*, Plenum Press, New York.

    a. Vol. 1 *Amino Acids*, 1974.

    b. Vol. 2 *Amines*, 1975.

    c. Vol. 3 *Other Organic Ligands*, 1977.

    d. Vol. 4 *Inorganic Complexes*, 1976.

    e. Vol. 5 *Supplement*, 1982.

    f. Vol. 6 *Supplement*, 1989.

136. Timmermans, J. *The Physico-chemical Constants of Binary Systems in Concentrated Solutions*, Interscience Publishers, New York 1959. Vol. 1–4.

4. Papers

137. Alberty, R. A. *Pure Appl. Chem. 73* (2001) 1349–1380. Use of Legendre transformations in chemical thermodynamics.

138. Anderson, J. G.; Brune, W. H.; Proffitt, M. H. *J. Geophys. Res. D 94* (1989) 11465–11479. Ozone destruction by chlorine radicals within the Antarctic Vortex: The spatial and temporal evolution of ClO-O_3 anticorrelation based on in situ ER-2 data.

139. Baker, J.; Bizzarro, M.; Wittig, N.; Connelly, J.; Haack, H. *Nature 436* (2005) 1127–1131. Early planetesimal melting from an age of 4.5662 Gyr for differentiated meteorites.

140. Balba, H. *J. Environ. Sci. Health, Part B, 42* (2007) 441–451. Review of strobilurin fungicide chemicals.

141. Bates, D. R.; Nicolet, M., *J. Geophys. Res. 55* (1950) 301. The photochemistry of atmospheric water vapour.

142. Baeyens, W.; Leermakers, M.; Papina, T.; Saprykin, A.; Brion, N.; Noyen, J.; De Gieter, M.; Elskens, M.; Goeyens, L. *Arch. Environm. Contamin. Toxicol. 45.4* (2003) 498–508. Bioconcentration and biomagnification of mercury and methylmercury in North Sea and Scheldt Estuary fish.

143. Berger, A.; Loutre, M. F. *Q. Sci. Rev. 10* (1991) 297–317. Insolation values for the climate of the last 10 million years.

144. Berger, A.; Loutre, M. F.; Mélice, J. L. *Clim. Past Discuss. 2* (2006) 519–513. Equatorial insolation: from precession harmonics to eccentricity frequencies.

145. Birchall, J. D.; Exley, C.; Chappell, J. S.; Phillips, M. J. *Nature 338* (1989) 146–148. Acute toxicity of aluminium to fish eliminated in silicon-rich acid waters.

146. Bjerrum, N. *Z. physik. Chem. 106* (1923) 219–242. Dissociations-konstanten von mehrbasischen Säuren und ihrer Anwendung zur Berechnung molekularer Dimensionen.

147. Bowman, A. F.; Lee, D. S.; Asman, W. A. H.; Dentener, F. J.; van der Hoek, K. W.; Olivier, J. G. J. 1997. *Global Biogeochem. Cycles 11* (1997) 561–587. A global high-resolution emission inventory for ammonia.

148. Brønsted, J. N.; Pedersen, K. J. *Z. physik. Chem. 108* (1924) 185–235. Die katalytische Zersetzung des Nitramids und ihre physikalish-chemische Bedeutung.

149. Cameron, A. G. W. *Space Sci. Rev. 15* (1973) 121–146. Abundance of the elements in the solar system.

150. Canup, R. M.; Asphaug, E. *Nature 412* (2001) 708–712. Origin of the Moon in a giant impact near the end of the Earth's formation.

151. Chen, Y. H.; Prinn, R. G. *J. Geophys. Research Atm. 111* (2006) D10307. Estimation of atmospheric methane emissions between 1996 and 2001 using a three-dimensional global chemical transport model.

151a. Chapman, S. *Mem. Roy. Meterol. Soc. 3* (1930) 103–105. A theory of upper-atmosphere ozone.

152. Charlson, R. J.; Lovelock, J. E.; Andreae, M. O.; Warren, S. G. *Nature 326* (1987) 655–661. Oceanic phytoplankton, atmospheric sulfur, cloud albedo and climate.

153. Ciavatta, L.; Ferri, D.; Grenthe, I.; Salvatore, F. *Inorg. Chem. 20* (1981) 463–467. The first acidification step of the tris(carbonato) dioxouranate(VI) $UO_2(CO_3)_3{}^{4-}$.

154. Cooper, G.; Kimmich, N.; Belisle, W.; Sarinana, J.; Brabham, K.; Garrel, L. *Nature 414* (2001) 879–883. Carbonaceous meteorites as a source of sugar-related organic compounds for the early Earth.

155. Cox, J. D. *Pure Appl. Chem. 54* (1982) 1239–1250. Standard states.

156. Dahl, T. W.; Stevenson, D. J. *Earth Planet. Sci. Lett. 295* (2010) 177–186. Turbulent mixing of metal and silicate during planet accretion – and interpretation of the Hf–W chronometer.

157. Dansgaard, W. *Tellus 16* (1964) 436–468. Stable isotopes in precipitation.

158. Dockery, D. W.; Cunningham, J.; Damokosh, A. I.; Neas, L. M.; Spengler, J. D.; Koutrakis, P.; Ware, J. H.; Raizenne, M.; Speizer, F. E. *Environ. Health Perspect. 104* (1996) 500–505. Health effects of acid aerosols on North American children: respiratory symptoms.

159. Dyrssen, D.; Sillén, L. G. *Tellus 19* (1967) 113–121. Alkalinity and total carbonate in sea water. A plea for *p*-*T*-independent data.

160. Ehde, P. M.; Andersson, I.; Pettersson, L. *Acta Chem. Scand. 43* (1989) 136–143. Multicomponent polyanions. 43. A study of aqueous equilibria in the vanadocitrate system.

161. Everett, D. H. *Pure Appl. Chem. 31* (1972) 578–638. Manual of symbols and terminology for physicochemical quantities and units. Apendix II: Definitions, terminology and symbols in colloid and surface chemistry.

162. Ewing, M. B.; Lilley, T. H.; Olofsson, G. M.; Rätzsch, M. T.; Somsen, G. *Pure Appl. Chem. 66* (1994) 533–552. Standard quantities in chemical thermodynamics. Fugacities, activities, and equilibrium constants for pure and mixed phases.

162a. Farman, J. C.; Gardiner, B. G.; Shanklin J. D. *Nature 315* (1985) 207. Large losses of total ozone in Antarctica reveal seasonal ClO*x*/NO*x* interaction.

163. Ferri, D.; Grenthe, I.; Salvatore, F. *Acta Chem. Scand. A35* (1981) 165–168. Dioxouranium(VI) carbonate complexes in neutral and alkaline solutions.

164. Fauvarque, J. *Pure Appl. Chem. 68* (1996) 1713–1720. The chlorine industry.

165. Galloway, J. N.; Dentener, F. J.; Capone, D. G.; Boyer, E. W.; Howarth, R. W.; Seitzinger, S. P.; Anser, G. P.; Cleveland, C. C.; Green, P. A.; Holland, E. A.; Karl, D. M.; Michaels, A. F.; Porter, J. H.; Townsend A. R.; Voeroesmarty, C. J. *Biogeochemistry 70* (2004) 153–226. Nitrogen cycles: past, present, and future.

166. Gamsjäger, H.; Lorimer, J. W.; Salomon, M.; Shaw, D. G.; Tomkins, R. P. T. *J. Phys. Chem. Ref. Data 39* (2010) 1–13. The IUPAC-NIST solubility data series: A guide to preparation and use of compilations and evaluations.

167. Giauque, W. F.; Clayton, J. O. *J. Am. Chem. Soc. 55* (1933) 4875–4889. The heat capacity and entropy of nitrogen. Heat of vaporization. Vapor pressures of solid and liquid. The reaction $\frac{1}{2}N_2 + \frac{1}{2}O_2 = NO$ from spectroscopic data.

168. Goldschmidt, V. M. *J. Chem. Soc.* (1937) 655–673. The principles of distribution of chemical elements in minerals and rocks.

169. Graedel, T. E.; Keene, W. C. *Pure Appl. Chem. 68* (1996) 1689–1697. The budget and cycle of Earth's natural chlorine.

170. Grenthe, I.; Plyasunov A. *Pure Appl. Chem. 69* (1997) 951–958. On the use of semiempirical electrolyte theories for the modeling of solution chemical data.

171. Gribble, G. W. *Pure Appl. Chem. 68* (1996) 1699–1712. The diversity of natural organochlorines in living organisms.

172. Grossman, L. *Geochim. Cosmoschim. Acta 36* (1972) 597–619. Condensation in the primitive solar nebula.

173. Guenther, A.; et al. *J. Geophys. Res. D 100* (1995) 8873–8892. A global model of natural volatile organic compound emissions.

174. Hansen, J.; Nazarenko, L. *Proc. Natl. Acad. Sci. USA 101* (2004) 423–428. Soot climate forcing via snow and ice albedos.

175. Hansson, I. *Acta Chem. Scand. 27* (1973) 931–944. The determination of dissociation constants of carbonic acid in synthetic sea water in the salinity range of 20-40 ‰ and temperature range of 5-30 °C.

176. Hofmann, D. J.; Harder, J. W.; Rosen, J. M.; Hereford, J. V.; Carpenter, J. R. *J. Geophys Res. D 94* (1989) 16527–16535. Ozone profile measurements at McMurdo Station, Antarctica, during the spring of 1987.

177. Imbrie, J.; Imbrie, J. Z. *Science 207* (1980) 943–953. Modeling the climatic response to orbital variations.

178. IUPAC *Pure Appl. Chem. 56* (1984) 595–624. Nomenclature and symbolism for amino acids and peptides.

179. IUPAC *Pure Appl. Chem.* (1960–2000). Pesticide reports no. 1–40.

180. Jawald, M.; Ingman, F.; Hay Liem, D.; Wallin, T. *Acta Chem. Scand. A32* (1978) 7–14. Solvent extraction studies of complex formation between methylmercury(II) and bromide, chloride and nitrate.

181. Jin, Z.; Charlock, T; Smith, W., Jr.; Rutledge, K. *Geophys. Res. Let. 31* (2004) L22301. A parameterization of ocean surface albedo.

182. Johnston, H. S. *Rev. Geophys. Space Phys. 13* (1975) 637. Global ozone balance in the natural stratosphere.

183. Jones, G.; Bradshaw, B. C. *J. Am. Chem. Soc. 55* (1933) 1780–1800. The measurements of conductivities of electrolytes. V.

184. Kirschner, E. M. *Chem. Eng. News* (1995) April 10, 16–20. Production of top 50 chemicals increased substantially in 1994.

185. Koppenol, W. H. *Pure Appl. Chem. 72* (2000) 437–446. Names for inorganic radicals.

186. Kowal, W.; Beattie, O. B.; Baadsgaard, H.; Krahn, P. M. *Nature 343* (1990) 319. Source identification of lead found in tissues of sailors from the Franklin arctic expedition of 1845.

187. Kulharni, R. S.; Kanekar, P. P. *Curr. Microbiol. 37* (1998) 191–194. Production of different proteases from fish gut micro flora utilizing Tannery fleshing.

188. Laidler, K. J. *J. Chem. Educ. 61* (1984) 494–498. The development of the Arrhenius equation.

189. Lyklema, J. *Pure Appl. Chem. 63* (1991) 895–906. Electrified interfaces in aqueous dispersions of solids.

190. Marion, G. M.; Babcock, K. L. *Soil Sci. 122* (1976) 181–244. Predicting specific conductance and salt concentration in dilute aqueous solutions.

191. Martin, R.-P.; Martens, G. D., Eds. *Pure Appl. Chem. 68* (1996) 1683–1824. IUPAC White Book on Chlorine.

192. Mattauch, J. *J. Am. Chem. Soc. 80* (1958) 4125–4126. The rational choice of a unified scale for atomic weights and nuclidic masses.

193. McDuff, R. E.; Morel, F. M. M. *Environm. Sci. Techn. 10* (1980) 1182–1186. The geochemical control of seawater (Sillén revisited).

194. Moss, G. P. *Pure Appl. Chem. 68* (1996) 2193–2222. Basic terminology of stereochemistry.

195. Müller, R. A.; McDonald, G. J. *Science 277* (1997) 215–218. Glacial cycles and astronomical forcing.

196. Möller, K.; Bretzke, C.; Huhnerfuss, H.; Kallenborn, R.; Kinkel, J. N.; Kopf, J.; Rimkus, G. *Angew. Chem. Int. Ed. 33* (1994) 882–884. The absolute configuration of $(+)$-α-1,2,3,4,5,6-hexachlorocyclohexane, and its permeation through the seal blood-brain barrier.

197. Mønsted, O. *Determination of acid dissociation constants*, Department of Chemistry, University of Copenhagen, Copenhagen 1996.

198. Papp, R. *Pure Appl. Chem. 68* (1996) 1801–1808. Organochlorine waste management.

199. Pearson, R. G. *J. Am. Chem. Soc. 85* (1963) 3533–3539. Hard and soft acids and bases.

200. Petit, J. R.; Jouzel, J.; Raynaud, D.; Barkov, N. I.; Barnola, J.-M.; Basile, I.; Bender, M.; Chappellaz, J.; Davis, M.; Delaygue, G.; Delmotte, M.; Kotlyakov, V. M.; Legrand, M.; Lipenkov, V. Y.; Lorius, C.; Pepin, L.; Ritz, C.; Saltzman, E.; Stievenard, M. *Nature 399* (1999) 429–436. Climate and atmospheric history of the past 420,000 years from the Vostok ice core, Antarctica.

201. Raupach, M. R.; Marland, G.; Ciais, P.; Quéré, C. L.; Canadell, J. G.; Klepper, G.; Field, C. B. *Proc. Natl. Acad. Sci. 104* (2007) 10288–10293. Global and regional drivers of accelerating CO_2 emissions.

202. Robino, C. V. *Metallurg. Materials Trans. 27B* (1996) 65–69. Representation of mixed reactive gases on free energy (Ellingharn-Richardson) diagrams.

203. Rouhi, A. M. *Chem. Eng. News 80* (2002) Issue 51, 29–38. Olefin metathesis: Big-deal reaction.

204. Scatchard, G.; Wood, S. E.; Mochel, J. M. *J. Am. Chem. Soc. 68* (1946) 1960–1963. Vapor-liquid equilibrium. VII. Carbon tetrachloride-methanol mixtures.

205. Schmidt, J. E.; Ahring, B. K. *Pure Appl. Chem. 69* (1997) 2447–2452. Treatment of waste water from a multi product food-processing company, in upflow anaerobic sludge blanket (UASB) reactors: The effect of seasonal variation.

206. Sillén, L. G. *Science 156* (1967) 1189–1197. The ocean as a chemical system.

207. Sillman, S.; Logan, J. A.; Wofsy, S. C. *J. Geophys. Res. D 95* (1990) 1837–1852. The sensitivity of ozone to nitrogen oxides and hydrocarbons in regional ozone episodes.

208. Skattebøl, L.; Stenstrøm, Y. *Acta Chem. Scand. B39* (1985) 291–304. Synthesis of (±)-lineatin, an aggregation pheromone component of *Trypodendron lineatum*.

209. Sugimura, Y.; Suzuki, Y.; Miyake, Y. *J. Oceanogr. Soc. Japan 32* (1976) 235–241. The content of selenium and its chemical form in sea water.

210. Sukhno, I. V.; Buzko, V. Y.; Pettit, L. D. *Chem. Intnl. 27*.3 (2005). Equilibrium in solution: a software aid.

211. Trantirek, L.; et al. *J. Biol. Chem. 276* (2000) 7734–7740. Reaction mechanism and stereochemistry of γ-hexachlorocyclohexane dehydrochlorinase LinA.

212. Ueno, Y.; Johnson, M. S.; Danielache, S. O.; Eskebjerg; C.; Pandey, A.; Yoshida, N. *Proc. Natl. Acad. Sci. 106* (2009) 14784–14785. Geological sulfur isotopes indicate elevated OCS in the Archean atmosphere, solving faint young sun paradox.

213. Wang, Y.; Jacob, D. J. *J. Geophys. Res. Atm. 103* (1998) 31123–31135. Anthropogenic forcing on tropospheric ozone and OH since preindustrial times.

214. Westöö, G. *Acta Chem. Scand. 20* (1966) 2131–2137. Determination of methylmercury compounds in foodstuffs.

215. Wiesser, M. E.; Coplen, T. B. *Pure Appl. Chem. 83* (2011) 359–396. Atomic weights of the elements 2009.

216. Yates, D. E.; James, R. O.; Healy, T. W. *J.C.S: Faraday I, 76* (1980) 1–8. Titanium dioxide-electrolyte interface part 1. Gas adsorption and tritium exchange studies.

217. Yates, D. E.; Healy, T. W. *J.C.S: Faraday I, 76* (1980) 9–18. Titanium dioxide-electrolyte interface part 2. Surface charge (titration) studies.

218. Zhao, J.; Zhang, R. *Adv. Quantum Chem. 55* (2008) 177–213. Theoretical investigation of atmospheric oxidation of biogenic hydrocarbons: A critical review.

5. Miscellaneous

219. Allégre, C. J.; Schneider, S. H. *Sci. Am.* (Oct. 1994) *66*, cited in ref. 12.

220. American Meteorological Society; www.ametsoc.org.

221. Beattie, O.; Geiger, J. *Frozen in Time*, Bloomsbury Publishing, London 1987.

222. British Petroleum. *Statistical Review of World Energy*, 2009; http://www.bp.com.

223. DeMore, W. B.; Sander, S. P.; Golden, D. M.; Hampson, R. F.; Kurylo, M. J.; Howard, C. J.; Ravishankara, A. R.; Kolb, C. E.; Molina, M. J., *Chemical kinetics and photochemical data for use in stratospheric modeling*, Evaluation number 12, JPL Publication 97–4, NASA, Pasadena, California, 1997.

224. EPA (U.S. Environmental Protection Agency), 2008; http://www.epa.gov/greenchemistry.

225. FAO (Food and Agricultural Organization of the United Nations), 2008; http://www.fao.org.

226. Finnish Forrest Industries, 2006; http://www.forestindustries.fi.

227. http://nobelprize.org/nobel_prizes/chemistry/laureates/2001; http://nobelprize.org/nobel_prizes/chemistry/laureates/2005.

228. http://www.surgeongeneral.gov/pressreleases/sg01132005.html.

229. ICSU (International Council for Science): www.icsu.org.

230. IPCC (Intergovernmental Panel on Climate Change) *Climate Change 2001: The Scientific Basis. Contribution of Working Group I to the Third Assessment Report of the IPCC*, Houghton, J. T.; Ding, Y.; Griggs, D. J.; Noguer, M.; van der Linden, P. J.; Dai, X.; Maskell, K.; Johnson, C. A., Eds. Cambridge University Press, New York 2001.

231. IPCC (Intergovernmental Panel on Climate Change) *Climate Change 2007: The Physical Science Basis. Contribution of Working Group I to the Fourth Assessment Report of the IPCC*, Solomon, S.; Qin, D.; Manning, M.; Chen, Z.; Marquis, M,; Averyt, K. B.; Tignor, M.; Miller, H. L., Eds. Cambridge University Press, Cambridge 2007.

232. IUGS (International Union of Geological Sciences) *A Geologic Time Scale 2004*, Gradstein, F. M.; Ogg, J. G.; Smith, A. G., Ed. Cambridge University Press, Cambridge 2005.

233. IUPAC http://www.iupac.org.

234. IUPAC Periodic Table of the Elements (November 2003). http://www.iupac.org/reports/periodic_table.

235. IUPAC Project: Ionic Strength Corrections for Stability Constants (2005) http://www.iupac.org/web/ins/2000-003-1-500.

236. IUPAC Summary of Evaluated Kinetic and Photochemical Data for Atmospheric Chemistry; http://www.iupac-kinetic.ch.cam.ac.uk.

237. IUPAC, *CHEMRAWN VII, The Chemistry of the Atmosphere: Its Impact on Global Change.* IUPAC and The American Chemical Society 1992.

238. IUPAC, *CHEMRAWN IX, The Role of Advanced Materials in Sustainable Development*, The Korean Chemical Society, Seoul 1996.

239. IUPAC, *CHEMRAWN XIV, Pure Appl. Chem. 73* (2001) 1229–1348. Toward environmentally benign processes and products.

240. Keeling, R. F.; Piper, S. C.; Bollenbacher, A. F.; Walker J. S., 2010; http://cdiac.ornl.gov/trends/co2/sio-mlo.html.

241. MCM (Master Chemical Mechanism), University of Leeds Master Chemical Mechanism for the Atmosphere; http://www.chem.leeds.ac.uk/Atmospheric/MCM/mcmproj.html.

242. NASA JPL (National Aeronautics and Space Administration, Jet Propulsion Laboratory, USA) Chemical kinetics and photochemical data for use in atmospheric studies; http://jpldataeval.jpl.nasa.gov.

243. NIST (U.S. National Institute of Standards and Technology)
    a. Chemistry Webbook, 2008; http://webbook.nist.gov/chemistry.
    b. Chemical Kinetics Database; http://kinetics.nist.gov/kinetics/index.jsp.

244. Novelli, P. C.; et al. *J. Geophys. Res. D 104* (1999) 30427.

245. Shanklin, J. D., 2010: http://www.antarctica.ac.uk/met/jds/ozone/data/ ZOZ5699.DAT.

246. SCUP Report 2009, www.sriconsulting.com/SCUP/Public/Reports/OVSPE000, 2011.

247. Soil Science Society of America, 2010: https://soils.org.

248. USDA (U.S. Department of Agriculture)
    a. http://www.fas.usda.gov.
    b. http://www.invasive.org.
    c. *Soil Taxonomy*, Agricultural Handbook Number 436, 1999.

249. U.S. Department of Commerce: National Climatic Data Center, 2009.
    a. http://www.ncdc.noaa.gov/oa/climate/research/anomalies/index.html.
    b. http://www.ncdc.noaa.gov/paleo/icecore/antarctica/vostok/vostok_data. html.

250. U.S. Department of Commerce; National Oceanic & Atmospheric Administration, 2010: Earth System Research Laboratory; Global Monitoring Division. http://www.esrl.noaa.gov/gmd.

251. Grube, A., Donaldson, D., Kiely, T., Wu, Li *Pesticides Industry Sales and Usage. 2006 and 2007 Market Estimates*. U.S. Environmental Protection Agency, Washington, DC 2011.

252. U.S. Geological Survey, 2008. http://minerals.usgs.gov.

253. SCOPE 21 *The Major Biogeochemical Cycles and Their Interactions*, 1983. http://www.icsu-scope.org/downloadpubs/scope21.

254. Stockholm Convention on Persistent Organic Pollutants http://chm.pops.int/ Programmes/NewPOPs/The9newPOPs/tabid/672/language/en-US/Default. aspx.

255. WCC (World Chlorine Council), 2010. http://worldchlorine.com.

256. WMO (World Meteorological Organization). Scientific assessment of ozone depletion, 1994.

Name index

Subject index

	1	2	3	4	5	6	7	8	9	10	11	12	13	14	15	16	17	18
1	H																	He
2	Li	Be											B	C	N	O	F	Ne
3	Na	Mg											Al	Si	P	S	Cl	Ar
4	K	Ca	Sc	Ti	V	Cr	Mn	Fe	Co	Ni	Cu	Zn	Ga	Ge	As	Se	Br	Kr
5	Rb	Sr	Y	Zr	Nb	Mo	Tc	Ru	Rh	Pd	Ag	Cd	In	Sn	Sb	Te	I	Xe
6	Cs	Ba	La	Hf	Ta	W	Re	Os	Ir	Pt	Au	Hg	Tl	Pb	Bi	Po	At	Rn
7	Fr	Ra	Ac	Rf	Db	Sg	Bh	Hs	Mt	Ds	Rg	Cn						
6				Ce	Pr	Nd	Pm	Sm	Eu	Gd	Tb	Dy	Ho	Er	Tm	Yb	Lu	
7				Th	Pa	U	Np	Pu	Am	Cm	Bk	Cf	Es	Fm	Md	No	Lr	

The periodic table of the elements[234]

	1	2	3	4	5	6	7	8	9	10	11	12	13	14	15	16	17	18
1	H 2.2																	
2	Li 1.0	Be 1.6											B 2.0	C 2.6	N 3.0	O 3.4	F 4.0	
3	Na 0.9	Mg 1.3											Al 1.6	Si 1.9	P 2.2	S 2.6	Cl 3.2	
4	K 0.8	Ca 1.0	Sc 1.4	Ti 1.5	V 1.6	Cr 1.9	Mn 1.6	Fe 1.8	Co 1.9	Ni 1.8	Cu 2.0	Zn 1.7	Ga 1.8	Ge 2.0	As 2.2	Se 2.6	Br 3.0	Kr 2.9
5	Rb 0.8	Sr 0.9	Y 1.2	Zr 1.3	Nb 1.6	Mo 2.2	Tc 1.9	Ru 2.2	Rh 2.2	Pd 2.2	Ag 1.9	Cd 1.7	In 1.8	Sn 2.0	Sb 2.1	Te 2.1	I 2.7	Xe 2.6
6	Cs 0.8	Ba 0.9	Ln 1.1	Hf 1.3	Ta 1.5	W 2.4	Re 1.9	Os 2.2	Ir 2.2	Pt 2.3	Au 2.5	Hg 2.0	Tl 2.0	Pb 2.3	Bi 2.0	Po 2.0	At 2.2	

Electronegativities of the elements (Pauling electronegativities[37a])